Physics2000

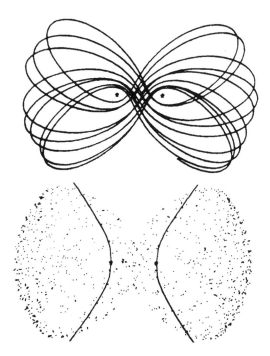

Student project by Bob Piela explaining the hydrogen molecule ion.

by E. R. Huggins
Department of Physics
Dartmouth College
Hanover, New Hampshire

ISBN 0-9707836-0-4 (*Physics 2000* 2–PART SET + CD)
ISBN 0-9707836-1-2 (*Physics 2000* PART 1 + CD)
ISBN 0-9707836-2-0 (*Physics 2000* PART 2 + CD)
ISBN 0-9707836-3-9 (*Physics 2000* CD)
ISBN 0-9707836-4-7 (*Calculus 2000* + CD)
ISBN 0-9707836-5-5 (*Physics 2000* Solutions)
ISBN 0-9707836-6-3 (*Physics 2000* 2–PART SET + *Calculus 2000* + CD)

Preface

ABOUT THE COURSE

Physics2000 is a calculus based, college level introductory physics course that is designed to include twentieth century physics throughout. This is made possible by introducing Einstein's special theory of relativity in the first chapter. This way, students start off with a modern picture of how space and time behave, and are prepared to approach topics such as mass and energy from a modern point of view.

The course, which was developed during 30 plus years working with premedical students at Dartmouth College, makes very gentle assumptions about the student's mathematical background. All the calculus needed for studying Physics2000 is contained in a supplementary chapter which is the first chapter of a *Physics Based Calculus* text. We can cover all the necessary calculus in one reasonable length chapter because the concepts are introduced in the physics text and the calculus chapter only needs to handle the formalism. (The remaining chapters of the calculus text introduce the mathematical tools and concepts used in advanced introductory courses for physics and engineering majors. These chapters will be available at *www.physics2000.com* in late 2000.)

In the physics text, the concepts of velocity and acceleration are introduced through the use of strobe photographs in Chapter 3. How these definitions can be used to predict motion is discussed in Chapter 4 on calculus and Chapter 5 on the use of the computer.

Students themselves have made major contributions to the organization and content of the text. Student's enthusiasm for the use of Fourier analysis to study musical instruments led to the development of the MacScope™ program. The program makes it easy to use Fourier analysis to study such topics as the normal modes of a coupled aircart system and how the energy-time form of the uncertainty principle arises from the particle-wave nature of matter.

Most students experience difficulty when they first encounter abstract concepts like vector fields and Gauss' law. To provide a familiar model for a vector field, we begin the section on electricity and magnetism with a chapter on fluid dynamics. It is easy to visualize the velocity field of a fluid, and Gauss' law is simply the statement that the fluid is incompressible. We then show that the electric field has mathematical properties similar to those of the velocity field.

The format of the standard calculus based introductory physics text is to put a chapter on special relativity following Maxwell's equations, and then put modern physics after that, usually in an extended edition. This format suggests that the mathematics required to understand special relativity may be even more difficult than the integral-differential equations encountered in Maxwell's theory. Such fears are enhanced by the strangeness of the concepts in special relativity, and are driven home by the fact that relativity appears at the end of the course where there is no time to comprehend it. This format is a disaster.

Special relativity does involve strange ideas, but the mathematics required is only the Pythagorean theorem. By placing relativity at the beginning of the course you let the students know that the mathematics is not difficult, and that there will be plenty of time to become familiar with the strange ideas. By the time students have gone through Maxwell's equations in *Physics2000*, they are thoroughly familiar with special relativity, and are well prepared to study the particle-wave nature of matter and the foundations of quantum mechanics. This material is not in an extended edition because there is time to cover it in a comfortably paced course.

ABOUT THE *PHYSICS2000* CD

The *Physics2000* CD contains the complete color version of the *Physics2000* text in Acrobat™ form along with a supplementary chapter covering all the calculus needed for the text. Included on the CD is the 36 minute motion picture *Time Dilation - An Experiment With Mu-Mesons*, and short movie segments of various physics demonstrations. Also a short cookbook on several basic dishes of Caribbean cooking.

The CD is available, for $10 postpaid, at the web site
www.physics2000.com
The black and white printed copy of the text, with the calculus chapter, is also available at the web site at a cost of $25. That includes the CD and shipping within the United States.

Use of the Text Material

Because we are trying to change the way physics is taught, Chapter 1 on special relativity, although copyrighted, may be used freely (except for the copyrighted photograph of Andromeda and frame of the muon film). All chapters may be printed and distributed to a class on a non profit basis.

ABOUT THE AUTHOR

E. R. Huggins has taught physics at Dartmouth College since 1961. He was an undergraduate at MIT and got his Ph.D. at Caltech. His Ph.D. thesis under Richard Feynman was on aspects of the quantum theory of gravity and the non uniqueness of energy momentum tensors. Since then most of his research has been on superfluid dynamics and the development of new teaching tools like the student-built electron gun and MacScope™. He wrote the non calculus introductory physics text *Physics1* in 1968 and the computer based text *Graphical Mechanics* in 1973. The *Physics2000* text, which summarizes over thirty years of experimenting with ways to teach physics, was written and class tested over the period from 1990 to 1998. All the work of producing the text was done by the author, and his wife, Anne Huggins. The text layout and design was by the author's daughter Cleo Huggins who designed eWorld™ for Apple Computer and the Sonata™ music font for Adobe Systems.

The author's eMail address is
lish.huggins@dartmouth.edu
The author welcomes any comments.

Table of Contents

PART 1

CHAPTER 18 ENTROPY

CHAPTER 19   THE ELECTRIC INTERACTION

CHAPTER 20   NUCLEAR MATTER

PART 2

CHAPTER ON GEOMETRICAL OPTICS

Calculus 1   INTRODUCTION TO CALCULUS

Chapter 23
Fluid Dynamics

Since the earth is covered by two fluids, air and water, much of our life is spent dealing with the dynamic behavior of fluids. This is particularly true of the atmosphere where the weather patterns are governed by the interaction of large and small vortex systems, that sometime strengthen into fierce systems like tornados and hurricanes. On a smaller scale our knowledge of some basic principles of fluid dynamics allows us to build airplanes that fly and sailboats that sail into the wind.

In this chapter we will discuss only a few of the basic concepts of fluid dynamics, the concept of the velocity field, of streamlines, Bernoulli's equation, and the basic structure of a well-formed vortex. While these topics are interesting in their own right, the subject is being discussed here to lay the foundation for many of the concepts that we will use in our discussion of electric and magnetic phenomena. This chapter is fairly easy reading, but it contains essential material for our later work. It is not optional.

The Current State of Fluid Dynamics

The ideas that we will discuss here were discovered well over a century ago. They are simple ideas that provide very good predictions in certain restricted circumstances. In general, fluid flows can become very complicated with the appearance of turbulent motion. Only in the twentieth century have we begun to gain confidence that we have the correct equations to explain fluid motion. Solving these equations is another matter and one of the most active research topics in modern science. Fluid theory has been the test bed of the capability of modern super computers as well as the focus of attention of many theorists. Only a few years ago, from the work of Lorenz it was discovered that it was not possible, even in principle, to make accurate long-range forecasts of the behavior of fluid systems, that when you try to predict too far into the future, the chaotic behavior of the system destroys the accuracy of the prediction.

Relative to the current work on fluid behavior, we will just barely touch the edges of the theory. But even there we find important basic concepts such as a vector field, streamlines, and voltage, that will be important throughout the remainder of the course. We are introducing these concepts in the context of fluid motion because it is much easier to visualize the behavior of a fluid than some of the more exotic fields we will discuss later.

THE VELOCITY FIELD

Imagine that you are standing on a bridge over a river looking down at the water flowing underneath you. If it is a shallow stream the flow may be around boulders and logs, and be marked by the motion of fallen leaves and specks of foam. In a deep, wide river, the flow could be quite smooth, marked only by the eddies that trail off from the bridge abutments or the whipping back and forth of small buoys.

Although the motion of the fluid is often hard to see directly, the moving leaves and eddies tell you that the motion is there, and you know that if you stepped into the river, you would be carried along with the water.

Our first step in constructing a theory of fluid motion is to describe the motion. At every point in the fluid, we can think of a small "particle" of fluid moving with a velocity $\vec{v}$. We have to be a bit careful here. If we picture too small a "particle of fluid", we begin to see individual atoms and the random motion between atoms. This is too small. On the other hand, if we think of too big a "particle", it may have small fluid eddies inside it and we can't decide which way this little piece of fluid is moving. Here we introduce a not completely justified assumption, namely that there is a scale of distance, a size of our particle of fluid, where atomic motions are too small to be seen and any eddies in the fluid are big enough to carry the entire particle with it. With this idealization, we will say that the velocity $\vec{v}$ of the fluid at some point is equal to the velocity of the particle of fluid that is located at that point.

We have just introduced a new concept which we will call the "velocity field". At every point in a fluid we define a vector $\vec{v}$ which is the velocity vector of the fluid particle at that point. To formalize the notation a bit, consider the point labeled by the coordinates (x, y, z). Then the velocity of the fluid at that point is given by the vector $\vec{v}$ (x, y, z), where $\vec{v}$ (x, y, z) changes as we go from one point to another, from one fluid particle to another.

As an example of what we will call a ***velocity field***, consider the bathtub vortex shown in Figure (1a). From the top view the water is going in a nearly circular motion around the vortex core as it spirals down the funnel. We have chosen Points A, B, C and D, and at each of the points drawn a velocity vector to represent the velocity of the fluid particle at that point. The velocity vectors are tangent to the circular path of the fluid and vary in size depending on the speed of the fluid. In a typical vortex the fluid near the core of the vortex moves faster than the fluid out near the edge. This is represented in Figure (1b) by the fact that the vector at Point D, in near the core, is much longer than the one at Point A, out near the edge.

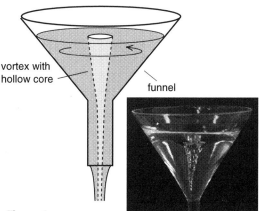

Figure 1a
The "bathtub" vortex is easily seen by filling a glass funnel with water, stirring the water, and letting the water flow out of the bottom.

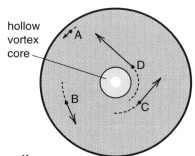

Figure 1b
Looking down from the top, we see the water moving around in a circular path, with the water near the core moving faster. The velocity vectors, drawn at four different points, get longer as we approach the core.

The Vector Field

The velocity field, illustrated in Figure (1) is our first example of a more general concept called a *vector field*. The idea of a vector field is simply that at every point in space there is a vector with an explicit direction and magnitude. In the case of the velocity field, the vector is the velocity vector of the fluid particle at that point. The vector $\vec{v}\,(x, y, z)$ points in the direction of motion of the fluid, and has a magnitude equal to the speed of the fluid.

It is not hard to construct other examples of vector fields. Suppose you took a 1 kg mass hung on the end of a spring, and carried it around to different parts of the earth. At every point on the surface where you stopped and measured the gravitational force $\vec{F} = m\vec{g} = \vec{g}$ (for m = 1) you would obtain a force vector that points nearly toward the center of the earth, and has a magnitude of about 9.8 m/sec² as illustrated in Figure (2). If you were ambitious and went down into tunnels, or up on very tall buildings, the vectors would still point toward the center of the earth, but the magnitude would

vary a bit depending how far down or up you went. (Theoretically the magnitude of $\vec{g}$ would drop to zero at the center of the earth, and drop off as $1/r^2$ as we went out away from the earth). This quantity $\vec{g}$ has a magnitude and direction at every point, and therefore qualifies as a vector field. This particular vector field is called the *gravitational field* of the earth.

It is easy to describe how to construct the gravitational field $\vec{g}$ at every point. Just measure the magnitude and direction of the gravitational force on a non-accelerated 1 kg mass at every point. What is not so easy is to picture the result. One problem is drawing all these vectors. In Figure (2) we drew only about five $\vec{g}$ vectors. What would we do if we had several million measurements?

The gravitational field is a fairly abstract concept—the result of a series of specific measurements. You have never seen a gravitational field, and at this point you have very little intuition about how gravitational fields behave (do they "behave"? do they do things?). Later we will see that they do.

In contrast you have seen fluid motion all your life, and you have already acquired an extensive intuition about the behavior of the velocity field of a fluid. We wish to build on this intuition and develop some of the mathematical tools that are effective in describing fluid motion. Once you see how these mathematical tools apply to an easily visualized vector field like the velocity field of a fluid, we will apply these tools to more abstract concepts like the gravitational field we just mentioned, or more importantly to the electric field, which is the subject of the next nine chapters.

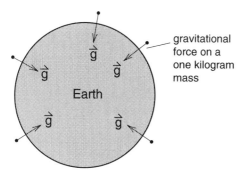

Figure 2
We can begin to draw a picture of the earth's gravitational field by carrying a one kilogram mass around to various points on the surface of the earth and drawing the vector $\vec{g}$ representing the force on that unit mass (m = 1) object.

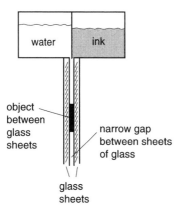

(a) Edge-view of the so called Hele-Shaw cell

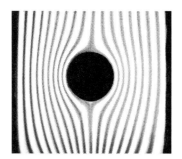

(b) Flow around a circular object.

(c) Flow around airplane wing shapes.

Figure 3
In a Hele-Shaw cell, bands of water and ink flow down through a narrow gap between sheets of glass. With this you can observe the flow around different shaped objects placed in the gap. The alternate black and clear bands of water and ink mark the streamlines of the flow.

Streamlines

We have already mentioned one problem with vector fields—how do you draw or represent so many vectors? A partial answer is through the concept of streamlines illustrated in Figure (3). In that figure we have two plates of glass separated by a narrow gap with water flowing down through the gap. In order to see the path taken by the flowing water, there are two fluid reservoirs at the top, one containing ink and the other clear water. The ink and water are fed into the gap in alternate bands producing the streaks that we see. Inside the gap are a plastic cut-out of both a cylinder and a cross section of an airplane wing, so that we can visualize how the fluid flows past these obstacles.

The lines drawn by the alternate bands of clear and dark water are called ***streamlines***. Each band forms a separate stream, the clear water staying in clear streams and the inky water in dark streams. What these streams or streamlines tell us is the direction of motion of the fluid. Because the streams do not cross and because the dark fluid does not mix with the light fluid, we know that the fluid is moving along the streamlines, not perpendicular to them. In Figure (4) we have sketched a pair of streamlines and drawn the velocity vectors $\vec{v}_1$, $\vec{v}_2$, $\vec{v}_3$ and $\vec{v}_4$ at four points along one of the streams. What is obvious is that the velocity vector at some point must be parallel to the streamline at that point, for that is the way the fluid is flowing. The streamlines give us a map of the directions of the fluid flow at the various points in the fluid.

Figure 4
Velocity vectors in a streamline. Since the fluid is flowing along the stream, the velocity vectors are parallel to the streamlines. Where the streamlines are close together and the stream becomes narrow, the fluid must flow faster and the velocity vectors are longer.

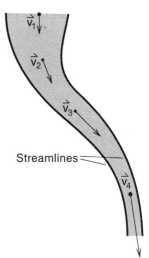

Continuity Equation

When we have a set of streamlines such as that in Figure (4), we have a good idea of the directions of flow. We can draw the direction of the velocity vector at any point by constructing a vector parallel to the streamline passing through that point. If the streamline we have drawn or photographed does not pass exactly through that point, then we can do a fairly good job of estimating the direction from the neighboring streamlines.

But what about the speed of the fluid? Every vector has both a magnitude and a direction. So far, the streamlines have told us only the directions of the velocity vectors. Can we determine or estimate the fluid speed at each point so that we can complete our description of the velocity field?

When there is construction on an interstate highway and the road is narrowed from two lanes to one, the traffic tends to go slowly through the construction. This makes sense for traffic safety, but it is just the wrong way to handle an efficient fluid flow. The traffic should go faster through the construction to make up for the reduced width of the road. (Can you imagine the person with an orange vest holding a sign that says "Fast"?) Water, when it flows down a tube with a constriction, travels faster through the constriction than

in the wide sections. This way, the same volume of water per second gets past the constriction as passes per second past a wide section of the channel. Applying this idea to Figure (4), we see why the velocity vectors are longer, the fluid speed higher, in the narrow sections of the streamline channels than in the wide sections.

It is not too hard to go from the qualitative idea that fluid must flow faster in the narrow sections of a channel, to a quantitative result that allows us to calculate how much faster. In Figure (5), we are considering a section of streamline or flow tube which has an entrance area A_1, and exit area A_2 as shown. In a short time Δt, the fluid at the entrance travels a distance $\Delta x_1 = v_1 \Delta t$, while at the exit the fluid goes a distance $\Delta x_2 = v_2 \Delta t$.

The volume of water that entered the stream during the time Δt is the shaded volume at the left side of the diagram, and is equal to the area A_1 times the distance Δx_1 that the fluid has moved

$$\left.\begin{array}{c} \text{Volume of water} \\ \text{entering in } \Delta t \end{array}\right\} = A_1 \Delta x_1 = A_1 v_1 \Delta t \qquad (1)$$

The volume of water leaving the same amount of time is

$$\left.\begin{array}{c} \text{Volume of water} \\ \text{leaving during } \Delta t \end{array}\right\} = A_2 \Delta x_2 = A_2 v_2 \Delta t \qquad (2)$$

If the water does not get squeezed up or compressed inside the stream between A_1 and A_2, if we have an *incompressible fluid*, which is quite true for water and in many cases even true for air, then the volume of fluid entering and the volume of the fluid leaving during the time Δt must be equal. Equating Equations (1) and (2) and cancelling the Δt gives

$$\boxed{A_1 v_1 = A_2 v_2} \qquad \textit{continuity equation} \qquad (3)$$

Equation (3) is known as the ***continuity equation*** for incompressible fluids. It is a statement that we do not squeeze up or lose any fluid in the stream. It also tells us that the velocity of the fluid is inversely proportional

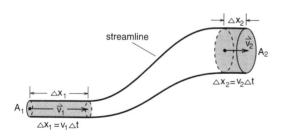

Figure 5
During the time Δt, water entering the small section of pipe travels a distance $v_1 \Delta t$, while water leaving the large section goes a distance $v_2 \Delta t$. Since the same amount of water must enter as leave, the entrance volume $A_1 \Delta x_1$ must equal the exit volume $A_2 \Delta x_2$. This gives $A_1 v_1 \Delta t = A_2 v_2 \Delta t$, or the result $A_1 v_1 = A_2 v_2$ which is one form of the continuity equation.

to the cross sectional area of the stream at that point. If the cross sectional area in a constriction has been cut in half, then the speed of the water must double in order to get the fluid through the constriction.

If we have a map of the streamlines, and know the entrance speed v_1 of the fluid, then we can determine the magnitude and direction of the fluid velocity $\vec{v}_2$ at any point downstream. The direction of $\vec{v}_2$ is parallel to the streamline at Point (2), and the magnitude is given by $v_2 = v_1 (A_1/A_2)$. Thus a careful map of the fluid streamlines, combined with the continuity equation, give us almost a complete picture of the fluid motion. The only additional information we need is the entrance speed.

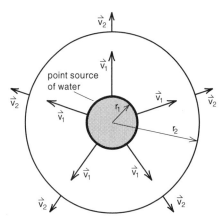

Figure 6

Point source of water. Imagine that water molecules are created inside the small sphere and flow radially out through its surface at a speed v_1. The same molecules will eventually flow out through the larger sphere at a lesser speed v_2. If no water molecules are created or destroyed outside the small sphere, then the continuity equation $A_1v_1 = A_2v_2$ requires that $\left(4\pi r_1^2\right)v_1 = \left(4\pi r_2^2\right)v_2$.

Velocity Field of a Point Source

This is an artificial example that shows us how to apply the continuity equation in a somewhat unexpected way, and leads to some ideas that will be very important in our later discussion of electric fields.

For this example, imagine a small magic sphere that creates water molecules inside and lets the water molecules flow out through the surface of the sphere. (Or there may be an unseen hose that supplies the water that flows out through the surface of the sphere.)

Let the small sphere have a radius r_1, area $4\pi r_1^2$ and assume that the water is emerging radially out through the small sphere at a speed v_1 as shown in Figure (6). Also let us picture that the small sphere is at the center of a huge swimming pool full of water, that the sides of the pool are so far away that the water continues to flow radially outward at least for several meters. Now conceptually construct a second sphere of radius $r_2 > r_1$ centered on the small sphere as in Figure (6).

During one second, the volume of water flowing out of the small sphere is v_1A_1, corresponding to $\Delta t = 1$ sec in Equation (1). By the continuity equation, the volume of water flowing out through the second sphere in one second, v_2A_2, must be the same in order that no water piles up between the spheres. Using the fact that $A_1 = 4\pi r_1^2$ and $A_2 = 4\pi r_2^2$, we get

$$v_1A_1 = v_2A_2 \qquad \begin{array}{l}\textit{continuity}\\\textit{equation}\end{array}$$

$$v_14\pi r_1^2 = v_24\pi r_2^2$$

$$v_2 = \frac{1}{r_2^2}\left(v_1r_1^2\right) \tag{4a}$$

Equation 4a tells us that as we go out from the "magic sphere", as the distance r_2 increases, the velocity v_2 drops off as the inverse square of r_2, as $1/r_2^2$. We can write this relationship in the form

$$v_2 \propto \frac{1}{r_2^2} \qquad \begin{array}{l}\textit{The symbol}\\\propto \textit{ means}\\\textit{"proportional to"}\end{array} \tag{4b}$$

A small spherical source like that shown in Figure (6) is often called a ***point source***. We see that a point source of water produces a $1/r^2$ velocity field.

Velocity Field of a Line Source

One more example which we will often use later is the *line source*. This is much easier to construct than the point source where we had to create water molecules. Good models for a line source of water are the sprinkling hoses used to water gardens. These hoses have a series of small holes that let the water flow radially outward.

For this example, imagine that we have a long sprinkler hose running down the center of an immense swimming pool. In Figure (7) we are looking at a cross section of the hose and see a radial flow that looks very much like Figure (6). The side view, however, is different. Here we see that we are dealing with a line rather than a point source of water.

Consider a section of the hose and fluid of length L. The volume of water flowing in one second out through this section of hose is $v_1 A_1$ where $A_1 = L$ times (the circumference of the hose) $= L(2\pi r_1)$.

$$\left.\begin{array}{l}\text{Volume of}\\\text{water / sec}\\\text{from a section}\\\text{L of hose}\end{array}\right\} = v_1 A_1 = v_1 L 2\pi r_1$$

If the swimming pool is big enough so that this water continues to flow radially out through a cylindrical area A_2 concentric with and surrounding the hose, then the volume of water per second (we will call this the "flux" of water) out through A_2 is

$$\left.\begin{array}{l}\text{Volume of}\\\text{water / sec out}\\\text{through } A_2\end{array}\right\} = v_2 A_2 = v_2 \left(2\pi r_2 L\right)$$

Using the continuity equation to equate these volumes of water per second gives

$$v_1 A_1 = v_2 A_2$$

$$v_1 L 2\pi r_1 = v_2 L 2\pi r_2 ; \quad v_1 r_1 = v_2 r_2$$

$$\boxed{v_2 = \frac{\left(v_1 r_1\right)}{r_2} \propto \frac{1}{r_2}} \tag{5}$$

We see that the velocity field of a line source drops off as $1/r$ rather than $1/r^2$ which we got from a point source.

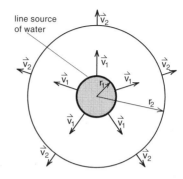

a) End view of line source

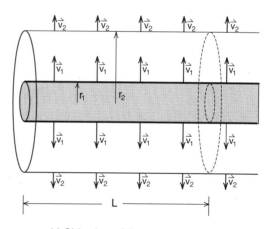

b) Side view of line source

Figure 7
Line source of water. In a line source, the water flows radially outward through a cylindrical area whose length we choose as L and whose circumference is $2\pi r$.

FLUX

Sometimes simply changing the name of a quantity leads us to new ways of thinking about it. In this case we are going to use the word *flux* to describe the *amount of water flowing per second* out of some volume. From the examples we have considered, the flux of water out through volumes V_1 and V_2 are given by the formulas

$$\left.\begin{array}{l}\text{Flux of}\\\text{water}\\\text{out of }V_1\end{array}\right\} \equiv \left\{\begin{array}{l}\text{Volume of}\\\text{water flowing}\\\text{per second}\\\text{out of }V_1\end{array}\right\} = v_1 A_1$$

$$\left.\begin{array}{l}\text{Flux of}\\\text{water}\\\text{out of }V_2\end{array}\right\} = v_2 A_2$$

The continuity equation can be restated by saying that the flux of water out of V_1 must equal the flux out of V_2 if the water does not get lost or compressed as it flows from the inner to the outer surface.

So far we have chosen simple surfaces, a sphere and a cylinder, and for these surfaces the flux of water is simply the fluid speed v times the area out through which it is flowing. Note that for our cylindrical surface shown in Figure (8), no water is flowing out through the ends of the cylinder, thus only the outside area $(2\pi rL)$ counted in our calculation of flux. A more general way of stating how we calculate flux is to say that it is the fluid speed v times the *perpendicular area* $A_\perp$ through which the fluid is flowing. For the cylinder, the perpendicular area $A_\perp$ is the outside area $(2\pi rL)$; the ends of the cylinder are parallel to the flow and therefore do not count.

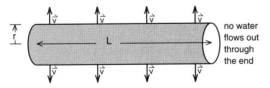

no water
flows out
through
the end

Figure 8
With a line source, all the water flows through the cylindrical surface surrounding the source and none through the ends. Thus $A_\perp$, the perpendicular area through which the water flows is $(2\pi r) \times L$.

The concept of flux can be generalized to irregular flows and irregularly shaped surfaces. To handle that case, break the flow up into a bunch of small flow tubes separated by streamlines, construct a perpendicular area for each flow tube as shown in Figure (9), and then calculate the total flux by adding up the fluxes from each flow tube.

$$\text{Total Flux} = v_1 A_1 + v_2 A_2 + v_2 A_2 + ... \\ = \sum_i v_i A_\perp^i \tag{6}$$

In the really messy cases, the sum over flow tubes becomes an integral as we take the limit of a large number of infinitesimal flow tubes.

For this text, we have gone too far. We will not work with very complicated flows. We can learn all we want from the simple ones like the flow out of a sphere or a cylinder. In those cases the perpendicular area is obvious and the flux easy to calculate. For the spherical flow of Figure (6), we see that the velocity field dropped off as $1/r^2$ as we went out from the center of the sphere. For the cylinder in Figure (7) the velocity field dropped off less rapidly, as $1/r$.

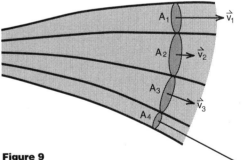

Figure 9
To calculate the flux of water in an arbitrarily shaped flow break up the flow into many small flux tubes where the fluid velocity is essentially uniform across the small tube as shown. The flux through the i-th tube is simply $v_i A_i$, and the total flux is the sum of the fluxes $\Sigma_i v_i A_i$ through each tube.

BERNOULLI'S EQUATION

Our discussion of flux was fairly lengthy, not so much for the results we got, but to establish concepts that we will use extensively later on in our discussion of electric fields. Another topic, Bernoulli's law, has a much more direct application to the understanding of fluid flows. It also has some rather surprising consequences which help explain why airplanes can fly and how a sailboat can sail up into the wind.

Bernoulli's law involves an energy relationship between the pressure, the height, and the velocity of a fluid. The theorem assumes that we have a constant density fluid moving with a steady flow, and that viscous effects are negligible, as they often are for fluids such as air and water.

Consider a small tube of flow bounded by streamlines as shown in Figure (10). In a short time Δt a small volume of fluid enters on the left and an equal volume exits on the right. If the exiting volume has more energy than the entering volume, the extra energy had to come from the work done by pressure forces acting on the fluid in the flow tube. Equating the work done by the pressure forces to the increase in energy gives us Bernoulli's equation.

To help visualize the situation, imagine that the streamline boundaries of the flow tube are replaced by frictionless, rigid walls. This would have no effect on the flow of the fluid, but focuses our attention on the ends of the tube where the fluid is flowing in on the left, at what we will call Point (1), and out on the right at Point (2).

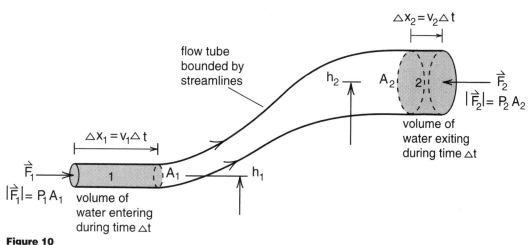

Figure 10

Derivation of Bernoulli's equation. Select a flow tube bounded by streamlines. For the steady flow of an incompressible fluid, during a time Δt the same volume of fluid must enter on the left as leave on the right. If the exiting fluid has more energy than the entering fluid, the increase must be a result of the net work done by the pressure forces acting on the fluid.

As a further aid to visualization, imagine that a small frictionless cylinder is temporarily inserted into the entrance of the tube as shown in Figure (11a), and at the exit as shown in Figure (11b). Such cylinders have no effect on the flow but help us picture the pressure forces.

At the entrance, if the fluid pressure is P_1 and the area of the cylinder is A_1, then the external fluid exerts a net force of magnitude

$$\left|\vec{F}_1\right| = P_1 A_1 \qquad (7a)$$

directed perpendicular to the surface of the cylinder as shown. We can think of this force $\vec{F}_1$ as the pressure force that the outside fluid exerts on the fluid inside the flow tube. At the exit, the external fluid exerts a pressure force $\vec{F}_2$ of magnitude

$$\left|\vec{F}_2\right| = P_2 A_2 \qquad (7b)$$

directed perpendicular to the piston, i.e., back toward the fluid inside the tube.

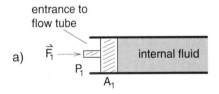

a)   entrance to flow tube
$\vec{F}_1$ → internal fluid
P_1
A_1

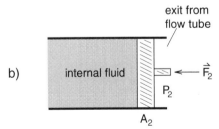

b)   exit from flow tube
internal fluid   ← $\vec{F}_2$
P_2
A_2

Figure 11
The flow would be unchanged if we temporarily inserted frictionless pistons at the entrance and exit.

Thus the fluid inside the flow tube is subject to external pressure forces, $\vec{F}_1$ in from the left and $\vec{F}_2$ in from the right. During a time Δt, the fluid at the entrance moves a distance $\Delta x_1 = v_1 \Delta t$ as shown in Figure (10). While moving this distance, the entering fluid is subject to the pressure force $\vec{F}_1$, thus the work ΔW_1 done by the pressure force at the entrance is

$$\Delta W_1 = \vec{F}_1 \cdot \Delta \vec{x}_1 = \left(P_1 A_1\right)\left(v_1 \Delta t\right) \qquad (8a)$$

At the exit, the fluid moves out a distance $\Delta x_2 = v_2 \Delta t$, while the external force pushes back in with a pressure force $\vec{F}_2$. Thus the pressure forces do **negative** work on the inside fluid, with the result

$$\Delta W_2 = \vec{F}_2 \cdot \Delta \vec{x}_2 = -\left(P_2 A_2\right)\left(v_2 \Delta t\right) \qquad (8b)$$

The net work ΔW done during a time Δt by external pressure forces on fluid inside the flow tube is therefore

$$\begin{aligned}\Delta W &= \Delta W_1 + \Delta W_2 \\ &= P_1\left(A_1 v_1 \Delta t\right) - P_2\left(A_2 v_2 \Delta t\right)\end{aligned} \qquad (9)$$

Equation (9) can be simplified by noting that $A_1 v_1 \Delta t = A_1 \Delta x_1$ is the volume ΔV_1 of the entering fluid. Likewise $A_2 v_2 \Delta t = A_2 \Delta x_2$ is the volume ΔV_2 of the exiting fluid. But during Δt, the same volume ΔV of fluid enters and leaves, thus $\Delta V_1 = \Delta V_2 = \Delta V$ and we can write Equation (9) as

$$\left.\begin{array}{l}\text{work done} \\ \text{by external} \\ \text{pressure} \\ \text{forces} \\ \text{on fluid} \\ \text{inside flow} \\ \text{tube}\end{array}\right\} \Delta W = \Delta V\left(P_1 - P_2\right) \qquad (10)$$

The next step is to calculate the change in energy of the entering and exiting volumes of fluid. The energy ΔE_1 of the entering fluid is its kinetic energy $1/2\left(\Delta m\right)v_1^2$ plus its gravitational potential energy $\left(\Delta m\right)gh_1$, where Δm is the mass of the entering fluid. If the fluid has a density ρ, then $\Delta m = \rho \Delta V$ and we get

$$\Delta E_1 = \frac{1}{2}\left(\rho\Delta V\right)v_1^2 + \left(\rho\Delta V\right)gh_1$$

$$= \Delta V\left(\frac{1}{2}\rho v_1^2 + \rho gh_1\right) \quad \text{(11a)}$$

At the exit, the same mass and volume of fluid leave in time Δt, and the energy of the exiting fluid is

$$\Delta E_2 = \Delta V\left(\frac{1}{2}\rho v_2^2 + \rho gh_2\right) \quad \text{(11b)}$$

The change ΔE in the energy in going from the entrance to the exit is therefore

$$\Delta E = \Delta E_2 - \Delta E_1$$

$$= \Delta V\left(\frac{1}{2}\rho v_2^2 + \rho gh_2 - \frac{1}{2}\rho v_1^2 - \rho gh_1\right) \quad \text{(12)}$$

Equating the work done, Equation (10) to the change in energy, Equation (12) gives

$$\Delta V(P_1 - P_2)$$

$$= \Delta V\left(\frac{1}{2}\rho v_2^2 + \rho gh_2 - \frac{1}{2}\rho v_1^2 - \rho gh_1\right) \quad \text{(13)}$$

Not only can we cancel the ΔVs in Equation (13), but we can rearrange the terms to make the result easier to remember. We get

$$\boxed{P_1 + \frac{1}{2}\rho v_1^2 + \rho gh_1 = P_2 + \frac{1}{2}\rho v_2^2 + \rho gh_2} \quad \text{(14)}$$

In this form, an interpretation of Bernoulli's equation begins to emerge. We see that the quantity $\left(P + \rho gh + \frac{1}{2}\rho v^2\right)$ has the same numerical value at the entrance, Point (1), as at the exit, Point (2). Since we can move the starting and ending points anywhere along the flow tube, we have the more general result

$$\boxed{\left(P + \rho gh + \frac{1}{2}\rho v^2\right) = \left\{\begin{array}{l}\text{constant anywhere} \\ \text{along a flow tube} \\ \text{or streamline}\end{array}\right.}$$

$$\text{(15)}$$

Equation (15) is our final statement of Bernoulli's equation. In words it says that for the steady flow of an incompressible, non viscous, fluid, the quantity $\left(P + \rho gh + 1/2\rho v^2\right)$ has a constant value along a streamline.

The restriction that $\left(P + \rho gh + 1/2\rho v^2\right)$ is constant *along a streamline* has to be taken seriously. Our derivation applied energy conservation to a plug moving along a small flow tube whose boundaries are streamlines. We did not consider plugs of fluid moving in different flow tubes, i.e., along different streamlines. For some special flows, the quantity $\left(P + \rho gh + 1/2\rho v^2\right)$ has the same value throughout the entire fluid. But for most flows, $\left(P + \rho gh + 1/2\rho v^2\right)$ has different values on different streamlines. Since we haven't told you what the special flows are, play it safe and assume that the numerical value of $\left(P + \rho gh + 1/2\rho v^2\right)$ can change when you hop from one streamline to another.

APPLICATIONS OF BERNOULLI'S EQUATION

Bernoulli's equation is a rather remarkable result that some quantity $\left(P + \rho gh + 1/2\rho v^2\right)$ has a value that doesn't change as you go along a streamline. The terms inside, except for the P term, look like the energy of a unit volume of fluid. The P term came from the work part of the energy conservation theorem, and cannot strictly be interpreted as some kind of pressure energy. As tempting as it is to try to give an interpretation to the terms in Bernoulli's equation, we will put that off for a while until we have worked out some practical applications of the formula. Once you see how much the equation can do, you will have a greater incentive to develop an interpretation.

Hydrostatics

Let us start with the simplest application of Bernoulli's equation, namely the case where the fluid is at rest. In a sense, all the fluid is on the same streamline, and we have

$$P + \rho gh = \begin{cases} \text{constant} \\ \text{throughout} \\ \text{the fluid} \end{cases} \qquad (16)$$

Suppose we have a tank of water shown in Figure (12). Let the pressure be atmospheric pressure at the surface, and set $h = 0$ at the surface. Therefore at the surface

$$P_{at} + \rho g(0) = \text{constant}$$

and the constant is P_{at}. For any depth $y = -h$, we have

$$P - \rho gy = \text{constant} = P_{at}$$

$$\boxed{P = P_{at} + \rho gy} \qquad (17)$$

We see that the increase in pressure at a depth y is ρgy, a well-known result from hydrostatics.

Exercise 1

The density of water is $\rho = 10^3 \text{Kg/m}^3$ and atmospheric pressure is $P_{at} = 1.0 \times 10^5 \text{N/m}^2$. At what depth does a scuba diver breath air at a pressure of 2 atmospheres? (At what depth does $\rho gy = P_{at}$?) (Your answer should be 10.2m or 33 ft.)

Exercise 2

What is the pressure, in atmospheres, at the deepest part of the ocean? (At a depth of 8 kilometers.)

Leaky Tank

For a slightly more challenging example, suppose we have a tank filled with water as shown in Figure (13). A distance h below the surface of the tank we drill a hole and the water runs out of the hole at a speed v. Use Bernoulli's equation to determine the speed v of the exiting water.

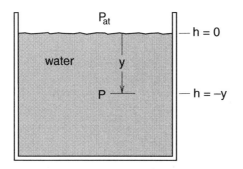
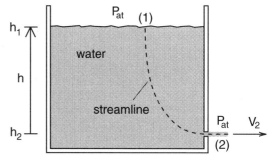

Figure 12
Hydrostatic pressure at a depth y is atmospheric pressure plus ρgy.

Figure 13
Water squirting out through a hole in a leaky tank. A streamline connects the leak at Point (2) with some Point (1) on the surface. Bernoulli's equation tells us that the water squirts out at the same speed it would have if it had fallen a height h.

Solution: Somewhere there will be a streamline connecting the free surface of the water (1) to a Point (2) in the exiting stream. Applying Bernoulli's equation to Points (1) and (2) gives

$$P_1 + \rho g h_1 + \frac{1}{2}\rho v_1^2 = P_2 + \rho g h_2 + \frac{1}{2}\rho v_2^2$$

Now $P_1 = P_2 = P_{at}$, so the Ps cancel. The water level in the tank is dropping very slowly, so that we can set $v_1 = 0$. Finally $h_1 - h_2 = h$, and we get

$$\frac{1}{2}\rho v_2^2 = \rho g(h_1 - h_2) = \rho g h \qquad (18)$$

The result is that the water coming out of the hole is moving just as fast as it would if it had fallen freely from the top surface to the hole we drilled.

Airplane Wing

In the example of a leaky tank, Bernoulli's equation gives a reasonable, not too exciting result. You might have guessed the answer by saying energy should be conserved. Now we will consider some examples that are more surprising than intuitive. The first explains how an airplane can stay up in the air.

Figure (14) shows the cross section of a typical airplane wing and some streamlines for a typical flow of fluid around the wing. (We copied the streamlines from our demonstration in Figure 3).

The wing is purposely designed so that the fluid has to flow farther to get over the top of the wing than it does to flow across the bottom. To travel this greater distance, the fluid has to move faster on the top of the wing (at Point 1), than at the bottom (at Point 2).

Arguing that the fluid at Point (1) on the top and Point (2) on the bottom started out on essentially the same streamline (Point 0), we can apply Bernoulli's equation to Points (1) and (2) with the result

$$P_1 + \frac{1}{2}\rho v_1^2 + \cancel{\rho g h_1} = P_2 + \frac{1}{2}\rho v_2^2 + \cancel{\rho g h_2}$$

We have crossed out the ρgh terms because the difference in hydrostatic pressure ρgh across the wing is negligible for a light fluid like air.

Here is the important observation. Since the fluid speed v_1 at the top of the wing is higher than the speed v_2 at the bottom, *the pressure P_2 at the bottom must be greater than P_1 at the top* in order that the sum of the two terms $\left(P + 1/2\rho v^2\right)$ be the same. The extra pressure on the bottom of the wing is what provides the lift that keeps the airplane up in the air.

There are two obvious criticisms of the above explanation of how airplanes get lift. What about stunt pilots who fly upside down? And how do balsa wood gliders with flat wings fly? The answer lies in the fact that the shape of the wing cross-section is only one of several important factors determining the flow pattern around a wing.

Figure (15) is a sketch of the flow pattern around a flat wing flying with a small angle of attack θ. By having an angle of attack, the wing creates a flow pattern where the streamlines around the top of the wing are longer than those under the bottom. The result is that the fluid flows faster over the top, therefore the pressure must be lower at the top (higher at the bottom) and we still get lift. The stunt pilot flying upside down must fly with a great enough angle of attack to overcome any downward lift designed into the wing.

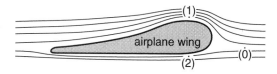

Figure 14
Streamline flow around an airplane wing. The wing is shaped so that the fluid flows faster over the top of the wing, Point (1) than underneath, Point (2). As a result the pressure is higher beneath Point (2) than above Point (1).

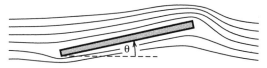

Figure 15
A balsa wood model plane gets lift by having the wing move forward with an upward tilt, or angle of attack. The flow pattern around the tilted wing gives rise to a faster flow and therefore reduced pressure over the top.

Sailboats

Sailboats rely on Bernoulli's principle not only to supply the "lift" force that allows the boat to sail into the wind, but also to create the "wing" itself. Figure (16) is a sketch of a sailboat heading at an angle θ off from the wind. If the sail has the shape shown, it looks like the airplane wing of Figure (14), the air will be moving faster over the outside curve of the sail (Position 1) than the inside (Position 2), and we get a higher pressure on the inside of the sail. This higher pressure on the inside both pushes the sail cloth out to give the sail an airplane wing shape, and creates the lift force shown in the diagram. This lift force has two components. One pulls the boat forward. The other component, however, tends to drag the boat sideways. To prevent the boat from slipping sideways, sailboats are equipped with a centerboard or a keel.

The operation of a sailboat is easily demonstrated using an air cart, glider and fan. Mount a small sail on top of the air cart glider (the light plastic shopping bags make excellent sail material) and elevate one end of the cart as shown in Figure (17) so that the cart rests at the low end. Then mount a fan so that the wind blows down and across as shown. With a little adjustment of the angle of the fan and the tilt of the air cart, you can observe the cart sail up the track, into the wind.

If you get the opportunity to sail a boat, remember that it is the Bernoulli effect that both shapes the sail and propels the boat. Try to adjust the sail so that it has a good airplane wing shape, and remember that the higher speed wind on the outside of the sail creates a low pressure that sucks the sailboat forward. You'll go faster if you keep these principles in mind.

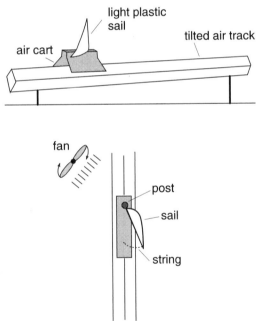

Figure 17
Sailboat demonstration. It is easy to rig a mast on an air cart, and use a small piece of a light plastic bag for a sail. Place the cart on a tilted air track so that the cart will naturally fall backward. Then turn on a fan as shown, and the cart sails up the track into the wind.

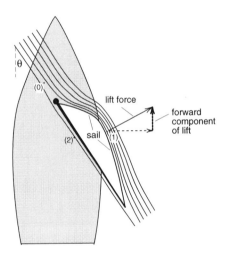

Figure 16
A properly designed sail takes on the shape of an airplane wing with the wind traveling faster, creating a lower pressure on the outside of the sail (Point 1). This low pressure on the outside both sucks the canvas out to maintain the shape of the sail and provides the lift force. The forward component of the lift force moves the boat forward and the sideways component is offset by the water acting on the keel.

The Venturi Meter

Another example, often advertised as a simple application of Bernoulli's equation, is the Venturi meter shown in Figure (18). We have a tube with a constriction, so that its cross-sectional area A_1 at the entrance and the exit, is reduced to A_2 at the constriction. By the continuity equation (3), we have

$$v_1 A_1 = v_2 A_2; \quad v_2 = \frac{v_1 A_1}{A_2}$$

As expected, the fluid travels faster through the constriction since $A_1 > A_2$.

Now apply Bernoulli's equation to Points (1) and (2). Since these points are at the same height, the $\rho g h$ terms cancel and Bernoulli's equation becomes

$$P_1 + \tfrac{1}{2}\rho v_1^2 = P_2 + \tfrac{1}{2}\rho v_2^2$$

Since $v_2 > v_1$, the pressure P_2 in the constriction must be less than the pressure P_1 in the main part of the tube. Using $v_2 = v_1 A_1/A_2$, we get

$$\left.\begin{matrix}\text{Pressure} \\ \text{drop in} \\ \text{constriction}\end{matrix}\right\} = \left(P_1 - P_2\right)$$

$$= \tfrac{1}{2}\rho\left(v_2^2 - v_1^2\right)$$

$$= \tfrac{1}{2}\rho\left(v_1^2 A_1^2/A_2^2 - v_1^2\right) \quad (19)$$

$$= \tfrac{1}{2}\rho v_1^2\left(A_1^2/A_2^2 - 1\right)$$

To observe the pressure drop, we can mount small tubes (A) and (B) as shown in Figure (18), to act as barometers. The lower pressure in the constriction will

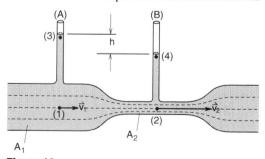

Figure 18
Venturi meter. Since the water flows faster through the constriction, the pressure is lower there. By using vertical tubes to measure the pressure drop, and using Bernoulli's equation and the continuity equation, you can determine the flow speeds v_1 and v_2.

cause the fluid level in barometer (B) to be lower than in the barometer over the slowly moving, high pressure stream. The height difference h means that there is a pressure difference

$$\left.\begin{matrix}\text{Pressure} \\ \text{difference}\end{matrix}\right) = (P_1 - P_2) = \rho g h \quad (20)$$

If we combine Equations (19) and (20), ρ cancels and we can solve for the speed v_1 of the fluid in the tube in terms of the quantities g, h, A_1 and A_2. The result is

$$v_1 = \sqrt{\frac{2gh}{\left(A_1^2/A_2^2 - 1\right)}} \quad (21)$$

Because we can determine the speed v_1 of the main flow by measuring the height difference h of the two columns of fluid, the setup in Figure (18) forms the basis of an often used meter to measure fluid flows. A meter based on this principle is called a *Venturi* meter.

Exercise 3

Show that all the terms in Bernoulli's equation have the same dimensions. (Use MKS units.)

Exercise 4

In a classroom demonstration of a venturi meter shown in Figure (18a), the inlet and outlet pipes had diameters of 2 cm and the constriction a diameter of 1 cm. For a certain flow, we noted that the height difference h in the barometer tubes was 7 cm. How fast, in meters/sec, was the fluid flowing in the inlet pipe?

Figure 18a
Venturi demonstration. We see about a 7cm drop in the height of the barometer tubes at the constriction.

The Aspirator

In Figure (18), the faster we move the fluid through the constriction (the greater v_1 and therefore v_2), the greater the height difference h in the two barometer columns. If we turn v_1 up high enough, the fluid is moving so fast through section 2 that the pressure becomes negative and we get suction in barometer 2. For even higher speed flows, the suction at the constriction becomes quite strong and we have effectively created a crude vacuum pump called an ***aspirator***. Typically aspirators like that shown in Figure (19) are mounted on cold water faucets in chemistry labs and are used for sucking up various kinds of fluids.

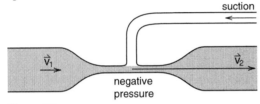

Figure 19a
If the water flows through the constriction fast enough, you get a negative pressure and suction in the attached tube.

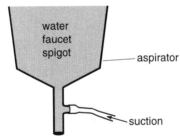

Figure 19b
If the constriction is placed on the end of a water faucet as shown, you have a device called an aspirator that is often used in chemistry labs for sucking up fluids.

Care in Applying Bernoulli's Equation

Although the Venturi meter and aspirator are often used as simple examples of Bernoulli's equation, considerable care must be used in applying Bernoulli's equation in these examples. To illustrate the trouble you can get into, suppose you tried to apply Bernoulli's equation to Points (3) and (4) of Figure (20). You would write

$$P_3 + \rho g h_3 + \frac{1}{2}\rho v_3^2 = P_4 + \rho g h_4 + \frac{1}{2}\rho v_4^2$$

$$(22)$$

Now $P_3 = P_4 = P_{atmosphere}$ because Points (3) and (4) are at the liquid surface. In addition the fluid is at rest in tubes (3) and (4), therefore $v_3 = v_4 = 0$. Therefore Bernoulli's equation predicts that

$$\rho g h_3 = \rho g h_4$$

or that $h_3 = h_4$ and there should be no height difference!

What went wrong? The mistake results from the fact that no streamlines go from position (3) to position (4), and therefore Bernoulli's equation does not have to apply. As shown in Figure (21) the streamlines flow across the bottom of the barometer tubes but do not go up into them. It turns out that we cannot apply Bernoulli's equation across this break in the streamlines. It requires some experience or a more advanced knowledge of hydrodynamic theory to know that you can treat the little tubes as barometers and get the

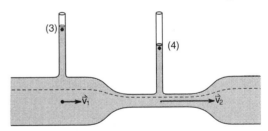

Figure 20
If you try to apply Bernoulli's equation to Points (3) and (4), you predict, incorrectly, that Points (3) and (4) should be at the same height. The error is that Points (3) and (4) do not lie on the same streamline, and therefore you cannot apply Bernoulli's equation to them.

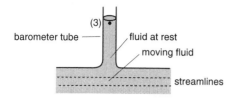

Figure 21
The water flows past the bottom of the barometer tube, not up into the tube. Thus Point (3) is not connected to any of the streamlines in the flow. The vertical tube acts essentially as a barometer, measuring the pressure of the fluid flowing beneath it.

correct answer. Most texts ignore this complication, but there are always some students who are clever enough to try to apply Bernoulli's equation across the break in the flow at the bottom of the small tubes and then wonder why they do not get reasonable answers.

There is a remarkable fluid called ***superfluid helium*** which under certain circumstances will not have a break in the flow at the base of the barometer tubes. (Superfluid helium is liquefied helium gas cooled to a temperature below $2.17°\,K$). As shown in Figure (22) the streamlines actually go up into the barometer tubes, Points (3) and (4) are connected by a streamline, Bernoulli's equation should apply and we should get no height difference. This experiment was performed in 1965 by Robert Meservey and the heights in the two barometer tubes were just the same!

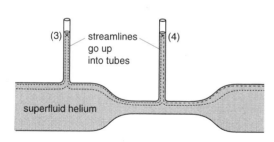

Figure 22
In superfluid helium, the streamlines actually go up into the barometer tubes and Bernoulli's equation can be applied to Points (3) and (4). The result is that the heights of the fluid are the same as predicted. (Experiment by R. Meservey, see Physics Of Fluids, July 1965.)

Hydrodynamic Voltage

When we studied the motion of a projectile, we found that the quantity $(1/2\,mv^2 + mgh)$ did not change as the ball moved along its parabolic trajectory. When physicists discover a quantity like $(1/2\,mv^2 + mgh)$ that does not change, they give that quantity a name, in this case *"the ball's total energy"*, and then say that they have discovered a new law, namely "the ball's total energy is conserved as the ball moves along its trajectory".

With Bernoulli's equation we have a quantity $\left(P + \rho gh + 1/2\rho v^2\right)$ which is constant along a streamline when we have the steady flow of an incompressible, non viscous fluid. Here we have a quantity $\left(P + \rho gh + 1/2\rho v^2\right)$ that is conserved under special circumstances; perhaps we should give this quantity a name also.

The term ρgh is the gravitational potential energy of a unit volume of the fluid, and $1/2\rho v^2$ is the same volume's kinetic energy. Thus our Bernoulli term has the dimensions and characteristics of the energy of a unit volume of fluid. But the pressure term, which came from the work part of the derivation of Bernoulli's equation, is not a real energy term. There is no pressure energy P stored in an incompressible fluid, and Bernoulli's equation is not truly a statement of energy conservation for a unit volume of fluid.

However, as we have seen, the Bernoulli term is a useful concept, and deserves a name. Once we name it, we can say that " " is conserved along a streamline under the right circumstances. Surprisingly there is not an extensive tradition for giving the Bernoulli term a name so that we have to concoct a name here. At this point our choice of name will seem a bit peculiar, but it is chosen with later discussions in mind. We will call the Bernoulli term ***hydrodynamic voltage***

$$\text{Hydrodynamic Voltage} \equiv \left(P + \rho gh + \frac{1}{2}\rho v^2\right) \quad (23)$$

and Bernoulli's equation states that ***the hydrodynamic voltage of an incompressible, non viscous fluid is constant along a streamline when the flow is steady.***

We obviously did not invent the word voltage; the name is commonly used in discussing electrical devices like high voltage wires and low voltage batteries. It turns out that there is a precise analogy between the concept of voltage used in electricity theory, and the Bernoulli term we have been discussing. To emphasize the analogy, we are naming the Bernoulli term hydrodynamic voltage. The word "hydrodynamic" is included to remind us that we are missing some of the electrical terms in a more general definition of voltage. We are discussing hydrodynamic voltage before electrical voltage because hydrodynamic voltage involves fluid concepts that are more familiar, easier to visualize and study, than the corresponding electrical concepts.

Town Water Supply

One of the familiar sights in towns where there are no nearby hills is the water tank somewhat crudely illustrated in Figure (23). Water is pumped from the reservoir into the tank to fill the tank up to a height h as shown.

For now let us assume that all the pipes attached to the tanks are relatively large and frictionless so that we can neglect viscous effects and apply Bernoulli's equation to the water at the various points along the water system. At Point (1), the pressure is simply atmospheric pressure P_{at}, the water is essentially not flowing, and the hydrodynamic voltage consists mainly of P_{at} plus the gravitational term gh_1

$$\left.\begin{array}{l}\text{Hydrodynamic}\\\text{Voltage}\end{array}\right\}_1 = P_{at} + gh_1$$

By placing the tank high up in the air, the gh_1 term can be made quite large. We can say that the tank gives us "high voltage" water.

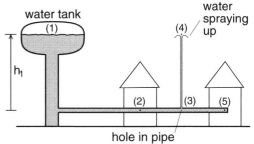

Figure 23
The pressure in the town water supply may be maintained by pumping water into a water tank as shown. If the pipes are big enough we can neglect the viscous effect and apply Bernoulli's equation throughout the system, including the break in the water pipe at Point (3), and the top of the fountain, Point (4).

Bernoulli's equation tells us that the hydrodynamic voltage of the water is the same at all the points along the water system. The purpose of the water tank is to ensure that we have high voltage water throughout the town. For example, at Point (2) at one of the closed faucets in the second house, there is no height left ($h_2 = 0$) and the water is not flowing. Thus all the voltage shows up as high pressure at the faucet.

$$\left. \begin{array}{l} \text{Hydrodynamic} \\ \text{Voltage} \end{array} \right\}_2 = P_2$$

At Point (3) we have a break in the pipe and water is squirting up. Just above the break the pressure has dropped to atmospheric pressure and there is still no height. At this point the voltage appears mainly in the form of kinetic energy.

$$\left. \begin{array}{l} \text{Hydrodynamic} \\ \text{Voltage} \end{array} \right\}_3 = P_{at} + \frac{1}{2}\rho v^2$$

Finally at Point (4) the water from the break reaches its maximum height and comes to rest before falling down again. Here it has no kinetic energy, the pressure is still atmospheric, and the hydrodynamic voltage is back in the form of gravitational potential energy. If no voltage has been lost, if Bernoulli's equation still holds, then the water at Point (4) must rise to the same height as the water at the surface in the town water tank.

In some sense, the town water tank serves as a huge "battery" to supply the hydrodynamic voltage for the town water system.

Viscous Effects

We said that the hydrodynamic analogy of voltage involves familiar concepts. Sometimes the concepts are too familiar. Has your shower suddenly turned cold when someone in the kitchen drew hot water for washing dishes; or turned hot when the toilet was flushed? Or been reduced to a trickle when the laundry was being washed? In all of these cases there was a pressure drop at the shower head of either the hot water, the cold water, or both. A pressure drop means that you are getting lower voltage water at the shower head than was supplied by the town water tank (or by your home pressure tank).

The hydrodynamic voltage drop results from the fact that you are trying to draw too much water through small pipes, viscous forces become important, and Bernoulli's equation no longer applies. Viscous forces always cause a drop in the hydrodynamic voltage. This voltage drop can be seen in a classroom demonstration, Figure (24), where we have inserted a series of small barometer tubes in a relatively small flow tube. If we run a relatively high speed stream of water through the flow tube, viscous effects become observable and the pressure drops as the water flows down the tube. The pressure drop is made clear by the decreasing heights of the water in the barometer tubes as we go downstream.

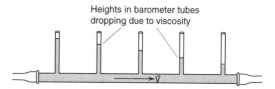

Heights in barometer tubes dropping due to viscosity

Figure 24
If we have a fairly fast flow in a fairly small tube, viscosity causes a pressure drop, or as we are calling it, a "hydrodynamic voltage" drop down the tube. This voltage drop is seen in the decreasing heights of the water in the barometer tubes. (In our Venturi demonstration of Figure (18a), the heights are lower on the exit side than the entrance side due to viscosity acting in the constriction.)

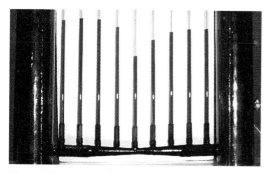

Figure 18a repeated
Venturi demonstration.

VORTICES

The flows we have been considering, water in a pipe, air past a sailboat sail, are tame compared to a striking phenomena seen naturally in the form of hurricanes and tornados. These are examples of a fluid motion called a *vortex*. They are an extension, to an atmospheric scale, of the common bathtub vortex like the one we created in the funnel seen in Figure (25).

Vortices have a fairly well-defined structure which is seen most dramatically in the case of the tornado (see Figures 29 and 30). At the center of the vortex is the core. The core of a bathtub vortex is the hollow tube of air that goes down the drain. In a tornado or water spout, the core is the rapidly rotating air. For a hurricane it is the eye, seen in Figures (27) and (28), which can be amazingly calm and serene considering the vicious winds and rain just outside the eye.

Figure 25
Bathtub vortex in a funnel. We stirred the water before letting it drain out.

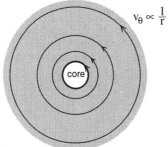

Figure 26
Vortices tend to have a circular velocity field about the core, a velocity field v_θ whose strength tends to drop off as 1/r as you go out from the core.

Outside the core, the fluid goes around in a circular pattern, the speed decreasing as the distance from the center increases. It turns out that viscous effects are minimized if the fluid speed drops off as 1/r where r is the radial distance from the center of the core as shown in Figure (26). At some distance from the center, the speed drops to below the speed of other local disturbances and we no longer see the organized motion.

The tendency of a fluid to try to maintain a 1/r velocity field explains why vortices have to have a core. You cannot maintain a 1/r velocity field down to r = 0, for then you would have infinite velocities at the center. To avoid this problem, the vortex either throws the fluid out of the core, as in the case of the hollow bathtub vortex, or has the fluid in the core move as a solid rotating object ($v_\theta = r\,\omega$) in the case of a tornado, or has a calm fluid when the core is large (i.e., viscous effects of the land are important) as in the case of a hurricane.

While the tornado is a very well organized example of a vortex, it has been difficult to do precise measurements of the wind speeds in a tornado. One of the best measurements verifying the 1/r velocity field was when a tornado hit a lumber yard, and a television station using a helicopter recorded the motion of sheets of 4' by 8' plywood that were scattered by the tornado. (Using doppler radar, a wind speed of 318 miles per hour was recorded in a tornado that struck Oklahoma city on May 3, 1999--a world wind speed record!)

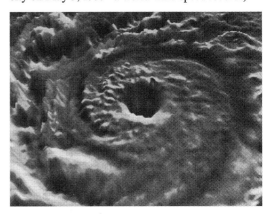

Figure 27
Eye of hurricane Allen viewed from a satellite. (Photograph courtesy of A. F. Haasler.)

Figure 29
Tornado in Kansas.

Figure 30
A tornado over water is called a water spout.

Figure 28
Hurricane approaching the east coast of the U.S.

Quantized Vortices in Superfluids

For precision, nothing beats the "quantized" vortex in superfluid helium. We have already mentioned that superfluid helium flows up and down the little barometer tubes in a Venturi meter, giving no height difference and nullifying the effectiveness of the device as a velocity meter. This happened because superfluid helium has **NO** viscosity (absolutely none as far as we can tell) and can therefore flow into tiny places where other fluids cannot move.

More surprising yet is the structure of a vortex in superfluid helium. The vortex has a core that is about one atomic diameter across (you can't get much smaller than that), and a precise 1/r velocity field outside the core. Even more peculiar is the fact that the velocity field outside the core is given by the formula

$$ v_\theta = \frac{\kappa}{2\pi r} \; ; \; \kappa = \frac{h}{m_{He}} \tag{24} $$

where κ, called the "circulation of the vortex", has the precisely known value h/m_{He}, where m_{He} is the mass of a helium atom, and h is an atomic constant known as **Planck's constant**. The remarkable point is that the strength of a helium vortex has a precise value determined by atomic scale constants. (This is why we say that vortices in superfluid helium are quantized.) When we get to the study of atoms, and particularly the Bohr theory of hydrogen, we can begin to explain why helium vortices have precisely the strength $\kappa = h/m_{He}$. For now, we are mentioning vortices in superfluid helium as examples of an ideal vortex with a well-defined core and a precise 1/r velocity field outside.

Quantized vortices of a more complicated structure also occur in superconductors and play an important role in the practical behavior of a superconducting material. The superconductors that carry the greatest currents, and are the most useful in practical applications, have quantized vortices that are pinned down and cannot move around. One of the problems in developing practical applications for the new high temperature superconductors is that the quantized vortices tend to move and cause energy losses. Pinning these vortices down is one of the main goals of current engineering research.

Exercise 5

This was an experiment, performed in the 1970s to study how platelets form plaque in arteries. The idea was that platelets deposit out of the blood if the flow of blood is too slow. The purpose of the experiment was to design a flow where one could easily see where the plaque began to form and also know what the velocity of the flow was there.

The apparatus is shown in Figure (31). Blood flows down through a small tube and then through a hole in a circular plate that is suspended a small distance d above a glass plate. When the blood gets to the glass it flows radially outward as indicated in Figure (31c). As the blood flowed radially outward, its velocity decreases. At a certain radius, call it r_p, platelets began to deposit on the glass. The flow was photographed by a video camera looking up through the glass.

For this problem, assume that the tube radius was $r_t = .4mm$, and that the separation d between the circular plate and the glass was d = .5 mm. If blood were flowing down the inlet tube at a rate of half a cubic centimeter per second, what is the average speed of the blood

a) inside the inlet tube?

b) at a radius $r_p = 2cm$ out from the hole in the circular plate?

(By average speed, we mean neglect fluid friction at walls, and assume that the flow is uniform across the radius of the inlet pipe and across the gap as indicated in Figure (31d).

Exercise 6

A good review of both the continuity equation and Bernoulli's equation, is to derive on your own, without looking back at the text, the formula

$$ v_1 = \sqrt{\frac{2gh}{A_1^2/A_2^2 - 1}} \tag{21} $$

for the flow speed in a venturi meter. The various quantities v_1, h, A_1 and A_2 are defined in Figure (18) reproduced on the opposite page. (If you have trouble with the derivation, review it in the text, and then a day or so later, try the derivation again on your own.

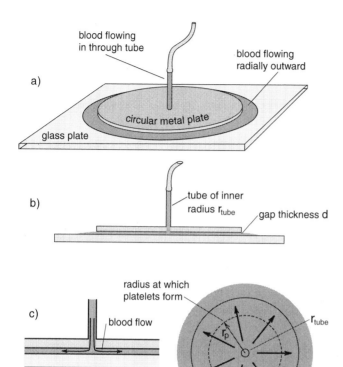

a) blood flowing in through tube

blood flowing radially outward

circular metal plate

glass plate

b) tube of inner radius r$_{tube}$

gap thickness d

c) radius at which platelets form

blood flow

r$_p$

r$_{tube}$

Figure 31 a,b,c
Experiment to measure the blood flow velocity at which platelets stick to a glass plate. This is an application of the continuity equation.

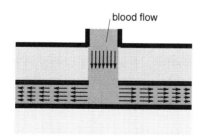

blood flow

Figure 31d
Neglect fluid friction at walls, and assume that the flow is uniform across the radius of the inlet pipe and across the gap

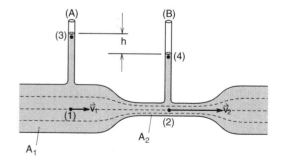

(A)   (B)

(3)   h   (4)

(1) $\vec{v}_1$ (2) $\vec{v}_2$

A$_1$ A$_2$

Figure 18
Venturi meter. Since the water flows faster through the constriction, the pressure is lower there. By using vertical tubes to measure the pressure drop, and using Bernoulli's equation and the continuity equation, you can determine the flow speeds v_1 and v_2.

Chapter 24

Coulomb's Law and Gauss' Law

In our discussion of the four basic interactions we saw that the electric and gravitational interaction had very similar $1/r^2$ force laws, but produced very different kinds of structures. The gravitationally bound structures include planets, solar systems, star clusters, galaxies, and clusters of galaxies. Typical electrically bound structures are atoms, molecules, people, and redwood trees. Although the force laws are similar in form, the differences in the structures they create result from two important differences in the forces. Gravity is weaker, far weaker, than electricity. On an atomic scale gravity is so weak that its effects have not been seen. But electricity has both attractive and repulsive forces. On a large scale, the electric forces cancel so completely that the weak but non-cancelling gravity dominates astronomical structures.

COULOMB'S LAW

In Chapter 18 we briefly discussed Coulomb's electric force law, primarily to compare it with gravitational force law. We wrote Coulomb's law in the form

$$\left| \vec{F} \right| = \frac{KQ_1Q_2}{r^2} \qquad \begin{array}{l} Coulomb's \\ Law \end{array} \qquad (1)$$

where Q_1 and Q_2 are two charges separated by a distance r as shown in Figure (1). If the charges Q_1 and Q_2 are of the same sign, the force is repulsive, if they are of the opposite sign it is attractive. The strength of the electric force decreases as $1/r^2$ just like the gravitational force between two masses.

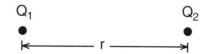

Figure 1
Two particles of charge Q_1 and Q_2, separated by a distance r.

CGS Units

In the CGS system of units, the constant K in Equation (1) is taken to have the numerical value 1, so that Coulomb's law becomes

$$\left|\vec{F}_e\right| = \frac{Q_1 Q_2}{r^2} \qquad (2)$$

Equation (2) can be used as an experimental definition of charge. Let Q_1 be some accepted standard charge. Then any other charge Q_2 can be determined in terms of the standard Q_1 by measuring the force on Q_2 when the separation is r. In our discussion in Chapter 18, we took the standard Q_1 as the charge on an electron.

This process of defining a standard charge Q_1 and using Equation (1) to determine other charges, is easy in principle but almost impossible in practice. These so-called *electrostatic* measurements are subject to all sorts of experimental problems such as charge leaking away due to a humid atmosphere, redistribution of charge, static charge on the experimenter, etc. Charles Coulomb worked hard just to show that the electric force between two charges did indeed drop as $1/r^2$. As a practical matter, Equation (2) is not used to define electric charge. As we will see, the more easily controlled magnetic forces are used instead.

MKS Units

When you buy a 100 *watt* bulb at the store for use in your home, you will see that it is rated for use at 110–120 *volts* if you live in the United States or Canada, or 220–240 volts most elsewhere. The circuit breakers in your house may allow each circuit to carry up to 15 or 20 *amperes* of current in each circuit. The familiar quantities *volts*, *amperes*, *watts* are all MKS units. The corresponding quantities in CGS units are the totally unfamiliar *statvolts*, *statamps*, and *ergs per second*. Some scientific disciplines, particularly plasma and solid state physics, are conventionally done in CGS units but the rest of the world uses MKS units for describing electrical phenomena.

Although we work with the familiar quantities volts, amps and watts using MKS units, there is a price we have to pay for this convenience. In MKS units the constant K in Coulomb's law is written in a rather peculiar way, namely

$$K = \frac{1}{4\pi\varepsilon_0}$$
$$\varepsilon_0 = 8.85 \times 10^{-12}\ \text{farads/meter} \qquad (3)$$

and Coulomb's law is written as

$$\left|\vec{F}_e\right| = \frac{Q_1 Q_2}{4\pi\varepsilon_0 r^2} \qquad (4)$$

where Q_1 and Q_2 are the charges measured in coulombs, and r is the separation measured in meters, and the force $\left|\vec{F}_e\right|$ is in newtons.

Before we can use Equation (4), we have to know how big a unit of charge a coulomb is, and we would probably like to know why there is a 4π in the formula, and why the proportionality constant ε_0 ("*epsilon naught*") is in the denominator.

As we saw in Chapter 18, nature has a basic unit of charge (e) which we call the charge on the electron. A coulomb of charge is 6.25×10^{18} times larger. Just as a liter of water is a large convenient collection of water molecules, 3.34×10^{25} of them, the coulomb can be thought of as a large collection of electron charges, 6.25×10^{18} of them.

In practice, the coulomb is defined experimentally, not by counting electrons, and not by the use of Coulomb's law, but, as we said, by a magnetic force measurement to be described later. For now, just think of the coulomb as a convenient unit made up of 6.25×10^{18} electron charges.

Once the size of the unit charge is chosen, the proportionality constant in Equation (4) can be determined by experiment. If we insist on putting the proportionality constant in the denominator and including a 4π, then ε_0 has the value of 8.85×10^{-12}, which we will often approximate as 9×10^{-12}.

Why is the proportionality constant ε_0 placed in the denominator? The kindest answer is to say that this is a historical choice that we still have with us. And why include the 4π? There is a better answer to this question. By putting the 4π now, we get rid of it in another law that we will discuss shortly, called *Gauss' law*. If you work with Gauss' law, it is convenient to have the 4π buried in Coulomb's law. But if you work with Coulomb's law, you will find the 4π to be a nuisance.

Checking Units in MKS Calculations

If we write out the units in Equation (4), we get

$$\vec{F}(\text{newtons}) = \frac{Q_1(\text{coul})Q_2(\text{coul})}{4\pi\varepsilon_0 r^2(\text{meter})^2} \qquad (4a)$$

In order for the units in Equation (3a) to balance, the proportionality constant ε_0 must have the dimensions

$$\varepsilon_0 \frac{\text{coulombs}^2}{\text{meter}^2\text{newton}} \qquad (5)$$

In earlier work with projectiles, etc., it was often useful to keep track of your units during a calculation as a check for errors. In MKS electrical calculations, it is almost impossible to do so. Units like *coul²/(meter²newton)* are bad enough as they are. But if you look up ε_0 in a textbook, you will find its units are listed as *farads/meter*. In other words the combination *coul²/(newton meter)* was given the name *farad*. With naming like this, you do not stand a chance of keeping units straight during a calculation. You have to do the best you can to avoid mistakes without having the reassurance that your units check.

Summary

The situation with Coulomb's law is not really that bad. We have a $1/r^2$ force law like gravity, charge is measured in coulombs, which is no worse than measuring mass in kilograms, and the proportionality constant just happens, for historical reasons, to be written as $1/4\pi\varepsilon_0$.

The units are incomprehensible, so do not worry too much about keeping track of units. After a bit of practice, Coulomb's law will become quite natural.

Example 1 *Two Charges*

Two positive charges, each 1 coulomb in size, are placed 1 meter apart. What is the electric force between them?

Solution: The force will be repulsive, and have a magnitude

$$|\vec{F}_e| = \frac{Q_1 Q_2}{4\pi\varepsilon_0 r^2} = \frac{1}{4\pi\varepsilon_0}$$
$$= \frac{1}{4\pi\times 9 \times 10^{-12}} = 10^{10} \text{ newtons}$$

From the answer, 10^{10} newtons, we see that a coulomb is a huge amount of charge. We would not be able to assemble two 1 coulomb charges and put them in the same room. They would tear the room apart.

Example 2 *Hydrogen Atom*

In a classical model of a hydrogen atom, we have a proton at the center of the atom and an electron traveling in a circular orbit around the proton. If the radius of the electron's orbit is $r = .5 \times 10^{-10}$ meters, how long does it take the electron to go around the proton once?

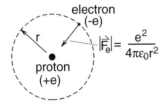

Solution: This problem is more conveniently handled in CGS units, but there is nothing wrong with using MKS units. The charge on the proton is (+e), on the electron (−e), thus the electrical force is attractive and has a magnitude

$$\left|\vec{F}_e\right| = \frac{(e)(e)}{4\pi\varepsilon_0 r^2} = \frac{e^2}{4\pi\varepsilon_0 r^2}$$

With $e = 1.6 \times 10^{-19}$ coulombs and $r = .5 \times 10^{-10}$m we get

$$\left|\vec{F}_e\right| = \frac{\left(1.6 \times 10^{-19}\right)^2}{4\pi \times \left(9 \times 10^{-12}\right) \times \left(.5 \times 10^{-10}\right)^2}$$

$$= 9 \times 10^{-8} \text{ newtons}$$

Since the electron is in a circular orbit, its acceleration is v^2/r pointing toward the center of the circle, and we get

$$\left|\vec{a}\right| = \frac{v^2}{r} = \frac{\left|\vec{F}\right|}{m} = \frac{F_e}{m}$$

With the electron mass m equal to 9.11×10^{-31} kg, we have

$$v^2 = \frac{rF_e}{m} = \frac{\left(.5 \times 10^{-10}\right) \times 9 \times 10^{-8}}{9.11 \times 10^{-31} \text{ kg}}$$

$$= 4.9 \times 10^{12} \frac{m^2}{s^2}$$

$$v = 2.2 \times 10^6 \frac{m}{s}$$

To go around a circle of radius r at a speed v takes a time

$$T = \frac{2\pi r}{v} = \frac{2\pi \times .5 \times 10^{-10}}{2.2 \times 10^6}$$

$$= 1.4 \times 10^{-16} \text{seconds}$$

In this calculation, we had to deal with a lot of very small or large numbers, and there was not much of an extra burden putting the $1/4\pi\varepsilon_0$ in Coulomb's law.

Exercise 1

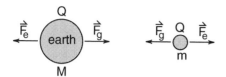

(a) Equal numbers of electrons are added to both the earth and the moon until the repulsive electric force exactly balances the attractive gravitational force. How many electrons are added to the earth and what is their total charge in coulombs?

(b) What is the mass, in kilograms of the electrons added to the earth in part (a)?

Exercise 2

Calculate the ratio of the electric to the gravitational force between two electrons. Why does your answer not depend upon how far apart the electrons are?

Exercise 3

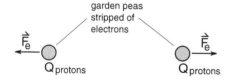

Imagine that we could strip all the electrons out of two garden peas, and then placed the peas one meter apart. What would be the repulsive force between them? Express your answer in newtons, and metric tons. (One metric ton is the weight of 1000 kilograms.) (Assume the peas each have about one Avogadro's number, or gram, of protons.)

Exercise 4

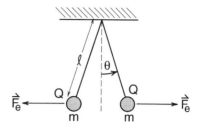

Two styrofoam balls covered by aluminum foil are suspended by equal length threads from a common point as shown. They are both charged negatively by touching them with a rubber rod that has been rubbed by cat fur. They spread apart by an angle 2θ as shown. Assuming that an equal amount of charge Q has been placed on each ball, calculate Q if the thread length is $\ell = 40\,cm$, the mass m of the balls is m=10 gm, and the angle is $\theta = 5°$. Use Coulomb's law in the form $\left|\vec{F_e}\right| = Q_1 Q_2 / 4\pi\varepsilon_0 r^2$, and remember that you must use MKS units for this form of the force law.

FORCE PRODUCED BY A LINE CHARGE

In our discussion of gravitational forces, we dealt only with point masses because most practical problems deal with spherical objects like moons, planets and stars which can be treated as point masses, or spacecraft which are essentially points.

The kind of problem we did not consider is the following. Suppose an advanced civilization constructed a rod shaped planet shown in Figure (2) that was 200,000 kilometers long and had a radius of 10,000 km. A satellite is launched in a circular orbit of radius 20,000 km, what is the period of the satellite's orbit?

We did not have problems like this because no one has thought of a good reason for constructing a rod shaped planet. Our spherical planets, which can be treated as a point mass, serve well enough.

In studying electrical phenomena, we are not restricted to spherical or point charges. It is easy to spread an electric charge along a rod, and one might want to know what force this charged rod exerted on a nearby point charge. In electricity theory we have to deal with various distributions of electric charge, not just the simple point concentrations we saw in gravitational calculations.

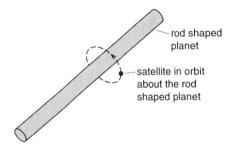

rod shaped planet

satellite in orbit about the rod shaped planet

Figure 2
Imagine that an advanced civilization creates a rod-shaped planet. The problem, which we have not encountered earlier, is to calculate the period of a satellite orbiting the planet.

We will see that there is a powerful theorem, discovered by Frederick Gauss, that considerably simplifies the calculation of electric forces produced by extended distribution of charges. But Gauss's law involves several new concepts that we will have to develop. To appreciate this effort, to see why we want to use Gauss's law, we will now do a brute force calculation, using standard calculus steps to calculate the force between a point charge and a line of charge. It will be hard work. Later we will use Gauss's law to do the same calculation and you will see how much easier it is.

The setup for our calculation is shown in Figure (3). We have a negative charge Q_T located a distance r from a long charged rod as shown. The rod has a positive charge density λ coulombs per meter spread along it. We wish to calculate the total force $\vec{F}$ exerted by the rod upon our negative test particle Q_T. For simplicity we may assume that the ends of the rod are infinitely far away (at least several feet away on the scale of the drawing).

To calculate the electric force exerted by the rod, we will conceptually break the rod into many short segments of length dx, each containing an amount of charge $dq = \lambda dx$. To preserve the left/right symmetry, we will calculate the force between Q_T and pairs of dq, one on the left and one on the right as shown. The left directional force $\vec{dF_1}$ and the right directed force $\vec{dF_2}$ add up to produce an upward directional force $\vec{dF}$. Thus, when we add up the forces from pairs of dq, the $\vec{dF}$'s are all upward directed and add numerically.

The force $\vec{dF_1}$ between Q_T and the piece of charge dq_1 is given by Coulomb's law as

$$\left| \vec{dF_1} \right| = \frac{KQ_Tdq_1}{R^2} = \frac{KQ_Tdq_1}{x^2 + r^2}$$

$$= \frac{K\lambda Q_Tdx}{x^2 + r^2} \qquad (6)$$

where, for now, we will use K for $1/4\pi\varepsilon_0$ to simplify the formula. The separation R between Q_T and dq_1 is given by the Pythagorean theorem as $R^2 = x^2 + r^2$.

The component of $d\vec{F}_1$ in the upward direction is $dF_1 \cos(\theta)$, so that $d\vec{F}$ has a magnitude

$$|d\vec{F}| = 2\,dF_1 \cos(\theta)$$

$$= 2\,\frac{K\lambda Q_T\,dx}{x^2 + r^2}\,\frac{r}{\sqrt{x^2 + r^2}} \qquad (7)$$

The factor of 2 comes from the fact that we get equal components from both $d\vec{F}_1$ and $d\vec{F}_2$.

To get the total force on Q_T, we add up the forces produced by all pairs of dq starting from x = 0 and going out to x = ∞. The result is the definite integral

$$|\vec{F}_{Q_T}| = \int_0^\infty \frac{2K\lambda Q_T r}{(x^2 + r^2)^{3/2}}\,dx$$

$$= 2K\lambda Q_T r \int_0^\infty \frac{dx}{(x^2 + r^2)^{3/2}} \qquad (8)$$

where r, the distance from the charge to the rod, is a constant that can be taken outside the integral.

The remaining integral $\int dx/(x^2 + r^2)^{3/2}$ is not a common integral whose result you are likely to have memorized, nor is it particularly easy to work out.

Instead we look it up in a table of integrals with the result

$$\int_0^A \frac{dx}{(x^2 + r^2)^{\frac{3}{2}}} = \frac{2(2x)}{4r^2\sqrt{r^2 + x^2}}\Bigg|_0^A$$

$$= \frac{A}{r^2\sqrt{r^2 + A^2}} - 0 \qquad (9)$$

For A >> r (very long rod), we can set $\sqrt{r^2 + A^2} \approx A$ and we get the results

$$\int_0^{A \gg r} \frac{dx}{(x^2 + r^2)^{\frac{3}{2}}} = \frac{A}{r^2 A} = \frac{1}{r^2} \qquad (10)$$

Using Equation (10) in Equation (8) we get

$$|\vec{F}_{Q_T}| = 2K\lambda Q_T \frac{r}{r^2}$$

$$\boxed{F_{Q_T} = \frac{2K\lambda Q_T}{r}} \qquad (11)$$

The important point of the calculation is that *the force between a point charge and a line charge drops off as $1/r$ rather than $1/r^2$*, as long as Q_T stays close enough to the rod that the ends appear to be very far away.

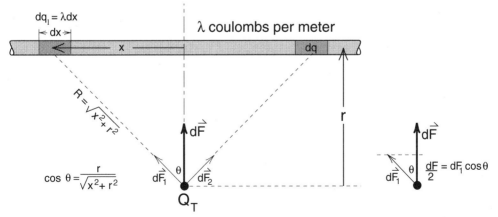

$dq_1 = \lambda dx$

λ coulombs per meter

$R = \sqrt{x^2 + r^2}$

$\cos\theta = \frac{r}{\sqrt{x^2 + r^2}}$

Q_T

$\frac{dF}{2} = dF_1 \cos\theta$

Figure 3
Geometry for calculating the electric force between a point charge Q_T a distance r from a line of charge with λ coulombs per meter.

One of the rules of thumb in doing physics is that if you have a simple result, there is probably an intuitive derivation or explanation. In deriving the answer in Equation (11), we did too much busy work to see anything intuitive. We had to deal with an integral of $(x^2 + r^2)^{-3/2}$, yet we got the simple answer that the force dropped off as $1/r$ rather than the $1/r^2$. We will see, when we repeat this derivation using Gauss's law, that the change from $1/r^2$ to a $1/r$ force results from the change from a three to a two dimensional problem. This basic connection with geometry is not obvious in our brute force derivation.

Exercise 5

Back to science fiction. A rod shaped planet has a mass density λ kilograms per unit length as shown in Figure (4). A satellite of mass m is located a distance r from the rod as shown. Find the magnitude of the gravitational force $\vec{F}_g$ exerted on the satellite by the rod shaped planet. Then find a formula for the period of the satellite in a circular orbit.

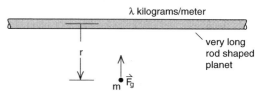

Figure 4
Rod shaped planet and satellite.

Short Rod

Our brute force calculation does have one advantage, however. If we change the problem and say that our charge is located a distance r from the center of a finite rod of length 2L as shown in Figure (5), then Equation (8) of our earlier derivation becomes

$$F_{Q_T} = 2K\lambda Q_T r \int_0^L \frac{dx}{(x^2 + r^2)^{\frac{3}{2}}} \qquad (12)$$

The only difference is that the integral stops at L rather than going out to infinity.

From Equation (9) we have

$$\int_0^L \frac{dx}{(x^2 + r^2)^{\frac{3}{2}}} = \frac{L}{r^2\sqrt{r^2 + L^2}} \qquad (13)$$

And the formula for the force on Q_T becomes

$$\boxed{F_Q = \frac{K(2L\lambda)Q_T}{r\sqrt{r^2 + L^2}}} \qquad (14)$$

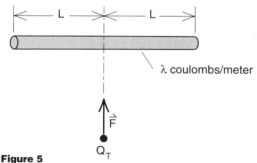

Figure 5
A harder problem is to calculate the force F exerted on Q_T by a rod of finite length 2 L.

Equation (14) has the advantage that it can handle both limiting cases of a long rod (L >> r), a short rod or point charge (L << r), or anything in between.

For example, if we are far away from a short rod, so that L << r, then $\sqrt{r^2 + L^2} \approx r$. Using the fact that $2L\lambda = Q_R$ is the total charge on the rod, Equation (14) becomes

$$F_{Q_T} = \frac{KQ_RQ_T}{r^2} \quad (L << r) \qquad (15b)$$

which is just Coulomb's law for point charges. The more general result, Equation (14), which we obtained by the brute force calculation, cannot be obtained with simple arguments using Gauss's law. This formula was worth the effort.

Exercise 6

Show that if we are very close to the rod, i.e. r << L, then Equation (14) becomes the formula for the force exerted by a line charge.

THE ELECTRIC FIELD

Our example of the force between a point charge and a line of charge demonstrates that even for simple distributions of charge, the calculation of electric forces can become complex. We now begin the introduction of several new concepts that will allow us to simplify many of these calculations. The first of these is the ***electric field***, a concept which allows us to rely more on maps, pictures and intuition, than upon formal calculations.

To introduce the idea of an electric field, let us start with the simple distribution of charge shown in Figure (6). A positive charge of magnitude Q_A is located at Point A, and a negative charge of magnitude $-Q_B$ is located at Point B. We will assume that these charges Q_A and Q_B are fixed, nailed down. They are our *fixed charge distribution*.

We also have a positive ***test charge*** of magnitude Q_T that we can move around in the space surrounding the fixed charges. Q_T will be used to test the strength of the electric force at various points, thus the name "test charge".

In Figure (6), we see that the test charge is subject to the repulsive force $\vec{F}_A$ and attractive force $\vec{F}_B$ to give a net force $\vec{F} = \vec{F}_A + \vec{F}_B$. The individual forces $\vec{F}_A$ and $\vec{F}_B$ are given by Coulomb's law as

$$\left|\vec{F}_A\right| = \frac{KQ_TQ_A}{r_A^2}$$

$$\left|\vec{F}_B\right| = \frac{KQ_TQ_B}{r_B^2}$$

(16)

where r_A is the distance from Q_T to Q_A and r_B is the distance from Q_T to Q_B. For now we are writing $1/4\pi\varepsilon_0 = K$ to keep the formulas from looking too messy.

In Figure (7), we have the same distribution of fixed charge as in Figure (6), namely Q_A and Q_B, but we are using a smaller test charge Q'_T. For this sketch, Q'_T is about half as big as Q_T of Figure (6), and the resulting force vectors point in the same directions but are about half as long;

$$\left|\vec{F}'_A\right| = \frac{KQ'_TQ_A}{r_A^2}$$

$$\left|\vec{F}'_B\right| = \frac{KQ'_TQ_B}{r_B^2}$$

(17)

The only difference between Equations (16) and (17) is that Q_T has been replaced by Q'_T in the formulas. You can see that Equations (17) can be obtained from Equations (16) by multiplying the forces by Q'_T/Q_T. Thus if we used a standard size test charge Q_T to calculate the forces, i.e. do all the vector additions, etc., then we can find the force on a different sized test charge Q'_T by multiplying the net force by the ratio Q'_T/Q_T.

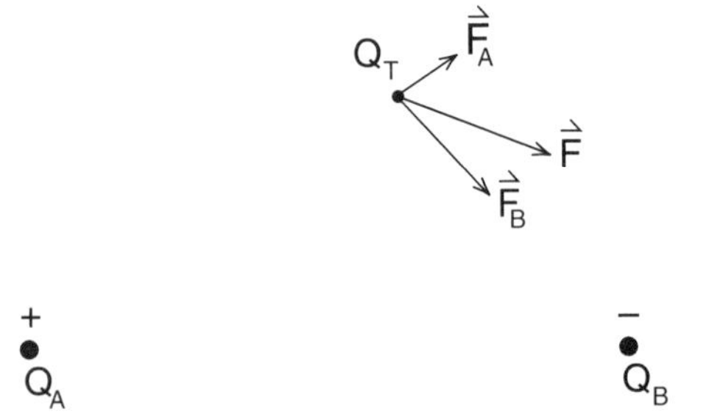

Figure 6
Forces exerted by two fixed charges
Q_A and Q_B on the test particle Q_T.

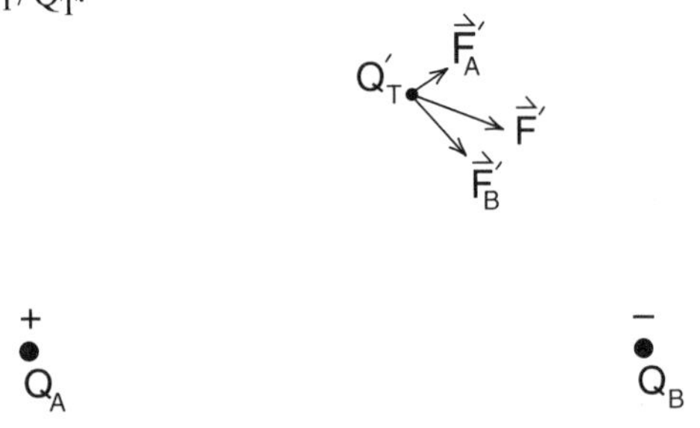

Figure 7
If we replace the test charge Q_T by a smaller
test charge Q'_T, everything is the same except
that the force vectors become shorter.

Unit Test Charge

The next step is to decide what size our standard test charge Q_T should be. Physically Q_T should be small so that it does not disturb the fixed distribution of charge. After all, Newton's third law requires that Q_T pull on the fixed charges with forces equal and opposite to the forces shown acting on Q_T.

On the other hand a simple mathematical choice is $Q_T = 1$ coulomb, what we will call a *"unit test charge"*. If $Q_T = 1$, then the force on another charge Q'_T is just Q'_T times larger $(\vec{F} = \vec{F} Q'_T / Q_T = \vec{F} Q'_T / 1 = \vec{F} Q'_T)$. The problem is that in practice, a coulomb of charge is enormous. Two point charges, each of strength +1 coulomb, located one meter apart, repel each other with a force of magnitude

$$\left|\vec{F}\right| = \frac{K \times 1 \text{ coulomb} \times 1 \text{ coulomb}}{1^2 \text{m}^2}$$

$$= K = \frac{1}{4\pi\varepsilon_0} = 9 \times 10^9 \text{ newtons} \quad (17)$$

a result we saw in Example 1. A force of nearly 10 billion newtons is strong enough to destroy any experimental structure you are ever likely to see. In practice the coulomb is much too big a charge to serve as a realistic test particle.

The mathematical simplicity of using a unit test charge is too great to ignore. Our compromise is the following. We use a unit test charge, but think of it as a "small unit test charge". Conceptually think of using a charge about the size of the charge of an electron, so that the

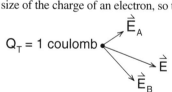

Figure 8
When we use a unit test charge $Q_T = 1$ coulomb, then the forces on it are called "electric field" vectors, $\vec{E}_A$, $\vec{E}_B$ and $\vec{E}$ as shown.

forces it exerts do not disturb anything, but mathematically treat the test charge as having a magnitude of 1 coulomb. In addition, we will always use a *positive unit* test charge. If we want to know the force on a negative test charge, simply reverse the direction of the force vectors.

Figure (8) is the same as our Figures (6) and (7), except that we are now using a unit test particle equal to 1 coulomb, to observe the electric forces surrounding our fixed charge distribution of Q_A and Q_B. The forces acting on $Q_T = 1$ coulomb are

$$\left|\vec{F}_A\right| = \frac{KQ_A \times 1 \text{ coulomb}}{r_A^2}$$
$$\left|\vec{F}_B\right| = \frac{KQ_B \times 1 \text{ coulomb}}{r_B^2} \quad (18)$$

To emphasize that these forces are acting on a unit test charge, we will use the letter E rather than F, and write

$$\left|\vec{E}_A\right| = \frac{KQ_A}{r_A^2}$$
$$\left|\vec{E}_B\right| = \frac{KQ_B}{r_B^2} \quad (19)$$

If you wished to know the force $\vec{F}$ on some charge Q located where our test particle is, you would write

$$\left|\vec{F}_A\right| = \frac{KQ_A Q}{r_A^2} = \left|\vec{E}_A\right| Q \quad (21a)$$

$$\left|\vec{F}_B\right| = \frac{KQ_B Q}{r_B^2} = \left|\vec{E}_B\right| Q \quad (21b)$$

$$\vec{F} = Q\vec{E}_A + Q\vec{E}_B = Q\left(\vec{E}_A + \vec{E}_B\right)$$
$$\vec{F} = Q\vec{E} \quad (22)$$

where $\vec{E} = \vec{E}_A + \vec{E}_B$. Equation (22) is an important result. It says that *the force on any charge Q is Q times the force $\vec{E}$ on a unit test particle.*

ELECTRIC FIELD LINES

The force $\vec{E}$ on a unit test particle plays such a central role in the theory of electricity that we give it a special name – the *electric field* $\vec{E}$.

$$\left. \begin{array}{c} \text{electric} \\ \text{field } \vec{E} \end{array} \right\} \equiv \left\{ \begin{array}{c} \text{force on a} \\ \text{unit test} \\ \text{particle} \end{array} \right. \tag{23}$$

Once we know the electric field $\vec{E}$ at some point, then the force $\vec{F}_Q$ on a charge Q located at that point is

$$\boxed{\vec{F} = Q\vec{E}} \tag{24}$$

If Q is negative, then $\vec{F}$ points in the direction opposite to $\vec{E}$.

From Equation (24), we see that the electric field $\vec{E}$ has the dimensions of newtons/coulomb, so that $Q\vec{E}$ comes out in newtons.

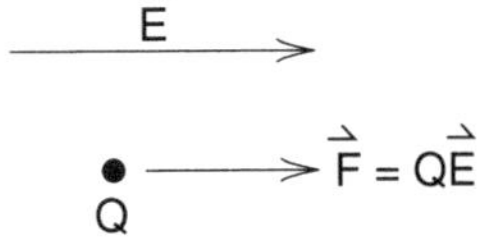

$$\xrightarrow{\quad\vec{E}\quad}$$

$$\underset{Q}{\bullet} \xrightarrow{\qquad} \vec{F} = Q\vec{E}$$

Figure 9
Once we know the electric field $\vec{E}$ at some point, we find the force $\vec{F}$ acting on a charge Q at that point by the simple formula $\vec{F} = Q\vec{E}$.

Mapping the Electric Field

In Figure (10), we started with a simple charge distribution +Q and -Q as shown, placed our unit test particle at various points in the region surrounding the fixed charges, and drew the resulting force vectors $\vec{E}$ at each point. If we do the diagram carefully, as in Figure (10), a picture of the electric field begins to emerge. Once we have a complete picture of the electric field $\vec{E}$, once we know $\vec{E}$ at every point in space, then we can find the force on any charge q by using $\vec{F}_Q = Q\vec{E}$. The problem we wish to solve, therefore, is how to construct a complete map or picture of the electric field $\vec{E}$.

Exercise 7

In Figure (10), we have labeled 3 points (1), (2), and (3). Sketch the force vectors $\vec{F}_Q$ on :

(a) a charge Q = 1 coulomb at Point (1)

(b) a charge Q = –1 coulomb at Point (2)

(c) a charge Q = 2 coulombs at Point (3)

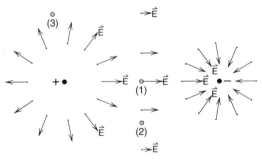

Figure 10
If we draw the electric field vectors $\vec{E}$ at various points around our charge distribution, a picture or map of the electric field begins to emerge.

In part (c) of Exercise (7), we asked you to sketch the force on a charge Q = 2 coulombs located at Point (3). The answer is $\vec{F}_3 = Q\vec{E}_3 = 2\vec{E}_3$, but the problem is that we have not yet calculated the electric field $\vec{E}_3$ at Point (3). On the other hand we have calculated $\vec{E}$ at some nearby locations. From the shape of the map that is emerging from the $\vec{E}$ vectors we have drawn, we can make a fairly accurate guess as to the magnitude and direction of $\vec{E}$ at Point (3) without doing the calculation. With a map we can build intuition and make reasonably accurate estimates without calculating $\vec{E}$ at every point.

In Figure (10) we were quite careful about choosing where to draw the vectors in order to construct the picture. We placed the points one after another to see the flow of the field from the positive to the negative charge. In Figure (11), we have constructed a similar picture for the electric field surrounding a single positive charge.

The difficulty in drawing maps or pictures of the electric field is that we have to show both the magnitude and direction at every point. To do this by drawing a large number of separate vectors quickly becomes cumbersome and time consuming. We need a better way to draw these maps, and in so doing will adopt many of the conventions developed by map makers.

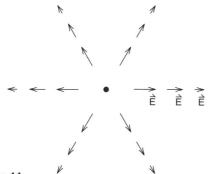

Figure 11
Electric field of a point charge.

Field Lines

As a first step in simplifying the mapping process, let us concentrate on showing the direction of the electric force in the space surrounding our charge distribution. This can be done by connecting the arrows in Figures (10) and (11) to produce the line drawings of Figures (12a) and (12b) respectively. The lines in these drawings are called *field lines.*

In our earlier discussion of fluid flow we saw diagrams that looked very much like the Figures (12a and 12b). There we were drawing stream lines for various flow patterns. Now we are drawing electric field lines. As illustrated in Figure (13), a streamline and an electric field line are similar concepts. At every point on a streamline, the velocity field $\vec{v}$ is parallel to the streamline, while at every point on an electric field line, the electric field $\vec{E}$ is parallel to the electric field line.

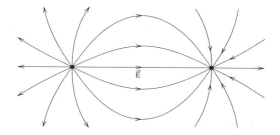

Figure 12a
We connected the arrows of Figure 10 to create a set of field lines for 2 point charges.

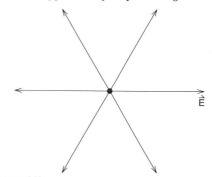

Figure 12b
Here we connected the arrows in Figure 11 to draw the field lines for a point charge.

Continuity Equation for Electric Fields

Figure (14) is our old diagram (23-6) for the velocity field of a point source of fluid (a small sphere that created water molecules). We applied the continuity equation to the flow outside the source and saw that the velocity field of a point source of fluid drops off as $1/r^2$.

Figure (15) is more or less a repeat of Figure (12b) for the electric field of a point charge. By Coulomb's law, the strength of the electric field drops off as $1/r^2$ as we go out from the point charge. We have the same field structure for a point source of an incompressible fluid and the electric field of a point charge. Is this pure coincidence, or is there something we can learn from the similarity of these two fields?

The crucial feature of the velocity field that gave us a $1/r^2$ flow was the continuity equation. Basically the idea is that all of the water that is created in the small sphere must eventually flow out through any larger sphere surrounding the source. Since the area of a sphere, $4\pi r^2$, increases as r^2, the speed of the water has to decrease as $1/r^2$ so that the same volume of water per second flows through a big sphere as through a small one.

If we think of the electric field as some kind of an incompressible fluid, and think of a point charge as a source of this fluid, then the continuity equation applied to this electric field gives us the correct $1/r^2$ dependence of the field. In a sense we can replace Coulomb's law by a continuity equation. Explicitly, we will use streamlines or field lines to map the direction of the field, and use the continuity equation to calculate the magnitude of the field. This is our general plan for constructing electric field maps; we now have to fill in the details.

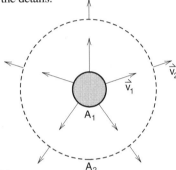

Figure 14
This is our old Figure 23-6 for the velocity field of a point source of water. The continuity equation $v_1 A_1 = v_2 A_2$, requires that the velocity field $\vec{v}$ drops off as $1/r^2$ because the area through which the water flows increases as πr^2.

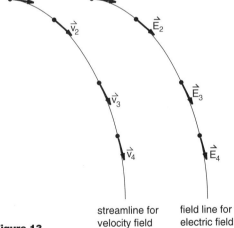

streamline for   field line for
velocity field   electric field

Figure 13
Comparison of the streamline for the velocity field and the field line for an electric field. Both are constructed in the same way by connecting successive vectors. The streamline is easier to visualize because it is the actual path followed by particles in the fluid.

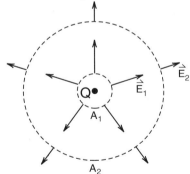

Figure 15
From Coulomb's law, $|\vec{E}| = KQ/r^2$, we see that the electric field of a point source drops off in exactly the same way as the velocity field of a point source. Thus the electric field must obey the same continuity equation $E_1 A_1 = E_2 A_2$ as does the velocity field.

Flux

To see how the continuity equation can be applied to electric fields, let us review the calculation of the $1/r^2$ velocity field of a point source of fluid, and follow the same steps to calculate the electric field of a point charge.

In Figure (16) we have a small sphere of area A_1 in which the water is created. The volume of water created each second, which we called the *flux* of the water and will now designate by the greek letter Φ, is given by

$$\left.\begin{array}{l}\text{volume of water} \\ \text{created per second} \\ \text{in the small sphere}\end{array}\right\} \equiv \Phi_1 = v_1 A_1 \qquad (25)$$

The flux of water out through a larger sphere of area A_2 is

$$\left.\begin{array}{l}\text{volume of water} \\ \text{flowing per second} \\ \text{out through a} \\ \text{larger sphere}\end{array}\right\} \equiv \Phi_2 = v_2 A_2 \qquad (26)$$

The continuity equation $v_1 A_1 = v_2 A_2$ requires these fluxes be equal

$$\Phi_1 = \Phi_2 \equiv \Phi \qquad \textit{continuity equation} \qquad (27)$$

Using Equations (26) and (27), we can express the velocity field v_2 out at the larger sphere in terms of the flux of water Φ created inside the small sphere

$$\boxed{v_2 = \frac{\Phi}{A_2} = \frac{\Phi}{4\pi r_2^2}} \qquad (28)$$

Let us now follow precisely the same steps for the electric field of a point charge. Construct a small sphere of area A_1 and a large sphere A_2 concentrically surrounding the point charge as shown in Figure (17). At the small sphere the electric field has a strength E_1, which has dropped to a strength E_2 out at A_2.

Let us define $E_1 A_1$ as the flux of our electric fluid flowing out of the smaller sphere, and $E_2 A_2$ as the flux flowing out through the larger sphere

$$\Phi_1 = E_1 A_1 \qquad (29)$$

$$\Phi_2 = E_2 A_2 \qquad (30)$$

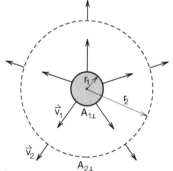

Figure 16
The total flux Φ_1 of water out of the small sphere is $\Phi_1 = v_1 A_{1\perp}$, where $A_{1\perp}$ is the perpendicular area through which the water flows. The flux through the larger sphere is $\Phi_2 = v_2 A_{2\perp}$. Noting that no water is lost as it flows from the inner to outer sphere, i.e., equating Φ_1 and Φ_2, gives us the result that the velocity field drops off as $1/r^2$ because the perpendicular area increases as r^2.

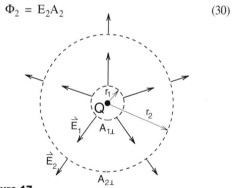

Figure 17
The total flux Φ_1 of the electric field out of the small sphere is $\Phi_1 = E_1 A_{1\perp}$, where $A_{1\perp}$ is the perpendicular area through which the electric field flows. The flux through the larger sphere is $\Phi_2 = E_2 A_{2\perp}$. Noting that no flux is lost as it flows from the inner to outer sphere, i.e., equating Φ_1 and Φ_2, gives us the result that the electric field drops off as $1/r^2$ because the perpendicular area increases as r^2.

Applying the continuity equation $E_1 A_1 = E_2 A_2$ to this electric fluid, we get

$$\Phi_1 = \Phi_2 = \Phi \tag{31}$$

Again we can express the field at A_2 in terms of the flux

$$E_2 = \frac{\Phi}{A_2} = \frac{\Phi}{4\pi r_2^2} \tag{32}$$

Since A_2 can be any sphere outside, but centered on the point charge, we can drop the subscript 2 and write

$$\boxed{E(r) = \frac{\Phi}{4\pi r^2}} \tag{33}$$

In Equation (33), we got the correct $1/r^2$ dependence for the electric field, but what is the appropriate value for Φ? How much electric flux Φ flows out of a point charge?

To find out, start with a fixed charge Q as shown in Figure (18), place our unit test charge a distance r away, and use Coulomb's law to calculate the electric force E on our unit test charge. The result is

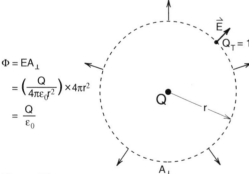

$\Phi = EA_\perp$

$= \left(\dfrac{Q}{4\pi\varepsilon_0 r^2}\right) \times 4\pi r^2$

$= \dfrac{Q}{\varepsilon_0}$

Figure 18
Using Coulomb's law for the electric field of a point charge Q, we calculate that the total flux out through any centered sphere surrounding the point charge is $\Phi = Q/\varepsilon_0$.

$$|\vec{E}| = \frac{Q}{4\pi\varepsilon_0 r^2} \qquad \begin{array}{l}\textit{Coulomb's}\\\textit{Law}\end{array} \tag{34}$$

where now we are explicitly putting in $1/4\pi\varepsilon_0$ for the proportionality constant K.

Comparing Equations (33) and (34), we see that if we choose

$$\boxed{\Phi = \frac{Q}{\varepsilon_0}} \qquad \begin{array}{l}\text{Flux emerging}\\\text{from a charge Q}\end{array} \tag{35}$$

then the continuity equation (33) and Coulomb's law (34) give the same answer. Equation (35) is the key that allows us to apply the continuity equation to the electric field. If we say that a point charge Q creates an electric flux $\Phi = Q/\varepsilon_0$, then applying the continuity equation gives the same results as Coulomb's law. (You can now see that by putting the 4π into Coulomb's law, there is no 4π in our formula (35) for flux.)

Negative Charge

If we have a negative charge $-Q$, then our unit test particle Q_T will be attracted to it as shown in Figure (19). From a hydrodynamic point of view, the electric fluid is flowing **into** the charge $-Q$ and being destroyed there. Therefore a generalization of our rule about electric flux is that a positive charge creates a positive, outward flux of magnitude Q/ε_0, while a negative charge destroys the electric flux, it has a negative flux $-Q/\varepsilon_0$ that flows into the charge and disappears.

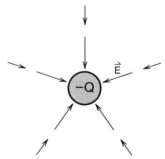

Figure 19
The electric field of a negative charge flows into the charge. Just as positive charge creates flux, negative charge destroys it.

Flux Tubes

In our first pictures of fluid flows like Figure (23-5) reproduced here, we saw that the streamlines were little tubes of flow. The continuity equation, applied to a streamline was $\Phi = v_1 A_1 = v_2 A_2$. This is simply the statement that the flux of a fluid along a streamline is constant. We can think of the streamlines as small tubes of flux. By analogy *we will think of our electric field lines as small tubes of electric flux*.

Conserved Field Lines

When we think of the field line as a small flux tube, the continuity equation gives us a very powerful result, namely the flux tubes must be continuous, must maintain their strength in any region where the fluid is neither being created nor destroyed. For the electric fluid, the flux tubes or field lines are created by, start at, positive charge. And they are destroyed by, or stop at, negative charge. But in between the electric fluid is conserved and the field lines are continuous. We will see that this continuity of the electric field lines is a very powerful tool for mapping electric fields.

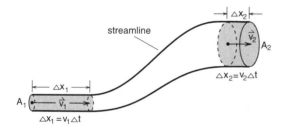

Figure 23-5
Flux tube in the flow of water.

A Mapping Convention

If an electric field line represents a small flux tube, the question remains as to how much flux is in the tube? Just as we standardized on a unit test charge $Q_T = 1$ coulomb for the definition of the electric field $\vec{E}$, we will standardize on a unit flux tube as the amount of flux represented by one electric field line. With this convention, we should therefore draw $\Phi = Q/\varepsilon_0$ field lines or unit flux tubes coming out of a positive charge $+Q$, or stopping on a negative charge $-Q$. Let us try a few examples to see what a powerful mapping convention this is.

In Figure (20) we have a positive charge $Q/\varepsilon_0 = +5$, and a negative charge $Q/\varepsilon_0 = -3$, located as shown. By our new mapping convention we should draw 5 unit flux tubes or field lines out of the positive charge, and we should show 3 of them stopping on the negative charge. Close to the positive charge, the negative charge is too far away to have any effect and the field lines must go radially out as shown. Close to the negative charge, the lines must go radially in because the positive charge is too far away.

$Q/\varepsilon_0 = -3$ $Q/\varepsilon_0 = 5$

Figure 20
We begin a sketch of the electric field by drawing the field lines in close to the charges, where the lines go either straight in or straight out. Here we have drawn 3 lines into the charge $-3\varepsilon_0$ and 5 lines out of the charge $+5\varepsilon_0$. To make a symmetric looking picture, we oriented the lines so that one will go straight across from the positive to the negative charge.

Now we get to the interesting part; what happens to the field lines out from the charges? The basic rule is that *the lines can start on positive charge, stop on negative charge, but must be continuous in between*. A good guess is that 3 of the lines starting on the positive charge go over to the negative charge in roughly the way we have drawn by dotted lines in Figure (21). There is no more room on the minus charge for the other two lines, so that all these two lines can do is to continue on out to infinity.

Let us take Figure (21), but step far back, so that the + and the – charge look close together as shown in Figure (22). Between the charges we still have the same heart-shaped pattern, but we now get a better view of the two lines that had nowhere to go in Figure (21). To get a better understanding for Figure (22) draw a sphere around the charges as shown. The net charge inside this sphere is

$$\left(\frac{Q_{net}}{\varepsilon_0}\right)_{inside\,sphere} = +5 - 3 = 2$$

Thus by our mapping convention, two field lines should emerge from this sphere, and they do. Far away

where we cannot see the space between the point charges, it looks like we have a single positive charge of magnitude $Q/\varepsilon_0 = 2$.

Summary

When we started this chapter with the brute force calculus calculation of the electric field of a line charge, you may have thought that the important point of this chapter was how to do messy calculations. Actually, exactly the opposite is true! **We want to learn how to avoid doing messy calculations.** The sketches shown in Figures (21) and (22) are an important step in this process. From what we are trying to get out of this chapter, it is far more important that you learn how to do sketches like Figures (21) and (22), than calculations illustrated by Figure (1).

With a little experience, most students get quite good at sketching field patterns. The basic constraints are that Q/ε_0 lines start on positive charges, or stop on negative charges. Between charges the lines are unbroken and should be smooth, and any lines left over must either go to or come from infinity as they did in Figure (22).

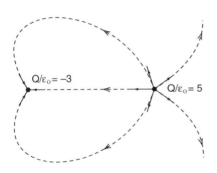

Figure 21
Once the in close field lines have been drawn, we can sketch in the connecting part of the lines as shown above. Three of the lines starting from the positive charge must end on the negative one. The other must go out to infinity. Using symmetry and a bit of artistic skill, you will become quite good at drawing these sketches.

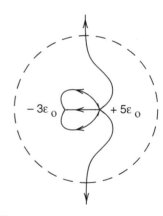

Figure 22
Distant view of our charge distribution. If we step way back from the charge distribution of Figure (21), we see a small object whose net charge is $-3\varepsilon_0 + 5\varepsilon_0 = +2\varepsilon_0$. Thus 2 lines must finally emerge from this distribution as shown.

A Computer Plot

Figure (23) is a computer plot of the electric field lines for the +5, -3 charge distribution of Figures (21) and (22). The first thing we noted is that the computer drew a lot more lines than we did. Did it violate our mapping convention that the lines represent unit flux tubes, with Q/ε_o lines starting or stopping on a charge Q? Yes. The computer drew a whole bunch of lines so that we could get a better feeling for the shape of the electric field. Notice, however, that the ratio of the number of lines starting from the positive charge to the number ending on the negative charge is still 5/3. One of the standard tricks in map making is to change your scale to make the map look as good as possible. Here the computer drew 10 lines per unit flux tube rather than 1. We will see that the only time we really have to be careful with the number of lines we draw is when we are using a count of the number of lines to estimate the strength of the electric field.

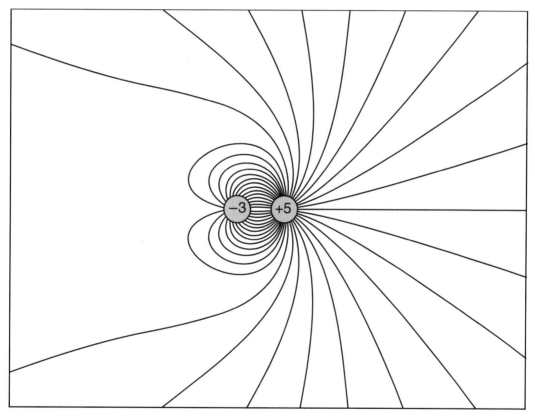

Figure 23
Computer plot of the field lines of a –3 and +5 charge distribution. Rather than drawing Q/ε_0 out of a positive charge or into a negative charge, the computer is programmed to draw enough lines to make the shape of the electric field as clear as possible.

GAUSS' LAW

The idea of using the continuity equation to map field lines was invented by Frederick Gauss and is known as *Gauss' law*. A basic statement of the law is as follows. Conceptually construct a closed surface (often called a *Gaussian surface*) around a group of sources or sinks, as shown in Figure (24). In that figure we have drawn the Gaussian surface around three sources and one sink. Then calculate the total flux Φ_{tot} coming from these sources. For Figure (24), we have

$$\Phi_{tot} = \Phi_1 + \Phi_2 + \Phi_3 + \Phi_4 \qquad (36)$$

where Φ_2 happens to be negative. Then if the fluid is incompressible, or we have an electric field, the total flux flowing out through the Gaussian surface must be equal to the amount of flux Φ_{tot} being created inside.

Gauss' law applies to any closed surface surrounding our sources and sinks. But the law is useful for calculations when the Gaussian surface is simple enough in shape that we can easily write the formula for the flux flowing through the surface. To illustrate the way we use Gauss' law, let us, one more time, calculate the electric field of a point charge.

In Figure (25), we have a point charge Q, and have drawn a spherical Gaussian surface around the charge. The flux produced by the point charge is

$$\Phi_{tot} = \frac{Q}{\varepsilon_o} \qquad \begin{array}{l}\textit{flux produced}\\ \textit{by the point}\\ \textit{charge}\end{array} \qquad (37)$$

At the Gaussian surface there is an electric field $\vec{E}$ (which we wish to calculate), and the surface has an area $A = 4\pi r^2$. Therefore the electric flux flowing out through the sphere is

$$\Phi_{out} = EA$$
$$= E \times 4\pi r^2 \qquad \begin{array}{l}\textit{flux flowing}\\ \textit{out through}\\ \textit{the sphere}\end{array} \qquad (38)$$

Equating the flux created inside (Equation 37) to the flux flowing out (Equation 38) gives

$$E \times 4\pi r^2 = \frac{Q}{\varepsilon_o}$$

$$\boxed{E = \frac{Q}{4\pi\varepsilon_o r^2}} \qquad \begin{array}{l}\textit{old result}\\ \textit{obtained}\\ \textit{new way}\end{array} \qquad (39)$$

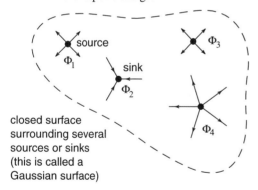

closed surface surrounding several sources or sinks (this is called a Gaussian surface)

Figure 24
If these were sources and sinks in a fluid, it would be obvious that the total flux of fluid out through the closed surface is equal to the net amount of fluid created inside. The same concept applies to electric flux. The net flux Φ_{tot} out through the Gaussian surface is the sum of the fluxes $\Phi_1 + \Phi_2 + \Phi_3 + \Phi_4$ created inside.

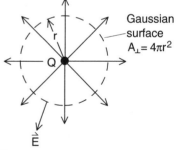

Figure 25
Calculating the electric field of a point charge by equating the flux Q/ε_0 created by the point charge to the flux $\Phi = EA_\perp$ flowing through the Gaussian surface. This gives

$$Q/\varepsilon_0 = E \times 4\pi r^2$$

$$E = Q/4\pi\varepsilon_0 r^2$$

as we expect.

Electric Field of a Line Charge

As a real test of Gauss' law, let us calculate the electric field of a line charge and compare the result with our brute force calculus calculation. In the calculus derivation, we found that the force on a test charge Q_T a distance r from our line charge was (from Equation (11))

$$\vec{F} = \frac{2K\lambda Q_T}{r} = \frac{2\lambda Q_T}{4\pi\varepsilon_o r}$$

where λ is the charge density on the rod as shown in Figure (26). Setting $Q_T = 1$ coulomb, $\vec{F}$ becomes the force $\vec{E}$ on a unit test charge:

$$\boxed{\left|\vec{E}\right| = \frac{\lambda}{2\pi\varepsilon_o r}} \qquad \begin{array}{l}\textit{calculus derived}\\ \textit{formula for the}\\ \textit{electric field of}\\ \textit{a line charge}\end{array} \qquad (40)$$

To apply Gauss' law we first construct a cylindrical Gaussian surface that surrounds a length L of the charge as shown in Figure (27). Since the charge density is λ, the total charge Q_{in}, inside our Gaussian surface is

$$Q_{in} = \lambda L$$

This amount of charge creates an amount of flux

$$\Phi_{inside} = \frac{Q_{in}}{\varepsilon_o} = \frac{\lambda L}{\varepsilon_o} \qquad (41)$$

Now we can see from Figure (27) that because the electric field lines go radially outward from the line charge, they only flow out through the curved outer surface of our Gaussian cylinder and not through the flat ends. This cylindrical surface has an area $A = (2\pi r)L$ (circumference $\times$ length), and the electric field out at a distance r is E(r). Thus the flux out through the Gaussian cylinder is

$$\Phi_{out} = E(r)\,2\pi r\,L \qquad (42)$$

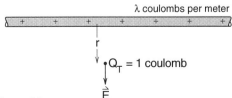

Figure 26
The force $\vec{E}$ on a unit test charge near a line charge.

Equating the flux created inside (Equation 41) to the flux flowing out through the cylindrical surface (Equation 42) gives

$$E(r)\,2\pi r L = \frac{\lambda L}{\varepsilon_o}$$

The L's cancel and we get

$$\boxed{E(r) = \frac{\lambda}{2\pi\varepsilon_o r}} \qquad (43)$$

Voilà! We get the same result. Compare the calculus derivation with its integral of $\left(x^2 + r^2\right)^{3/2}$, to the simple steps of Equations (41) and (42). We noted that in physics, a simple answer, like the 1/r dependence of the electric field of a line charge, should have an easy derivation. The easy derivation is Gauss' law. The simple idea is that for a line charge the flux is flowing out through a cylindrical rather than a spherical surface. The area of a cylindrical surface increases as r, rather than as r^2 for a sphere, therefore the electric field drops off as 1/r rather than as $1/r^2$ as it did for a point charge.

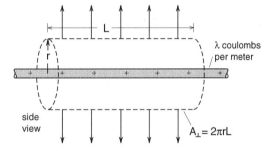

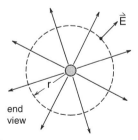

Figure 27
Using Gauss' law to calculate the electric field of a line charge. Draw the Gaussian surface around a section of the rod. The flux all flows out through the cylindrical surface.

Flux Calculations

In our calculations of the flux through a Gaussian surface, Equation (38) for a spherical surface and Equation (42) for a cylindrical surface, we multiplied the strength of the electric field times the area A through which the field was flowing. In both cases we were careful to construct the area perpendicular to the field lines, for flux is equal to the strength of the field times cross sectional or perpendicular area through which it is flowing. It would be more accurate to write the formula for flux as

$$\boxed{\Phi = E\,A_\perp} \tag{44}$$

where the $\perp$ sign reminds us that $A_\perp$ is the perpendicular area.

Area as a Vector

A more formal way to present the formula for flux is to turn the area A into a vector. To illustrate the procedure, consider the small flow tube shown in Figure (28). We have sliced the tube with a plane, and the intersection of the tube and the plane gives us an area A as shown. We turn A into the vector $\vec{A}$ by drawing an arrow perpendicular to the plane, and of length A.

To show why we bothered turning A into a vector, in Figure (29) we have constructed a cross-sectional area $\vec{A}_\perp$ as well as the area $\vec{A}$ of Figure (28). The cross-sectional area is the smallest area we can construct

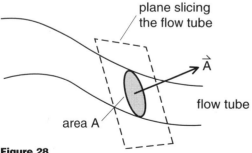

Figure 28
Definition of the area as a vector. Slice a flow tube by a plane. The area A is the area of the region in the plane bounded by the flow tube. We define the direction of $\vec{A}$ as pointing perpendicular to the plane as shown.

across the tube. The other areas are bigger by a factor $1/\cos(\theta)$ where θ is the angle between $\vec{A}$ and $\vec{A}_\perp$. I.e.,

$$A_\perp = A\cos(\theta) \tag{45}$$

Now the flux in the flow tube is the fluid speed v times the cross-sectional area $A_\perp$

$$\Phi = vA_\perp \tag{46}$$

Using Equation (45) for $A_\perp$ in Equation (46) gives

$$\Phi = vA_\perp = vA\cos(\theta)$$

But $vA\cos(\theta)$ is just the vector dot product of the velocity vector $\vec{v}$ (which points in the direction $\vec{A}_\perp$) and $\vec{A}$, thus we have the more general formula

$$\Phi = vA_\perp = \vec{v}\cdot\vec{A} \tag{47a}$$

By analogy, the electric flux through an area $\vec{A}$ is

$$\boxed{\Phi = EA_\perp = \vec{E}\cdot\vec{A}} \tag{47b}$$

In extreme cases where the Gaussian surface is not smooth, you may have to break up the surface into small pieces, calculate the flux $d\Phi_i = \vec{E}\cdot d\vec{A}_i$ for each piece $d\vec{A}_i$ and then add up all the contributions from each piece to get the total flux Φ_{out}. The result is called a **surface integral** which we will discuss later. For now we will make sure that our Gaussian surfaces are smooth and perpendicular to the field, so that we can use the simple form of Equation (47).

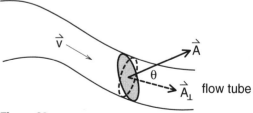

Figure 29
The cross-sectional area, which we have been calling $\vec{A}_\perp$, is the smallest area that crosses the entire tube. The tilted area is larger than $\vec{A}_\perp$ by a factor of $1/\cos(\theta)$, where θ is the angle between $\vec{A}$ and $\vec{A}_\perp$. As a result we can write
$$A_\perp = A\cos(\theta)$$
$$\Phi = vA_\perp = vA\cos(\theta) = \vec{v}\cdot\vec{A}$$

GAUSS' LAW FOR THE GRAVITATIONAL FIELD

In our earliest work with gravitational force problems, such as calculating the motion of the moon or artificial earth satellites, we got the correct answer by replacing the extended spherical earth by a point mass M_e located at the center of the earth. It is surprising that the gravitational force exerted on you by every rock, mountain, body of water, the earth's iron core, etc. all adds up to be equivalent to the force that would be exerted by a point mass M_e located 6,000 km beneath you. A simplified version of history is that Isaac Newton delayed his publication of the theory of gravity 20 years, and invented calculus, in order to show that the gravitational force of the entire earth was equivalent to the force exerted by a point mass located at the center.

One can do a brute force calculus derivation to prove the above result, or one can get the result almost immediately from Gauss' law. With Gauss' law, we can also find out, almost by inspection, how the gravitational force decreases as we go down inside the earth.

Since gravity and electricity are both $1/r^2$ forces, Gauss' law also applies to gravity, and we can get the formulas for the gravitational version by comparing the constants that appear in the force laws. Defining the gravitational field $\vec{g}$ as the force on a unit mass (note that this is also the acceleration due to gravity), we have for a point charge and a point mass

$$\begin{matrix} \text{electric} \\ \text{field} \end{matrix} \quad \vec{E} = \frac{Q}{4\pi\varepsilon_0 r^2} \qquad (48)$$

$$\begin{matrix} \text{gravitational} \\ \text{field} \end{matrix} \quad \vec{g} = \frac{GM}{r^2} \qquad (49)$$

Aside from the fact that gravitational forces are always attractive (therefore the field lines always go into a mass m) the only other difference is that $1/4\pi\varepsilon_0$ is replaced by G

$$G \Leftrightarrow \frac{1}{4\pi\varepsilon_0} \qquad (50)$$

For the electric forces, a point charge Q produces an amount of flux

$$\Phi_E = \frac{Q}{\varepsilon_0} = \frac{1}{4\pi\varepsilon_0} \times 4\pi Q \qquad (51)$$

Replacing $1/4\pi\varepsilon_0$ by G and Q by M, we expect that a mass M destroys (rather than creates) an amount of flux Φ_G given by

$$\Phi_G = G(4\pi M) = 4\pi GM \qquad (52)$$

Gravitational Field of a Point Mass

Let us check if we have the correct flux formula by first calculating the gravitational field of a point mass. In Figure (30) we have a point mass M surrounded by a Gaussian spherical surface of radius r. The flux flowing into the point mass is given by Equation (52) as $4\pi GM$. The flux flowing in through the Gaussian surface is

$$\Phi_G \begin{pmatrix} \textit{flux in} \\ \textit{through} \\ \textit{Gaussian} \\ \textit{surface} \end{pmatrix} = \vec{g} \cdot \vec{A} = gA_\perp = g\left(4\pi r^2\right)$$

$$(53)$$

where the area of the sphere is $4\pi r^2$. Equating the flux in through the sphere (53) to the flux into M (52) gives

$$g\left(4\pi r^2\right) = 4\pi GM$$

$$g = \frac{GM}{r^2} \qquad (54)$$

which is the gravitational force exerted on a unit mass.

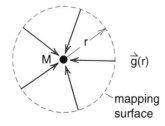

Figure 30
Calculating the gravitational field $\vec{g}$ of a point mass using Gauss' law.

Gravitational Field of a Spherical Mass

Now let us model the earth as a uniform sphere of mass as shown in Figure (31). By symmetry the gravitational field lines must flow radially inward toward the center of the sphere. If we draw a Gaussian surface of radius r outside the earth, we have a total flux flowing in through the sphere given by

$$\Phi_{in} = \vec{g} \cdot \vec{A} = gA_{\perp} \\ = g4\pi r^2 \tag{55}$$

Now the total amount of mass inside the sphere is M_e, so that the total amount of flux that must stop somewhere inside the Gaussian surface is

$$\Phi_G = 4\pi GM_e \tag{56}$$

Since Equations (56) and (55) are identical to Equations (52) and (53) for a point mass, we must get the same answer. Therefore the gravitational field outside a spherically symmetric mass M_e is the same as the field of a point mass M_e located at the center of the sphere.

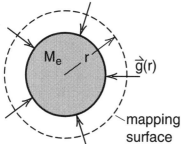

Figure 31
Since $4\pi GM_e$ lines of flux go in through the mapping surface in both Figures 30 and 31, the field at the mapping surface must be the same for both. This is why, when we are above the surface of the earth, we can treat the earth as a point mass located at the center of the earth.

Gravitational Field Inside the Earth

The gravitational field outside a spherical mass is so easy to calculate using Gauss' law, that you might suspect that we really haven't done anything. The calculation becomes more interesting when we go down inside the earth and the field is no longer that of a point mass M_e. By determining how the field decreases as we go inside the earth, we can gain some confidence in the calculational capabilities of Gauss' law.

In Figure (32), we are representing the earth by a uniform sphere of matter of mass M_e and radius R_e. Inside the earth we have drawn a Gaussian surface of radius r, and assume that the gravitational field has a strength g(r) at this radius. Thus the total flux in through the Gaussian surface Φ_{in} is given by

$$\Phi_{in} = \vec{g}(r) \cdot \vec{A} = g(r)A_{\perp} \\ = g(r)4\pi r^2 \tag{57}$$

Now the amount of mass inside our Gaussian surface is no longer M_e, but only the fraction of M_e lying below the radius r. That amount is equal to M_e times the ratio

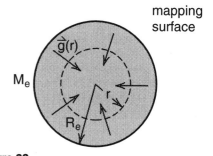

Figure 32
To calculate the gravitational field inside the earth, we draw a mapping surface inside, at a radius r less than the earth radius R_e. The amount of mass M_{in} inside the mapping surface is equal to the earth mass M_e times the ratio of the volume of the mapping sphere to the volume of the earth.

of the volume of a sphere of radius r to the volume of the entire earth.

$$M_r \left(\begin{matrix} mass \\ below\,r \end{matrix} \right) = M_e \frac{{}^{4}\!/_{3}\pi r^3}{{}^{4}\!/_{3}\pi R_e^{\,3}}$$

$$= M_e \frac{r^3}{R_e^{\,3}} \qquad (58)$$

The amount of flux that the mass M_r absorbs is given by Equation (52) as

$$\Phi_G = 4\pi G M_r = 4\pi G M_e \frac{r^3}{R_e^{\,3}} \qquad (59)$$

Equating the flux flowing in through the Gaussian surface (Equation (57) to the flux absorbed by M_r (Equation 59) gives

$$g(r)4\pi r^2 = 4\pi G M_e \frac{r^3}{R_e^{\,3}}$$

or

$$g(r) = G M_e \frac{r}{R^3} \qquad \begin{matrix} \text{Gravitational field} \\ \text{inside the earth} \end{matrix} \qquad (60)$$

This result, which can be obtained by a much more difficult calculus calculation, shows the earth's gravitational field dropping linearly (proportional to r), going to zero as r goes to 0 at the center of the earth.

Figure (32) gives an even more general picture of how the earth's gravitational field changes as we go down inside. The flux going in past our Gaussian surface is determined entirely by the mass inside the surface. *The mass in the spherical shell outside the Gaussian surface has no effect at all!* If we are down inside the earth, a distance R_i from the center, we can accurately determine the gravitational force on us by assuming that all the mass below us ($r < R_i$) is located at a point at the center of the earth, and all the mass above ($r > R_i$) does not exist.

Exercise 8

As shown in Figure (33) a plastic ball of radius R has a total charge Q uniformly distributed throughout it. Use Gauss' law to:

a) Calculate the electric field $\vec{E}(r)$ outside the sphere ($r > R$). How does this compare with the electric field of a point charge?

b) Calculate the electric field inside the plastic sphere ($r < R$).

(Try to do this now. The solution is on the next page as an example.)

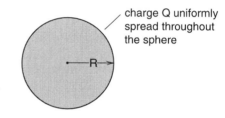

charge Q uniformly spread throughout the sphere

Figure 33
Diagram for exercises 8&10. A plastic sphere of radius R has a charge Q spread uniformly throughout. The problem is to calculate the electric field inside and outside.

Solving Gauss' Law Problems

Using Gauss' law to solve for electric fields can be handled in a relatively straightforward way using the following steps:

1) Carefully sketch the problem.

2) Draw a mapping surface that passes through the point where you want to solve for the field. Construct the surface so that any field lines going through the surface are perpendicular to the surface. This way you can immediately spot the perpendicular area $A_\perp$.

3) Identify Q_{in}, the amount of electric charge inside your mapping surface.

4) Solve for E using the Equation $\Phi = EA_\perp = Q_{in}/\varepsilon_o$.

5) Check that your answer is reasonable.

As an example, let us follow these steps to solve part (b) of Exercise 8, i.e., find the electric field E inside a uniform ball of charge.

1) Sketch the problem. The sphere has a radius R, and total charge +Q. By symmetry the electric field must go radially outward for a positive charge (or radially inward for a negative charge).

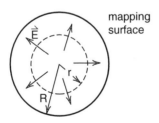

mapping surface

2) Since we want the field inside the charged sphere, we will use a spherical mapping surface of radius $r < R$. Because the electric field is everywhere perpendicular to the mapping surface, the area of the mapping surface is $A_\perp = 4\pi r^2$.

3) The simplest way to calculate the amount of charge Q_{in} inside our mapping surface is to note that since the charge is uniformly spread throughout the sphere, Q_{in} is equal to the total charge Q times the ratio of the volume inside the mapping surface to the total volume of the charged sphere; i.e.,

$$Q_{in} = Q \times \frac{(4/3)\pi r^3}{(4/3)\pi R^3} = \frac{Qr^3}{R^3}$$

4) Now use Gauss' law to calculate E:

$$\Phi = EA_\perp = \frac{Q_{in}}{\varepsilon_o}$$

$$E \times 4\pi r^2 = \frac{Qr^3}{\varepsilon_o R^3}$$

$$\boxed{E = \frac{Qr}{4\pi\varepsilon_o R^3}}$$

5) Check to see if the answer is reasonable. At $r = 0$, we get $E = 0$. That is good, because at the center of the sphere, there is no unique direction for $\vec{E}$ to point. At $r = R$, our formula for E reduces to $E = Q/(4\pi\varepsilon_o R^2)$, which is the field of a point charge Q when we are a distance R away. This agrees with the idea that once we are outside a spherical charge, the electric field is the same as if all the charge were at a point at the center of the sphere.

Exercise 9

As shown in Figure (34), the inside of the plastic sphere has been hollowed out. The total charge on the sphere is Q. Use Gauss' law to

a) Determine the strength of the electric field inside the hollow cavity.

b) Calculate the strength of the electric field inside the plastic.

c) Calculate the strength of the electric field outside the plastic.

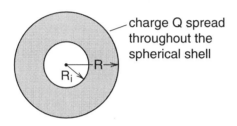

charge Q spread throughout the spherical shell

Figure 34
Diagram for Exercises 9 & 11. The charge is now spread throughout a spherical shell.

Exercise 10

Repeat *Exercise 8*, assuming that Figure (33) represents the end view of a very long charged plastic rod with a charge of λ coulombs per meter. (A section of length L will thus have a charge Q = λ L.)

Exercise 11

Repeat *Exercise 9*, assuming that Figure (34) represents the end view of a charged hollow plastic rod with a charge of λ coulombs per meter.

Exercise 12

A hydrogen atom (H atom) consists of a proton with an electron moving about it. The classical picture is that the electron orbits about the proton much like the earth orbits the sun. A model that has its origins in quantum mechanics and is more useful to chemists, is to picture the electron as being smeared out, forming a ball of negative charge surrounding the proton. This ball of negative charge is called an "electron cloud ".

For this problem, assume that the electron cloud is a *uniform* sphere of negative charge, a sphere of radius R centered on the proton as shown in Figure (35). The total negative charge (–e) in the electron cloud just balances the positive charge (+e) on the proton, so that the net charge on the H atom is zero.

a) Sketch the electric field for this model of the H atom. Show the electric field both inside and outside the electron cloud.

b) Calculate the magnitude of the electric field for both r < R (inside the cloud) and r > R (outside the cloud).

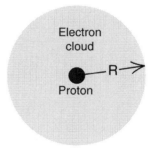

Figure 35
Picture a hydrogen atom as a proton (of charge +e), surrounded by an electron cloud. Think of the cloud as a uniform ball of negative charge, with a net charge -e.

Exercise 13

A butterfly net with a circular opening of radius R, is in a uniform electric field of magnitude E as shown in Figure (36). The opening is perpendicular to the field. Calculate the net flux of the electric field through the net itself. (The amount of flux through each hole in the net is $\vec{E} \cdot d\vec{A}$ where $d\vec{A}$ is the area of the hole.) (This is one of our favorite problems from Halliday and Resnick.)

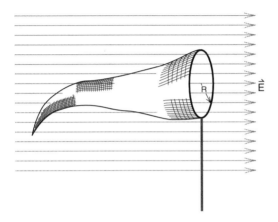

Figure 36
Butterfly net in a uniform electric field. With Gauss' law, you can easily calculate the flux of the electric field through the net itself.

Exercise 14

Electric fields exist in the earth's atmosphere. (You get lightning if they get too strong.)

On a particular day, it is observed that at an altitude of 300 meters, there is a downward directed electric field of magnitude

$$E\left(300\,m\right) = 70\,\frac{newton}{coulomb}$$

Down at an altitude of 200 meters, the electric field still points down, but the magnitude has increased to

$$E\left(200\,m\right) = 100\,\frac{newton}{coulomb}$$

How much electric charge is contained in a cube 100 meters on a side in this region of the atmosphere?

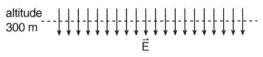

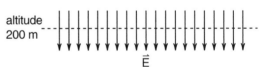

Figure 37
Electric field at two different altitudes.

PROBLEM SOLVING

One can devise a number of Gauss' law problems where one plugs various numerical values into the formulas we have derived. But that is not the point of this chapter. Here we are interested, not so much in the answers, as in the concepts and techniques used to derive them.

In this chapter we have introduced two new concepts. One is electric flux $\Phi = EA_\perp$, and the other is that the total flux out through a closed surface is equal to Q_{inside}/ε_0. These two concepts allow us to easily solve for the electric field in certain special cases. Those cases are where $A_\perp$ is a sphere, a cylinder, or a plane.

What you need to get from this discussion is the beginning of an intuitive picture of electric flux, how it is related to the flow of an incompressible fluid, and how this concept can be used to handle the few but important examples where $A_\perp$ is either a sphere, cylinder, or plane. Numerical applications can come later, now is the time to develop intuition.

Most students have some difficulty handling Gauss's law problems the first time they see them because the concepts involved are new and unfamiliar. Then when the problems are solved in a homework session, a common reaction is, "Oh, those are not so hard after all". One gets the feeling that by just watching the problem solved, and seeing that it is fairly easy after all, they understand it. The rude shock comes at an exam where suddenly the problem that looked so easy, has become unsolvable again.

There is a way to study to avoid this rude shock. Pick one of the problems you could not solve on your own, a problem you saw solved in class or on an answer sheet. A problem that looked so easy after you saw it solved.

Wait a day or two after you saw the solution, clean off your desk, take out a blank sheet of paper, and try solving the problem.

Something awful may happen. That problem that looked so easy in the homework review session is now impossible again. You can't see how to do it, and it looked so easy two days ago. You feel really bad—but don't, it happens to everyone.

Instead, if you cannot get it, just peek at the solution to see what point you missed, then put the solution away and solve it on your own. You may have to peek a couple of times, but that is OK.

If you had to peek at the solution, then wait another day or so, clean off your desk, and try again. Soon you will get the solution without looking, and you will not forget how to solve that problem. You will get more out of this technique than solving 15 numerical examples.

When you are studying a new topic with new, unfamiliar concepts, the best way to learn the subject is to thoroughly learn a few, well chosen worked out examples. By learn, we mean problems you can work on a blank sheet of paper without looking at a solution.

Pick examples that are relatively simple but clearly illustrate the concepts involved. For this chapter, one could pick the example of calculating the electric field inside and outside a uniform ball of charge. If you can do that problem on a clean desk, you can probably do most of the other problems in this chapter without too much difficulty.

Why learn a sample problem for each new topic? The reason is that if you know one worked example you will find it easy to remember the entire topic. That worked example reminds you immediately how that concept works, how it functions.

In this text, Chapters 24-32 on electric and magnetic fields involve many new concepts. Concepts you will not have seen unless you have already taken the course. As we go along, we will suggest sample problems, what we call "clean desk problems", which serve as a good example of the way the new concept is used. You may wish to choose different sample problems, but the best way to learn this topic is to develop a repertory of selected sample problems you understand cold.

At this point, go back to some of the problems in this chapter, particularly Exercises 9 through 14 and see if you can solve them on a clean desk. If you can, you are ready for the next chapter.

Chapter 25

Field Plots and Electric Potential

*Calculating the electric field of any but the simplest distribution of charges can be a challenging task. Gauss' law works well where there is considerable symmetry, as in the case of spheres or infinite lines of charge. At the beginning of Chapter (24), we were able to use a brute force calculus calculation to determine the electric field of a short charged rod. But to handle more complex charge distributions we will find it helpful to apply the techniques developed by map makers to describe complex terrains on a flat map. This is the technique of the contour map which works equally well for mapping electric fields and mountain ranges. Using the contour map ideas, we will be lead to the concept of a **potential** and **equipotential lines** or **surfaces**, which is the main topic of this chapter.*

THE CONTOUR MAP

Figure (1) is a contour map of a small island. The contour lines, labeled 0, 10, 20, 30 and 40 are lines of equal height. Anywhere along the line marked 10 the land is 10 meters above sea level. (You have to look at some note on the map that tells you that height is measured in meters, rather than feet or yards.)

You can get a reasonable understanding of the terrain just by looking at the contour lines. On the south side of the island where the contour lines are far apart, the land slopes gradually upward. This is probably where the beach is located. On the north side where the contour lines are close together, the land drops off sharply. We would expect to see a cliff on this side of the island.

Figure 1
Contour map of a small island with a beach on the south shore, two hills, and a cliff on the northwest side. The slope of the island is gradual where the lines are far apart, and steep where the lines are close together. If you were standing at the point labeled (A) and the surface were slippery, you would start to slide in the direction of the arrow.

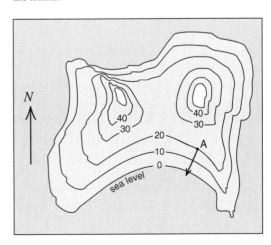

Although we would rather picture this island as being in the south seas, imagine that it is in the North Atlantic and a storm has just covered it with a sheet of ice. You are standing at the point labeled A in Figure (1), and start to slip. If the surface is smooth, which way would you start to slip?

A contour line runs through Point A which we have shown in an enlargement in Figure (2). You would not start to slide along the contour line because all the points along the contour line are at the same height. Instead, you would start to slide in the steepest down-hill direction, which is perpendicular to the contour line as shown by the arrow.

If you do not believe that the direction of steepest descent is perpendicular to the contour line, choose any smooth surface like the top of a rock, mark a horizontal line (an equal height line) for a contour line, and carefully look for the directions that are most steeply sloped down. You will see that all along the contour line the steepest slope is, in fact, perpendicular to the contour line.

Skiers are familiar with this concept. When you want to stop and rest and the slope is icy, you plant your skis along a contour line so that they will not slide either forward or backward. The direction of steepest descent is now perpendicular to your skis, in a direction that ski instructors call the *fall line*. The fall line is the direction you will start to slide if the edges of your skis fail to hold.

In Figure (3), we have redrawn our contour map of the island, but have added a set of perpendicular lines to show the directions of steepest descent, the direction of the net force on you if you were sitting on a slippery surface. These lines of steepest descent, are also called ***lines of force***. They can be sketched by hand, using the rule that the lines of force must always be perpendicular to the contour lines.

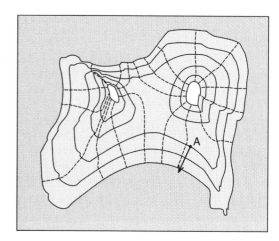

Figure 3
You can sketch in the lines of steepest descent by drawing a set of lines that are always perpendicular to the contour lines. These lines indicate the direction a ball would <u>start</u> to roll if placed at a point on the line.

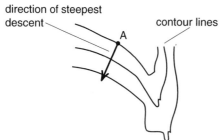

Figure 2
Along a contour line the land is level. The direction of steepest slope or descent is perpendicular to the contour line.

In Figure (4) we have the same island, but except for the zero height contour outlining the island, we show only the lines of force. The exercise here, which you should do now, is sketch in the contour lines. Just use the rule that the contour lines must be drawn perpendicular to the lines of force. The point is that you can go either way. Given the contour lines you can sketch the lines of force, or given the lines of force you can sketch the contour lines. This turns out to be a powerful technique in the mapping of any complex physical or mathematical terrain.

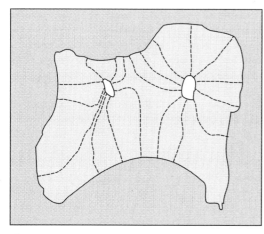

Figure 4
Here we have removed the contour lines leaving only the lines of steepest descent and the outline of the island. By drawing a set of lines perpendicular to the lines of steepest descent, you can more or less reconstruct the contour lines. The idea is that you can go back and forth from one set of lines to the other.

Figure 6
The gravitational potential energy of two masses separated by a distance r is $-GMm/r$.

EQUIPOTENTIAL LINES

On a contour map of an island, the contour lines are lines of equal height. If you walk along a contour line, your height h, and therefore your gravitational potential energy mgh, remains constant. As a result, we can call these the lines of constant or equal potential energy, *equipotential* energy lines for short.

Let us apply these mapping concepts to the simpler situation of a spherical mass M shown in Figure (5). As in Chapter 24, we have drawn the gravitational field lines, which point radially inward toward the center of M. We determined these field lines by placing a test particle of mass m in the vicinity of M as shown in Figure (6). The potential energy of this test mass m is given by our old formula (see Equation 10-50a)

$$PE = -\frac{GMm}{r}$$

attractive forces
have negative   (1)
potential energy

In our discussion of electric and gravitational fields, we defined the fields as the force on a unit test particle. Setting m = 1 for a unit test mass, we get as the formula for the potential energy of our unit test mass

$$\left.\begin{array}{c}\text{Potential energy of}\\ \text{a unit test mass}\end{array}\right\} = -\frac{GM}{r} \quad (2)$$

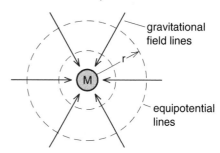

Figure 5
The gravitational field lines for a spherical mass point radially inward. The lines of constant potential energy are circles of equal height above the mass. The equipotential lines are everywhere perpendicular to the field lines, just as, in the map of the island, the contour lines were everywhere perpendicular to the lines of steepest descent.

From Equation (2) we see that if we stay a constant distance r out from M, if we are one of the concentric circles in Figure (5), then the potential energy of the unit test mass remains constant. These circles, drawn perpendicular to the lines of force, are again equal potential energy lines.

There is a convention in physics to use the word *potential* when talking about the potential energy of a unit mass or unit charge. With this convention, then – GM/r in Equation (2) is the formula for the gravitational potential of a mass M, and the constant radius circles in Figure (5) are lines of constant potential. Thus the name *equipotential lines* for these circles is fitting.

Negative and Positive Potential Energy

In Figure (7) we have drawn the electric field lines of a point charge Q and drawn the set of concentric circles perpendicular to the field lines as shown. From the close analogy between the electric and gravitational force, we expect that these circles represent lines of constant electric potential energy, that they are the electric equipotential lines.

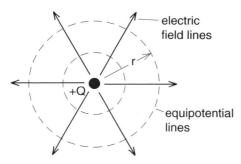

Figure 7
The <u>electric potential</u> is the potential energy of a positive unit test charge q_{test} = + 1 coulomb. Because a positive charge + Q and a positive test charge repel this potential energy is positive.

But there is one important difference between Figures (5) and (7). In Figure (5) the gravitational force on our unit test mass is attractive, in toward the mass M. In Figure (7), the force on our unit positive test particle is out, away from Q if Q is positive. *When we have an attractive force as in Figure (5), the potential energy is negative as in Equation (1). But when the force is repulsive, as in Figure (7), the potential energy is positive.* Let us briefly review the physical origin for this difference in the sign of the potential energy.

In any discussion of potential energy, it is necessary to define the zero of potential energy, i.e. to say where the floor is. In the case of satellite motion, we defined the satellite's potential energy as being zero when the satellite was infinitely far away from the planet. If we release a satellite at rest a great distance from the planet, it will start falling toward the planet. As it falls, it gains kinetic energy, which it must get at the expense of gravitational potential energy. Since the satellite started with zero gravitational potential energy when far out and loses potential energy as it falls in, it must end up with negative potential energy when it is near the planet. This is the physical origin of the minus sign in Equation (1). Using the convention that potential energy is zero at infinity, then attractive forces lead to negative potential energies.

If the force is repulsive as in Figure (7), then we have to do work on our test particle in order to bring it in from infinity. The work we do against the repulsive force is stored up as positive potential energy which could be released if we let go of the test particle (and the test particle goes flying out). Thus the convention that potential energy is zero at infinity leads to positive potential energies for repulsive forces like that shown in Figure (7).

ELECTRIC POTENTIAL OF A POINT CHARGE

Using the fact that we can go from the gravitational force law to Coulomb's law by replacing GMm by $Qq/4\pi\varepsilon_0$ (see Exercise 1), we expect that the formula for the electric potential energy of a charge q a distance r from Q is

$$\left.\begin{array}{l}\text{electric potential energy}\\\text{of a charge q}\end{array}\right\} = +\frac{Qq}{4\pi\varepsilon_0 r} \quad (3)$$

The + sign in Equation (3) indicates that for positive Q and q we have a repulsive force and positive potential energy. (If Q is negative, but q still positive, the force is attractive and the potential energy must be negative.)

To determine the potential energy of a *unit test charge*, we set q = 1 in Equation (3) to get

$$\left.\begin{array}{l}\text{electric potential}\\\text{energy of a unit}\\\text{test charge}\end{array}\right\} \equiv \left(\begin{array}{l}\text{electric}\\\text{potential}\end{array}\right) = \frac{Qq}{4\pi\varepsilon_0 r}$$

$$(4)$$

Following the same convention we used for gravity, we will use the name *electric potential* for the potential energy of a unit test charge. Thus Equation (4) is the formula for the electric potential in the region surrounding the charge Q. As expected, the lines of *equal potential*, the *equipotential lines* are the circles of constant radius seen in Figure (7).

Exercise 1

Start with Newton's gravitational force law, replace GMm by $Qq/4\pi\varepsilon_0$, and show that you end up with Coulomb's electrical force law.

CONSERVATIVE FORCES

(This is a formal aside to introduce a point that we will treat in much more detail later.)

Suppose we have a fixed charge Q and a small test particle q as shown in Figure (8). The potential energy of q is defined as zero when it is infinitely far away from Q. If we carry q in from infinity to a distance r, we do an amount of work on the particle

$$\text{Work we do}\Big\} = \int_{\infty}^{r} \vec{F}_{us}\cdot d\vec{x} \quad (5)$$

If we apply just enough force to overcome the electric repulsive force, if $\vec{F}_{us} = -q\vec{E}$, then the work we do should all be stored as electric potential energy, and Equation (5), with $\vec{F}_{us} = -q\vec{E}$ should give us the correct electric potential energy of the charge q.

But an interesting question arises. Suppose we bring the charge q in along two different paths, paths (1) and (2) shown in Figure (8). Do we do the same amount of work, store the same potential energy for the two different paths?

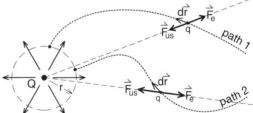

Figure 8
If we bring a test particle q in from infinity to a distance r from the charge Q, the electric potential energy equals $Qq/4\pi\varepsilon_o r$. But this potential energy is the work we do in bringing q in from infinity:

$$\int_{\infty}^{r} \vec{F}_{us}\cdot d\vec{r} = \frac{Qq}{4\pi\varepsilon_o r}$$

This answer does not depend upon the path we take bringing q in.

If we lift an eraser off the floor up to a height h, and hold it still, then it does not matter what path we took, the net amount of work we did was mgh and this is stored as gravitational potential energy. When the work we do against a force depends only on the initial and final points, and not on the path we take, we say that the force is conservative.

In contrast, if we move the eraser over a horizontal table from one point to another, the amount of work we do against friction depends very much on the path. The longer the path the more work we do. As a result we cannot define a **friction potential energy** because it has no unique value. Friction is a *non-conservative force*, and non-conservative forces do not have unique potential energies.

The gravitational fields of stationary masses and the electric fields of stationary charges all produce conservative forces, and therefore have unique potential energies. We will see however that moving charges can produce electric fields that are not conservative! When that happens, we will have to take a very careful look at our picture of electric potential energy. But in dealing with the electric fields of static charges, as we will for a few chapters, we will have unique electric potential energies, and maps of equipotential lines will have an unambiguous meaning.

ELECTRIC VOLTAGE

In our discussion of Bernoulli's equation, we gave the collection of terms $(P + \rho gh + 1/2\rho v^2)$ the name *hydrodynamic voltage*. The content of Bernoulli's equation is that this hydrodynamic voltage is constant along a stream line when the fluid is incompressible and viscous forces can be neglected. Two of the three terms, ρgh and $1/2\rho v^2$ represent the energy of a unit volume of the fluid, thus we see that our hydrodynamic voltage has the dimensions of energy per unit volume.

Electric voltage is a quantity with the dimensions of *energy per unit charge* that in different situations is represented by a series of terms like the terms in Bernoulli's hydrodynamic voltage. There is the potential energy of an electric field, the chemical energy supplied by a battery, even a kinetic energy term, seen in careful studies of superconductors, that is strictly analogous to the $1/2\rho v^2$ term in Bernoulli's equation. In other words, electric voltage is a complex concept, but it has one simplifying feature. Electric voltages are measured by a common experimental device called a voltmeter. In fact we will take as the definition of electric voltage, that quantity which we measure using a voltmeter.

This sounds like a nebulous definition. Without telling you how a voltmeter works, how are you to know what the meter is measuring? To overcome this objection, we will build up our understanding of what a voltmeter measures by considering the various possible sources of voltage one at a time. Bernoulli's equation gave us all the hydrodynamic voltage terms at once. For electric voltage we will have to dig them out as we find them.

Our first example of an electric voltage term is the ***electric potential energy of a unit test charge***. This has the dimensions of energy per unit charge which in the MKS system is joules/coulomb and called volts.

$$\boxed{1\ \frac{\text{joule}}{\text{Coulomb}} \equiv 1\ \text{volt}} \qquad (6)$$

In Figure (9), which is a repeat of Figure (8) showing the electric field lines and equipotential lines for a point charge Q, we see from Equation (4) that a unit test particle at Point (1) has a potential energy, or voltage V_1 given by

$$V_1 = \frac{Q}{4\pi\varepsilon_0 r_1} \qquad \begin{array}{l} \textit{electric potential or} \\ \textit{voltage at Point (1)} \end{array}$$

At Point (2), the electric potential or voltage V_2 is given by

$$V_2 = \frac{Q}{4\pi\varepsilon_0 r_2} \qquad \begin{array}{l} \textit{electric potential or} \\ \textit{voltage at Point (2)} \end{array}$$

Voltmeters have the property that they only measure the *difference in voltage* between two points. Thus if we put one lead of a voltmeter at Point (1), and the other at Point (2) as shown, then we get a voltage reading V given by

$$\begin{array}{l} \text{voltmeter} \\ \text{reading} \end{array} \quad V \equiv V_2 - V_1 = \frac{Q}{4\pi\varepsilon_0}\left(\frac{1}{r_2} - \frac{1}{r_1}\right)$$

If we put the two voltmeter leads at points equal distances from Q, i.e. if $r_1 = r_2$, then the voltmeter would read zero. Since the voltage difference between any two points on an equipotential line is zero, the voltmeter reading must also be zero when the leads are attached to any two points on an equipotential line.

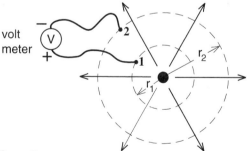

Figure 9
A voltmeter measures the difference in electrical voltage between two points.

This observation suggests an experimental way to map equipotential lines or surfaces. Attach one lead of the voltmeter to some particular point, call it Point (A). Then move the other lead around. Whenever you get a zero reading on the voltmeter, the second lead must be at another point of the same equipotential line as Point (A). By marking all the points where the meter reads zero, you get a picture of the equipotential line.

The discussion we have just given for finding the equipotential lines surrounding a point charge Q is not practical. This involves electrostatic measurements that are extremely difficult to carry out. Just the damp air from your breath would affect the voltages surrounding a point charge, and typical voltmeters found in the lab cannot make electrostatic measurements. Sophisticated meters in carefully controlled environments are required for this work.

But the idea of potential plotting can be illustrated nicely by the simple laboratory apparatus illustrated in Figure (10). In that apparatus we have a tray of water (slightly salty or dirty, so that it is somewhat conductive), and two metal cylinders attached by wire leads to a battery as shown. There are also two probes consisting of a bent, stiff wire attached to a block of wood and adjusted so that the tips of the wires stick down in the water. The other end of the probes are attached to a voltmeter so we can read the voltage difference between the two points (A) and (B), where the probes touch the water.

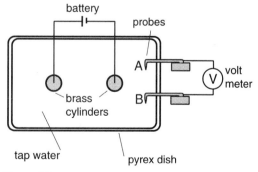

Figure 10
Simple setup for plotting fields. You plot equipotentials by placing one probe (A) at a given position and moving the other (B) around. Whenever the voltage V on the voltmeter reads zero, the probes are at points of equipotential.

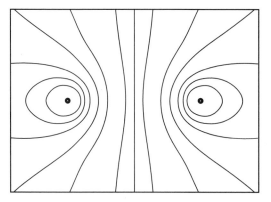

Figure 11
*Plot of the equipotential lines from a student project by
B. J. Grattan. Instead of a tray of water, Grattan used a
sheet of conductive paper, painting two circles with
aluminum paint to replace the brass cylinders. (The
conductive paper and the tray of water give similar
results.) We used the Adobe Illustrator program to
draw the lines through Grattan's data points.*

If we keep Probe (A) fixed and move Probe (B) around,
whenever the voltmeter reads zero, Probe (B) will be on
the equipotential line that goes through Point (A).
Without too much effort, one can get a complete plot of
the equipotential line. Each time we move Probe (A)
we can plot a new equipotential line. A plot of a series
of equipotential lines is shown in Figure (11).

Once we have the equipotential lines shown in Figure
(11), we can sketch the lines of force by drawing a set
of lines perpendicular to the equipotential as we did in
Figure (12). With a little practice you can sketch fairly
accurate plots, and the beauty of the process is that you
did not have to do any calculations!

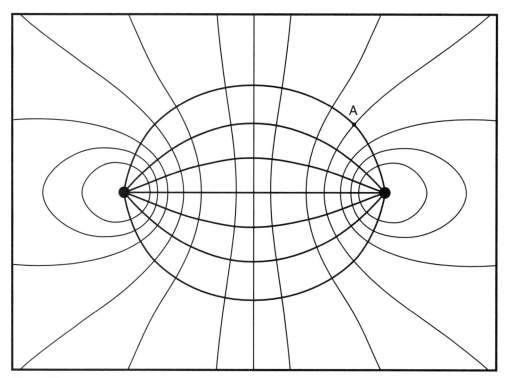

Figure 12
*It does not take too much practice to sketch in the field lines. Draw smooth lines, always
perpendicular to the equipotential lines, and maintain any symmetry that should be there.*

Exercise 2

The equipotential plot of Figure (11) and the field lines of Figure (12) were taken from a student project. The field lines look like the field of two point charges +Q and -Q separated by a distance r. But who knows what is happening in the shallow tank of water (or a sheet of conducting paper)? Perhaps the field lines more nearly represent the field of two line charges +λ and -λ separated by a distance r.

The field of a point charge drops off as $1/r^2$ while the field of a line charge drops off as $1/r$. The point of the exercise is to decide whether the field lines in Figure (12) (or your own field plot if you have constructed one in the lab) more closely represent the field of a point or a line charge.

Hint—Look at the electric field at Point A in Figure (12), enlarged in Figure (13). We know that the field $\vec{E}$ at Point A is made up of two components, $\vec{E}_1$ directed away from the left hand cylinder, and $\vec{E}_2$ directed toward the right hand cylinder, and the net field $\vec{E}$ is the vector sum of the two components. If the field is the field of point charges then $\vec{E}_1$ drops off as $1/r_1^2$ and $\vec{E}_2$ as $1/r_2^2$. But if the field is that of line charges, $\left|\vec{E}_1\right|$ drops off as $1/r_1$ and $\left|\vec{E}_2\right|$ as $1/r_2$. We have chosen Point (A) so that r_1, the distance from (A) to the left cylinder is quite a bit longer than the distance r_2 to the right cylinder. As a result, the ratio of $\left|\vec{E}_1\right|$ to $\left|\vec{E}_2\right|$ and thus the direction of $\vec{E}$, will be quite different for $1/r$ and $1/r^2$ forces. This difference is great enough that you can decide, even from student lab results, whether you are looking at the field of point or line charges. Try it yourself and see which way it comes out.

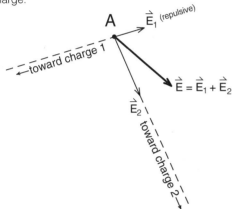

Figure 13
Knowing the direction of the electric field at Point (A) allows us to determine the relative magnitude of the fields $\vec{E}_1$ and $\vec{E}_2$ produced by charges 1 and 2 alone. At Point (A), construct a vector $\vec{E}$ of convenient length parallel to the field line through (A). Then decompose $\vec{E}$ into component vectors $\vec{E}_1$ and $\vec{E}_2$, where $\vec{E}_1$ lies along the line from charge 1 to Point (A), and $\vec{E}_2$ along the line toward charge 2. Then adjust the lengths of $\vec{E}_1$ and $\vec{E}_2$ so that their vector sum is $\vec{E}$.

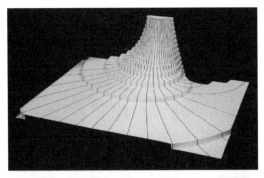

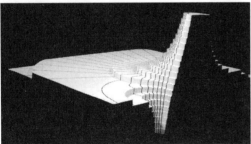

Figure 14
Model of the electric field in the region of two point charges $Q_+ = +3$, $Q_- = -1$. Using the analogy to a topographical map, we cut out plywood slabs in the shape of the equipotentials from the computer plot of Figure 15, and stacked the slabs to form a three dimensional surface. The field lines, which are marked with narrow black tape on the model, always lead in the direction of steepest descent on the surface.

A Field Plot Model

The analogy between a field plot and a map maker's contour plot can be made even more obvious by constructing a plywood model like that shown in Figure (14).

To construct the model, we made a computer plot of the electric field of charge distribution consisting of a charge +3 and –1 seen in Figure (15). We enlarged the computer plot and then cut out pieces of plywood that had the shapes of the contour lines. The pieces of plywood were stacked on top of each other and glued together to produce the three dimensional view of the field structure.

In this model, each additional thickness of plywood represents one more equal step in the electric potential or voltage. The voltage of the positive charge $Q = +3$ is represented by the fat positive spike that goes up toward $+\infty$ and the negative charge $q = -1$ is represented by the smaller hole that heads down to $-\infty$. These spikes can be seen in the back view in Figure (14), and the potential plot in Figure (16).

In addition to seeing the contour lines in the slabs of plywood, we have also marked the lines of steepest descent with narrow strips of black tape. These lines of steepest descent are always perpendicular to the contour lines, and are in fact, the electric field lines, when viewed from the top as in the photograph of Figure (15).

Figure (17) is a plywood model of the electric potential for two positive charges, $Q = +5$, $Q = +2$. Here we get two hills.

Figure 16
Potential plot along the line of the two charges +3, –1. The positive charge creates an upward spike, while the negative charge makes a hole.

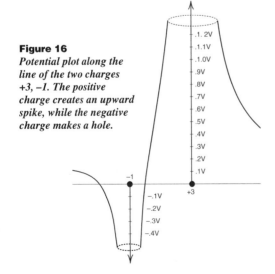

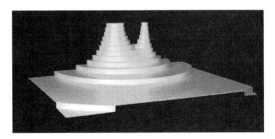

Figure 17
Model of the electric potential in the region of two point charges $Q = +5$ and $Q = +2$.

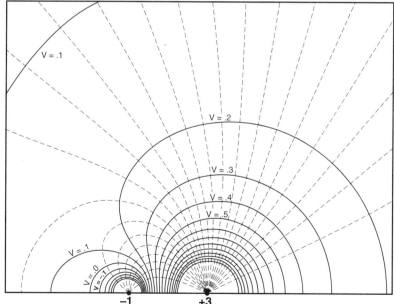

Figure 15

Computer plot of the field lines and equipotentials for a charge distribution consisting of a positive charge + 3 and a negative charge − 1. These lines were then used to construct the plywood model.

Computer Plots

There are now many excellent programs that have personal computers draw out field plots for various charge distributions. In most of these programs you enter an array of charges and the computer draws the field and equipotential lines. You should practice with one of these programs in order to develop an intuition for the field structures various charge distributions produce. In particular, try the charge distribution shown in Figure (18) and (19). In Figure (18), we wish to see the field of oppositely charged plates (a positive plate on the left and a negative one on the right). This charge distribution will appear in the next chapter in our discussion of the parallel plate capacitor.

In Figure (19) we are modeling the field of a circle or in 3-dimensions a hollow sphere of charge. Something rather remarkable happens to the electric field lines in this case. Try it and see what happens!

Exercise 3

If you have a computer plotting program available, plot the field lines for the charge distributions shown in Figures (18,19), and explain what the significant features of the plot are.

Figure 18
The idea is to use the computer to develop an intuition for the shape of the electric field produced by various distributions of electric charge. Here the parallel lines of charge simulates two plates with opposite charge.

Figure 19
We have placed + charges around a circle to simulate a cylinder or sphere of charge. You get interesting results when you plot the field lines for this distribution of charge.

Exercise 4

Figures (20a) and (20b) are computer plots of the electric field of opposite charges. One of the plots represents the $1/r^2$ field of 3 dimensional point charges. The other is the end view of the $1/r$ field of line charges. You are to decide which is which, explaining how you can tell.

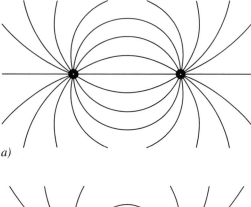

a)

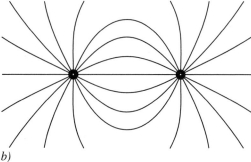

b)

Figure 20
Computer plots of $1/r$ and $1/r^2$ (two dimensional and three dimensional) fields of equal and opposite charges. You are to figure out which is which.

Chapter 26

Electric Fields and Conductors

In this chapter we will first discuss the behavior of electric fields in the presence of conductors, and then apply the results to three practical devices, the Van de Graaff generator, the electron gun, and the parallel plate capacitor. Each of these examples provides not only an explanation of a practical device, but also helps build an intuitive picture of the concept of electric voltage.

ELECTRIC FIELD INSIDE A CONDUCTOR

If we have a piece of metal a few centimeters across as illustrated in Figure (1), and suddenly turn on an electric field, what happens? Initially the field goes right through the metal. But within a few pico seconds (1 pico second = 10^{-12} seconds) the electrons in the metal redistribute themselves inside the metal creating their own field that soon cancels the external applied electric field, as indicated in Figure (2).

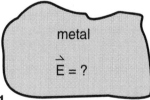

Figure 1
What is the electric field inside a chunk of metal? Metals have conduction electrons that are free to move. If there were an electric field inside the metal, the conduction electrons would be accelerated by the field.

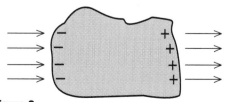

Figure 2
If you place a chunk of metal in an external electric field, the electrons move until there is no longer a force on them.

The very concept of an electrical conductor requires that, in the steady state, there be no electric field inside. To see why, imagine that there is a field inside. Since it is a conductor, the electrons in the conductor are free to move. If there is a field inside, the field will exert a force on the electrons and the electrons will move. They will continue to move until there is no force on them, i.e., until there is no field remaining inside. The electrons must continue to move until the field they create just cancels the external field you applied.

Surface Charges

Where does the redistributed charge have to go in order to create an electric field that precisely cancels the applied electric field? Gauss' law provides a remarkably simple answer to this question. The redistributed charge must reside on the *surface of the conductor.* This is because Gauss' law requires that there be no net charge inside the volume of a conductor.

To see why, let us assume that a charge Q is inside a conductor as shown in Figure (3). Draw a small Gaussian $\vec{\text{surface}}$ around Q. Then by Gauss' law the flux $\Phi = \vec{E} \cdot \vec{A}$ coming out through the Gaussian surface must be equal to Q_{in}/ε_0 where Q_{in} is the net charge inside the Gaussian surface. But if there is no field inside the conductor, if $\vec{E} = 0$, then the flux $\vec{E} \cdot \vec{A}$ out through the Gaussian surface must be zero, and therefore the charge Q_{in} must be zero.

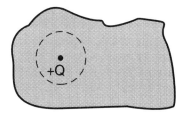

Figure 3
Is there any electric charge inside a conductor? To find out, draw a Gaussian surface around the suspected charge. Since there is no electric field inside the conductor, there is no flux out through the surface, and therefore no charge inside.

If there is no charge inside the conductor, then the only place any charge can exist is in the surface. If there is a redistribution of charge, the redistributed charge must lie on the surface of the conductor.

Figure (4) is a qualitative sketch of how surface charge can create a field that cancels the applied field.

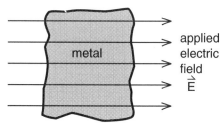

Figure 4a
An external field is applied to a block of metal.

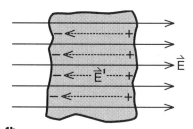

Figure 4b
In response to the electric field, the electrons move to the left surface of the metal, leaving behind positive charge on the right surface. These two surface charges have their own field $\vec{E}'$ that is oppositely directed to $\vec{E}$.

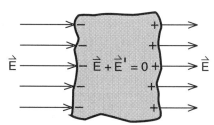

Figure 4c
Inside the block of metal the fields cancel. The result is that the external field on the left stops on the negative surface charge. The field on the right starts again on the positive surface charge.

In Figure (4a) we see the electric field just after it has been turned on. Since the electrons in the metal are negatively charged $(q = -e)$, the force on the electrons $\vec{F} = (-e)\,\vec{E}$ is opposite to $\vec{E}$ and directed to the left.

In Figure (4b), electrons have been sucked over to the left surface of the metal, leaving positive charge on the right surface. The negative charge on the left surface combined with the positive charge on the right produced the left directed field $\vec{E}'$ shown by the dotted lines. The oppositely directed fields $\vec{E}$ and $\vec{E}'$ cancel in Figure (4c) giving no net field inside the metal.

Surface Charge Density

When a field $\vec{E}$ impinges on the surface of a conductor, it must be oriented at right angles to the conductor as shown in Figure (5). The reason for this is that if $\vec{E}$ had a component $\vec{E}_{||}$ parallel to the surface, $\vec{E}_{||}$ would pull the movable charge along the surface and change the charge distribution. The only direction the surface charge cannot be pulled is directly out of the surface of the conductor, thus for a stable setup the electric field at the surface must be perpendicular as shown.

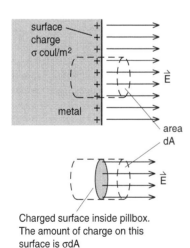

Charged surface inside pillbox. The amount of charge on this surface is σdA

Figure 5
To calculate the surface charge density, we draw a small cylindrical pill box of cross-sectional area dA. We then equate the flux of electric field out through the right surface of the pill box to $1/\varepsilon_0$ times the charge inside the pill box.

Gauss' law can be used to calculate how much charge must be at the surface if a field of strength $\vec{E}$ is impinging as shown in Figure (5). In that figure we have drawn a small pill box shaped Gaussian surface, with one end in the conductor and the other outside in the field $\vec{E}$. If the area of the end of the pill box is dA, then the flux out of the pill box on the right is
$$\Phi_{out} = EdA.$$

Let σ coulombs/meter^2 be the charge density on the surface. The amount of the conductor's surface surrounded by the pill box is dA, thus the amount of charge inside the pill box is

$$\left.\begin{array}{r}\text{Amount of charge} \\ \text{inside the} \\ \text{Gaussian surface}\end{array}\right\} \equiv Q_{in} = \sigma dA$$

By Gauss' law, the flux $\Phi_{out} = EdA$ must equal $1/\varepsilon_0$ times the total charge inside the Gaussian surface and we get

$$\Phi_{out} = \frac{Q_{in}}{\varepsilon_0} = \frac{\sigma dA}{\varepsilon_0} = EdA$$

The dA's cancel and we are left with

$$E = \frac{\sigma}{\varepsilon_0} \qquad \begin{array}{l} E = electric\ field\ at\ the\ conductor \\ \sigma = charge\ density\ at\ the\ surface \end{array} \quad (1)$$

Equation (1) gives a simple relation between the strength of the electric field at the surface of a conductor, and the surface charge density σ at that point. Just remember that the field $\vec{E}$ must be perpendicular to the surface of the conductor. (If the applied field was not originally perpendicular to the surface, surface charges will slide along the surface, reorienting the external field to make it perpendicular.)

To appreciate how far we have come with the concepts of fields and Gauss' law, just imagine trying to derive Equation (1) from Coulomb's law. We wouldn't even know how to begin.

We will now work an example and assign a few exercises to build an intuition for the behavior of fields and conductors. Then we will apply the results to some practical devices.

Example: Field
in a Hollow Metal Sphere

Suppose we have the hollow metal sphere shown in Figure (6). A total charge Q is placed on the sphere. What are the electric fields outside and inside the sphere?

One key to solving this problem is to realize that since the sphere is symmetric, the fields it produces must also be symmetric. We are not interested in fields that do one thing on the left side and something else on the right, for we do not have any physical cause for such an asymmetry.

In Figure (7) we have drawn a Gaussian surface surrounding the metal sphere as shown. Since there is a net charge +Q on the sphere, and therefore inside the Gaussian surface, there must be a net flux Q/ε_0 out through the surface. Since the Gaussian surface has an area $4\pi r^2$, Gauss' law gives

$$\Phi = E_{out}A_\perp = E_{out} \times 4\pi r^2 = \frac{Q}{\varepsilon_0}$$

$$E_{out} = \frac{Q}{4\pi\varepsilon_0 r^2} \qquad (2)$$

which happens to be the field of a point charge.

In Figure (8) we have drawn a Gaussian surface inside the metal at a radius r_i. Since there is no field inside the metal, $EA_\perp = 0$ and there is no flux flowing out through the Gaussian surface. Thus by Gauss' law there can be *no net charge inside the Gaussian surface*. Explicitly this means that there is no surface charge on the inside of the conductor. The charge Q we spread on the conducting sphere *all went to the outside surface*!

Finally in Figure (9) we have drawn a Gaussian surface inside the hollow part of the hollow sphere. Since there is no charge—only empty space inside this Gaussian surface, there can be no flux out through the surface, and the *field* $\vec{E}$ *inside the hollow part of the sphere is exactly zero*. This is a rather remarkable result considering how little effort was required to obtain it.

Figure 6
We place a charge Q on a hollow metal sphere. Where do the charge and the field lines go?

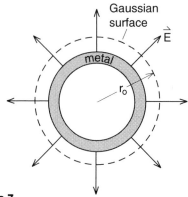

Figure 7
If we place a Gaussian surface around and outside the sphere, we know that the charge Q must be inside the Gaussian surface, and therefore Q/ε_o lines must come out through the surface

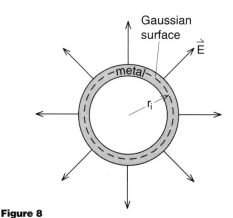

Figure 8
If we place our Gaussian surface inside the metal where $\vec{E} = 0$, no lines come out through the Gaussian surface and therefore there must be no net charge Q inside the Gaussian surface. The fact that there is no charge within that surface means all the charge we placed on the sphere spreads to the outside surface.

Exercise 1

A positive charge +Q is surrounded concentrically by a conducting sphere with an inner radius r_a and outer radius r_b as shown in Figure (10). The conducting sphere has no net charge. Using Gauss' law, find the electric field inside the hollow section $(r < r_a)$, inside the conducting sphere $(r_a < r < r_b)$ and outside the sphere $(r > r_b)$. Also calculate the surface charge densities on the inner and outer surfaces of the conducting sphere. Show that Equation (1) applies to the charge densities you calculate.

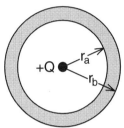

Figure 10
Start with an uncharged hollow metal sphere and place a charge +Q inside. Use Gauss' law to determine the electric field and the surface charges throughout the region.

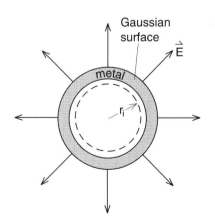

Figure 9
If the Gaussian surface is drawn inside the hollow cavity as shown, then there is no charge inside the Gaussian surface. Thus no field lines emerge through the Gaussian surface, and $\vec{E}$ must be zero inside the cavity.

Exercise 2

A chunk of metal has an irregularly shaped cavity inside as shown in Figure (11). There are no holes and the cavity is completely surrounded by metal.

The metal chunk is struck by lightning which produces huge electric fields and deposits an unknown amount of charge on the metal, but does not burn a hole into the cavity. Show that the lightning does not create an electric field inside the cavity. *(For a time on the order of pico seconds, an electric field will penetrate into the metal, but if the metal is a good conductor like silver or copper, the distance will be very short.)*

(What does this problem have to do with the advice to stay in a car during a thundershower?)

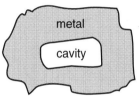

Figure 11
A chunk of metal with a completely enclosed hollow cavity inside is struck by lightning.

Exercise 3

A positive charge +Q placed on a conducting sphere of radius R, produces the electric field shown.

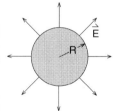

a) What is the charge density σ on the surface of the sphere?

b) Use Equation (1) to find a formula for the magnitude of the electric field $\vec{E}$ produced by the surface charge density σ.

c) How does the field calculated in part *b)* compare with the strength of the electric field a distance R from a point charge Q?

Exercise 4

Repeat Exercise 1 assuming that the conducting sphere has a net charge of $-Q$. Does the charge on the conducting sphere have any effect on the fields inside the sphere? Why is there no field outside the sphere?

VAN DE GRAAFF GENERATOR

The Van de Graaff generator is a conceptually straight-forward device designed to produce high voltages. A sketch of the apparatus is shown in Figure (12), where we have a hollow metal sphere with a hole in the bottom, and a conveyer belt whose purpose is to bring charge up into the sphere. The belt is driven by a motor at the bottom.

The first step is to get electric charge onto the belt. This is done electrostatically by having an appropriate material rub against the belt. For example, if you rub a rubber rod with cat fur, you leave a negative charge on the rubber rod. If you rub a glass rod with silk, a positive charge will be left on the glass rod. I do not know what sign of charge is left on a comb when you run it through your hair on a dry day, but enough charge can be left on the comb to pick up small pieces of paper. We will leave the theory of creating electrostatic charges to other texts. For our discussion, it is sufficient to visualize that some kind of rubbing of the belt at the bottom near the motor deposits charge on the belt. (As an example of charging by rubbing, run a comb through your hair several times. The comb becomes electrically charged and will pick up small pieces of paper.)

Acting like a conveyor belt, the motorized belt carries the charge up and into the inside of the hollow metal sphere. If there is already charge on the sphere, then, as we have seen in Example (1), there will be an electric field outside the sphere as shown in Figure (13). (For this example we are assuming that the belt is carrying positive charge.) But inside the sphere there will be no field. (The hole in the bottom of the sphere lets a small amount of electric field leak inside, but not enough to worry about.)

As the charge is being carried up by the belt, the electric field outside the sphere pushes back on the charge, and the belt has to do work to get the charge up to the sphere. The more charge that has built up on the sphere, the stronger the electric field $\vec{E}$, and the more work the belt has to do. In a typical Van de Graaff generator used in lecture demonstration, you can hear the motor working harder when a large charge has built up on the sphere.

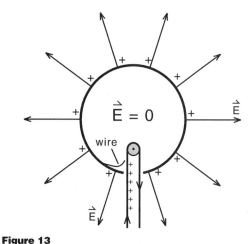

Figure 13
*It takes work to carry the charge up to the sphere against the electric field that is pushing down on the charge. But once inside the sphere where there is almost no field, the charge freely moves off the belt, onto the wire, charging up the sphere. The more charge on the sphere, the stronger the electric field **E** outside the sphere, and the more work required to bring new charge up into the sphere. (In the demonstration model, you can hear the motor slow down as the sphere becomes charged up.)*

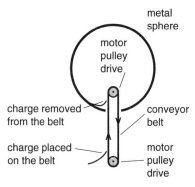

Figure 12
The Van de Graaff generator. Electric charge is carried up the belt and dumped inside the hollow metal sphere. Since there are no electric fields inside the sphere, the electric charge freely flows off the belt to the sphere, where it then spreads evenly to the <u>outside</u> surface of the sphere.

When the charge gets to the sphere how do we get it off the belt onto the sphere? When the sphere already has a lot of positive charge on it, why would the positive charge on the belt want to flow over to the sphere? Shouldn't the positive charge on the belt be repelled by the positive sphere?

Here is where our knowledge of electric fields comes in. As illustrated in Figure (13), there may be very strong electric fields outside the sphere, but inside there are none. Once the conveyor belt gets the charge inside the sphere, the charge is completely free to run off to the sphere. All we need is a small wire that is attached to the inside of the sphere that rubs against the belt. In fact, the neighboring + charge on the belt helps push the charge off the belt onto the wire.

Once the charge is on the wire and flows to the inside of the sphere, it must immediately flow to the outside of the sphere where it helps produce a stronger field $\vec{E}$ shown in Figure (13).

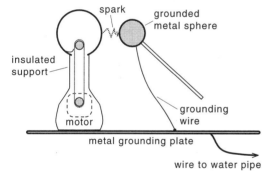

Figure 14
We can discharge the Van de Graaff generator by bringing up a grounded sphere as shown. Since about 100,000 volts are required to make a spark one inch long, we can use the maximum length of sparks to estimate the voltage produced by the Van de Graaff generator.

Electric Discharge

When a large amount of charge has accumulated on the metal sphere of the Van de Graaff generator, we can produce some very strong fields and high voltages. We can estimate the voltage by bringing a grounded sphere up to the Van de Graaff generator as shown in Figure (14). A voltage of about 100,000 volts is required to make a spark jump about an inch through air. Thus if we get a spark about 2 inches long between the Van de Graaff generator and the grounded sphere, we have brought enough charge onto the generator sphere to create a voltage of about 200,000 volts. (The length of the sparks acts as a crude voltmeter!)

As an exercise, let us estimate how many coulombs of charge must be on the Van de Graaff generator sphere to bring it up to a voltage of 200,000 volts.

Outside the Van de Graaff generator sphere, the electric field is roughly equal to the electric field of a point charge. Thus the voltage or electric potential of the sphere should be given by Equation (25-4) as

$$V = \frac{Q}{4\pi\varepsilon_0 r} \tag{25-4}$$

where r is the radius of the Van de Graaff generator sphere. *(Remember that r is not squared in the formula for potential energy or voltage.)*

Let us assume that r = 10 cm or .1 m, and that the voltage V is up to 200,000 volts. Then Equation (25-4) gives

$$Q = 4\pi\varepsilon_0 r V$$
$$= 4\pi \times 9 \times 10^{-12} \times .1 \times 200,000$$

$$Q \approx 2 \times 10^{-6} \text{ coulombs}$$

A couple millionth's of a coulomb of charge is enough to create 200,000 volt sparks. As we said earlier, a whole coulomb is a huge amount of charge!

Grounding

The grounded sphere in Figure (14) that we used to produce the sparks, provides a good example of the way we use conductors and wires.

Beneath the Van de Graaff generator apparatus we have placed a large sheet of aluminum called a grounding plane that is attached to the metal pipes and the electrical ground in the room. (Whenever we have neglected to use this grounding plane during a demonstration we have regretted it.) We have attached a copper wire from the grounding plane to the "grounded" sphere as shown.

Thus in Figure (14), the grounding plane, the room's metal pipes and electrical ground wires, and the grounded sphere are all attached to each other via a conductor. Now there can be no electric field inside a conductor, therefore all these objects are at the same electric potential or voltage. (If you have a voltage difference between two points, there must be an electric field between these two points to produce the voltage difference.) It is common practice in working with electricity to define the voltage of the water pipes (or a metal rod stuck deeply into the earth) as zero volts or "ground". (The ground wires in most home wiring are attached to the water pipes.) Any object that is connected by a wire to the water pipes or electrical ground wire is said to be **grounded**. The use of the earth as the definition of the zero of electric voltage is much like using the floor of a room as the definition of the zero of the gravitational potential energy of an object.

In Figure (14), when the grounded sphere is brought up to the Van de Graaff generator and we get a 2 inch long spark, the spark tells us that the Van de Graaff sphere had been raised to a potential of at least 200,000 volts above ground.

Van de Graaff generators are found primarily in two applications. One is in science museums and lecture demonstration to impress visitors and students. The other is in physics research. Compared to modern accelerators, the 200,000 volts or up to 100 million volts that Van de Graaff generators produce is small. But the voltages are very stable and can be precisely controlled. As a result the Van de Graaff's make excellent tools for studying the fine details of the structure of atomic nuclei.

THE ELECTRON GUN

In Figure (15) we have a rough sketch of a television tube with an electron gun at one end to create a beam of electrons, deflection plates to move the electron beam, and a phosphor screen at the other end to produce a bright spot where the electrons strike the end of the tube.

Figure (16) illustrates how a picture is drawn on a television screen. The electron beam is swept horizontally across the face of the tube, then the beam is moved down one line and swept horizontally again. An American television picture has about 500 horizontal lines in one picture.

As the beam is swept across, the brightness of the spot can be adjusted by changing the intensity of the electron beam. In Figure (16), line 3, the beam starts out bright, is dimmed when it gets to the left side of the letter A, shut off completely when it gets to the black line, then turned on to full brightness to complete the line. In a standard television set, one sweep across the tube takes about 60 microseconds. To draw the fine details you see on a good television set requires that the intensity of the beam can be turned up and down in little more than a tenth of a microsecond.

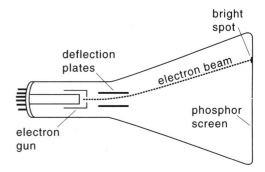

Figure 15
Cathode ray tubes, like the one shown above, are commonly used in television sets, oscilloscopes, and computer monitors. The electron beam (otherwise known as a "cathode ray") is created in the electron gun, is aimed by the deflection plates, and produces a bright spot where it strikes the phosphor screen.

The heart of this system is the electron gun which creates the electron beam. The actual electron gun in a television tube is a complex looking device with indirect heaters and focusing rings all mounted on the basic gun. What we will describe instead is a student-built gun which does not produce the fine beam of a commercial gun, but which is easy to build and easy to understand.

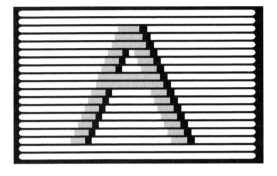

Figure 16
The letter A on a TV screen. To construct an image the electron beam is swept horizontally, and turned up where the picture should be bright and turned down when dark. The entire image consists of a series of these horizontal lines, evenly spaced, one below the other.

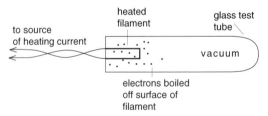

Figure 17
Source of the electrons. The tungsten filament is heated by an electric current. When it becomes red-hot, electrons boil out through the surface. The white coating on the filament makes it easier for the electrons to escape.

The Filament

As shown in Figure (17), the source of the electrons in an electron gun is the **filament**, a piece of wire that has been heated red-hot by the passage of an electric current. At these temperatures, some of the electrons in the filament gain enough thermal kinetic energy to evaporate out through the surface of the wire. The white coating you may see on a filament reduces the amount of energy an electron needs to escape out through the metal surface, and therefore helps produce a more intense beam of electrons. At standard temperature and pressure, air molecules are about 10- molecular diameters apart as indicated in Figure (18). Therefore if the filament is in air, an electron that has evaporated from the filament can travel, at most, a few hundred molecular diameters before striking an air molecule. This is why the red-hot burner on an electric stove does not emit a beam of electrons. The only way we can get electrons to travel far from the filament is to place the filament in a vacuum as we did in Figure (17). The better the vacuum, the farther the electrons can travel.

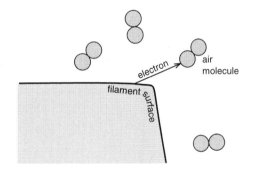

Figure 18
Whenever we heat a metal to a high enough temperature, electrons boil out of the surface. But if there is air at standard pressure around, the electrons do not get very far before striking an air molecule.

Accelerating Field

Once the electrons are out of the filament we use an electric field to accelerate them. This is done by placing a metal cap with a hole in the end over the end of the filament as shown in Figure (19). The filament and cap are attached to a battery as shown in Figure (20) so that the cap is positively charged relative to the filament.

Intuitively the gun works as follows. The electrons are repelled by the negatively charged filament and are attracted to the positively charged cap. Most of the electrons rush over, strike, and are absorbed by the cap as shown in Figure (21). But an electron headed for the hole in the cap discovers too late that it has missed the cap and goes on out to form the electron beam.

A picture of the resulting electron beam is seen in Figure (22). The beam is visible because some air remains inside the tube, and the air molecules glow when they are struck by an electron.

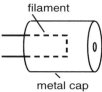

filament

0

metal cap

Figure 19
To create a beam of electrons, we start by placing a metal cap with a hole in it, over the filament.

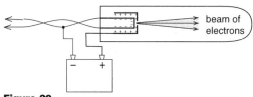

beam of
electrons

−   +

Figure 20
We then attach a battery to the metal cap so that the cap has a positive voltage relative to the filament.

A Field Plot

A field plot of the electric field lines inside the electron gun cap gives a more precise picture of what is happening. Figure (23) is a computer plot of the field lines for a cylindrical filament inside a metal cap. We chose a cylindrical filament rather than a bent wire filament because it has the cylindrical symmetry of the cap and is therefore much easier to calculate and draw. But the fields for a wire filament are not too different.

First notice that the field lines are perpendicular to both metal surfaces. This agrees with our earlier discussion that an electric field at the surface of a conductor cannot have a parallel component for that would move the charge in the conductor. The second thing to note is that due to the unfortunate fact that the charge on the electron is negative, the electric field points oppositely to the direction of the force on the electrons. The force is in the direction of $-\vec{E}$.

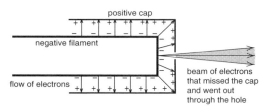

positive cap

negative filament

flow of electrons

beam of electrons
that missed the cap
and went out
through the hole

Figure 21
Electrons flow from the negative filament to the positive cap. The beam of electrons is formed by the electrons that miss the cap and go out through the hole.

Figure 22
Resulting electron beam.

The electrons, however, do not move along the $-\vec{E}$ field lines. If they boil out of the filament with a negligible speed they will start moving in the direction $-\vec{E}$. But as the electrons gain momentum, the force $-e\vec{E}$ has less and less effect. (Remember, for example, that for a satellite in a circular orbit, the force on the satellite is down toward the center of the earth. But the satellite moves around the earth in an orbit of constant radius.) In Figure (23), the dotted lines show a computer plot of the trajectories of the electrons at several points. The most important trajectories for our purposes are those that pass through the hole in the cap and go out and form the electron beam.

Exercise 5

Describe two other examples where an object does not move in the direction of the net force acting on it.

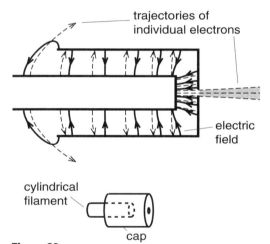

trajectories of
individual electrons

electric
field

cylindrical
filament

cap

Figure 23
Plot of the electric field in the region between the filament and the cap. Here we assume that we have a cylindrical filament heated by a wire inside.

Equipotential Plot

Once we know the field lines, we can plot the equipotential lines as shown in Figure (24). The lines are labeled assuming that the filament is grounded (0 volts) and that the cap is at 100 volts. The shape of the equipotentials, shown by dashed lines, does not change when we use different accelerating voltages, only the numerical value of the equipotentials changes.

The reason that the equipotential lines are of such interest in Figure (24) is that they can also be viewed as a map of the electron's kinetic energy.

Remember that the voltage V is the potential energy of a unit positive test charge. A charge q has a potential energy qV, and an electron, with a charge $-e$, has an electric potential energy $-eV$.

In our electron gun, the electrons evaporate from the filament with very little kinetic energy, call it zero. By the time the electrons get to the 10-volt equipotential, their electric potential energy has dropped to $(-e \times 10)$ joules, and by conservation of energy, their kinetic energy has gone up to $(+e \times 10)$ joules. At the 50 volt equipotential the electron's kinetic energy has risen to $(e \times 50)$ joules, and when the electrons reach the 100 volt cap, their energy is up to $(e \times 100)$ joules. Thus the equipotential lines in Figure (24) provide a map of the kinetic energy of the electrons.

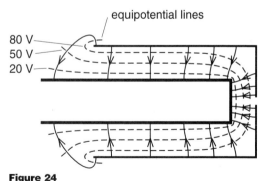

equipotential lines

80 V
50 V
20 V

Figure 24
Equipotential plot. We see that by the time the electrons have reached the hole in the cap, they have crossed the same equipotential lines and therefore have gained as much kinetic energy as the electrons that strike the cap. (From a student project by Daniel Leslie and Elad Levy.)

ELECTRON VOLT
AS A UNIT OF ENERGY

What is perhaps most remarkable about the electron gun is that every electron that leaves the filament and strikes the cap gains precisely the same kinetic energy. If we use a battery that produces 100 volt accelerating voltage, then every electron gains precisely $(e \times 100)$ joules of kinetic energy. This is also true of the electrons that miss the cap and go out and form the electron beam.

The amount of energy gained by an electron that falls through a 1 volt potential is $(e \times 1 \text{ volt}) = 1.6 \times 10^{-19}$ joules. This amount of energy is called an *electron volt* and designated by the symbol eV.

$$1 \text{eV} = \begin{cases} \text{energy gained by an electron} \\ \text{falling through a 1 volt potential} \end{cases}$$

$$= (e \text{ coulombs}) \times (1 \text{ volt}) \qquad (3)$$

$$= 1.6 \times 10^{-19} \text{ joules}$$

The dimensions in Equation (3) make a bit more sense when we realize that the volt has the dimensions of joule/coulomb, so that

$$1 \text{eV} = \left(e \text{ coulombs}\right) \times \left(1 \frac{\text{joule}}{\text{coulomb}}\right)$$

$$= (e) \text{ joules} \qquad (3a)$$

The electron volt is an extremely convenient unit for describing the energy of electrons produced by an electron gun. If we use a 100 volt battery to accelerate the electrons, we get 100 eV electrons. Two hundred volt batteries produce 200 eV electrons, etc.

To solve problems like calculating the speed of a 100 eV electron, you need to convert from eV to joules. The conversion factor is

$$\boxed{1.6 \times 10^{-19} \frac{\text{joules}}{\text{eV}}} \qquad \begin{array}{l} conversion \\ factor \end{array} \qquad (4)$$

For example, if we have a 100 eV electron, its kinetic energy $1/2 \text{ mv}^2$ is given by

$$KE = 1/2 \text{ mv}^2$$

$$= 100 \text{ eV} \times 1.6 \times 10^{-19} \frac{\text{joules}}{\text{eV}} \qquad (5)$$

Using the value $m = 9.11 \times 10^{-31}$ kg for the electron mass in Equation (5) gives

$$v = \sqrt{\frac{2 \times 100 \times 1.6 \times 10^{-19}}{9.11 \times 10^{-31}}}$$

$$\qquad (6)$$

$$= 6 \times 10^6 \frac{\text{meters}}{\text{sec}}$$

which is 2% the speed of light.

In studies involving atomic particles such as electrons and protons, the electron volt is both a convenient and very commonly used unit. If the electron volt is too small, we can measure the particle energy in MeV (millions of electron volts) or GeV (billions of electron volts or *Gigavolts*).

$$1 \text{ MeV} \equiv 10^6 \text{ eV}$$

$$1 \text{ GeV} \equiv 10^9 \text{ eV} \qquad (6)$$

For example, if you work the following exercises, you will see that the rest energies $m_0 c^2$ of an electron and a proton have the values

$$\boxed{\begin{array}{l} \text{electron rest energy} = .51 \text{ MeV} \\ \text{proton rest energy} = .93 \text{ GeV} \end{array}} \qquad (7)$$

The reason that it is worth remembering that an electron's rest energy is about .5 MeV and a proton's about 1 GeV, is that when a particle's kinetic energy gets up toward its rest energy, the particle's speed becomes a significant fraction of the speed of light and nonrelativistic formulas like $1/2 \text{ mv}^2$ for kinetic energy no longer apply.

Example

Calculate the rest energy of an electron in eV.

Solution:

$$E = \frac{m_0 c^2 \text{ joules}}{1.6 \times 10^{-19} \frac{\text{joules}}{\text{eV}}}$$

$$= \frac{9.11 \times 10^{-31} \times \left(3 \times 10^8\right)^2}{1.6 \times 10^{-19}}$$

$$= .51 \times 10^6 \text{ eV}$$

Exercise 6

Calculate the rest energy of a proton in eV and GeV.

Exercise 7

What accelerating voltage must be used in an electron gun to produce electrons whose kinetic energy equals their rest energy?

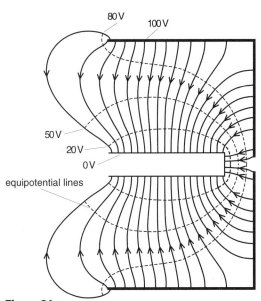

Figure 24a

Another field plot by Leslie and Levy, showing the electric field and equipotential lines in a gun with a shorter cap.

About Computer Plots

One final note in our discussion of the electron gun. You might feel that by using the computer plots in Figures (23) and (24) we have cheated a bit. We haven't done the work ourselves, we let somebody (or something) else do the calculations for us and we are just using their answers. Yes and no!

First of all, with a little bit of practice you can learn to draw sketches that are quite close to the computer plots. Use a trick like noting that field lines must be perpendicular to the surface of a conductor where they touch the conductor. If two conductors have equal and opposite charge – if they were charged by a battery – all the field lines that start on the positive conductor will stop on the negative one. Use any symmetry you can find to help sketch the field lines and then sketch the equipotential lines perpendicular to the field lines. Some places it is easier to visualize the equipotential lines, e.g., near the surface of a conductor, and then draw in the perpendicular field lines.

The other point is that, for a number of practical problems the geometry of the conductors is complicated enough that only by using a computer can we accurately plot the field lines and equipotentials. But once a computer plot is drawn, we do not have to worry about how it was calculated. Like a hiker in a new territory, we can use the computer plot as our contour map to tell us the shape and important features of the terrain. For example in our field plots of the electron gun, we see that there is virtually no field out in front of the hole where the electrons emerge, therefore from the time the electrons leave the hole they coast freely at constant speed and energy down the tube.

THE PARALLEL PLATE CAPACITOR

Our final example in this chapter of fields and conductors is the parallel plate capacitor. Here we will work with a much simpler field structure than for the electron gun, and will therefore be able to calculate field strengths and voltages. The parallel plate capacitor serves as the prototype example of a capacitor, a device used throughout physics and electrical engineering for storing electric fields and electric energy.

Suppose we take two circular metal plates of area A, separate them by a distance d, and attach a battery as shown in Figure (25). This setup is called a parallel plate capacitor, and the field lines and equipotential for this setup are shown in the computer plot of Figure (26).

Except at the edges of the plates, the field lines go straight down from the positive to the negative plate, and the equipotentials are equally spaced horizontal lines parallel to the plates. If the plate separation d is small compared to the diameter D of the plates, then we can neglect the fringing of the field at the edge of the plates. The result is what we will call an ideal parallel plate capacitor whose field structure is shown in Figure (27). The advantage of working with this ideal capacitor is that we can easily derive the relationship between the charging voltage V, and the charge Q.

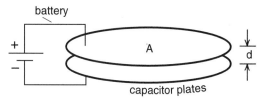

Figure 25
The parallel plate capacitor. The capacitor is charged up by connecting a battery across the plates as shown.

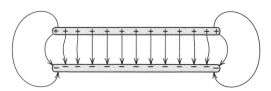

Figure 26
The electric field between and around the edge of the capacitor plates.

Let us take a close look at what we have in Figure (27). The electric field lines $\vec{E}$ leave the positively charged top plate and go straight down to the negatively charged bottom plate. Since all the lines starting at the top plate stop at the bottom one, there must be an equal and opposite charge +Q and -Q on the two plates. There is no net charge on the capacitor, only a separation of charge. And because the field lines go straight down, nowhere do they get closer together or farther apart, the field must have a uniform strength E between the plates.

We can use Gauss' law to quickly calculate the field strength E. The top plate has a charge Q, therefore the total flux out of the top plate must be $\Phi = Q/\varepsilon_0$. But we also have a field of strength E flowing out of a plate of area A. Thus flux of E flowing between the plates is $\Phi = EA$. Equating these two formulas for flux gives

$$\Phi = EA = \frac{Q}{\varepsilon_0}$$

$$\boxed{E = \frac{Q}{\varepsilon_0 A}} \tag{8}$$

We can relate the voltage V and the field strength E by remembering that *E is the force on a unit test charge and V is the potential energy of a unit test charge. If I lift a unit positive test charge from the bottom plate a distance d up to the top one, I have to exert an upward force of strength E for a distance d and therefore do an amount of work $E \times d$. This work is stored as the electric potential energy of the unit test charge, and is therefore the voltage V:*

$$\boxed{V = E\,d} \tag{9}$$

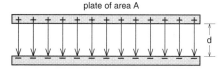

Figure 27
In our idealized parallel plate capacitor the field lines go straight from the positive to the negative plate, and the field is uniform between the plates.

It may seem surprising, but V is also the voltage of the battery (see Figure 25) used to charge up the capacitor.

There is also a simple relationship between the charge Q on the capacitor plates and the voltage difference V between them. Substituting the value of E from Equation (8) into Equation (9) gives

$$V = \left(\frac{d}{\varepsilon_0 A}\right)Q \qquad (10)$$

Equation (10) makes an interesting prediction. If we have a fixed charge Q on the capacitor (say we charged up the capacitor and removed the battery), then if we increase the separation d between the plates, the voltage V will increase.

One problem with trying to measure this increase in voltage is that if we attach a common voltmeter between the plates to measure V, the capacitor will quickly discharge through the voltmeter. In order to see this effect we must use a special voltmeter called an **electrometer** that will not allow the capacitor to discharge. The classic electrometer, used in the 1800's, is the gold leaf electrometer shown in Figures (28) and (29). When the top plate of the electrometer is charged, some of the charge flows to the gold leaves, forcing the leaves apart. The greater the voltage, the greater the charge and the greater the force separating the leaves. Thus the separation of the leaves is a rough measure of the voltage.

In Figure (28), we see a gold leaf electrometer attached to two metal capacitor plates. When the plates are charged, the gold leaves separate, indicating that there is a voltage difference between the plates.

In Figures (29a,b), we are looking through the electrometer at the edge of the capacitor plates. In going from (29a) to (29b), we moved the plates apart without changing the charge on the plates. We see that when the plates are farther apart, the gold leaves are more separated, indicating a greater voltage as predicted by equation (10).

Figure 28
Gold leaf electrometer attached to a parallel plate capacitor.

Figure 29a
Looking through the electrometer at the edge of the charged capacitor plates.

Figure 29b
Without changing the charge, the plates are moved further apart. The increased separation of the gold leaves shows that the voltage difference between the capacitor plates has increased.

Exercise 8

Two circular metal plates of radius 10 cm are separated by microscope slide covers of thickness d = .12 mm. A voltage difference of 5 volts is set up between the plates using a battery as shown in Figure (25). What is the charge Q on the plates?

Deflection Plates

A fitting conclusion to this chapter is to see how the fields in parallel plate capacitor can be used to deflect the beam of electrons produced by an electron gun. In Figure (30) the beam of electrons from an electron gun is aimed between the plates of a parallel plate capacitor. The upward directed electric field $\vec{E}$ produces a downward directed force $-e\vec{E}$ on the electrons, so that when the electrons emerge from the plates, they have been deflected downward by an angle θ as shown. We wish to calculate this angle θ which depends on the strength of the deflection voltage V_p, the length D of the plates, and on the speed v of the electrons.

While the electrons are between the plates, their acceleration is given by

$$\vec{A} = \frac{\vec{F}}{m_e} = \frac{-e\vec{E}}{m_e}$$

where m_e is the electron mass. This acceleration is constant and directed downward, just as in our old projectile motion studies. Using Equation (9) E = V/d for the magnitude of $\vec{E}$, we find that the downward acceleration $\vec{A}$ of the electrons has a magnitude

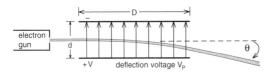

Figure 30
To deflect the beam of electrons, we place what is essentially a parallel plate capacitor in the path of the beam as shown. The electrons are deflected by the electric field between the capacitor plates.

$$|\vec{A}| = \frac{eE}{m_e} = \frac{eV_p}{m_e d}$$

where V_p, is the voltage and d the separation of the deflection plates.

If a particle is subjected to a downward acceleration for a time T, and initially has no downward velocity, its final downward velocity v_{fy} is from the constant acceleration formulas as

$$v_{fy} = A_y T = \frac{eV_p}{m_e d} T \tag{11}$$

If the electrons emerge from the electron gun at a speed v, then the time T it takes them to pass between the plates is

$$T = \frac{D}{v} \tag{12}$$

The tangent of the deflection angle θ is given by the ratio v_{fy}/v which we can get from Equations (11) and (12):

$$v_{fy} = \frac{eV_p}{m_e d} \times \frac{D}{v}$$

$$\tan\theta = \frac{v_{fy}}{v} = \frac{eV_p D}{m_e d v^2} \tag{13}$$

The final step is to note that the speed v of the electrons is determined by the electron gun accelerating voltage V_{acc} by the relationship

$$\frac{1}{2}m_e v^2 = eV_{acc} \quad or \quad v^2 = \frac{2eV_{acc}}{m_e} \tag{14}$$

Equations (13) and (14) finally give

$$\tan\theta = \frac{eV_p D}{m_e d\left(2eV_{acc}/m_e\right)} = \frac{1}{2}\frac{D}{d}\frac{V_p}{V_{acc}} \tag{15}$$

which is a fairly simple result considering the steps we went through to get it. It is reassuring that $\tan\theta$ comes out as a dimensionless ratio, which it must.

Exercise 9

In an electron gun, deflection plates 5 cm long are separated by a distance $d = 1.2$ cm. The electron beam is produced by a 75 volt accelerating voltage. What deflection voltage V_p is required to bend the beam 10 degrees?

Exercise 10

In what is called the Millikan oil drop experiment, shown in Figure (31), a vapor of oil is sprayed between two capacitor plates and the oil drops are electrically charged by radioactive particles.

Consider a particular oil drop of mass m that has lost one electron and therefore has an electric charge q = + e. (The mass m of the drop was determined by measuring its terminal velocity in free fall in the air. We will not worry about that part of the experiment, and simply assume that the drop's mass m is known.) To measure the charge q on the oil drop, and thus determine the electron charge e, an upward electric field $\vec{E}$ is applied to the oil drop. The strength of the field E is adjusted until the upward electric force just balances the downward gravitational force. When the forces are balanced, the drop, seen through a microscope, will be observed to come to rest due to air resistance.

The electric field $\vec{E}$ that supports the oil drop is produced by a parallel plate capacitor and power supply that can be adjusted to the desired voltage V. The separation between the plates is d.

a) Reproduce the sketch of Figure (31), Then put a + sign beside the positive battery terminal and a – sign beside the negative one.

b) Find the formula for the voltage V required to precisely support the oil drop against the gravitational force. Express your answer in terms of the geometry of the capacitor (plate separation d, area A, etc.) the drop's mass m, the acceleration due to gravity g, and the electron charge e.

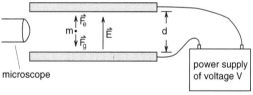

Figure 31
Millikan oil drop apparatus

Chapter 27
Basic Electric Circuits

*In the modern age (post 1870) we have been sur-
rounded by electric circuits. House wiring is our most
familiar example, but we have become increasingly
familiar with electric circuits in radio and television
sets, and even the digital watch you may be wearing. In
this chapter we will discuss the basic electric circuits in
order to introduce the concepts of electric current,
resistance, and voltage drops around the circuit. We
will restrict ourselves to devices like batteries, resistors, light bulbs, and capacitors. The main purpose is
to develop the background needed to work with electric
circuits and electronic measuring equipment in the
laboratory.*

ELECTRIC CURRENT

An electric current in a wire is conceptually somewhat like the current of water in a river. We can define the current in a river as the amount of water per second flowing under a bridge. The amount of water could be defined as the number of water molecules, but a more convenient unit would be gallons, liters, or cubic meters.

An electric current in a wire is usually associated with the flow of electrons and is measured as the amount of charge per second flowing past some point or through some cross-sectional area of the wire, as illustrated in Figure (1). We could measure the amount of charge by counting the number of electrons crossing the area, but it is more convenient to use our standard unit of charge, the coulomb, and define an electric current as the number of coulombs per second passing the cross-sectional area. The unit of current defined this way is called an ***ampere***.

$$1 \text{ ampere } = \left\{ \begin{array}{l} 1 \text{ coulomb per} \\ \text{second passing} \\ \text{a cross--sectional} \\ \text{area of wire} \end{array} \right. \qquad (1)$$

From your experience with household wiring you should already be familiar with the ampere (amp) as a unit of current. A typical light bulb draws between 1/2 and 1 ampere of current, and so does the typical motor in an electric appliance (drill, eggbeater, etc.). A microwave oven and a toaster may draw up to 6 amps, and hair dryers and electric heaters up to 12 amps. Household wiring is limited in its capability of carrying electric current. If you try to carry too much current in a wire, the wire gets hot and poses a fire hazard.

Household wiring is protected by fuses or circuit breakers that shut off the current if it exceeds 15 or 20 amps. (You can see why you do not want to run a hair dryer and an electric heater on the same circuit.)

There is a common misconception that the electrons in a wire travel very fast when a current is flowing in the wire. After all when you turn on a wall light switch the light on the other side of the room appears to turn on instantly. How did the electrons get there so fast?

The answer can be seen by an analogy to a garden hose. When you first attach an empty hose to a spigot and turn on the water, it takes a while before the hose fills up with water and water comes out of the other end. But when the hose is already full and you turn on the spigot, water almost instantly comes out of the other end. Not the water that just went in, but the water that was already in the hose.

A copper wire is analogous to the hose that is already full of water; the electrons are already there. When you turn on the light switch, the light comes on almost instantly because all the "electric fluid" in the wire starts moving almost at once.

To help build an intuition, let us estimate how fast the electrons must move in a copper wire with a 1 millimeter cross-sectional area carrying an electric current of one ampere. This is not an unreasonable situation for household wiring.

A copper atom has a nucleus containing 29 protons surrounded by a cloud of 29 electrons. Of the 29 electrons, 27 are tightly bound to the nucleus and 2 are in an outer shell, loosely bound. (All metal atoms have one, two, and sometimes 3 loosely bound outer electrons.) When copper atoms are collected together to form a copper crystal, the 27 tightly bound electrons remain with their respective nuclei, but the two loosely bound electrons are free to wander throughout the crystal. In a metal crystal or wire, it is the loosely bound electrons (called ***conduction electrons***) that form the electric fluid that makes the wire a conductor.

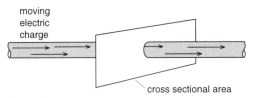

Figure 1
An electric current is defined as the amount of <u>charge per second</u> flowing past a cross-sectional area.

Copper has an atomic weight of 63.5, thus there are 63.5 grams of copper in a mole. And the density of copper is 9 gm /cm³, thus a mole of copper has a volume

$$\text{volume of one} \atop \text{mole of copper} \Bigg\} = \frac{63.5\,\text{gm / mole}}{9\,\text{gm / cm}^3} = 7\frac{\text{cm}^3}{\text{mole}}$$

Since a mole of a substance contains an Avogadro's number 6×10^{23} of particles of that substance, and since there are 2 conduction electrons per copper atom, 7 cm³ of copper contain 12×10^{23} conduction electrons. Dividing by 7, we see that there are 1.7×10^{23} conduction electrons in every cubic centimeter of copper and 1.7×10^{20} in a cubic millimeter. Converting this to coulombs, we get

$$\left. \begin{array}{l} \text{number of} \\ \text{coulombs of} \\ \text{conduction} \\ \text{electrons} \\ \text{in 1mm}^3 \\ \text{of copper} \end{array} \right\} = \frac{1.7 \times 10^{20}\,\text{electrons/mm}^3}{6.25 \times 10^{18}\,\text{electrons/coulomb}}$$

$$= 27\,\text{coulombs/mm}^3$$

In our 1 millimeter cross-sectional area wire, if the electrons flowed at a speed of 1 millimeter per second, 27 coulombs of charge would flow past any point in the wire per second, and we would have a current of 27 amperes. To have a current of 1 ampere, the electrons would have to move only 1/27 as fast, *or 1/27 of a millimeter per second!* This slow speed results from the huge density of conduction electrons.

Positive and Negative Currents

If you are using a hose to fill a bucket with water, there is not much question about which way the current of water is flowing—from the hose to the bucket. But with electric current, because there are two kinds of electric charge, the situation is not that simple. As shown in Figure (2), there are two ways to give an object a positive charge, add positive charge or remove negative charge. If a wire connected to the object is doing the charging, it may be difficult to tell whether there is a current of positive charge into the object or a current of negative charge out of the object. Both have essentially the same effect.

You may argue that at least for copper wires a current of positive charge doesn't make sense because the electric current is being carried by the negative conduction electrons. But a simple model of an electric current will clearly demonstrate that a positive current flowing one way is essentially equivalent to a negative current flowing the other way.

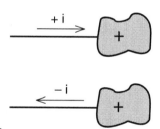

Figure 2
A current of positive charge into an object, or a current of negative charge out, leaves the object positively charged.

In Figure (3) we have tried to sketch a picture of a copper wire in which the conduction electrons are moving to the left producing a left directed negative current. The problem with Figure (3) is that it is hard to show the conduction electrons flowing through the lattice of stationary positive copper nuclei. The picture is difficult to draw, and Figure (3) is not particularly informative.

To more clearly show that the positive charge is at rest and that it is the negative charge that is moving, we have in Figure (4) constructed a model of a copper wire in which we have two separate rods, one moving and one at rest. The stationary rod has the positive copper nuclei and the moving rod has the negative conduction electrons. This model is not a very good representation of what is going on inside the copper wire, but it does remind us clearly that the positive charge is at rest, and that the current is being carried by the moving negative charge.

When you see this model, which we will use again in later discussions, think of the two rods as merged together. Picture the minus charge as flowing through the lattice of positive charge. Remember that the only reason that we drew them as separate rods was to clearly show which charge was carrying the electric current.

Using the results of the previous section, we can make our model of Figure (4) more specific by assuming it represents a copper wire with a 1 millimeter cross section carrying a current of one ampere. In that example the average speed of the conduction electrons was 1/27 of a millimeter per second, which we will take as the speed v of the moving negative rod in Figure (4).

Figure (5a) is the same as Figure (4), except we have drawn a stick figure representing a person walking to the left at a speed v. The person and the negatively charged rod are both moving to the left at the same speed.

Figure (5b) is the same situation from the point of view of the stick figure person. From her point of view, the negative rod is at rest and it is the positive rod that is moving to the right. *Our left directed negative current in Figure (5a) is seen by the moving observer to be a right directed positive current (Figure 5b).* Whether we have a left directed negative current or a right directed positive current just depends upon the point of view of the observer.

But how fast was our moving observer walking? If Figure (5) is a model of a 1 mm^2 copper wire carrying a current of 1 ampere, *the speed v in Figure (5) is 1/27 of a millimeter per second*. This is about 2 millimeters per minute! Although faster than the continental drift, this motion should certainly have little effect on what we see. If the wire is leading to a toaster, the toast will come out the same whether or not we walk by at a speed of 2 mm per minute. For most purposes, we can take a left directed negative current and a right directed positive current as being equivalent. Relatively sophisticated experiments, such as those using the Hall effect (to be discussed later) are required to tell the difference.

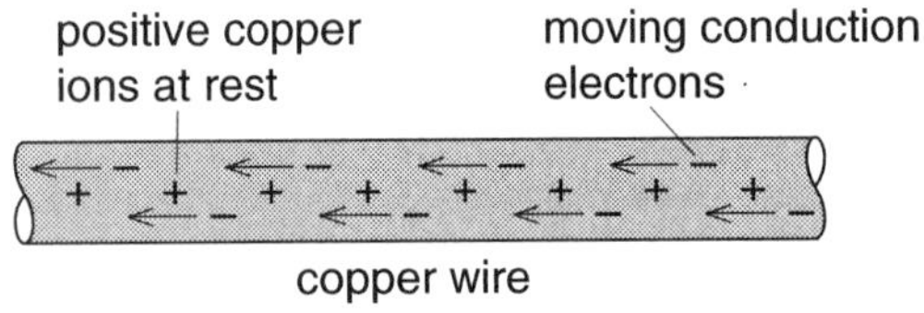

positive copper   moving conduction
ions at rest   electrons

Figure 3
A copper wire at rest with the conduction electrons moving to the left. This gives us a left-directed negative current.

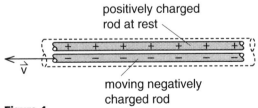

positively charged
rod at rest

moving negatively
charged rod

Figure 4
Model of a copper wire carrying an electric current. We are representing the positive copper ions by a positively charged rod at rest, and the conduction electrons by a moving, negatively charged rod.

A Convention

It was Ben Franklin who made the assignment of positive and negative charge. The charge left on a glass rod rubbed by silk was defined as positive, and that left on a rubber rod rubbed by cat fur as negative. This has often been considered a tragic mistake, for it leaves the electron, the common carrier of electric current, with a negative charge. It also leads to the unfortunate intuitive picture that an atom that has lost some electrons ends up with a positive charge.

Some physics textbooks written in the 1930s redefined the electron as being positive, but this was a disaster. We cannot undo over two centuries of convention that leads to the electron as being negative.

The worst problem with Franklin's convention comes when we try to handle the minus signs in problems involving the flow of electrons in a wire. But we have just seen that the flow of electrons in one direction is almost completely equivalent to the flow of positive charge in the other. If we do our calculations for positive currents, then we know that the electrons are simply moving in the opposite direction.

In order to maintain sanity and not get tangled up with minus signs, in this text we will, whenever possible, talk about the flow of positive currents, and talk about the force on positive test charges. If the problem we are working on involves electrons, we will work everything assuming positive charges and positive currents, and only at the end of the problem we will take into account the negative sign of the electron. With some practice, you will find this an easy convention to use.

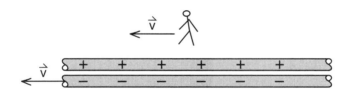

a) observer walking along with the moving negatively charged rod

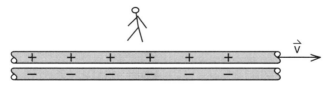

b) from the observer's point of view the negative rod is at rest and the positive charge is moving to the right

Figure 5 a, b
In (a) we have a left directed negative current, while in (b) we have a right directed positive current. The only difference is the perspective of the observer. (You can turn a negative current into an oppositely flowing positive one simply by moving your head.)

CURRENT AND VOLTAGE

Students first studying electricity can have difficulty conceptually distinguishing between the concepts of current and voltage. This problem can be handled by referring back to our hydrodynamic analogy of Chapter 23.

In Chapter 23 we were discussing Bernoulli's equation which stated that the quantity $(P + \rho gh + 1/2\ \rho v^2)$ was constant along a stream line if we could neglect viscous effects in the fluid. Because of the special nature of this collection of terms, we gave them the name *hydrodynamic voltage*.

$$\left.\begin{array}{r}\text{hydrodynamic}\\ \text{voltage}\end{array}\right\} = P + \rho gh + \frac{1}{2}\rho v^2 \quad (23\text{-}23)$$

(The second and third terms in the hydrodynamic voltage are the potential energy of a unit volume of fluid and the kinetic energy. The pressure term, while not a potential energy, is related to the work required to move fluid into a higher pressure region.)

Many features of hydrodynamic voltage should already be familiar. If you live in a house with good water pressure, when you turn on the faucet the water comes out rapidly. But if someone is running the washing machine in the basement or watering the garden, the water pressure may be low, and the water just dribbles out of the faucet. We will think of the high pressure water as *high voltage* water, and the low pressure water as *low voltage* water.

Let us look more carefully at high voltage water in a faucet. When the faucet is shut off, the water is at rest but the pressure is high, and the main contribution to the hydrodynamic voltage is the P term. When the faucet is on, the water that has just left the faucet has dropped back to atmospheric pressure but it is moving rapidly. Now it is the $1/2\ \rho v^2$ that contributes most to the hydrodynamic voltage. If the water originally comes

from a town water tank, when the water was at the top of the tank it was at atmospheric pressure and not moving, but was at a great height h. In the town water tank the hydrodynamic voltage comes mainly from the ρgh term.

Let us focus our attention on the high pressure in a faucet that is shut off. In this case we have high voltage water but no current. We can get a big current if we turn the faucet on, but the voltage is there whether or not we have a current.

In household wiring, the electrical outlets may be thought of as faucets for the electrical fluid in the wires. The high voltage in these wires is like the high pressure in the water pipes. You can have a high voltage at the outlet without drawing any current, or you can connect an appliance and draw a current of this high pressure electrical fluid.

Resistors

In an electric heater the electrical energy supplied by the power station is converted into heat energy by having electric current flow through a dissipative or resistive material. The actual process by which electrical energy is turned into heat energy is fairly complex but not unlike the conversion of mechanical energy to heat through friction. One can think of resistance as an internal friction encountered by the electric current.

In our discussion of Bernoulli's equation we saw that the hydrodynamic voltage $P + \rho gh + 1/2\ \rho v^2$ was constant along a stream line if there were no viscous effects. But we also saw in Figure (23-24) that when there were viscous effects this hydrodynamic voltage dropped as we went along a stream line.

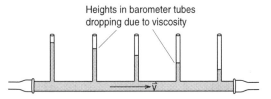

Heights in barometer tubes dropping due to viscosity

Figure 23-24
Hydrodynamic voltage drop due to viscous effects.

In fluid flows, we get the most dissipation where the fluid is moving rapidly through a narrow constriction. This is seen in our venturi demonstration of Figure (23-18), reproduced here in Figure (6). Here we have a large tube with a constriction. The glass barometer tubes show us that the pressure remains relatively constant before the constriction, but does not return to its original value afterward. There is a net pressure drop of ρgh, where h is the height drop indicated in the figure.

Consider the points in the fluid at the dots labeled (2) and (9), in the center of the stream below tubes 2 and 9. These points are at the same heights ($h_2 = h_9$), and the fluid velocities are the same ($v_2 = v_9$) because the flow tube has returned to its original size. Because of the pressure drop ($P_9 < P_2$), the hydrodynamic voltage ($P_9 + \rho g h_9 + 1/2 \rho v_9^2$) at point (9) is less than that at point (2) by an amount equal to $P_2 - P_9 = \rho g h$. The barometer tubes 2 and 9 are acting as *hydrodynamic voltmeters* showing us where the voltage drop occurs.

Just as in fluid flows, dissipation in electric currents are associated with voltage drops, in this case electrical voltage drops. In general, the amount of the voltage drop depends on the amount of current, the geometry of the flow path, on the material through which the current is flowing, and on the temperature of the material. But in a special device called a *resistor*, the voltage drop ΔV depends primarily on the current i through the resistor

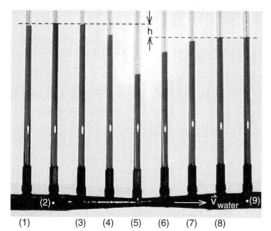

(1)　　　(3)　(4)　(5)　(6)　(7)　(8)

Figure 6
The hydrodynamic voltage, as measured by the barometer tubes, drops by an amount ρgh in going across the constriction from Point (2) to Point (9).

and is proportional to that current. ***When the voltage drop ΔV is proportional to the current i, the resistor is said to obey Ohm's law.*** This can be written as the equation

$$\boxed{\Delta V = iR} \qquad \textit{Ohm's law} \qquad (1)$$

The proportionality constant R is called the ***resistance R*** of the resistor. From Equation (1) you can see that R has the dimensions volt/amp. This unit is called an ***ohm,*** a name which is convenient in practice but which further complicates the problem of following dimensions in electrical calculations.

$$R = \frac{\Delta V}{i}\frac{\text{volts}}{\text{amps}} = \frac{\Delta V}{i}\text{ohms}$$

Resistors are the most common element in electronic circuits. They usually consist of a small cylinder with wire pigtails sticking out each end as shown in Figure (7). The material inside the cylinder which creates the voltage drop, which turns electrical energy into heat energy, is usually carbon.

The resistors you find in an electronics shop come in a huge selection of values, with resistances ranging from about 0.1 ohm up to around 10^9 ohms in a standard series of steps. The physical size of the resistor depends not on the value of the resistance but on the amount of electrical energy the resistor is capable of dissipating without burning up. The value of the resistance is usually indicated by colored stripes painted on the resistor, there being a standard color code so that you can read the value from the stripes.

(A light bulb is a good example of an electrical device that dissipates energy, in this case mostly in the form of heat and some light. The only problem with a light bulb is that as the filament gets hot, its resistance increases. If we wish to use Ohm's law, we have to add the qualification that the bulb's resistance R increases with temperature.)

Figure 7
The resistor, found in most electronic circuits. The purpose of the resistor is to cause an electric voltage drop analogous to the hydrodynamic voltage drop we saw in Figure 6 across the restriction in the flow tube.

A Simple Circuit

To get some intuition for how resistors are used, consider the circuit shown in Figure (8) containing a battery and a resistor connected by wires. In drawing circuits, it is convention to use a line ――――― for a wire, the symbol –\\/\\/\\/– for a resistor, and ――+|⊢―― for a battery. In the symbol for a battery, the short perpendicular line represents the negative terminal of the battery and the long side the positive terminal. When we have a current i flowing through the wire we draw an arrow indicating the direction of flow of positive charge ――i⇒―― and label the current with a letter such as i, i_1, etc.

In Figure (9), we have labeled the voltages V_1, V_2, V_3 and V_4 at four points around the circuit. By definition we will take the negative side of the battery as being zero volts, or what we call **ground**

$$V_4 = 0 \text{ volts} \qquad \textit{by definition} \qquad (2)$$

On the positive side of the battery, the voltage is up to the battery voltage V_b which is 1.5 volts for a common flashlight battery and up to 9 volts for many transistor radio batteries

$$V_1 = V_b \qquad \textit{the battery voltage} \qquad (3)$$

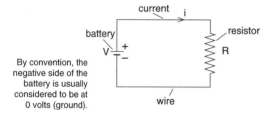

By convention, the negative side of the battery is usually considered to be at 0 volts (ground).

Figure 8
About the simplest electrical circuit consists of a battery connected to a resistor. If the resistor were a light bulb, you would have a flashlight.

Point (2) at the upper end of the resistor, is connected to the positive terminal of the battery, Point (1), by a wire. In our circuit diagrams we always assume that our wires are good conductors, having no electric fields inside them and therefore no voltage drops along them. Thus

$$V_2 = V_1 \; (= V_b) \qquad \begin{matrix} \textit{no voltage drop} \\ \textit{along a wire} \end{matrix} \qquad (4)$$

The bottom of the resistor is connected to the negative terminal of the battery by a wire, therefore

$$V_3 = V_4 \; (= 0) \qquad \begin{matrix} \textit{no voltage drop} \\ \textit{along a wire} \end{matrix} \qquad (5)$$

Equations (4) and (5) determine the voltage drop ΔV that must be occurring at the resistor

$$\Delta V = V_2 - V_3 = V_b \qquad (6)$$

And by Ohm's law, Equation (1), this voltage drop is related to the current i through the resistor by

$$\Delta V = iR = V_b \qquad \textit{Ohms law} \qquad (7)$$

Solving for the current i in the circuit gives

$$\boxed{i = \frac{V_b}{R}} \qquad (8)$$

In future discussions of circuits we will not write out all the steps as we have in Equations (2) through (8), but the first time through a circuit we wanted to show all the details.

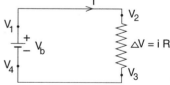

Figure 9
Voltages around the circuit.

Equation (8) is the one that really shows us how resistors are used in a circuit. We can see from Equation (8) that if we use a small resistor, we get a big current, and if we use a large resistor we get a small current. *In most applications resistors are used to control the flow of current.*

In modern electronics such as radios and computers, typical battery voltages are around 5 volts and typical currents a milliampere (10^{-3} amps). What size resistor R do we have to use in Equation (8) so that we get a one milliampere current from a 5 volt battery? The answer is

$$R = \frac{V_b}{i} = \frac{5 \text{ volts}}{10^{-3} \text{ amps}}$$
$$= 5000 \text{ ohms} \equiv 5000 \, \Omega \quad (9)$$

where we used the standard symbol Ω for ohms. Many of the resistors in electronics circuits have values like this in the $1,000 \, \Omega$ to $10,000 \, \Omega$ range.

The Short Circuit

Equation (8) raises an interesting problem. What if $R = 0$? The equation predicts an infinite current! We could try to make $R = 0$ by attaching a wire rather than a resistor from Points (2) to (3) in Figure (9). What would happen is that a very large current would start to flow and either melt the wire, start a fire, drain the battery, or destroy the power supply. (A power supply is an electronic battery.) When this happens, you have created what is called a *short circuit*. The common lingo is that you have *shorted* out the battery or power supply and this is not a good thing to do.

Power

As one of the roles of a resistor is electrical power dissipation, let us determine the power that is being dissipated when a current is flowing through a resistor. Recall that power is the amount of energy transferred or dissipated per unit time. In the MKS system power has the dimensions of joules per second which is called a watt

$$\text{Power} = \frac{\text{joules}}{\text{second}} = \text{watt} \quad (10)$$

Now suppose we have a current flowing through a resistor R as shown in Figure (10). The voltage drop across the resistor is V, from a voltage of V volts at the top to 0 volts at the bottom as shown.

Because V is the electric potential energy of a unit charge (the coulomb), every coulomb of charge flowing through the resistor loses V joules of electric potential energy which is changed to heat.

If we have a current i, then i coulombs flow through the resistor every second. Thus the energy lost per second is the number of coulombs (i) times the energy lost per coulomb (V) or (iV):

$$\text{Power} = i\frac{\text{coul}}{\text{sec}} \times V\frac{\text{joules}}{\text{coul}}$$
$$= iV\frac{\text{joules}}{\text{sec}} = iV \text{ watts} \quad (11)$$

Ohm's law, Equation (1), can be used to express the power in terms of R and either i or V

$$\text{Power} = iV = i^2R = \frac{V^2}{R} \quad (11a)$$

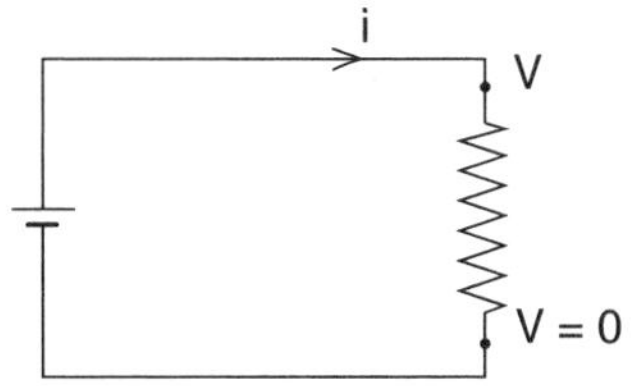

Figure 10
The voltage drops from V to 0 as the current i flows through the resistor. The power dissipated is the current i coulombs/second times the voltage drop V joules/coulomb, which is iV joules/second, or watts.

Exercise 1

These are some simple exercises to have you become familiar with the concepts of volts and amps.

a) Design a circuit consisting of a 9 volt battery and a resistor, where the current through the resistor is 25 milliamperes (25×10^{-3} amps).

b) A flashlight consists of a 1.5 volt battery and a 1 watt light bulb. How much current flows through the bulb when the flashlight is on?

c) When you plug a 1000 watt heater into a 120 volt power line, how much current goes through the heater? What is the resistance R of the heater when the filament is hot?

d) In most households, each circuit has a voltage of 120 volts and is fused for 20 amps. (The circuit breaker opens up if the current exceeds 20 amps). What is the maximum power you can draw from one circuit in your house?

e) An electric dryer requires 3000 watts of power, yet it has to be plugged into wires that can handle only 20 amps. What is the least voltage you can have on the circuit?

f) In many parts of the world, the standard voltage is 240 volts. The wires to appliances are much thinner. Explain why.

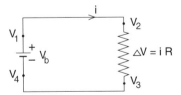

Figure 9 (redrawn)
Voltages around the circuit.

KIRCHOFF'S LAW

Imagine that you are going for an afternoon hike on a nearby mountain. You drive up to the base lodge, park your car, and start up the trail. The trail goes up over a ridge, down into a ravine, up to the peak of the mountain, down the other side and then around the mountain back to the base lodge. When you get back to your car, how much gravitational potential energy have you gained from the trip? The answer is clearly zero—you are right back where you started.

If you defined gh, which is the potential energy of a unit mass, as your gravitational voltage, then as you went up the ridge, there was a voltage rise as h increased. Going down into the ravine there was a voltage drop, or what we could call a negative voltage rise. The big voltage rise is up to the top of the mountain, and the big negative voltage rise is down the back side of the mountain. When you add up all the voltage rises for the complete trip, counting voltage drops as negative rises, the sum is zero.

Consider our Figure (9) redrawn here. If we start at Point (4) where the voltage is zero, and "walk" around the circuit in the direction of the positive current i, we first encounter a voltage rise up to $V = V_b$ due to the battery, then a voltage drop back to zero at the resistor. When we get back to the starting point, the sum of the voltage rises is zero just as in our trip through the mountains. Even in more complicated circuits with many branches and different circuit elements, it is usually true that the sum of the voltage rises around any complete path, back to your starting point, is zero. It turns out that this is a powerful tool for analyzing electric circuits, and is known as ***Kirchoff's law***. (Kirchoff's law can be violated, we can get a net voltage rise in a complete circuit, if changing magnetic fields are present. We will treat this phenomenon in a later chapter. For now we will discuss the usual situation where Kirchoff's law applies.)

Application of Kirchoff's Law

There are some relatively standard, cookbook like procedures that make it easy to apply Kirchoff's law to the analysis of circuits. The steps in the recipe are as follows:

(1) Sketch the circuit and use arrows to show the direction of the positive current in each loop as we did in Figure (11). Do not be too concerned about getting the correct direction for the current i. If you have the

Figure 11
Labeling the direction of the current.

arrow pointing the wrong way, then when you finish solving the problem, i will turn out to be negative.

(2) Label all the *voltage rises* in the circuit. Use arrows to indicate the direction of the voltage rise as we did in Figure (12). Note that if we go through the resistor in the direction of the current, we get a voltage drop. Therefore the arrow showing the voltage rise in a resistor must point back, opposite to the direction of the

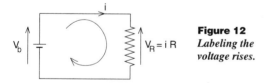

Figure 12
Labeling the voltage rises.

current i in the resistor. (The analogy is to a rock strewn waterfall where the water loses hydrodynamic voltage as it flows down through the rocks. The direction of the voltage rise is back up the waterfall, in a direction opposite to that of the current.)

(3) The final step is to *"walk"* around the loop in the direction of i (or any direction you choose), and set the sum of the voltage rises you encounter equal to zero. If you encounter an arrow that points in the direction you are walking, it counts as a positive voltage rise (like V_b in Figure 12). If the arrow points against you (like V_R), then it is a negative rise. Applying this rule to Figure (12) gives

$$\left. \begin{array}{l} \text{Sum of the voltage rises} \\ \text{going clockwise around} \\ \text{the circuit of Figure 12} \end{array} \right\} = V_b + V_R \quad (12)$$

$$= V_b + (-iR)$$

$$= 0$$

Equation (12) gives

$$i = \frac{V_b}{R} \quad (13)$$

which is the result we had back in Equation (8).

Series Resistors

By now we have beaten to death our simple battery resistor circuit. Let us try something a little more challenging—let us put in two resistors as shown in Figure (13). In that figure we have drawn the circuit and labeled the direction of the current (Step 1), and drawn in the arrows representing the voltage rises (Step 2). Setting the sum of the voltage rises equal to zero (Step 3) gives

$$V_b + (-iR_1) + (-iR_2) = 0 \quad (14)$$

$$i = \frac{V_b}{(R_1 + R_2)} \quad (15)$$

The two resistors in Figure (13) are said to be connected in series. Comparing Equation (13) for a single resistor and Equation (15) for the series resistors, we see that if

$$R_1 + R_2 = R \quad \text{(series resistors)} \quad (16)$$

then we get the same current i in both cases (if we use the same battery). We say that if $R_1 + R_2 = R$ then the series resistors are equivalent to the single resistor R.

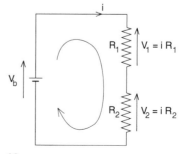

Figure 13
Two resistors in series.

Parallel Resistors

A bit more challenging is the circuit of Figure (14) where the resistors are wired in "parallel". In Step (1), we drew the circuit and labeled the currents. But here we have something new. When the current gets to the point labeled (A), it is like a fork in the stream and the current divides. We have labeled the two branch currents i_1 and i_2, and have the obvious subsidiary condition (conservation of current, if you like).

$$i_1 + i_2 = i \qquad (17)$$

There is no problem with Step (2), the voltage rises are V_b, i_1R_1 and i_2R_2 as shown. But we get something new when we try to write down Kirchoff's law for the sum of the voltage rises around a complete circuit. Now we have three different ways we can go around a complete circuit, as shown in Figures (15 a, b, c).

Applying Kirchoff's law to the path shown in Figure (15a) we get

$$V_b + (-i_1R_1) = 0 \qquad (18)$$

For Figure (15b) we get

$$(-i_2R_2) + (i_1R_1) = 0 \qquad (19)$$

and for Figure (15c) we get

$$V_b + (-i_2R_2) = 0 \qquad (20)$$

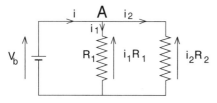

Figure 14
Two resistors in parallel.

The main problem with using Kirchoff's laws for complex circuits is that we can get more equations than we need or want. For our current example, if you solve Equation (18) for $V_b = i_1R_1$, then put that result in Equation (20), you get $i_1R_1 - i_2R_2 = 0$ which is Equation (19). In other words Equation (19) does not tell us anything that we did not already know from Equations (18) and (20). The mathematicians would say that Equations (18), (19), and (20) are not linearly independent.

Let us look at the situation from a slightly different point of view. To completely solve the circuit of Figure (15), we have to determine the currents i, i_1 and i_2. We have three unknowns, but four equations, Equations (17), (18), (19) and (20). It is well known that you need as many equations as unknowns to solve a system of equations, and therefore we have one too many equations.

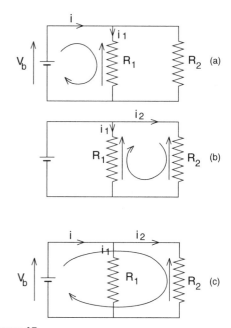

Figure 15
Three possible loops for analyzing the parallel resistance circuit. They give more equations than needed.

We cannot arbitrarily throw out one of the equations for the remaining three must be linearly independent. For example, if we threw out Equation (17) and tried to solve Equations (18), (19) and (20) for i_1, i_2, and i, we couldn't get an answer because Equation (19) contains no information not already in Equation (18) and (20). When you are working with a system of linear equations, the hardest problem is to decide which is a set of linear independent equations. Then you can use a standard set of procedures that mathematicians have for solving linear equations. These procedures involve determinants and matrices, which are easily handled on a computer, but are tedious to work by hand.

In our treatment of circuit theory we will limit our discussion to simple circuits where we can use grade school methods for solving the equations. Problems of linear independence, determinants and matrices will be left to other treatments of the topic.

To solve our parallel resistor circuit of Figure (14), we have from Equation (18)

$$i_1 = \frac{V_b}{R_1}$$

and from Equation (20)

$$i_2 = \frac{V_b}{R_2}$$

Substituting these values in Equation (17) gives

$$i = i_1 + i_2 = \frac{V_b}{R_1} + \frac{V_b}{R_2}$$
$$= V_b \left(\frac{1}{R_1} + \frac{1}{R_2}\right) \tag{21}$$

Comparing Equation (21) for parallel resistors, and Equation (13) for a single resistor

$$i = V_b \left(\frac{1}{R}\right) \tag{13}$$

We see that two parallel resistors R_1 and R_2 are equivalent to a single resistor R if they obey the relationship

$$\boxed{\frac{1}{R} = \frac{1}{R_1} + \frac{1}{R_2}}$$
equivalent parallel resistors (22)

Exercise 2

You are given a device, sealed in a box, with electrical leads on each end. (Such a device is often referred to as a "black box", the word black referring to our lack of knowledge of the contents, rather than the actual color of the device.) You use an instrument called an **ohmmeter** to measure the electrical resistance between the two terminals and find that it's resistance R is 470 ohms (470Ω).

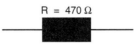
R = 470 Ω

a) Sketch a circuit, containing the black box and one resistor, where the total resistance of the circuit is 500Ω.

b) Sketch a circuit, containing the black box and one resistor, where the total resistance of the circuit is 400Ω.

Exercise 3 The Voltage Divider

We wish to measure the voltage V_b produced by a high voltage power supply, but our voltmeter has the limited range of +2 to -2 volts. To make the measurement we use the voltage divider circuit shown below, containing a big resistor R_1 and a small resistor R_2. If, for example, R_2 is 1000 times smaller than R_1, then the voltage across R_2 is 1000 times smaller than that across R_1. By measuring the small voltage across the small resistor we can use this result to determine the big voltage V_b.

a) What current i flows through the circuit. Express your answer in terms of V_b.

b) Find the formula for V_b in terms of V_2, the voltage measured *across* the small resistor.

c) Find a formula for V_b in terms of V_2, R_1 and R_2, assuming $R_1 >> R_2$, so that you can replace $(R_1 + R_2)$ by R_1 in the equation for i.

d) Our voltmeter reads $V_2 = .24$ volts. What was V_b?

Voltage divider circuit

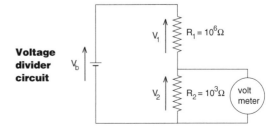

CAPACITANCE AND CAPACITORS

In addition to the resistor, another common circuit element is the capacitor. A resistor dissipates energy, causes a voltage drop given by Ohm's law $V = iR$, and is often used to limit the amount of current flowing in a section of a circuit. A capacitor is a device for storing electrical charge and maintains a voltage proportional to the charge stored. We have already seen one explicit example of a capacitor, the parallel plate capacitor studied in the last chapter. Here we will abstract the general features of capacitors, and see how they are used as circuit elements.

Hydrodynamic Analogy

Before focusing on the electrical capacitor, it is instructive to consider an accurate hydrodynamic analogy—the cylindrical water tank shown in Figure (16). If the tank is filled to a height h, then all the water in the tank has a hydrodynamic voltage

$$V_h = P + \rho gh + \frac{1}{2}\rho v^2 = \rho gh \qquad (23)$$

For water at the top of the tank, $y = h$, the voltage is all in the form of gravitational potential energy ρgh. (We will ignore atmospheric pressure.) At the bottom of the tank where $y = 0$, the voltage is all in the pressure term $P = \rho gh$. The dynamic voltage term $1/2\,\rho v^2$ does not play a significant role.

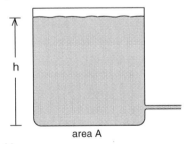

Figure 16
Analogy between a cylindrical tank of water and an electrical capacitor. In the tank, all the water in the tank is at a hydrodynamic voltage $V_h = \rho gh$, and the quantity Q of water in the tank, given by $Q = Ah = \left(A/\rho g\right)\rho gh = \left(A/\rho g\right)V_h$ is proportional to V_h.

Let us denote by the letter Q the quantity or volume of water stored in the tank. If we talk only about cylindrical tanks (of cross-sectional area A), then this volume is proportional to the height h and therefore the hydrodynamic voltage V_h

$$\left.\begin{array}{l}\text{Volume of water}\\ \text{in cylindrical tank}\end{array}\right\} \equiv Q = Ah = \left(\frac{A}{\rho g}\right)\rho gh$$

$$Q = \left(\frac{A}{\rho g}\right)V_h \qquad (24)$$

If we define the proportionality constant $A/\rho g$ in Equation (24) as the ***capacitance C*** of the tank

$$C = \frac{A}{\rho g} \equiv \left\{\begin{array}{l}\textbf{\textit{capacitance}} \text{ of}\\ \text{a cylindrical tank}\\ \text{with a cross}-\\ \text{sectional area A}\end{array}\right. \qquad (25)$$

then we get

$$\boxed{Q = CV_h} \qquad (26)$$

as the relation between the hydrodynamic voltage and volume Q of water in the tank.

Cylindrical Tank as a Constant Voltage Source

One of the main uses of a water storage tank is to maintain a water supply at constant hydrodynamic voltage.

Figure (17) is a schematic diagram of a typical town water supply. Water is pumped from the reservoir up into the water tank where a constant height h and therefore constant voltage ρgh is maintained. The houses in the town all draw constant voltage water from this tank.

Let us see what would happen if the water tank was too small. As soon as several houses started using water, the level h in the tank would drop and the pump at the reservoir would have to come on. The pump would raise the level back to h and shut off. Then the level would drop again and the pump would come on again. The result would be that the hydrodynamic voltage or water pressure supplied to the town would vary and customers might complain.

On the other hand if the town water tank has a large cross-sectional area and therefore large capacitance C, a few houses drawing water would have very little effect on the level h and therefore voltage ρgh of the water. The town would have a constant voltage water supply and the water company could pump water from the reservoir at night when electricity rates were low.

We will see that one of the important uses of electrical capacitors in electric circuits is to maintain constant or nearly constant electric voltages. There is an accurate analogy to the way the town water tank maintains constant voltage water. If we use too small a capacitor, the electrical voltage will also fluctuate when current is drawn.

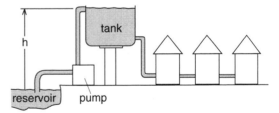

Figure 17
Town water supply. By maintaining a constant height h of water in the storage tank, all the water supplied to the town has a constant hydrodynamic voltage $V_h \rho gh$.

Electrical Capacitance

Figure (18) is a repeat of the sketches of the parallel plate capacitor discussed in Chapter 26. The important features of the capacitor are the following. We have two metal plates of area A separated by a distance d. The positive plate shown on top has a charge + Q, the bottom plate a charge – Q. Since the area of the plates is A, the surface charge density on the inside of the plate is

$$\sigma = \frac{Q}{A} \tag{27}$$

In Chapter 26, page 26-3, we saw that a charge density σ on the surface of a conductor produced an electric field of strength

$$E = \frac{\sigma}{\varepsilon_0} \tag{28}$$

perpendicularly out of the conductor. In Figure (18) this field starts at the positive charge on the inside of the upper plate and stops at the negative charge on the inside surface of the bottom plate.

Recall that one form of electric voltage is the electric potential energy of a unit test charge. To lift a positive unit test charge from the bottom plate to the top one requires an amount of work equal to the force E on a unit charge times the distance d the charge was lifted. This work $E*d$ is equal to the increase of the potential energy of the unit charge, and therefore to the increase in voltage in going from the bottom to the top plate. If we say that the bottom plate is at a voltage $V = 0$, then the voltage at the top plate is

$$V = E\,d \tag{29}$$

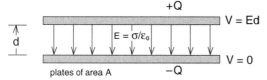

plates of area A

Figure 18
The parallel plate capacitor. If we place charges + Q and – Q on plates of area A, the charge density on the plates will be $\sigma = Q/A$, the electric field will be $E = \sigma/\varepsilon_0$ and the voltage between the plates $V = Ed$.

Using Equation (28) for E and Equation (27) for σ, we get the relationship

$$V = \frac{\sigma}{\varepsilon_0}\,d = \frac{Q}{\varepsilon_0 A}\,d$$

or

$$Q = \left(\frac{\varepsilon_0 A}{d}\right)V \tag{30}$$

which is our old Equation (26-10).

As in our hydrodynamic analogy, we see that the quantity of charge Q stored in the capacitor is proportional to the voltage V on the capacitor. Again we call the proportionality constant the capacitance C

$$Q = CV \qquad \begin{array}{l}\textit{definition of}\\ \textit{electrical}\\ \textit{capacitance}\end{array} \tag{31}$$

Comparing Equations (30) and (31) we see that the formula for the capacitance C of a parallel plate capacitor is

$$C = \frac{\varepsilon_0 A}{d} \qquad \begin{array}{l}\textit{capacitance of a}\\ \textit{parallel plate}\\ \textit{capacitor of area A,}\\ \textit{plate separation d}\end{array} \tag{32}$$

For both the parallel plate capacitor and the cylindrical water tank, the capacitance is proportional to the cross-sectional area A. The new feature for the electrical capacitor is that the capacitance increases as we make the plate separation d smaller and smaller.

Our parallel plate capacitor is but one example of many kinds of capacitors used in electronic circuits. In some, the geometry of the metal conductors is different, and in others the space between the conductors is filled with a material called a dielectric which increases the effective capacitance. But in all common capacitors the amount of charge Q is proportional to voltage V across the capacitor, i.e. $Q = CV$, where C is constant independent of the voltage V and in most cases independent of the temperature.

The dimensions of capacitance C are coulombs per volt, which is given the name **farad** in honor of Michael Faraday who pioneered the concept of an electric field. Although such an honor may be deserved, this is one more example of the excessive use of names in the MKS system that make it hard to follow the dimensions in a calculation.

To get a feeling for the size of a farad, suppose that we have two metal plates with an area $A = 0.1$ meter^2 and make a separation $d = 1$ millimeter $= 10^{-3}$ meters. These plates will have a capacitance C given by

$$C = \frac{\varepsilon_0 A}{d} = \frac{9 \times 10^{-12} \times .1}{10^{-3}}$$

$$= 9 \times 10^{-10} \text{ farads}$$

which is about one billionth of a farad. If you keep the separation at 1 millimeter you would need plates with an area of 100 million square meters (an area 10 kilometers on a side) to have a capacitance of 1 farad.

Commercial capacitors used in electronic circuits come in various shapes like those shown in Figure (19), and in an enormous range of values from a few farads down to 10^{-14} farads.

Our calculation of the capacitance of a parallel plate capacitor demonstrates that it is not an easy trick to produce capacitors with a capacitance of 10^{-6} farads or larger. One technique is to take two long strips of metal foil separated by an insulator, and roll them up into a small cylinder. This gives us a large plate area with a reasonably small separation, stuffed into a relatively small volume.

In a special kind of a capacitor called an electrolytic capacitor, the effective plate separation d is reduced to almost atomic dimensions. Only this way are we able to create the physically small 1 farad capacitor shown in Figure (19). The problem with electrolytic capacitors is that one side has to be positive and the other negative, as marked on the capacitor. If you reverse the voltage on an electrolytic capacitor, it will not work and may explode.

Exercise 4 - Electrolytic Capacitor

In an *electrolytic capacitor*, one of the plates is a thin aluminum sheet and the other is a conducting dielectric liquid surrounding the aluminum. A nonconducting oxide layer forms on the surface of the aluminum and plays the same role as the air gap in the parallel plate capacitors we have been discussing. The fact that the oxide layer is very thin means that you can construct a capacitor with a very large capacitance in a small container.

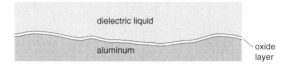

For this problem, assume that you have a dielectric capacitor whose total capacitance is 1 farad, and that the oxide layer acts like an air gap 10^{-7} meters thick in a parallel plate capacitor. From this, estimate the area of the aluminum surface in the capacitor.

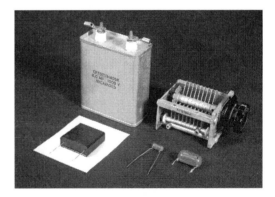

Figure 19
Examples of capacitors used in electronic circuits. The one on the right is a variable capacitor whose plate area is changed by turning the knob. The square black capacitor is a 4 farad electrolytic. Its capacitance is one million times greater than the tall regular capacitor behind it.

ENERGY STORAGE IN CAPACITORS

In physics, one of the important uses of capacitors is energy storage. The advantage of using capacitors is that large quantities of energy can be released in a very short time. For example, Figure (20) is a photograph of the Nova laser at the Lawrence Livermore National Laboratory. This laser produces short, but very high energy pulses of light for fusion research. The laser is powered by a bank of capacitors which, for the short length of time needed, can supply power at a rate about 200 times the power generating capacity of the United States.

The easiest way to determine the amount of energy stored in a capacitor is to calculate how much work is required to charge up the capacitor. In Figure (21) we have a capacitor of capacitance C that already has a charge $+Q$ on the positive plate and $-Q$ on the negative plate. The voltage V across the capacitor is related to Q by Equation (12), $Q = CV$.

Now let us take a charge dQ out of the bottom plate, leaving a charge $-(Q + dQ)$ behind, and lift it to the top plate, leaving $(Q + dQ)$ there. The work dW we do to lift the charge is equal to dQ times the work required to lift a unit test charge, namely dQ times the voltage V

$$dW = VdQ$$

or replacing V by Q/C, we have

$$dW = \frac{Q}{C}dQ \qquad (33)$$

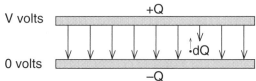

Figure 21
Charging up a capacitor. If the capacitor is already charged up to a voltage V, the amount of work required to lift an additional charge dQ from the bottom to top plate is dW = VdQ.

Figure 20
The Nova laser, powered by a bank of capacitors. While the laser is being fired, the capacitors supply 200 times as much power as the generating capacity of the United States.

You can see from Equation (33) that when the capacitor is uncharged and we lift the first dQ, no work is required because there is no field yet in the capacitor. However once there is a big charge on the capacitor, much work is required to lift an additional dQ. The total amount of work to charge the capacitor from zero charge to a final charge Q_f is clearly given by the integral

$$\text{Work} = \int dW = \int_0^{Q_f} \frac{Q}{C} dQ$$

The fact that the capacitance C is a constant, means that we can take it outside the integral and we get

$$W = \frac{1}{C} \int_0^{Q_f} Q dQ = \frac{Q_f^2}{2C} \tag{34}$$

Since it is easier to measure the final voltage V rather than the charge Q_f in a capacitor, we use $Q_f = CV$ to rewrite Equation (34) in the form

$$\left. \begin{array}{c} \text{Energy stored} \\ \text{in a capacitor} \end{array} \right\} = \frac{CV_f^2}{2} \tag{35}$$

The energy stored is proportional to the capacitance C of the capacitor, and the square of the voltage V.

Energy Density in an Electric Field

Equation 35 can be written in a form that shows that the energy stored in a capacitor is proportional to the square of the strength of the electric field. Substituting $V_f = E \times d$ and $C = \varepsilon_0 A/d$ into Equation 35 gives

$$\left. \begin{array}{c} \text{Energy stored} \\ \text{in a capacitor} \end{array} \right\} = \frac{CV_f^2}{2} = \frac{1}{2} \frac{\varepsilon_0 A}{d} \times E^2 d^2$$

$$= \frac{\varepsilon_0 E^2}{2} \times A d = \frac{\varepsilon_0 E^2}{2} \times \left(\begin{array}{c} \text{Volume} \\ \text{Inside} \\ \text{capacitor} \end{array} \right)$$

where we note that $A \times d$ is the volume inside the capacitor.

Since the energy stored in the capacitor is proportional to the volume occupied by the electric field, we see that the energy per unit volume, the energy density, is simply given by

$$\left. \begin{array}{c} \text{Energy} \\ \text{density} \end{array} \right\} = \frac{\varepsilon_0 E^2}{2} \tag{36}$$

This result, that the energy density in an electric field is proportional to the square of the strength of the field, turns out to be a far more general result than we might expect from the above derivation. It applies not just to the uniform electric field in an idealized capacitor, but to electric fields of arbitrary shape.

Exercise 5

A parallel plate capacitor consists of two circular aluminum plates with a radius of 11 cm separated by a distance of 1 millimeter. The capacitor is charged to a voltage of 5 volts.

a) What is the capacitance, in farads, of the capacitor?

b) Using Equation 35, calculate the energy stored in the capacitor.

c) What is the magnitude of the electric field E between the plates?

d) Using equation 36, calculate the energy density in the electric field.

e) What is the volume of space, in cubic meters, between the plates?

f) From your answers to parts d) and e), calculate the total energy in the electric field between the plates. Compare your answer with your answer to part b-

. g) Using Einstein's formula $E = mc^2$, calculate the mas , in kilograms, of the electric field between the plate

.h) The mass of the electric field is equal to the mass f how many electron

---

CAPACITORS AS CIRCUIT ELEMENTS

Figure (22) is a simple circuit consisting of a battery of voltage V_b and a capacitor of capacitance C. The standard circuit symbol for a capacitor is ——$|$|——, which is a sketch of a parallel plate capacitor.

When the battery is attached to the capacitor, the upper plate becomes positively charged and the lower one negatively charged as shown. The upper plate could actually become positively charged either by positive charge flowing into it or negative charge flowing out— it does not matter. We have followed our convention of always showing the direction of positive currents, thus we show i flowing into the positive plate and out of the negative one.

We have also followed our convention of labeling the voltage rises with an arrow pointing in the direction of the higher voltage. The voltage V_c on the capacitor is related to the charge Q stored by the definition of capacitance, $V_c = Q/C$.

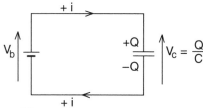

Figure 22
A battery and a capacitor in a circuit. We have drawn the diagram showing positive current flowing into the top plate and out of the bottom plate. The upper plate could have become positively charged by having a negative current flowing out of it. The arrow designating the voltage on the capacitor points in the direction of the voltage rise.

Applying Kirchoff's law to Figure (22), i.e., setting the sum of the voltage rises around the circuit equal to zero, we get

$$V_b + (-V_c) = V_b - Q/C = 0$$
$$Q = CV_b \tag{37}$$

Thus we get a relatively straightforward result for the amount of charge stored by the battery.

For something a little more challenging, we have connected two capacitors in parallel to a battery as shown in Figure (23). Because single wires go all the way across the top and across the bottom, the three voltages V_b, V_1 and V_2 must all be equal, and we get

$$Q_1 = C_1 V_b \qquad\qquad Q_2 = C_2 V_b$$

The total charge Q stored on the two capacitors in parallel is therefore

$$Q = Q_1 + Q_2 = (C_1 + C_2)V_b$$

Comparing this with Equation (37), we see that two capacitors in parallel store the same charge as a single capacitor C given by

$$\boxed{C = C_1 + C_2} \qquad \begin{array}{l}\textit{capacitors}\\ \textit{attached in}\\ \textit{parallel}\end{array} \tag{38}$$

Comparing this result with Equation (16), we find that for capacitors in parallel or resistors in series, the effective capacitance or resistance is just the sum of the values of the individual components.

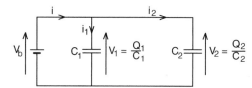

Figure 23
Capacitors connected in parallel. The three voltages V_b, V_1 and V_2 must all be level because the wires go all the way across the three elements.

In Figure (24) we have two capacitors in series. The trick here is to note that all the charge that flowed out of the bottom plate of C_1 flowed into the top plate of C_2, as indicated in the diagram. But if there is a charge $-Q$ on the bottom plate of C_1, there must be an equal and opposite charge $+Q$ on the top and we have $Q_1 = Q$. Similarly we must have $Q_2 = Q$.

To apply Kirchoff's law, we set the sum of the voltage rises to zero to get

$$V_b + \frac{-Q_1}{C_1} + \frac{-Q_2}{C_2} = 0$$

Setting $Q_1 = Q_2 = Q$ gives

$$V_b = Q\left(\frac{1}{C_1} + \frac{1}{C_2}\right) \tag{39}$$

Comparing Equation (39) with Equation (37) in the form $V_b = Q/C$ we see that

$$\boxed{\frac{1}{C} = \frac{1}{C_1} + \frac{1}{C_2}} \quad \begin{array}{l}\textit{capacitors}\\ \textit{attached in}\\ \textit{series}\end{array} \tag{40}$$

is the formula for the effective capacitance of capacitors connected in series. This is analogous to the formula for parallel resistors.

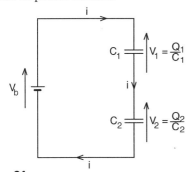

Figure 24
Capacitors in series. In this case the sum of V_1 and V_2 must be equal to the battery voltage V_b.

It is interesting to note that for storing charge, parallel capacitors are more efficient because the charge can flow into both capacitors as seen in Figure (23). When the capacitors are in series, charge flowing out of the bottom of one capacitor flows into the top of the next, and we get no enhancement in charge storage capability. What we do get from series capacitors is higher voltages, the total voltage rise across the pair is the sum of the voltage rise on each.

Exercise 6

You have a 5 microfarad (abbreviated 5μf) capacitor and a 10 μf capacitor. What are all the values of capacitor you can make from these two?

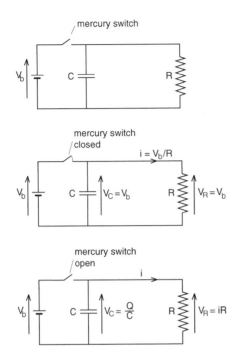

Figure 25
An RC circuit. When the mercury switch is closed, the capacitor quickly charges up to a voltage $V_C = V_b$. When the switch is opened, the capacitor discharges through the resistor.

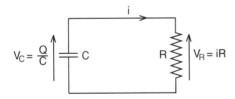

Figure 26
Capacitor discharge. When the switch is open, the only part of the circuit we have to look at is the capacitor discharging through the resistor.

THE RC CIRCUIT

The capacitor circuits we have discussed so far are not too exciting. When you are working with an electronic circuit you do not hitch capacitors together in series or parallel, you simply go to the parts drawer and select a capacitor of the desired value.

If we add a resistor to the circuit as shown in Figure (25), we begin to get some interesting results. The circuit is designed so that if the mercury switch is closed, the capacitor is charged up to a voltage V_b by the battery. Then, at a time we will call $t = 0$, the switch is opened, so that the capacitor will discharge through the resistor. During the discharge, the battery is disconnected and the only part of the circuit that is active is that shown in Figure (26). (The reason for using a mercury switch was to get a clean break in the current. Mechanical switches do not work well.)

Figure (27) shows the capacitor voltage just before and for a while after the switch was opened. We are looking at the experimental results of discharging a $C = 10^{-6}$ farad (one microfarad) capacitor through an $R = 10^4$ ohm resistor. We see that a good fraction of the capacitor voltage has decayed in about 10 milliseconds (10^{-2} seconds) .

To analyze the capacitor discharge, we apply Kirchoff's law to the circuit in Figure (26). Setting the sum of the voltage rises around the circuit equal to zero gives

$$V_C - V_R = 0$$

$$\frac{Q}{C} - iR = 0$$

which can be written in the form

$$-i + \frac{Q}{RC} = 0 \qquad (41)$$

The problem with Equation (41) is that we have two unknowns, i and Q, and only one equation. We need to find another relationship between these variables in order to predict the behavior of the circuit.

The additional relationship is obtained by noting that the current i is the number of coulombs per second flowing out of the capacitor. In a short time dt, an amount of charge dQ that leaves the capacitor is given by

$$dQ = idt \qquad (42)$$

Dividing Equation (42) through by dt, and including a minus sign to represent the fact that i is causing a decrease in the charge Q in the capacitor, we get

$$\frac{dQ}{dt} = -i \qquad \begin{array}{l}\textit{discharge of} \\ \textit{a capacitor}\end{array} \qquad (43)$$

Substituting Equation (43) in (41) gives one equation for the unknown Q

$$\frac{dQ}{dt} + \frac{Q}{RC} = 0 \qquad \begin{array}{l}\textit{equation for} \\ \textit{the discharge} \\ \textit{of a capacitor}\end{array} \qquad (44)$$

Exponential Decay

The next problem is that Equation (44) is a differential equation, of a type we have not yet discussed in the text. We met another kind of differential equation in our discussion of harmonic motion, a differential equation that involved second derivatives and had oscillating sinusoidal solutions. Equation (44) has only a first derivative, and produces a different kind of solution.

There are two principle ways of solving a differential equation. One is to use a computer, and the other, the so-called analytic method, is to guess the answer and then check to see if you have made the correct guess. We will first apply the analytic method to Equation (44). In the supplement we will show how a computer solution is obtained.

The important thing to remember about a differential equation is that the solution is a shape or a curve, not a number. The equation $x^2 = 4$ has the solutions $x = \pm 2$, the solution to Equation (44) is the curve shown in Figure (27). One of the advantages of working with electric circuits is that the theory gives you the differential equation, and the equipment in the lab allows you to look at the solution. The curve in Figure (27) is the voltage on the capacitor recorded by the computer based oscilloscope we used to record the motion of air carts and do the analysis of sound waves. We are now using the device as a voltmeter that draws a picture of the voltage.

The curve in Figure (27) is well known to scientists in many fields as an *exponential decay*. Exponential decays are best known in studies of radioactive decay and are associated with the familiar concept of a half life. Let us first write down the formula for an exponential decay, check that the formula is, in fact, a solution to Equation (44) and then discuss the special properties of the curve.

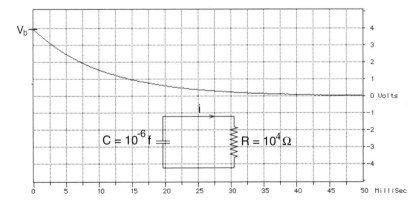

Figure 27
Experimental results from discharging the one microfarad $\left(10^{-6}f\right)$ capacitor through a 10 k ohm $\left(10^4\,\Omega\right)$ resistor. The switch, shown in Figure 26, is thrown at time t = 0.

If Figure (27) represents an exponential decay of the capacitor voltage V_c, then V_c must be of the form

$$V_C = V_0 e^{-\alpha t} \qquad (45)$$

where V_0 and α are constants to be determined. Since Equation (44) is in terms of Q rather than V_C, we can use the definition of capacitance Q = CV to rewrite Equation (45) as

$$CV_C = CV_0 e^{-\alpha t} \qquad (46)$$

$$Q = Q_0 e^{-\alpha t}$$

where $Q_0 = CV_0$.

Differentiating Equation (46) with respect to time gives

$$\frac{dQ}{dt} = -\alpha Q_0 e^{-\alpha t} = -\alpha Q \qquad (47)$$

This result illustrates one of the properties of an exponential decay, namely that the derivative of the function is proportional to the function itself (here $dQ/dt = -\alpha Q$).

Substituting Equation (47) into our differential Equation (44) gives

$$\frac{dQ}{dt} + \frac{Q}{RC} = 0 \qquad (44)$$

$$-\alpha Q + \frac{Q}{RC} = 0 \qquad (48)$$

The Qs cancel in Equation (48) and we get

$$\boxed{\alpha = \frac{1}{RC}} \qquad (49)$$

Thus the coefficient of the exponent is determined by the differential equation.

Exercise 7

Determine the constant V_0 in Equation (45) from Figure (27), by noting that at t = 0, $e^{-\alpha t} = e^0 = 1$.

The Time Constant RC

Substituting Equation (49) for α back into our formula for V_C gives

$$V_C = V_0 e^{-(t/RC)} \qquad (50)$$

Since the exponent (t/RC) must be dimensionless, the quantity RC in the denominator must have the dimensions of time. Since R is in ohms and C in farads, we must have

$$\boxed{\text{ohms} * \text{farads} = \text{seconds}} \qquad (51)$$

We have mentioned that units in electrical calculations are hard to follow, and this is a prime example. We leave it as a challenge to go back and actually show, from the definition of the ohm and of the farad, that the product ohms times farads comes out in seconds.

The quantity RC that appears in Equation (50) is known as the ***time constant*** for the decay. At the time t = RC, the voltage V_C has the value

$$V_C \,(\text{at t} = RC) = V_b e^{-RC/RC} \qquad (52)$$

$$= V_b e^{-1} = \frac{V_b}{e}$$

I.e., in one time constant RC, the voltage has decayed to 1/e = 1/2.7 of its initial value.

To see if this analysis works experimentally, we have gone back to Figure (27) and marked the time RC. In that experiment

$$R = 10^4 \text{ ohms}$$

$$C = 10^{-6} \text{ farads}$$

thus

$$RC = 10^{-2} \text{ ohm farads}$$

$$= 10^{-2} \text{ seconds} \qquad (53)$$

$$= 10 \text{ milliseconds}$$

We see that at a time T = RC, the voltage dropped from V_b = 4 volts to V(t=RC) = 1.5 volts, which is down by a factor 1/e = 1/2.7.

If we wait another time constant, until $t=2RC$, we have

$$V_C \text{ (at } t = 2RC) = V_b e^{-2} = \frac{V_b}{e^2}$$

and we get another factor of $e=2.7$ in the denominator. In Figure (28) the voltage is down to $4/(2.7*2.7) = .55$ volts at $t=2RC$. **After each succeeding time constant RC, the voltage drops by another factor of 1/e**.

Half-Lives

When you first studied radioactive decay, you learned about half-lives. A half-life was the time it took for half of the remaining radioactive particles to decay. Wait another half-life and half of those are gone. In our description of the exponential decay, the time constant RC is similar to a half-life, but just a bit longer. When we wait for a time constant, the voltage decays down to 1/2.7 of its initial value rather than 1/2 of its initial value. In Figure (29) we compare the half-life t 1/2 and the time constant RC. Although the half-life is easier to explain, we will see that the time constant RC provides a more convenient unit of time for the analysis of the exponential decay curve.

Initial Slope

One of the special features of a time constant is the fact that the initial slope of the curve intercepts the zero value one time constant later, as illustrated in Figure (30). It does not matter where we take our initial time to be. Pick any point on the curve, draw a tangent line at that point, and the tangent line intercepts the $V = 0$ line one time constant RC later. This turns out to be the most convenient way to determine the time constant from an experimental curve. Try it yourself in the following exercise.

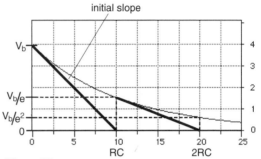

Figure 30
The initial slope of the discharge curve intersects the $V_C = 0$ origin at a time $t = RC$, one time constant later. This fact provides an easy way to estimate the time constant for an exponential curve.

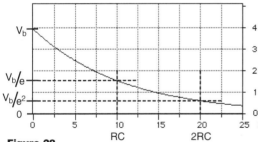

Figure 28
Exponential decay of the voltage in the capacitor. In a time $t = RC$ the voltage drops by a factor $1/e = 1/2.7$. In the next time interval RC, the voltage drops by another factor of 1/e.

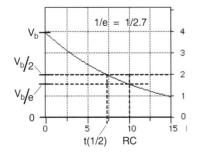

Figure 29
The time it takes the voltage to drop to half its initial value, what we could call the "half life" of the voltage, is a bit shorter than the time constant RC.

Exercise 8

In Figure (31) we have the experimental voltage decay for an RC circuit where R is known to be 10^5 ohms. Determine the time constant for this curve and from that find out what the value of the capacitance C must have been.

The Exponential Rise

Figure (32) is a circuit in which we use a battery to charge a capacitor through a resistance R. The experimental result, for our capacitor $C = 10^{-6}$ farads, $R = 10^4$ ohms is shown in Figure (33). Here the capacitor starts charging relatively fast, then the rate of charging slows until the capacitor voltage finally reaches the battery voltage V_b.

If the shape of the curve in Figure (33) looks vaguely familiar, it should. Turn the curve over and it looks like our exponential decay curve with time increasing toward the left. If that is true, then the initial slope of the charge up curve should intercept the $V_C = V_b$ line one time constant later as shown in Figure (34), and it does.

Because we have a battery in the circuit while the capacitor is charging up, the analysis is a bit more messy than for the capacitor discharge. (You should be able to repeat the analysis of the capacitor discharge on your own, and at least be able to follow the steps for analyzing the charge up.)

In Figure (32) we have drawn the circuit diagram and indicated the voltage rises around the circuit. Kirchoff's law there gives

$$V_b + (-V_R) + (-V_C) = 0$$

$$V_b - iR - \frac{Q}{C} = 0 \tag{54}$$

One difference between Figure (32) for the charge up and Figure (27) for the discharge, is that for the charge up, the current i is flowing into, rather than out of, the capacitor. Therefore in a time dt the charge in the capacitor **increases** by an amount dQ = idt, and we have

$$\frac{dQ}{dt} = +i \tag{55}$$

Using Equation (55) in (54), dividing through by R, and rearranging a bit, gives

$$\frac{dQ}{dt} + \frac{Q}{RC} = \frac{V_b}{R} \tag{56}$$

It is the term on the right that makes this differential equation harder to solve. A simple guess like the one we made in Equation (46) does not work, and we have to try a more complicated guess like

$$Q = A + Be^{-\alpha t} \tag{57}$$

When you take a course in solving differential equations, much of the time is spent learning how to guess the form of solutions. For now, let us just see if the guess in Equation (57) can be made to work for some value of the constants A, B, and α.

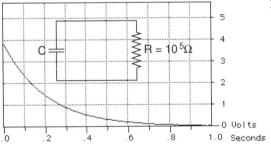

Figure 31
Experimental results for the discharge of a capacitor through a $10^5\ \Omega$ resistor.

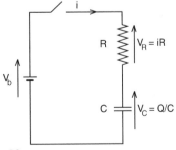

Figure 32
Charging up a capacitor through a resistor.

Differentiating Equation (57) with respect to time gives

$$\frac{dQ}{dt} = -\alpha B e^{-\alpha t} \tag{58}$$

Substituting Equations (57) and (58) into (56) gives

$$-\alpha B e^{-\alpha t} + \frac{A}{RC} + \frac{B}{RC} e^{-\alpha t} = \frac{V_b}{R} \tag{59}$$

The only way we can satisfy Equation (59) is have the two terms with an $e^{-\alpha t}$ cancel each other. I.e., we must have

$$-\alpha B + \frac{B}{RC} = 0$$

$$\alpha = \frac{1}{RC} \tag{60}$$

which is a familiar result.

The remaining terms in Equation (59) give

$$\frac{A}{RC} = \frac{V_b}{R} \qquad A = CV_b \tag{61}$$

Putting the values for A and α (Equations 61 and 60) back into our guess (Equation 57), we get

$$Q = CV_b + Be^{-t/RC} \tag{62}$$

The final step is to note that at time $t=0$, $Q=0$, so that

$$0 = CV_b + Be^0 \qquad B = -CV_b$$

thus our final result is

$$Q = CV_b\left(1 - e^{-t/RC}\right) \tag{63a}$$

We can express this result in terms of voltages if we divide through by C and use $V_C = Q/C$, we get

$$V_C = V_b\left(1 - e^{-t/RC}\right) \qquad \text{charging a capacitor} \tag{63}$$

(We warned you that the addition of just one more term to our differential equation would make it messier to solve.)

The answer, Equation (63) is the standard form for an exponential rise. It is in fact just our exponential decay curve turned upside down, and Figure (34) represents the easy way to determine the time constant RC from experimental data.

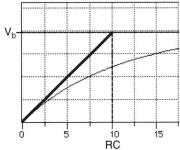

Figure 34
If you continue along the initial slope line of the charge-up curve, you intersect the final voltage V_b at a time t = RC, one time constant later. Turn this diagram upside down and it looks like Figure 31.

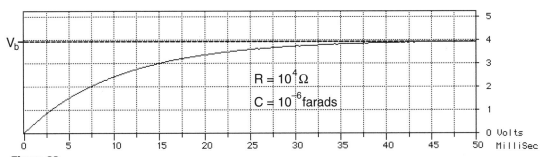

Figure 33
Plot of the capacitor voltage versus time for the charging up of a capacitor through a resistor. If you turn the diagram upside down, you get the curve for the discharge of a capacitor.

Exercise 9

Figure (35) shows the voltage across a capacitor C being charged through a resistance R. Given that $C = 1.0 \times 10^{-8}$ farads, estimate the value of R.

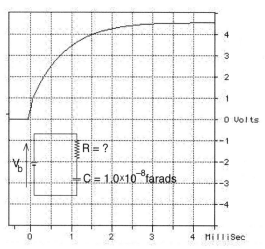

Figure 35
Given the experimental results for the charge-up of a capacitor, determine the value of the resistance R. (Answer R = 68K)

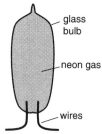

glass bulb

neon gas

wires

Figure 36
A neon bulb. When the voltage across the wires reaches a threshold value, typically around 100 volts, the neon gas starts to glow, and the gas suddenly changes from an insulator to a conductor.

THE NEON BULB OSCILLATOR

We will end this chapter on basic electric circuits with a discussion of an electronic device called a neon bulb oscillator. This is conceptually the world's simplest electronic device that does something useful—it oscillates and its frequency of oscillation can be adjusted. The device is not practical, for it is hard to adjust, its waveform is far from being a pure sinusoidal shape, and it requires a relatively high voltage power supply. But when you work with this apparatus, you will begin to get a feeling for the kind of tricks we pull in order to make useful apparatus.

The Neon Bulb

The new circuit element we will add to our neon oscillator circuit is the common neon bulb which glows orange and is often used as a night light. The bulb, which we will designate by the symbol —⊙— is simply a small glass tube with neon gas inside and two wires as shown in Figure (36). The bulb turns on when there is a large enough voltage difference between the wires that the neon gas becomes ionized and starts to glow. For typical neon bulbs, the glow starts when the voltage reaches approximately 100 volts. When the bulb is glowing the neon gas is a good conductor and the bulb is like a closed switch. When the bulb is not glowing, the gas is inert and the bulb is like an open switch.

A given neon bulb has a rather consistent voltage V_f (*firing voltage*) at which it turns on, and voltage V_q (*quenching voltage*) at which it shuts off. In a typical bulb V_f may be 100 volts and V_q equal to 40 volts. These numbers will, however, vary from bulb to bulb. When a neon bulb is included in a circuit, it acts like an automatic switch, closing (turning on) when the voltage across it reaches V_f and opening (shutting off) when the voltage drops to V_q.

The Neon Oscillator Circuit

We can make a neon oscillator using the circuit shown in Figure (37). The left hand part of the circuit is just the RC circuit we used in Figure (32) to charge up a capacitor. The only really new feature is the neon bulb in parallel with the capacitor. The recording voltmeter, indicated by the symbol —(V)— is there to record the capacitor voltage V_C.

The output of the neon oscillator circuit is shown in Figure (38). Initially the capacitor is charging up just as it did in Figure (33). During the charge up, the neon bulb is off; it is like an open switch and might as well not be there. The effective circuit is shown in Figure (39). When the voltage on the capacitor (and on the neon bulb) reaches the firing voltage V_f, the neon bulb turns on and acts like a short circuit as shown in Figure (40). It is not exactly a short circuit, the neon bulb and the wire leads have some small resistance. But the resistance is so small that the capacitor rapidly discharges through the bulb, and the capacitor voltage drops almost instantly.

When the capacitor and neon bulb voltage V_c drops to the bulb quenching voltage V_q, the bulb shuts off, and the capacitor starts charging up again. As seen in Figure (38), this process keeps repeating and we get the oscillating voltage shown.

For the last cycle in Figure (38), we opened a switch to disconnect the neon bulb, allowing the capacitor to charge up all the way to the power supply voltage V_b. This allowed us to display all three voltages V_b, V_f, and V_q on one experimental plot.

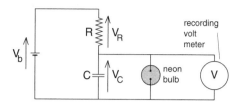

Figure 37
Neon bulb oscillator circuit.

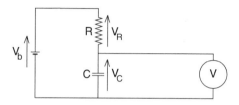

Figure 39
Effective circuit while the neon bulb is off.

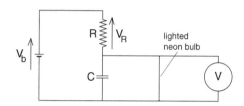

Figure 40
Effective circuit while the neon bulb is glowing.

Figure 38
Experimental output of a neon oscillator circuit. The capacitor charges up until the voltage reaches the neon bulb firing voltage V_f, at which point the neon bulb turns on and the voltage rapidly drops. When the voltage has fallen to the quench voltage V_q, the neon bulb shuts off and the capacitor voltage starts to rise again. On the last cycle, we opened a switch to disconnect the neon bulb, allowing the capacitor to charge up all the way to the power supply voltage V_b.

(A voltage divider was used to measure these high voltages. In the figure, the voltage scale has been corrected to represent the actual voltage on the capacitor.)

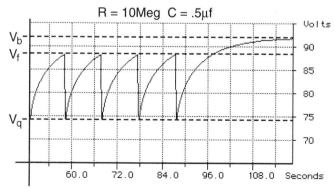

Period of Oscillation

To calculate the period of oscillation, we start with the diagram of Figure (41) showing a cycle of the oscillation superimposed upon the complete charge up curve which starts at $V_C = 0$ and goes up to $V_C = V_b$. This curve is given by the formula

$$V_C = V_b\left(1 - e^{-t/RC}\right) \tag{63}$$

What we want to calculate is the time $T = (t_2 - t_1)$ it takes for the capacitor to charge up from a voltage V_q to V_f.

At time t_1, $V_C = V_q$ and Equation (63) gives

$$V_q = V_b\left(1 - e^{-t_1/RC}\right) \tag{64}$$

At time $t = t_2$, $V_C = V_f$ and we have

$$V_f = V_b\left(1 - e^{-t_2/RC}\right) \tag{65}$$

Equation (64) and (65) can be rearranged to give

$$e^{-t_1/RC} = 1 - \frac{V_q}{V_b} \tag{66}$$

$$e^{-t_2/RC} = 1 - \frac{V_f}{V_b} \tag{67}$$

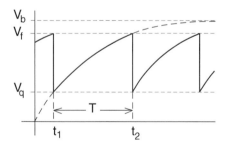

Figure 41
Determining the period T of the oscillation. The formulas for t_1 and t_2 are obtained from the capacitor charge up equations

$$V_f = V_b\left(1 - e^{-t_1/RC}\right)$$

$$V_q = V_b\left(1 - e^{-t_2/RC}\right)$$

The messy part is extracting the period $T = t_2 - t_1$ from these equations.

Dividing Equation (66) by Equation (67) gives

$$e^{(t_2-t_1)/RC} = \frac{1 - V_q/V_b}{1 - V_f/V_b} = \frac{V_b - V_q}{V_b - V_f} \tag{68}$$

where we used the fact that

$$\frac{e^{-\alpha}}{e^{-\beta}} = e^{\beta - \alpha}$$

Taking the logarithm of Equation (68), using $\ln(e^{\alpha}) = \alpha$, we have

$$\frac{t_2 - t_1}{RC} = \ln\left(\frac{V_b - V_q}{V_b - V_f}\right)$$

or using $T = t_2 - t_1$ we get for the period of oscillation

$$\boxed{T = (RC)\ln\left(\frac{V_b - V_q}{V_b - V_f}\right)} \quad \begin{array}{l} period\ of \\ neon\ oscillator \end{array} \tag{69}$$

Equation (69) was a bit messy to derive, and it is not a fundamental result that you need to memorize. You are unlikely to meet a neon oscillator except in an introductory physics lab. But we have used the theory developed in this chapter to make an explicit prediction that can be tested in the laboratory.

Exercise 10

See how well Equation (69) applies to the experimental data of Figure (38). (The marked values on resistors and capacitors are usually accurate only to within ± 10%.)

Equation (69) provides clear instructions on how to change the period or frequency of a neon oscillator. The easiest way to make major changes in the period is to change the time constant RC. Because of the ease with which we can select different values of R and C (typical values of R ranging from 10^2 ohms to 10^8 ohms, and typical values of C from 10^{-4} to 10^{-12} farads), a large range of time constants RC are available. However high frequencies are limited by the characteristics of the neon bulb. We found it difficult to get the circuit to oscillate faster than 30 cycles per second.

Adjusting the battery voltage V_b changes the shape of the neon oscillator wave and also allows fine adjustments in the period .

Experimental Setup

An experimental problem you face while working with the neon oscillator circuit, is that the voltages of interest range up to 100 volts or more. Modern oscilloscopes or recording voltmeters tend to operate in the range of +5 to –5 volts or less, and should not be attached to a voltage source of the order of 100 volts. This problem can be solved by using the voltage divider circuit discussed in Exercise (3), and shown in Figure (42).

For a standard laboratory experiment, we have found it convenient to mount, in one box, the voltage divider, neon bulb, and switch - the components shown inside the dotted rectangle of Figure (42). This reduces student exposure to high voltages and guarantees that voltmeter will be exposed to voltages 1000 times smaller than those across the capacitor.

In Figure (43), we have recorded the entire voltage range of the experiment, starting from an uncharged capacitor. We first opened the switch above the neon bulb in Figure (42), and let the capacitor charge up to the full voltage V_b . Then closing the switch allowed the capacitor voltage to oscillate between V_f and V_q as seen in Figure (38). While the actual capacitor voltage ranges from 0 to 100 volts, the recording voltmeter shows a range of 0 to 100 millivolts because of the 1000 to 1 voltage divider.

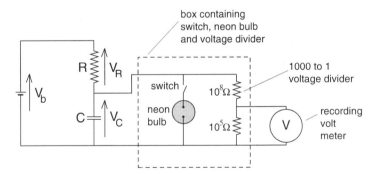

box containing
switch, neon bulb
and voltage divider

1000 to 1
voltage divider

recording
volt
meter

Figure 42
Neon oscillator circuit with voltage divider. The switch above the neon bulb allows us to disconnect the bulb from the circuit. It is convenient to mount, in a single box, the components within the dotted rectangle.

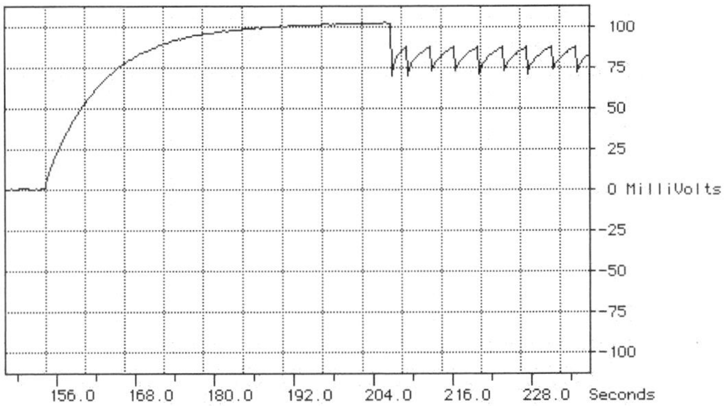

Figure 43
Full range of voltages from the neon oscillator circuit. The voltage scale is in millivolts because of the voltage divider.

Exercise 11 Review Problem

Figure (44a) shows the circuit used to observe the discharge of a capacitor. The capacitor is made from the two circular aluminum plates shown in Figure (44b). The plates have a diameter of 22 cm and are separated a distance (d) by small pieces of glass. In Figure (44c), we are observing the discharge of the capacitor through a 10kΩ $(10^4\Omega)$ resistor. For this discharge, what is the separation (d) of the plates?

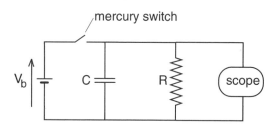

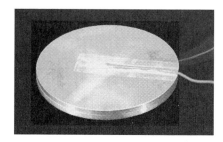

Figure 44a
Circuit for observing the discharge of a capacitor.

Figure 44b
The capacitor plates.

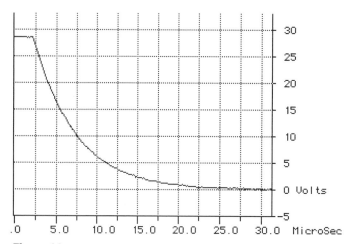

Figure 44c
Voltage during discharge.

Chapter 28
Magnetism

In our discussion of the four basic interactions, we saw that electric forces are very strong but in most circumstances tend to cancel. The strength of the forces are so great, but the cancellation is so nearly complete that the slightest imbalance in the cancellation leads to important effects such as molecular forces. As illustrated in Figure (18-6) reproduced here, a positively charged proton brought up to a neutral hydrogen atom experiences a net attractive force because the negative charge in the atom is pulled closer to the proton. This net force is the simplest example of the type of molecular force called a covalent bond.

In this chapter we will study another way that the precise balance between attractive and repulsive electric forces can be upset. So far in our discussion of electrical phenomena such as the flow of currents in wires, the charging of capacitors, etc., we have ignored the effects of special relativity. And we had good reason to. We saw that the conduction electrons in a wire move at utterly nonrelativistic speeds, like two

millimeters per minute. One would not expect phenomena like the Lorentz contraction or time dilation to play any observable role whatever in such electrical phenomena.

But, as we shall see, observable effects do result from the tiny imbalance in electric forces caused by the Lorentz contraction. Since these effects are not describable by Coulomb's law, they are traditionally given another name—magnetism. Magnetism is one of the consequences of requiring that the electrical force law and electric phenomena be consistent with the principle of relativity.

Historically this point of view is backwards. Magnetic effects were known in the time of the ancient Greeks. Hans Christian Oersted first demonstrated the connection between magnetic and electric forces in 1820 and James Clerk Maxwell wrote out a complete theory of electromagnetic phenomena in 1860. Einstein did not discover special relativity until 1905. In fact, Einstein used Maxwell's theory as an important guide in his discovery.

If you follow an historical approach, it appears that special relativity is a consequence of electricity theory, and a large number of physics texts treat it that way. Seldom is there a serious discussion of special relativity until after Maxwell's theory of electricity has been developed. This is considered necessary in order to explain the experiments and arguments that lead to the discovery of the special theory.

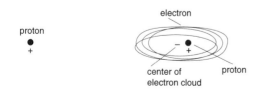

Figure 18-6
The net attraction between a positive charge and a neutral atom is caused by a redistribution of charge in the atom.

But as we know today, electricity is one of but several basic forces in nature, and all of them are consistent with special relativity. Einstein's famous theory of gravity called **general relativity** *can be viewed as a repair of Newton's theory of gravity to make it consistent with the principle of relativity. (This "repair" produced only minor corrections when applied to our solar system, but has sweeping philosophical implications.) If the principle of relativity underlies the structure of all forces in nature, if all known phenomena are consistent with the principle, then it is not especially necessary to introduce special relativity in the context of its historical origins in electromagnetic theory.*

In this chapter we are taking a non-historical point of view. We already know about special relativity (from chapter one), and have just studied Coulomb's electrical force law and some simple applications like the electron gun and basic circuits. We would now like to see if Coulomb's law is consistent with the principle of relativity. In some sense, we would like to do for Coulomb's law of electricity what Einstein did to Newton's law of gravity.

Two Garden Peas

In preparation for our discussion of relativistic effects in electricity theory, let us review a homely example that demonstrates both how strong electric forces actually are, and how complete the cancellation must be for the world to act the way it does. Suppose we had two garden peas, each with a mass of about 2 grams, separated by a distance of 1 meter. Each pea would contain about one mole (6×10^{23}) of protons in the atomic nuclei, an equal number of electrons surrounding the nuclei. Thus each pea has a total positive charge $+Q$ in the protons given by

$$
\left. \begin{array}{l} \text{total positive} \\ \text{charge in a} \\ \text{garden pea} \end{array} \right\} \approx 6 \times 10^{23} e
$$

$$
\begin{aligned}
&= 6 \times 10^{23} \times 1.6 \times 10^{-19} \\
&= 10^{5} \text{ coulombs}
\end{aligned}
\tag{1}
$$

and there is an equal and opposite amount of negative charge in the electrons.

When two peas are separated by a distance of 1 meter as shown in Figure (1), we can think of there being four pairs of electric forces involved. The positive charge in pea (1) repels the positive charge in pea (2) with a force of magnitude

$$\left.\begin{array}{l}\text{repulsive force}\\ \text{between positive}\\ \text{charge in}\\ \text{the two peas}\end{array}\right\} = \frac{QQ}{4\pi\varepsilon_0 r^2} \qquad (2)$$

which gives rise to one pair of repulsive forces. The negative charges in each pea also repel each other with a force of the same magnitude, giving rise to the second repulsive pair of electric forces. But the positive charge in Pea (1) attracts the negative charge in Pea (2), and the negative charge in Pea (1) attracts the positive charge in Pea (2). This gives us two pairs of attractive forces that precisely cancel the repulsive forces.

Let us put numbers into Equation (2) to see how big these cancelling electric forces are. Equation (2) can be viewed as giving the net force if we removed all the electrons from each garden pea, leaving just the pure positive charge of the protons. The result would be

$$|\vec{F}| = \frac{Q^2}{4\pi\varepsilon_0 r^2}$$

$$= \frac{(10^5 \text{ coulombs})^2}{4\pi \times 9 \times 10^{-12} \times (1)^2}$$

$$= 8.8 \times 10^{19} \text{ newtons} \qquad (3)$$

pea   pea

1   2

Figure 1
Electric forces between two garden peas. On pea #1, there is the attractive force between the protons in pea #1 and the electrons in pea #2, and between the electrons in pea #1 and the protons in pea #2. The two repulsive forces are between the electrons in the two peas and the protons in the two peas. The net force is zero.

To put this answer in a more recognizable form, note that the weight of one metric ton (1000 kg) of matter is

$$F_g \text{ (1 metric ton)} = mg$$

$$= 10^3 \text{ kg} \times 9.8 \frac{m}{sec^2}$$

$$= 9.8 \times 10^3 \text{ newtons}$$

Expressing the force between our two positively charged peas in metric tons we get

$$\left.\begin{array}{l}\text{repulsive force}\\ \text{between two}\\ \text{positive peas}\\ \text{1 meter apart}\end{array}\right\} = \frac{8.8 \times 10^{19} \text{ newtons}}{9.8 \times 10^3 \text{ newtons/ton}}$$

$$\approx \boxed{10^{15} \text{ tons !}}$$

$$(4)$$

If we stripped the electrons from two garden peas, and placed them one meter apart, they would repel each other with an electric force of 10^{15} tons!! Yet for two real garden peas, the attractive and repulsive electric force cancel so precisely that the peas can lie next to each other on your dinner plate.

Exercise 1

Calculate the strength of the gravitational force between the peas. How much stronger is the uncancelled electric force of Equation (3)?

With forces of the order of 10^{15} tons precisely cancelling in two garden peas, we can see that even the tiniest imbalance in these forces could lead to striking results. An imbalance of one part in 10^{15}, one part in a million billion, would leave a one ton residual electric force. This is still huge. We have to take seriously imbalances that are thousands of times smaller. One possible source of an imbalance is the Lorentz contraction, as seen in the following thought experiment.

A THOUGHT EXPERIMENT

In our previous discussion of electric currents, we had difficulty drawing diagrams showing the electrons flowing through the positive charge. To clarify the role of the positive and negative charge, we suggested a model of a copper wire in which we think of the positive and negative charge as being attached to separate rods as shown in Figure (27-5a) repeated here. In that model the rods have equal and opposite charge to represent the fact that the copper wire is electrically neutral, and the negative rod is moving to represent the electric current being carried by a flow of the negative conduction electrons.

The point of the model in Figure (27-5) was to show that a left directed negative current, seen in (a) is essentially equivalent to a right directed positive current seen in (b). In Figure (27-5a), we drew a stick figure diagram of a person walking to the left at the same speed v as the negative rod. Figure (27-5b) is the same setup from the point of view of the stick figure person. She sees the negative rod at rest and the positive rod moving to the right as shown.

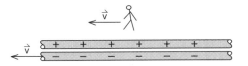

a) observer walking along with the moving negatively charged rod

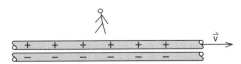

b) from the observer's point of view the negative rod is at rest and the positive charge is moving to the right

Figure 27-5 a,b
In (a) we have a left directed negative current, while in (b) we have a right directed positive current. The only difference is the perspective of the observer. (You can turn a negative current into an oppositely flowing positive one simply by moving your head.)

In another calculation, we saw that if a millimeter cross section copper wire carried a steady current of one ampere, the conduction electrons would have to move at the slow speed of 1/27 of a millimeter per second, a motion so slow that it would be hard to detect. As a result there should be no important physical difference between the two points of view, and a left directed negative current should be physically equivalent to a right directed positive current.

A closer examination of Figure (27-5) shows that we have left something out. The bottom figure, (27-5b) is not precisely what the moving observer sees. To show what has been left out, we have in Figure (2a) redrawn Figure (27-5a) and carefully labeled the individual charges. To maintain strict overall charge neutrality we have used charges +Q on the positive rod, charges -Q on the negative rod, and both sets of charges have equal separations of ℓ centimeters.

From the point of view of the moving observer in Figure (2b), the negative rod is at rest and the positive rod is moving to the right as we saw back in Figure (27-5b). But, *due to the Lorentz contraction, the spacing between the charges is no longer ℓ!* Since the positive rod was at rest and is now moving, the length of the positive spacing must be contracted to a distance $\ell\sqrt{1 - v^2/c^2}$ as shown.

On the other hand the negative rod was moving in Figure (2a), therefore the negative spacing must expand to $\ell/\sqrt{1 - v^2/c^2}$ when the negative rod comes to rest. *(Start with a spacing $\ell/\sqrt{1 - v^2/c^2}$ for the negative charges at rest in Figure (2b), and go up to Figure (2a) where the negative rod is moving at a speed v. There the spacing must contract by a factor $\sqrt{1 - v^2/c^2}$, and the new spacing is $\ell/\sqrt{1 - v^2/c^2} \times \sqrt{1 - v^2/c^2} = \ell$ as shown.)*

As a result of the Lorentz contraction, the moving observer will see that the positive charges on her moving rod are closer together than the negative charges on her stationary rod. (We have exaggerated this effect in our sketch, Figure (2b)). Thus the moving observer of Figure (2b) sees not only a right directed positive current, but also *a net positive charge density* on her two rods. The Lorentz contraction has changed a neutral wire in Figure (2a) into a positively charged one in Figure (2b)!

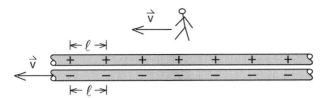

a) Observer walking along with the moving negatively charged rod.

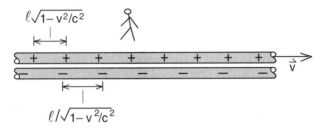

b) Charged rods from the observer's point of view. Now that the positive charge is moving, the spacing between positive charges has **contracted** from ℓ to $\ell\sqrt{1-v^2/c^2}$. The negative rod is now at rest, the Lorentz contraction is undone, and the negative spacing has **expanded** from ℓ to $\ell/\sqrt{1-v^2/c^2}$.

Figure 2
An electric current from two points of view.

Charge Density on the Two Rods

Our next step will be to calculate the net charge density λ on the pair of rods shown in Figure (2b). Somewhat messy algebra is required for this calculation, but the result will be used in much of the remainder of the text. The effort will be worth it.

If we have a rod with charges spaced a distance d apart as shown in Figure (3), then a unit length of the rod, 1 meter, contains 1/d charges. (For example, if d = .01 meter, then there will be 1/d = 100 charges per meter.) If each charge is of strength Q, then there is a total charge Q/d on each meter of the rod. Thus the charge density is λ = Q/d coulombs per meter. Applying this result to the positive rod of Figure (2b) gives us a positive charge density

$$\lambda_+ = \frac{Q}{d_+} = \frac{Q}{\ell\sqrt{1 - v^2/c^2}} \tag{5}$$

And on the negative rod the charge density is

$$\lambda_- = \frac{-Q}{d_-} = \frac{-Q}{\ell/\sqrt{1 - v^2/c^2}}$$

$$= \frac{-Q\sqrt{1 - v^2/c^2}}{\ell} \tag{6}$$

Multiplying the top and bottom of the right side of Equation (6) by $\sqrt{1 - v^2/c^2}$, we can write λ_- as

$$\lambda_- = \frac{-Q\sqrt{1 - v^2/c^2}}{\ell} \times \frac{\sqrt{1 - v^2/c^2}}{\sqrt{1 - v^2/c^2}}$$

$$= \frac{-Q}{\ell\sqrt{1 - v^2/c^2}}\left(1 - v^2/c^2\right) \tag{7}$$

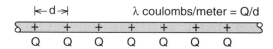

|←─d─→| λ coulombs/meter = Q/d

Figure 3
If the charges are a distance d apart, then there are 1/d charges per meter of rod. (If d = .1 meters, then there are 10 charges/meter.) If the magnitude of each charge is Q, then λ, the charge per meter is Q times as great, i.e., $\lambda = Q * (1/d)$.

The net charge density λ is obtained by adding λ_+ and λ_- of Equations (5) and (7) to get

$$\lambda = \lambda_+ + \lambda_- = \frac{Q}{\ell\sqrt{1 - v^2/c^2}}\left\{1 - (1 - v^2/c^2)\right\}$$

$$\lambda = \frac{Q}{\ell\sqrt{1 - v^2/c^2}}\left\{\frac{v^2}{c^2}\right\} = \lambda_+\frac{v^2}{c^2} \tag{8}$$

Equation (8) can be simplified by noting that the current i carried by the positive rod in Figure (2b) is equal to the charge λ_+ on 1 meter of the rod times the speed v of the rod

$$i = \lambda_+ v \qquad \begin{array}{l}\textit{current i}\\ \textit{carried by the}\\ \textit{positive rod}\end{array} \tag{9}$$

(In one second, v meters of rod move past any fixed cross-sectional area, and the charge on this v meters of rod is λ_+ v) Using Equation (9), we can replace λ_+ and one of the v's in Equation (8) by i to get the result

$$\boxed{\lambda = \frac{iv}{c^2}} \tag{10}$$

Due to the Lorentz contraction , the moving observer in Figure (2b) sees a net ***positive charge density*** $\lambda = iv/c^2$ on the wire which from our point of view, Figure (2a) ***was precisely neutral***.

Although Equation (10) may be formally correct, one has the feeling that it is insane to worry about the Lorentz contraction for speeds as slow as 2 millimeters per minute. But the Lorentz contraction changes a precisely neutral pair of rods shown in Figure (2a), into a pair with a net positive charge density $\lambda = iv/c^2$ in Figure (2b). We have unbalanced a perfect cancellation of charge which could lead to an imbalance in the cancellation of electrostatic forces. Since we saw from our discussion of the two garden peas that imbalances as small as one part in 10^{18} or less might be observable, let us see if there are any real experiments where the charge density λ is detectable.

A Proposed Experiment

How would we detect the charge imbalance in Figure (2b)? If there is a net positive charge density l on the two rods in Figure (2b), repeated here again in Figure (4), then the net charge should produce a radial electric field whose strength is given by the formula

$$E = \frac{\lambda}{2\pi\varepsilon_0 r} \qquad (11)$$

We derived this result in our very first discussions of Coulomb's law in Chapter 24. *(Remember that the two separate rods are our model for a single copper wire carrying a current. The rods are not physically separated as we have had to draw them, the negative conduction electrons and positive nuclei are flowing through each other.)*

We can test for the existence of the electric field produced by the positive charge density $\lambda = iv/c^2$ by placing a test particle of charge q a distance r from the wire as shown in Figure (4). This test particle should experience a force

$$\vec{F} = q\vec{E} \qquad (12)$$

which would be repulsive if the test particle q is positive and attractive if q is negative. Using Equations (10) for λ and (11) for E, Equation (12) gives for the predicted magnitude of $\vec{F}$:

$$\left|\vec{F}\right| = q\left|\vec{E}\right| = \frac{q\lambda}{2\pi\varepsilon_0 r} = \frac{q\,iv}{2\pi\varepsilon_0 rc^2} \qquad (13)$$

Rearranging the terms on the right side of Equation (13), we can write $\left|\vec{F}\right|$ in the form

$$\left|\vec{F}\right| = qv\times\left\{\frac{i}{2\pi r\varepsilon_0 c^2}\right\} \qquad (14)$$

Why we have written Equation (14) this way will become clear shortly.

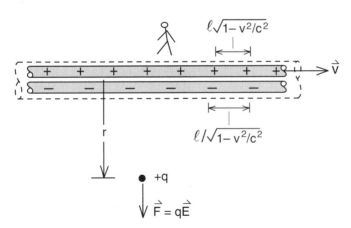

Figure 4
To test for the net charge density, as seen by the observer at rest relative to the minus charge, the observer places a test charge q a distance r from the wire as shown. If there is a net charge λ on the wire, the charge will produce an electric field $\vec{E}$, which will exert a force $\vec{F} = q\vec{E}$ on the test particle as shown.

Origin of Magnetic Forces

You might think that the next step is to put reasonable numbers into Equation (14) and see if we get a force $\vec{F}$ that is strong enough to be observed. But there is an important thought experiment we will carry out first. The idea is to look at the force on a test particle from two different points of view, one where the wire appears charged as in Figures (4 & 2b), and where the wire appears neutral as in Figure (2a). The two points of view are shown in Figure (5).

Figure (5b), on the left, is the situation as observed by the moving observer. She has a copper wire carrying a positive current directed to the right. Due to the Lorentz contraction, her copper wire has a charge density λ which creates an electric field $\vec{E}$. To observe $\vec{E}$, she mounts a test particle -q at one end of a spring whose other end is fixed, nailed to her floor. She detects the force $\vec{F} = -q\vec{E}$ by observing how much the spring has been stretched.

Our point of view is shown in Figure (5a). It is exactly the same setup, we have touched nothing! It is just viewed by someone moving to the right relative to her.

In our point of view, the moving observer, the negative rod, and the test particle are all moving to the left at a speed v. The positive rod is at rest, the Lorentz contractions are undone, and there is *no net charge* on our rods. All we have is a negative current flowing to the left.

We can also see the test particle. It is now moving to the left at a speed v, and it is still attached to the spring.

Here is the crucial point of this discussion. We also see that the spring is stretched. We also see that the end of the spring has been pulled beyond the mark indicating the unstretched length. We also detect the force $\vec{F}$ on the test particle!

Why do we see a force $\vec{F}$ on the test particle? Our copper wire is electrically neutral; we do not have an electric field $\vec{E}$ to produce the force $\vec{F}$. Yet $\vec{F}$ is there. If we cut the spring, the test particle would accelerate toward the copper wire, and both we and the moving observer would see this acceleration.

At this point, we have come upon a basic problem. Even if the Lorentz contraction is very small and the force $\vec{F}$ in Figure (5b) is very small, we at least predict that $\vec{F}$ exists. In Figure (5a) we predict that a neutral wire, that is carrying a current but *has absolutely no net charge on it*, exerts an attractive force on a moving negative charge as shown.

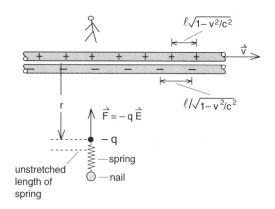

(b) her view   (a) our view

Figure 5
Two views of the same experiment. For the observer moving with the electrons, she sees a positively charged wire exerting an attractive force on the negative charge at rest. We see an electrically neutral wire carrying a negative current, and a moving negative charge. The spring is still stretched, meaning the attractive force is still there.

With a few modifications, the experiment shown in Figure (5a) is easy to perform and gives clear results. Instead of a negative test particle attached to a spring, we will use a beam of electrons in an electron gun as shown in Figure (6). In Figure (6a) we see the setup of our thought experiment. In Figure (6b) we have replaced the two charged rods with a neutral copper wire carrying a current -i, and replaced the test particle with an electron beam.

According to Equation (14), the force $\vec{F}$ on the test particle -q should have a strength proportional to the current i in the wire. Thus when we turn on a current (*shorting the wire on the terminals of a car storage battery to produce a healthy current*) we will see the electron beam deflected toward the wire if there is an observable force. The experimental result is shown in Figure (6c). There is a large, easily observed deflection. The force $\vec{F}$ is easily seen.

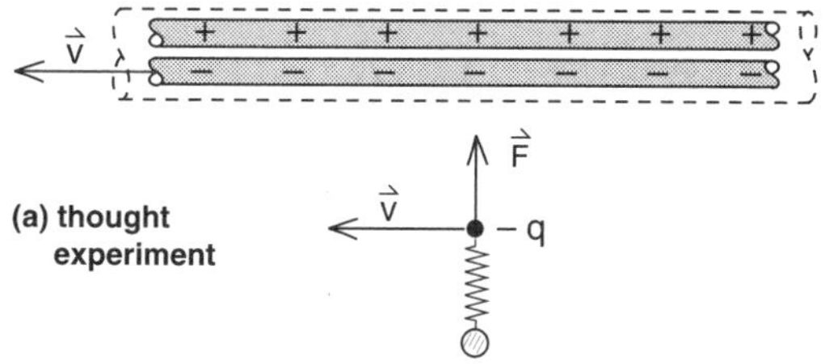

Exercise 2

In Figure (7) we reversed the direction of the current in the wire and observe that the electron beam is deflected away from the wire. Devise a thought experiment, analogous to the one shown in Figure (5a,b) that explains why the electron beam is repelled from the wire by this setup. (This is not a trivial problem; you may have to try several charge distributions on moving rods before you can imitate the situation shown in Figure (7a). But the effort is worth it because you will be making a physical prediction that is checked by the experimental results of Figure (7b).

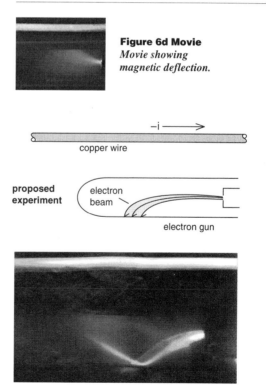

Figure 6d Movie
Movie showing magnetic deflection.

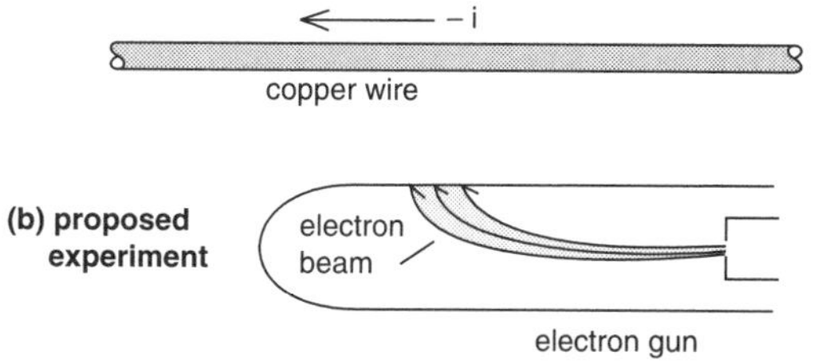

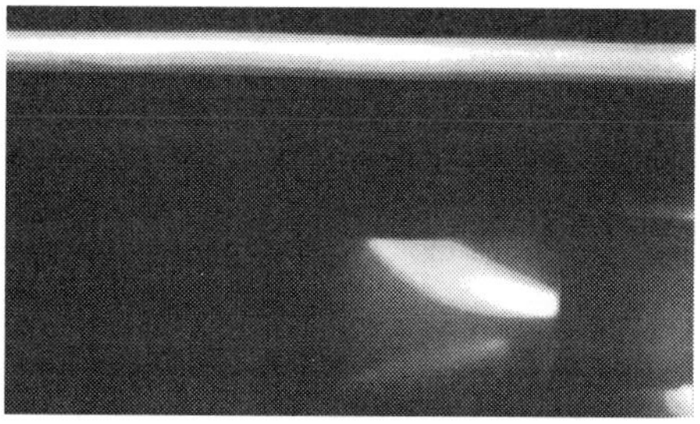

(a) thought experiment

(b) proposed experiment

electron beam

electron gun

copper wire

− i

Figure 6c
For an experimental test of the results of the thought experiment, we replace the moving negative charge with a beam of electrons in an electron gun. The electrons are attracted to the wire as predicted.

proposed experiment

electron beam

electron gun

copper wire

−i

Figure 7
If we reverse the direction of the current in the wire, the electrons in the beam are repelled

Magnetic Forces

Historically an electric force was defined as the force between charged particles and was expressed by Coulomb's law. The force in Figure (5a) between a moving test charge and an **uncharged** wire does not meet this criterion. You might say that for historical reasons, it is not eligible to be called an electric force.

The forces we saw in Figures (6c) and (7b), between a moving charge and a neutral electric current, were known before special relativity and were called **magnetic forces**. Our derivation of the magnetic force in Figure (5a) from the electric force seen in Figure (5b) demonstrates that *electric and magnetic forces in this example are the same thing just seen from a different point of view*.

When we go from Figure (5b) to (5a), which we can do by moving our head at a speed of 2 millimeters per minute, we see essentially no change in the physical setup but we have an enormous change in perspective. We go from a right directed positive current to a left directed negative current, and the force on the test particle changes from an electric to a magnetic force.

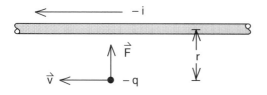

Figure 8
Force on a charge -q moving at a speed v parallel to a negative current -i a distance r away.

MAGNETIC FORCE LAW

From our Coulomb's law calculation of the electric force in Figure (5a), we were able to obtain the formula for the magnetic force in Figure (5b). The result, Equation (14) repeated here, is

$$\left|\vec{F}\right| = qv \times \left\{ \frac{i}{2\pi r \varepsilon_0 c^2} \right\} \tag{14}$$

where q is the charge on the test particle, i the current in the wire, and r the distance from the wire to the charge as shown in Figure (8). The only thing our derivation does not make clear is whether v in Equation (14) is the speed of the test charge or the speed of the electrons in the wire. We can't tell because we used the same speed v for both in our thought experiment. A more complex thought experiment will show that the v in Equation (14) is the speed of the test particle.

The Magnetic Field B

In Equation (14) we have broken the somewhat complex formula for the magnetic force into two parts. The first part qv is related to the test charge (q is its charge and v its speed), and the second part in the curly brackets, which we will designate by the letter B

$$B \equiv \frac{i}{2\pi r \varepsilon_0 c^2} \tag{15}$$

is related to the wire. The wire is carrying a current i and located a distance r away.

The quantity B in Equation (15) is called the magnitude of *the magnetic field of the wire*, and in terms of B the magnetic force becomes

$$\boxed{\left|\vec{F}_{magnetic}\right| = qvB} \tag{16}$$

Equation (16) is almost a complete statement of the magnetic force law. What we have left to do for the law is to assign a direction to B, i.e. turn it into the vector $\vec{B}$, and then turn Equation (16) into a vector equation for the force $\vec{F}_{magnetic}$.

There is one more definition. In the MKS system of units, it is traditional to define the constant μ_0 by the equation

$$\mu_0 = \frac{1}{\varepsilon_0 c^2} \qquad \textit{definition of } \mu_0 \qquad (17)$$

Using this definition of μ_0 in Equation (15) for B, we get

$$B = \frac{\mu_0 i}{2\pi r} \qquad \textit{magnetic field of a wire} \qquad (18)$$

as the formula for the magnetic field of a wire.

It turns out to be quite an accomplishment to get Equations (16), (17), and (18) out of one thought experiment. These equations will provide the foundation for most of the rest of our discussion of electric and magnetic (electromagnetic) theory.

Direction of the Magnetic Field

We will temporarily leave our special relativity thought experiment and approach magnetism in a more traditional way. Figure (9) is a sketch of the magnetic field of the earth. By convention the direction of the magnetic field lines are defined by the direction that a compass needle points. At the equator the magnetic field lines point north (as does a compass needle) and the field lines are parallel to the surface of the earth. As we go north from the equator the magnetic field lines begin to point down into the earth as well as north. At the north magnetic pole the magnetic field lines go straight down.

Figure (9) is drawn with the magnetic north pole at the top. The earth's rotational axis, passing through the true north pole, is at an angle of 11.5 degrees as shown. Over time the location of the earth's magnetic pole wanders, and occasionally flips down to the southern hemisphere. Currently the north magnetic pole is located in north central Canada.

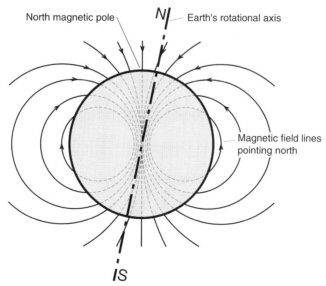

North magnetic pole — N — Earth's rotational axis

Magnetic field lines pointing north

S

Figure 9

Magnetic field of the earth. The magnetic field lines show that the direction a freely floating compass needle would point at any location outside the earth. For example, at the equator the compass needle would be parallel to the surface of the earth and point north. At the north magnetic pole, the compass needle would point straight down (and thus not be very useful for navigation).

As we mentioned, it is by long standing convention that the direction of the magnetic field is defined by the direction a compass needle points. We can therefore use a set of small compasses to map the direction of the magnetic field.

In 1820, while preparing a physics lecture demonstration for a class of students, Hans Christian Oersted discovered that an electric current in a wire could

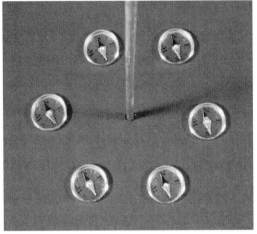

Figure 10a
With no current flowing in the wire,
all the compass needles point north.

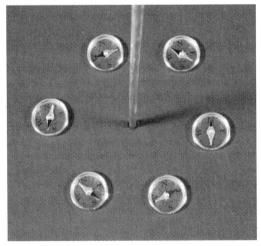

Figure 10b
When an upward directed current is turned on,
the compass needles point in a
counterclockwise circle about the wire.

deflect a compass needle. This was the first evidence of the connection between the subject of electricity with its charges and currents, and magnetism with its magnets and compasses.

The fact that a wire carrying a current deflects a compass needle means that the current must be producing a magnetic field. We can *use the deflected compass needles to show us the shape of the magnetic field* of a wire. This is done in Figures (10a,b) where we see a ring of compasses surrounding a vertical wire. In (10a) there is no current in the wire, and all the compass needles all point north (black tips). In (10b) we have turned on an upward directed current in the wire, and the compass needles point in a circle around the wire. *Using the north pole of the compass needle to define the direction of the magnetic field*, we see that the magnetic field goes in a counterclockwise circle around the wire.

In Figure (11) we have replaced the compasses in Figure (10) with a sprinkle of iron filings. When the current in the wire is turned on, the iron filings align themselves to produce the circular field pattern shown. What is happening is that each iron filing is acting as a small compass needle and is lining up parallel to the magnetic field. While we cannot tell which way is north with iron filings, we get a much more complete

Figure 11
Iron fillings sprinkled around a current form a circular
pattern. Each iron filing lines up like a compass needle,
giving us a map of the magnetic field.

picture or map of the direction of the magnetic field. Figure (11) is convincing evidence that the magnetic field surrounding a wire carrying a current is in a circular field, not unlike the circular flow pattern of water around the core of a vortex.

The use of iron filings turns out to be a wonderfully simple way to map magnetic field patterns. In Figure (12), a sheet of cardboard was placed on a bar magnet and iron filings sprinkled on the cardboard. The result, with two poles or points of focus resembles what is called a *dipole field*.

In Figure (13) we have thrown iron filings at an old iron magnet and created what one young observer called a "magnet plant." Here we see the three dimensional structure of the magnetic field, not only between the pole pieces but over the top half of the magnet.

The Right Hand Rule for Currents

Iron filings give us an excellent picture of the shape of the magnetic field, but do not tell us which way the field is pointing. For that we have to go back to compasses as in Figure (9), where $\vec{B}$ is defined as pointing in the direction of the north tip of the compass needle. In that figure we see that when a positive current i is flowing toward us, the magnetic field goes in a counter clockwise direction as illustrated in Figure (14).

The above description for the direction may be hard to remember. A more concise description is the following. *Point the thumb of your right hand in the direction of the current* as shown in Figure (14), then *your fingers will curl in the direction of the magnetic field*. This mnemonic device for remembering the direction of $\vec{B}$ is one of the *right hand rules*. (This is the version we used in Figure 2-37 to distinguish right and left hand threads.) If we had used compasses that pointed south, we would have gotten a left hand rule.

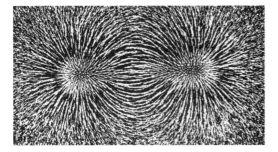

Figure 12
A sheet of cardboard is placed over the poles of a magnet and sprinkled with iron filings. From the pattern of the filings we see the shape of the more complex magnetic field of the magnet.

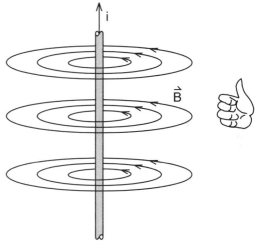

Figure 14
Right hand rule for the magnetic field of a current i. Point the thumb in the direction of the positive current and your fingers curl in the direction of the magnetic field.

Figure 13
You get a three dimensional picture of the magnetic field if you pour the iron filings directly on the magnet. Our young daughter called this a **Magnet Plant.**

Parallel Currents Attract

While we are in the business of discussing mnemonic rules, there is another that makes it easy to remember whether a charge moving parallel to a current is attracted or repelled. In Figure (6) we had a beam of negative electrons moving parallel to a negative current -i, and the electrons were attracted to the current. In Figure (7) the current was reversed and the electrons were repelled. One can work out a thought experiment similar to the ones we have done in this chapter to show that a positive charge moving parallel to a positive current as shown in Figure (15) is attracted.

The simple, yet general rule is that **parallel currents attract, opposite currents repel**. A positive charge moving in the direction of a positive current, or a negative charge moving along with a negative current are attracting parallel currents. When we have negative charges moving opposite to a negative current as in Figure (7) we have an example of opposite currents that repel.

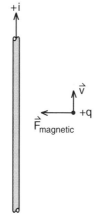

Figure 15
A positive charge, moving parallel to a positive current, is attracted by the current. Thinking of the moving positive charge as a positive upward directed current, we have the rule that parallel currents attract, opposite currents repel.

The Magnetic Force Law

Now that we have a direction assigned to the magnetic field $\vec{B}$ we are in a position to include directions in our formula for magnetic forces. In Figure (16) which is the same as (15) but also shows the magnetic field, we have a positive charge moving parallel to a positive current, and therefore an attractive force whose magnitude is given by Equation (16) as

$$\left| \vec{F}_{mag} \right| = q\,v\,B \qquad (16)$$

There are three different vectors in Equation (16), $\vec{F}_{mag}, \vec{v},$ and $\vec{B}$. Our problem is to see if we can combine these vectors in any way so that something like Equation (16) tells us both the magnitude and the direction of the magnetic force $\vec{F}_{mag}$. That is, can we turn Equation (16) into a vector equation?

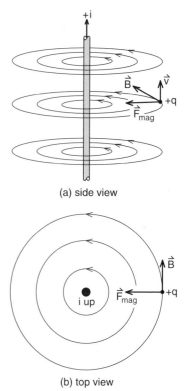

(a) side view

(b) top view

Figure 16
The directions of the vectors $\vec{F}_{mag}$, $\vec{v}$ and $\vec{B}$ for a positive charge moving parallel to a positive current.

The right hand side of Equation (16) involves the product of the vectors $\vec{v}$ and $\vec{B}$. So far in the text we have discussed two different ways of multiplying vectors; the dot product $\vec{A} \cdot \vec{B}$ which gives a scalar number C, and the cross product $\vec{A} \times \vec{B}$ which gives the vector $\vec{C}$. Since we want the product of $\vec{v}$ and $\vec{B}$ to give us the vector $\vec{F}_{mag}$, the cross product appears to be the better candidate, and we can try

$$\vec{F}_{mag} = q\vec{v} \times \vec{B} \qquad \begin{array}{l} \textit{magnetic} \\ \textit{force law} \end{array} \qquad (19)$$

as our vector equation.

To see if Equation (19) works, look at the three vectors $\vec{v}$, $\vec{B}$, and $\vec{F}_{mag}$ of Figure (16) redrawn in Figure (17). The force $\vec{F}_{mag}$ is perpendicular to the plane defined by $\vec{v}$ and $\vec{B}$ which is the essential feature of a vector cross product. To see if $\vec{F}_{mag}$ is in the correct direction, we use the cross product right hand rule. Point the fingers of your **right** hand in the direction of the first vector in the cross product, in this case $\vec{v}$, and curl them in the direction of the second vector, now $\vec{B}$. Then your thumb will point in the direction of the cross product $\vec{v} \times \vec{B}$. Looking at Figure (17), we see that the thumb of the right hand sketch does point in the direction of $\vec{F}_{mag}$, therefore the direction of $\vec{F}_{mag}$ is correctly given by the cross product $\vec{v} \times \vec{B}$. (If the direction had come out wrong, we could have used $\vec{B} \times \vec{v}$ instead.)

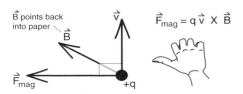

B points back into paper

$\vec{B}$

$\vec{v}$

$\vec{F}_{mag}$

$\vec{F}_{mag} = q\,\vec{v} \times \vec{B}$

+q

Figure 17
Right hand rule for the vector cross product $\vec{v} \times \vec{B}$. Point the fingers of your right hand in the direction of the first vector $\vec{v}$, and then curl them in the direction of the second vector $\vec{B}$. Your thumb ends up pointing in the direction of the vector $\vec{v} \times \vec{B}$.

Although the formula for $\vec{F}_{mag}$, Equation (19), was derived for a special case, the result is general. Whenever a particle of charge q is moving with a velocity $\vec{v}$ through a magnetic field $\vec{B}$, no matter what the relative directions of $\vec{v}$ and $\vec{B}$, the magnetic force is correctly given as $q\vec{v} \times \vec{B}$.

Exercise 3

Using the magnetic force law $\vec{F}_{mag} = q\vec{v} \times \vec{B}$ and the right hand rule for the magnetic field of a current, show that:

(a) An electron moving parallel to a negative current -i is attracted (Figure 6)

(b) An electron moving opposite to a negative current is repelled (Figure 7)

Lorentz Force Law

Since electric and magnetic forces are closely related, it makes sense to write one formula for both the electric and the magnetic force on a charged particle. If we have a charge q moving with a velocity $\vec{v}$ through an electric field $\vec{E}$ and a magnetic field $\vec{B}$, then the electric force is $q\vec{E}$, the magnetic force $q\vec{v} \times \vec{B}$, and the total "electromagnetic" force is given by

$$\boxed{\vec{F} = q\vec{E} + q\vec{v} \times \vec{B}} \qquad \begin{array}{l} \textit{Lorentz} \\ \textit{force law} \end{array} \qquad (20)$$

Equation (20), which is known as the **Lorentz force law**, is a complete description of the electric and magnetic forces on a charged particle, provided $\vec{E}$ and $\vec{B}$ are known.

Dimensions of the
Magnetic Field, Tesla and Gauss

The dimensions of the magnetic field can be obtained from the magnetic force law. In the MKS system we have

$$F\left(\text{newtons}\right) = q\left(\text{coulombs}\right) v\left(\frac{\text{meters}}{\text{second}}\right) \times B$$

which gives us B in units of newton seconds per coulomb meter. This set of dimensions is given the name *tesla*

$$\boxed{\frac{\text{newton second}}{\text{coulomb meter}} \equiv \text{tesla}} \quad \begin{array}{l} \textit{MKS units} \\ \textit{for} \\ \textit{magnetic} \\ \textit{fields} \end{array} \quad (21)$$

Although most MKS electrical quantities like the volt and ampere are convenient, the tesla is too large. Only the strongest electromagnets, or the new superconducting magnets used in particle accelerators or magnetic resonance imaging apparatus, can produce fields of the order of 1 tesla or more. Fields produced by coils of wire we use in the lab are typically 100 times weaker, and the earth's magnetic field is 100 times weaker still.

In the CGS system of units, magnetic fields are measured in gauss, where

$$\boxed{1 \text{ gauss} = 10^{-4} \text{ tesla}}$$

The gauss is so much more convenient a unit that there is a major incentive to work with CGS units when studying magnetic phenomena. For example the earth's magnetic field has a strength of about 1 gauss at the earth's surface, and the magnetic field that deflected the electrons in Figures (6) and (7) has a strength of about 30 gauss at the electron beam. Refrigerator magnets have comparable strengths.

We could be pedantic, insist on using only MKS units, and suffer with numbers like .00021 tesla in discussions of the earth's magnetic field. But if someone wants you to measure a magnetic field, they hand you a "gauss meter" not a tesla meter. Magnetic-type instruments are usually calibrated in gauss. If you worked only with tesla, you would have a hard time communicating with much of the scientific community. What we will do in this text is use either gauss or tesla depending upon which is the more convenient unit. When we come to a calculation, we will convert any gauss to tesla, just as we convert any distances measured in centimeters to meters.

Uniform Magnetic Fields

Using the magnetic force law $\vec{F}_{mag} = q\vec{v} \times \vec{B}$ to calculate magnetic forces is often the easy part of the problem. The hard part can be to determine the magnetic field $\vec{B}$. For a current in a straight wire, we were able to use a thought experiment and the Lorentz contraction to get Equation (20) for the strength of $\vec{B}$. But in more complicated situations, where we may have bent wires, thought experiments become too difficult and we need other techniques for calculating $\vec{B}$.

One of the other techniques, which we will discuss in the next chapter, is called *Ampere's law*. This law will give us the ability to calculate the magnetic field of simple current distribution much the same way that Gauss' law allowed us to calculate the electric field of simple charge distributions. But until we get to Ampere's law in the next chapter, we will confine our study of the magnetic force law to the simplest of all possible magnetic fields, the uniform magnetic field.

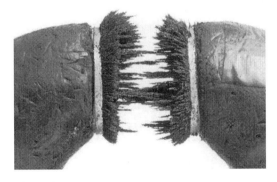

Figure 18
Between the poles of this magnet there is a relatively uniform magnetic field.

Working with uniform magnetic fields, fields that are constant in both magnitude and direction, is so convenient that physicists and engineers go to great lengths to construct them. One place to find a uniform field is between the flat pole pieces of a magnet, as seen in Figure (18) which is our "magnet plant" of Figure (13) with fewer iron filings.

If we bend a wire in a loop, then a current around the loop produces the fairly complex field pattern shown in Figure (19). When we use two loops as seen in Figure (20), the field becomes more complicated in some places but begins to be more uniform in the central region between the coils. With many loops, with the coil of wire shown in Figure (21a), we get a nearly uniform field inside. Such a coil is called a solenoid, and will be studied extensively in the next chapter. An iron filing map of the field of a large diameter, tightly wound solenoid is seen in Figure (21b).

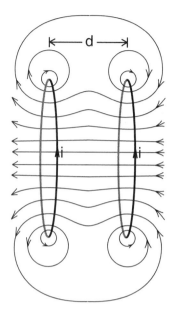

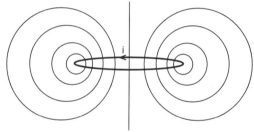

Figure 19
The magnetic field of a current loop is fairly complex.

Figure 20
The magnetic field in the region between a pair of coils is relatively uniform. We can achieve the greatest uniformity by making the separation d between the coils equal to the radius of the coils. Such a setup is called a pair of **Helmholtz coils.**

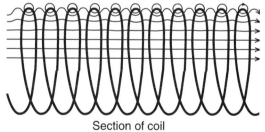

Section of coil

Figure 21a
Magnetic field in the upper half of a section of a coil of wire. When you have many closely spaced coils, the field inside can become quite uniform through most of the length of the coil.

Figure 21b
Iron filing map of the magnetic field of a large diameter coil. (Student project, Alexandra Lesk and Kirsten Teany.)

Helmholtz Coils

For now we will confine our attention to the reasonably uniform field in the central region between two coils seen in Figure (20). Helmholtz discovered that when the coils are spaced a distance d apart equal to the coil radius r (Figure 22), we get a maximally uniform field $\vec{B}$ between the coils. This arrangement, which is called a pair of *Helmholtz coils*, is commonly used in physics and engineering apparatus. Figure (23) shows a pair of Helmholtz coils we use in our undergraduate physics labs and which will be used for several of the experiments discussed later. An iron filing map of the field produced by these coils is seen in Figure (24a), and one of the experiments will give us a field plot similar to Figure (24b).

In our derivation of the magnetic field of a current in a straight wire, we saw that *the strength of the magnetic field was proportional to the current i in the wire*. This is true even if the wire is bent to form coils, or even twisted into a complex tangle. That means that once you have mapped the magnetic field for a given current (i) in a set of wires, doubling the current produces the same shape map with twice as strong a field.

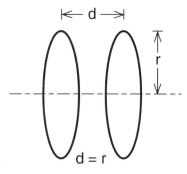

Figure 22
For Helmholtz coils, the separation d equals the coil radius r.

For the Helmholtz coils in Figure (23), it was observed that when a current of one amp flowed through the coils, the strength of the magnetic field in the central regions was 8 gauss. A current of 2 amps produced a 16 gauss field. Thus the field strength, **for these coils,** is related to the current i by

$$B(\text{gauss}) = 8i(\text{amps}) \qquad \begin{array}{l} \textit{for the Helmholtz} \\ \textit{coils of} \\ \textit{Figure (23) only} \end{array}$$

In the lab we measure the strength of B simply by reading (i) from an ampmeter and multiplying by 8. Of course, if you are using a different set of coils,(i) will be multiplied by a different number. *(Do not worry about the mixed units, remember that we convert gauss to tesla before doing MKS calculations.)*

One can derive a formula for an idealized set of Helmholtz coils. The derivation is complicated and the answer $B = 8\mu_0 N i/(5\sqrt{5}\ r)$ is rather a mess. (N is the number of turns in each coil.) The simple feature which we expected, is that B is proportional to the strength of the current (i) in the coils. Because another law, called Faraday's law, can be used to give us a more accurate calibration of real Helmholtz coils, we will leave the derivation of the Helmholtz formula above to other texts.

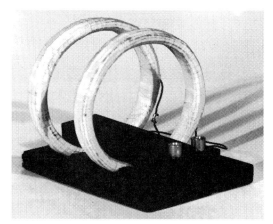

Figure 23
Helmholtz coils used in a number of lab experiments discussed in the text. Each coil consists of 60 turns of fairly heavy magnet wire.

Figure 24a
Iron filing map of the magnetic field of the Helmholtz coils. (Student project, Alexandra Lesk and Kirsten Teany.)

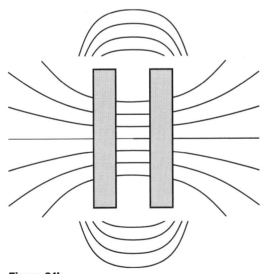

Figure 24b
Plot from a student experiment, of the magnetic field in the region between and around the coils.

MOTION OF CHARGED PARTICLES IN MAGNETIC FIELDS

In physics, one of the primary uses of magnetic fields is to control the motion of charged particles. When the magnetic field is uniform, the motion is particularly simple and has many practical applications from particle accelerators to mass spectrometers. Here we will discuss this motion and several of the applications.

The main feature of the magnetic force law,

$$\vec{F}_{magnetic} = q\vec{v} \times \vec{B} \tag{19}$$

is that because of the cross product $\vec{v} \times \vec{B}$, the magnetic force is *always* perpendicular to the velocity $\vec{v}$ of the charged particle. This has one important immediate consequence. *Magnetic forces do no work!* The formula for the power, i.e., the work done per second, is

$$\left. \begin{array}{l} \text{Work done} \\ \text{per second} \\ \text{by a force } \vec{F} \end{array} \right\} = \text{power} = \vec{F} \cdot \vec{v} \tag{22}$$

Since the magnetic force $\vec{F}_{mag}$ is always perpendicular to $\vec{v}$, we have

$$\left. \begin{array}{l} \text{Work done by a} \\ \text{magnetic force} \end{array} \right\} = \vec{F}_{magnetic} \cdot \vec{v}$$

$$= q\left(\vec{v} \times \vec{B}\right) \cdot \vec{v} \equiv 0 \tag{23}$$

magnetic fields do not change the energy of a particle, they simply change the direction of motion.

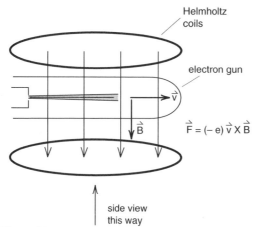

Helmholtz
coils

electron gun

$\vec{v}$

$\vec{B}$

$\vec{F} = (-e)\,\vec{v} \times \vec{B}$

side view
this way

Figure 25a
Top view looking down on the electron gun placed between the Helmholtz coils. The electrons in the beam move perpendicular to the magnetic field $\vec{B}$. The magnetic force $\vec{F} = (-e)\,\vec{v} \times \vec{B}$ is directed up, out of the paper in this drawing.

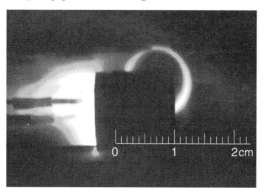

Figure 25b
Side view of the electron beam, as seen through the lower coil in Figure 25a. In this view the magnetic field is directed out of the paper toward the reader.

Figure 25d
Movie of the experiment.

Motion in a Uniform Magnetic Field

When we have a charged particle moving through a uniform magnetic field, we get a particularly simple kind of motion—the circular motion seen in Figure (25b). In Figure (25a) we sketched the experimental setup where an electron gun is placed between a pair of Helmholtz coils so that the magnetic field $\vec{B}$ is perpendicular to the electron beam as shown. Figure (25b) is a photograph of the electron beam deflected into a circular path. In Figure (25c) we have a sketch of the forces on an electron in the beam. The magnetic field B in this diagram is up out of the paper, thus $\vec{v} \times \vec{B}$ points radially out from the circle. But the electron has a negative charge, thus the magnetic force $\vec{F}_B$

$$\vec{F}_B = (-e)\,\vec{v} \times \vec{B}$$

points in toward the center of the circle as shown.

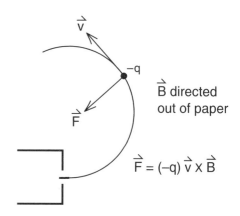

$\vec{v}$

$-q$

$\vec{B}$ directed
out of paper

$\vec{F}$

$\vec{F} = (-q)\,\vec{v} \times \vec{B}$

Figure 25c
As the electrons move along a curved path, the magnetic force $\vec{F} = -q\,\vec{v} \times \vec{B}$ always remains perpendicular to the velocity and therefore cannot change the speed of the electrons. The resulting motion is uniform circular motion where the force and the acceleration are directed toward the center of the circle.

To apply Newton's second law to the electrons in Figure (25), we note that a particle moving in a circle accelerates toward the center of the circle, the same direction as $\vec{F}_{mag}$ in Figure (25c). Thus $\vec{F}_B$ and $m\vec{a}$ are in the same direction and we can use the fact that for circular motion $a = v^2/r$ to get $\left|\vec{F}_B\right| = m\left|\vec{a}\right|$ or

$$qvB = mv^2/r \tag{24}$$

Solving for r, we predict from Equation (24) that the electron beam will be bent into a circle of radius r given by

$$\boxed{r = \frac{mv}{qB}} \tag{25}$$

Equation (25) is an important result that we will use often. But it is so easy to derive, and it is such good practice to derive it, that it may be a good idea not to memorize it.

Let us use the experimental numbers provided with Figure (25b) as an example of the use of Equation (25). In that figure, the strength of the magnetic field is B = 70 gauss, and the electrons were accelerated by an accelerating voltage of 135 volts. The constants m and q are the mass and charge of an electron.

The first step is to calculate the speed v of the electrons using the fact that the electrons have 135 eV of kinetic energy. We begin by converting from eV to joules using the conversion factor 1.6×10^{-19} joules per eV. This gives

$$\frac{1}{2}mv^2 = 135\,eV \times 1.6 \times 10^{-19} \frac{joules}{eV}$$

$$v^2 = \frac{2 \times 135 \times 1.6 \times 10^{-19}}{.911 \times 10^{-30}}$$

$$= 47.4 \times 10^{12} \frac{m^2}{s^2}$$

$$\boxed{v = 6.9 \times 10^6 \frac{m}{s}}$$

(26)

Next convert B from gauss to the MKS tesla

$$70\,gauss = 70 \times 10^{-4}\,tesla \tag{27}$$

Substituting Equations (26) and (27) in (25) gives

$$r = \frac{mv}{qB} = \frac{.911 \times 10^{-30} \times 6.9 \times 10^6}{1.6 \times 10^{-19} \times 7 \times 10^{-3}}$$

$$r = 5.6 \times 10^{-3}\,m = .56\,cm \tag{28}$$

Exercise 4

The scale of distance shown in Figure (25b) was drawn knowing the dimensions of the cap in the electron gun. Use this scale to estimate the radius of curvature of the electron beam and compare the result with the prediction of Equation (28).

Exercise 5

Use the experimental results shown in Figure (26) (B = 70 gauss) to estimate the accelerating voltage used for the electrons in the beam. (The experimental answer is included in the homework answer section.)

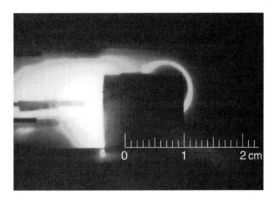

Figure 26
Use the fact that the magnetic field for this example was 70 gauss to estimate the accelerating voltage that produced the electron beam.

Particle Accelerators

Our knowledge of the structure of matter on a sub-atomic scale, where we study the various kinds of elementary particles, has come from our ability to accelerate particles to high energies in particle accelerators such as the synchrotron. In a synchrotron, an electric field $\vec{E}$ is used to give the particle's energy, and a magnetic field $\vec{B}$ is used to keep the particles confined to a circular track.

Figure (27a) is a schematic diagram of a small electron synchrotron. At the top is an electron gun that is used to produce a beam of electrons. In practice the gun is quickly turned on then off to produce a pulse of electrons.

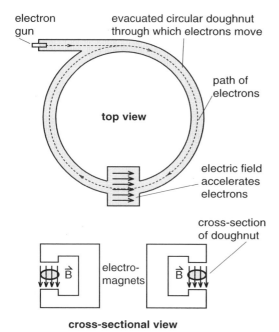

electron
gun

evacuated circular doughnut
through which electrons move

path of
electrons

top view

electric field
accelerates
electrons

cross-section
of doughnut

$\vec{B}$ electro- $\vec{B}$
magnets

cross-sectional view

Figure 27a
Diagram of a synchrotron, in which the electrons, produced by the electron gun, travel through a circular evacuated doughnut. The electrons are accelerated by an electric field, gaining energy on each trip around. The electrons are kept in a circular orbit by an increasingly strong magnetic field produced by the electromagnets.

The pulse of electrons enter an evacuated circular track shown in the top view. To keep the pulse of electrons moving in the circular track, large electromagnets shown in the cross-sectional view are used to provide a perpendicular magnetic field. In this example the magnetic field $\vec{B}$ points downward so that the magnetic force $q\vec{v} \times \vec{B} = -e\vec{v} \times \vec{B}$ points inward toward the center of the track.

We saw that a magnetic field cannot do any work on the electrons since the magnetic force is perpendicular to the particle's velocity. Therefore to give the electrons more energy, we use an electric field $\vec{E}$. This is done by inserting into one section of the path a device that produces an electric field so that the electric force $-e\vec{E}$ points in the direction of the motion of the electrons.

One might think of using a charged parallel plate capacitor to create the electric field $\vec{E}$, but that is not feasible. Later we will see that radio waves have an electric field $\vec{E}$ associated with them, and it is a radio wave electric field in a so-called "resonant cavity" that is used to produce the required strong fields. For now it does not matter how $\vec{E}$ is produced, it is this electric field that adds energy to the electrons.

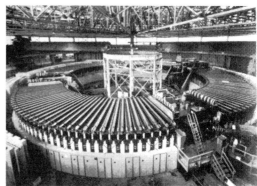

Figure 27b
The Berkeley synchrotron shown here, accelerated protons rather than electrons. It was the first machine with enough energy to create anti protons. After this machine was built, ways were devised for focusing the particle beam and using an evacuated doughnut with a much smaller cross-sectional area.

When electrons gain energy, their momentum $p = mv$ increases. Writing Equation (25) in the form

$$r = \frac{mv}{qB} = \frac{p}{qB} \tag{25a}$$

we see that an increase in the electron's momentum p will cause the orbital radius r to increase. The radius r will increase unless we compensate by increasing the strength B of the magnetic field. The rate at which we increase B must be synchronized with the rate at which we increase the particle's momentum p in order to keep r constant and keep the electrons in the circular path. Because of this synchronization, the device is called a *synchrotron*.

You can see that the amount of energy or momentum we can supply to the particles is limited by how strong a field B we can make. Iron electromagnets can create fields up to about 1 tesla (here the MKS unit is useful) or 10,000 gauss. The superconducting magnets, being used in the latest accelerator designs, can go up to around 5 tesla.

Noting that B is limited to one or a few tesla, Equation (25) tells us that to get more momentum or energy, we must use accelerators with a bigger radius r. This explains why particle accelerators are getting bigger and bigger. The biggest particle accelerator now operating in the United States is the proton accelerator at the Fermi National Accelerator Laboratory in Batavia, Illinois shown in Figures (28) and (29).

In Figure (28), we see a section of tunnel and the magnets that surround the 2 inch diameter evacuated pipe which carries the protons. Originally there was one ring using iron magnets (painted red and blue in the photograph). Later another ring with superconducting magnets was installed, in order to obtain stronger magnetic fields and higher proton energies. The ring of superconducting magnets (painted yellow) is beneath the ring of iron magnets.

Figure (29) is an aerial view showing the 4 mile circumference of the accelerator. Currently the largest accelerator in the world is at the European Center for Particle Physics (CERN). The 27 kilometer path of that accelerator is seen in Figure (30) on the next page.

Figure 28
The Fermi Lab accelerator has two accelerating rings, one on top of the other. In each, the evacuated doughnut is only 2 inches in diameter, and four miles in circumference. The bottom ring uses superconducting magnets (painted yellow), while the older upper ring has iron magnets (painted red and blue.)

Figure 29
Aerial view of the Fermi Lab particle accelerator.

RELATIVISTIC ENERGY AND MOMENTA

Even the smallest synchrotrons accelerate electrons and protons up to relativistic energies where we can no longer use the non relativistic formula $1/2 \, mv^2$ for kinetic energy. For any calculations involving the large accelerators we must use fully relativistic calculations like $E = mc^2$ for energy and $p = mv$ for momentum where $m = m_0/\sqrt{1 - v^2/c^2}$ is the relativistic mass.

Equation (25) or (25a) for a charged particle moving in a circular orbit of radius r, can be written in the form

$$p = qBr \qquad (25b)$$

where B is the strength of the uniform magnetic field and p the particle momentum. It turns out that Equation (25) is correct even at relativistic energies provided $p = mv$ is the relativistic momentum. Thus a knowl-

edge of the magnetic field and orbital radius immediately tells us the momentum of the particles in the large synchrotrons.

To determine the energy of the particles in these machines, we need a relationship between a particle's energy E and momentum p. The relationship can be obtained by writing out E and p in the forms

$$p = mv = \frac{m_0}{\sqrt{1 - v^2/c^2}} \, v \qquad (29)$$

$$E = mc^2 = \frac{m_0}{\sqrt{1 - v^2/c^2}} \, c^2 \qquad (30)$$

It is then straightforward algebraic substitution to show that

$$E^2 = p^2c^2 + m_0^2c^4 \qquad \text{An exact relationship} \qquad (31)$$

Figure 30
Path for the 8 kilometer circumference Super Proton Synchrotron (SPS, solid circle) and the 27 kilometer Large Electron-Positron collider (LEP, dashed circle) at CERN, on the border between France and Switzerland. The Geneva airport is in the foreground.

Exercise 6

Directly check Equation (31) by plugging in the values of p and E from Equations (29) and (30).

In the big particle accelerators the kinetic energy supplied by the accelerators greatly exceeds the particle's rest energy $m_0 c^2$, so that the $(m_0 c^2)^2$ term in Equation (31) is completely negligible. For these "highly relativistic" particles, we can drop the $m_0^2 c^4$ term in Equation (31) and we get the much simpler formula

$$\boxed{E \approx pc} \qquad \text{If } E \gg m_0 c^2 \qquad (32)$$

Equation (32) is an accurate relationship between energy and momentum for any particle moving at a speed so close to the speed of light that its total energy E greatly exceeds its rest energy $m_0 c^2$.

For the high energy particle accelerators we can combine Equations (25) and (32) to get

$$E = pc = qBrc \qquad (33)$$

Consider CERN's Super Proton Synchrotron or SPS, shown by the smaller solid circle in Figure (30), which was used to discover the particles responsible for the weak interaction. In this accelerator, the magnets produced fields of B = 1.1 tesla, and the radius of the ring was r = 1.3km (for a circumference of 8km). Thus we have

$$E = qBrc$$

$$= \left(1.6 \times 10^{-19} \text{ coulombs}\right) \times \left(1.1 \text{ tesla}\right)$$

$$\times \left(1.3 \times 10^3 \text{m}\right) \times \left(3 \times 10^8 \text{ m}/_{\text{s}}\right)$$

$$= 6.9 \times 10^{-8} \text{ joules}$$

Converting this answer to electron volts, we get

$$E = \frac{6.9 \times 10^{-8} \text{ joules}}{1.6 \times 10^{-19} \dfrac{\text{joules}}{\text{eV}}} = 430 \times 10^9 \text{ eV}$$

$$= 430 \text{GeV} \qquad (34)$$

How good was our approximation that we could neglect the particle's rest energy and use the simple Equation (32)? Recall that the rest energy of a proton is about 1GeV. Thus the SPS accelerator produced protons with a kinetic energy 430 times greater! For these particles it is not much of an error to neglect the rest energy.

Exercise 7

a) The Fermi lab accelerator, with its radius of 1 kilometer, uses superconducting magnets to produce beams of protons with a kinetic energy of 1000 GeV $\left(10^{12}\text{eV}\right)$. How strong a magnetic field is required to produce protons of this energy?

b) Iron electromagnets cannot produce magnetic fields stronger than 2 tesla, which is why superconducting magnets were required to produce the 1000 GeV protons discussed in part a). Before the ring with superconducting magnets was constructed, a ring using iron magnets already existed in the same tunnel. The iron magnets could produce 1.5 tesla fields. What was the maximum energy to which protons could be accelerated before the superconducting magnets were installed? (You can see both rings of magnets in Figure 28.)

Exercise 8

The large electron-positron (LEP) collider, being constructed at CERN, will create head on collisions between electrons and positrons. (Electrons will go around one way, and positrons, having the opposite charge, will go around the other.) The path of the LEP accelerator, which will have a circumference of 27 km, is shown in Figure (30), superimposed on the countryside north of Geneva, Switzerland.

a) Assuming that the LEP accelerator will use 3 tesla superconducting magnets, what will be the maximum kinetic energy, in eV, of the electrons and positrons that will be accelerated by this machine?

b) What will be the speed of these electrons and positrons? (How many 9's in v/c?)

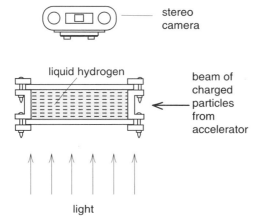

stereo
camera

liquid hydrogen

beam of
charged
particles
from
accelerator

light

Figure 31a
*Schematic diagram of the Berkeley
10-inch hydrogen bubble chamber.*

Figure 31b
*The 10-inch Bubble chamber at the Lawrence
Radiation Laboratory, University of California,
Berkeley. (Photograph copyright The Ealing
Corporation, Cambridge, Mass.)*

BUBBLE CHAMBERS

In the study of elementary particles, it is just as important to have adequate means of observing particles as it is to have accelerating machines to produce them. One of the more useful devices for this purpose is the bubble chamber invented by Donald Glaser in 1954.

It may not be true that Glaser invented the bubble chamber while looking at the streaks of bubbles in a glass of beer. But the idea is not too far off. When a charged particle like an electron, proton or some exotic elementary particle, passes through a container of liquid hydrogen, the charged particle tends to tear electrons from the hydrogen atoms that it passes, leaving a trail of ionized hydrogen atoms. If the pressure of the liquid hydrogen is suddenly reduced the liquid will start boiling if it has a "seed"—a special location where the boiling can start. The trail of ionized hydrogen atoms left by the charged particle provides a trail of seeds for boiling. The result is a line of bubbles showing where the particle went.

In a typical bubble chamber, a stereoscopic camera is used to record the three dimensional paths of the particles. It is impressive to look at the three dimensional paths in stereoscopic viewers, but unfortunately all we can conveniently do in a book is show a flat two dimensional image like the one in Figure (32). In that picture we see the paths of some of the now more common exotic elementary particles. In the interesting part of this photograph, sketched above, a negative π^- meson collides with a positive proton to create a neutral Λ^0 and a neutral K^0 meson. The neutral Λ^0 and K^0 do not leave tracks, but they are detected by the fact that the K^0 decayed into a π^+ and a π^- meson, and the Λ^0 decayed into a π^- and a proton p^+, all of which are charged particles that left tracks.

To analyze a picture like Figure (32) you need more information than just the tracks left behind by particles. You would also like to know the charge and the momentum or energy of the particles. This is done by placing the bubble chamber in a magnetic field so that positive particle tracks are curved one way and negatives ones the other. And, from Equation (25b), we see that the radii of the tracks tell us the momenta of the particles.

Another example of a bubble chamber photograph is Figure (33) where we see the spiral path produced by an electron. The fact that the path is spiral, that the radius of the path is getting smaller, immediately tells us that the electron is losing momentum and therefore energy as it moves through the liquid hydrogen. The magnetic field used for this photograph had a strength B = 1.17 tesla, and the initial radius of the spiral was 7.3 cm. From this we can determine the momentum and energy of the electron.

Exercise 9

Calculate the energy, in eV of the electron as it entered the photograph in Figure (33). Since you do not know off hand whether the particle was relativistic or not, use the exact relation

$$E^2 = p^2c^2 + m_0^2c^4 \qquad (31)$$

to determine E from p. From your answer decide whether you could have used the non relativistic formula $KE = 1/2\,m_0v^2$ or the fully relativistic formula E = pc, or whether you were in an intermediate range where neither approximation works well.

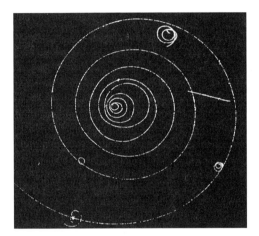

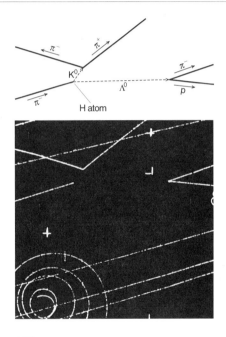

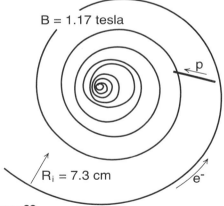

B = 1.17 tesla

p

R$_i$ = 7.3 cm

e⁻

Figure 33
Spiraling electron. An electron enters the chamber at the lower left and spirals to rest as it loses momentum. The spiral track is caused by the magnetic field applied to the chamber which deflects a charged particle into a curved path with a radius of curvature proportional to the particle's momentum. The straight track crossing the spiral is a proton recoiling from a collision with a stray neutron. Because the proton has much greater mass than the electron, its track is much less curved.

Figure 32
Bubble chamber photograph showing the creation of a K^0 meson and a Λ^0 particle, and their subsequent annihilations. We now know that the K meson is a quark/anti quark pair, and the Λ^0 particle contains 3 quarks as does a proton and a neutron. The K and Λ particles last long enough to be seen in a bubble chamber photograph because they each contain a strange quark which decays slowly via the weak interaction. (Photo copyright The Ealing Corporation Cambridge, Mass.)

The Mass Spectrometer

A device commonly seen in chemistry and geology labs is the *mass spectrometer* which is based on the circular orbits that a charged particle follows in a uniform magnetic field. Figure (34) is a sketch of a mass spectrometer which consists of a semi circular evacuated chamber with a uniform magnetic field $\vec{B}$ directed up out of the paper. The direction of $\vec{B}$ is chosen to deflect positive ions around inside the chamber to a photographic plate on the right side. The ions to be studied are boiled off a heated filament and accelerated by a negative cap in a reversed voltage electron gun shown in Figure (35). By measuring the position where the ions strike the photographic plate, we know the radius of the orbit taken by the ion. Combine this with the knowledge of the field B of the spectrometer, and we can determine the ion's momentum p if the charge q is known. The speed of the ion is determined by the accelerating voltage in the gun, thus knowing p gives us the mass m of the ions. Non relativistic formulas work well and the calculations are nearly identical to our analysis of the path of the electrons in Figure (25). (See Equations 26 to 28.)

Mass spectrometers are used to identify elements in a small sample of material, and are particularly useful in being able to separate different isotopes of an element. Two different isotopes of an element have different numbers of neutrons in the nucleus, everything else being the same. Thus ions of the two isotopes will have slightly different masses, and land at slightly different distances down the photographic plate. If an isotope is missing in one sample the corresponding line on the photographic plate will be absent. The analogy between looking at the lines identifying isotopes, and looking at a photographic plate showing the spectrum of light, suggested the name *mass spectrometer*.

Exercise 10

Suppose that you wish to measure the mass of an iodine atom using the apparatus of Figures (34) and (35). You coat the filament of the gun in Figure (35) with iodine, and heat the filament until iodine atoms start to boil off. In the process, some of the iodine atoms lose an electron and become positive ions with a charge +e. The ions are then accelerated in the gun by a battery of voltage V_b and then pass into the evacuated chamber.

(a) Assuming that V_b = 125 volts (accelerating voltage) and B = 1000 gauss (0.1 tesla), and that the iodine atoms follow a path of radius r = 18.2 cm, calculate the mass m of the iodine atoms.

(b) How many times more massive is the iodine ion than a proton? From the fact that protons and neutrons have about the same mass, and that an electron is 2000 times lighter, use your result to estimate how many nuclear particles (protons or neutrons) are in an iodine nucleus.

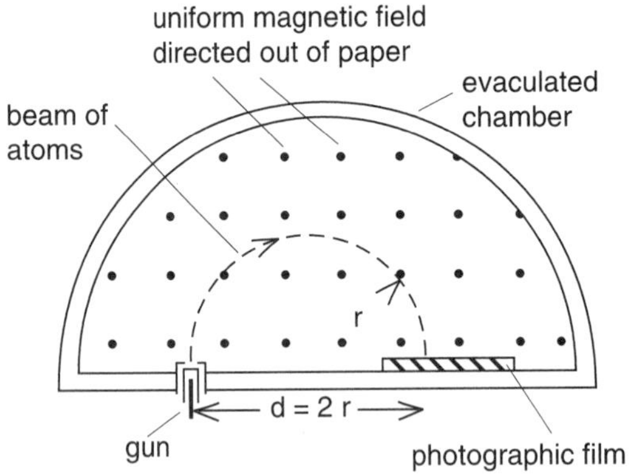

---

uniform magnetic field
directed out of paper

beam of
atoms

evaculated
chamber

r

d = 2 r

gun

photographic film

Figure 34
Top view of a mass spectrograph. A uniform magnetic field $\vec{B}$ rises directly up through the chamber. The beam of atoms is produced by the accelerating gun shown in Figure 35.

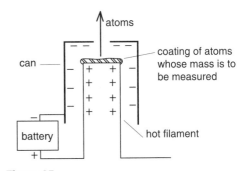

atoms

can

coating of atoms
whose mass is to
be measured

battery

hot filament

Figure 35
When the substance to be studied is heated by a filament, atoms evaporate and some lose an electron and become electrically charged positive ions. The ions are then accelerated by an electric field to produce a beam of ions of known kinetic energy.

Magnetic Focusing

In the magnetic force examples we have considered so far, the velocity $\vec{v}$ of the charged particle started out perpendicular to $\vec{B}$ and we got the circular orbits we have been discussing.

If we place an electron gun so that the electron beam is aimed down the axis of a pair of Helmholtz coils, as shown in Figure (36), the electron velocity $\vec{v}$ is parallel to $\vec{B}$, $\vec{v} \times \vec{B} = 0$ and there is no magnetic force.

Figure (36) is a bit too idealized for the student built electron gun we have been using in earlier examples. Some of the electrons do come out straight as shown in Figure (36), but many come out at an angle as shown in Figure (37a). In Figure (37b) we look at the velocity components $\vec{v}_\perp$ and $\vec{v}_\parallel$ of an electron emerging at an angle q. Because $\vec{v}_\parallel \times \vec{B} = 0$ only the perpendicular component $\vec{v}_\perp$ contributes to the magnetic force

$$\vec{F}_{mag} = q\vec{v}_\perp \times \vec{B} \tag{36}$$

This force is perpendicular to both $\vec{v}_\perp$ and $\vec{B}$ as shown in the end view of the electron gun, Figure (37b). In this end view, where we can't see $\vec{v}_\parallel$, the electron appears to travel around the usual circular path.

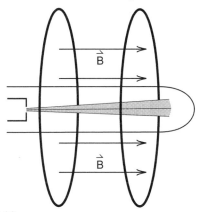

Figure 36
Electron gun inserted so that the beam of electrons moves parallel to the magnetic field of the coils. If the beam is truly parallel to $\vec{B}$, there will be no magnetic force on the electrons.

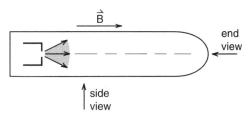

Figure 37a
In reality the electron beam spreads out when it leaves the cap. Most of the electrons are not moving parallel to $\vec{B}$, and there will be a magnetic force on them.

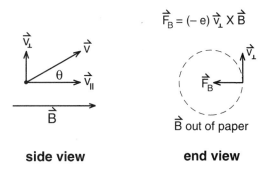

side view **end view**

Figure 37b
Consider an electron emerging from the cap at an angle θ from the center line as shown in the side view above. Such an electron has a component of velocity $\vec{v}_\perp$ perpendicular to the magnetic field. This produces a magnetic force $\vec{F}_B = (-e)\vec{v}_\perp \times \vec{B}$ which points toward the axis of the gun. The magnetic force $\vec{F}_B$ can be seen in the end view above. From the end view the electron will appear to travel in a circle about the axis of the gun. The stronger the magnetic field, the smaller the radius of the circle.

It is in the side view, Figure (38) that we see the effects of $\vec{v}_\parallel$. Since there is no force related to $\vec{v}_\parallel$, this component of velocity is unchanged and simply carries the electron at a constant horizontal speed down the electron gun. The quantity $\vec{v}_\parallel$ is often called the drift speed of the particle. (The situation is not unlike projectile motion, where the horizontal component v_x of the projectile's velocity is unaffected by the vertical acceleration a_y.)

When we combine the circular motion, seen in the end view of Figure (37c), with the constant drift speed $\vec{v}_\parallel$, down the tube seen in Figure (38a) the net effect is a helical path like a stretched spring seen in Figure (38b). The electron in effect spirals around and travels along the magnetic field line. The stronger the magnetic field, the smaller the circle in Figure (37c), and the tighter the helix.

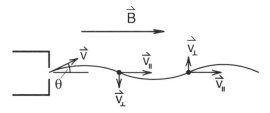

Figure 38a
In the side view of the motion of the electron, we see that $v_\parallel$ is unchanged, $\vec{v}_\parallel$ just carries the electron down the tube.

helical motion
of the electron

Figure 38b
Oblique view of the helical motion of the electron. When you combine the uniform motion of the electron down the tube with the circular motion around the axis of the tube, you get a helical motion with the same shape as the wire in a stretched spring.

The tightening of the helix is seen in Figure (39) where in (a) we see an electron beam with no magnetic field. The electrons are spraying out in a fairly wide cone. In (b) we have a 75 gauss magnetic field aligned parallel to the axis of the gun and we are beginning to see the helical motion of the electrons. In (c) the magnetic field is increased to 200 gauss and the radius of the helix has decreased considerably. As B is increased, the electrons are confined more and more closely to a path along the magnetic field lines. In our electron gun, the magnetic field is having the effect of focusing the electron beam.

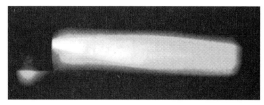

a) No magnetic field

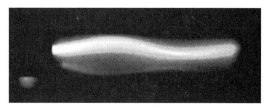

b) B = 75 gauss

c) B = 200 gauss

d) Movie

Figure 39
Focusing an electron beam with a parallel magnetic field. The beam travels along a helical path which becomes tighter as the strength of the magnetic field is increased.

SPACE PHYSICS

Even in non-uniform magnetic fields there is a tendency for a charged particle to move in a spiral path along a magnetic field line as illustrated in Figure (40). This is true as long as the magnetic field is reasonably uniform over a distance equal to the radius r of the spiral (from Equation (25), $r = mv_\perp/q\vec{v}_B$). Neglecting the spiral part of the motion, we see that the large scale effect is that charged particles tend to move or flow along magnetic field lines. This plays an important role in space physics phenomena which deals with charged particles emitted by the sun (the "solar wind") and the interaction of these particles with the magnetic field of the earth and other planets.

There are so many interesting and complex effects in the interaction of the solar wind with planetary magnetic fields that space physics has become an entire field of physics. Seldom are we aware of these effects unless a particularly powerful burst of solar wind particles disrupts radio communications or causes an Aurora Borealis to be seen as far south as the temperate latitudes. The Aurora are caused when particles from the solar wind spiral in along the earth's magnetic field lines and end up striking atoms in the upper atmosphere. The atoms struck by the solar wind particles emit light just like the residual air atoms struck by the electrons in an electron gun.

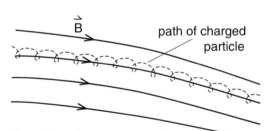

Figure 40
When charged particles from the sun enter the earth's magnetic field, they spiral around the magnetic field lines much like the electrons in the magnetic focusing experiment of Figure 39.

The Magnetic Bottle

If a magnetic field has the correct shape, if the field lines pinch together as shown at the left or the right side of Figure (41), then the magnetic force $\vec{F}_{mag}$ on a charged particle has a component that is directed back from the pinch. For charged particles with the correct speed, this back component of the magnetic force can reflect the particle and reverse $v_\parallel$. If the magnetic field is pinched at both ends, as in Figure (41) the charged particle can reflect back and forth, trapped as if it were in a magnetic bottle.

In the subject of plasma physics, one often deals with hot ionized gasses, particularly in experiments designed to study the possibility of creating controlled fusion reactions. These gases are so hot that they would melt and vaporize any known substance they touch. The only known way to confine these gases to do experiments on them is either do the experiments so fast that the gas does not have time to escape (inertial confinement), or use magnetic fields and devices like the magnetic bottle shown in Figure (41) (magnetic confinement).

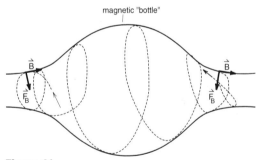

Figure 41
Magnetic bottle. When the magnetic field lines pinch together, the charged particles can be reflected back in a process called magnetic mirroring. (At the two ends of the magnetic bottle above, the magnetic force $\vec{F}_B$ has a component back into the bottle.)

Van Allen Radiation Belts

The earth's magnetic field shown in Figure (9) and repeated in Figure (42) forms magnetic bottles that can trap charged particles from the solar wind. The ends of the bottles are where the field lines come together at the north and south magnetic poles, and the regions where significant numbers of particles are trapped are called the Van Allen radiation belts shown in Figure (42). Protons are trapped in the inner belt and electrons in the outer one.

It is not feasible to do hand calculations of the motion of charged particles in non-uniform magnetic fields. The motion is just too complicated. But computer calculations, very similar to the orbit calculations discussed in Chapter 4, work well for electric and magnetic forces. As long as we have a formula for the shape of $\vec{E}$ or $\vec{B}$, we can use the Lorentz force law (Equation 20)

$$\vec{F} = q\vec{E} + q\vec{v} \times \vec{B}$$

as one of the steps in the computer program. The computer does not care how complicated the path is, but we might have trouble drawing and interpreting the results.

In Figure (43), a student, Jeff Lelek, started with the formula for a "dipole magnetic field", namely

$$\vec{B} = -\left(\frac{B_0}{R^3}\right) * \left(\hat{Z} - 3*(\hat{Z}\cdot\hat{R})*\hat{R}\right) \qquad (37)$$

which is a reasonably accurate representation of the earth's magnetic field, and calculated some electron orbits for this field. The result is fairly complex, but we do get the feeling that the electron is spiraling around the magnetic field lines and reflecting near the magnetic poles.

To provide a simpler interpretation of this motion, the student let the calculation run for a long time, saving up the particle coordinates at many hundreds of different points along the long orbit. These points are then plotted as the dot pattern shown in Figure (43). (In this picture, the latitude of the particle is ignored, the points are all plotted in one plane so we can see the extent of the radial and north-south motion of the particles.) The result gives us a good picture of the distribution of particles in a Van Allen radiation belt. This and similar calculations are discussed in the supplement on computer calculations with the Lorentz force law.

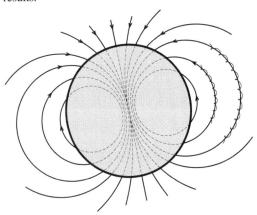

Figure 42
Charged particles, trapped by the earth's magnetic field, spiral around the magnetic field lines reflecting where the lines pinch together at the poles. The earth's magnetic field thus forms a magnetic bottle, holding the charged particles of the Van Allen radiation belts.

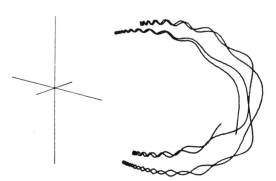

Figure 43
Computer plot of the motion of a proton in a dipole magnetic field. The formula for this field and the computer program used to calculate the motion of the proton are given in the Appendix. As you can see, the motion is relatively complex. Not only does the proton reflect back and forth between the poles, but also precesses around the equator.

Auroras

Most of the electrons and protons that come to the earth from the sun do not become captured in the radiation belts. Instead they follow magnetic field lines down into the earth's atmosphere. There they excite the atoms in the atmosphere, just as the electrons in our electron gun excited air atoms in the gun, illuminating the path of the electrons. The glow created by the electrons and protons coming down through the earth's atmosphere creates an *aurora*, or what is often known as the *northern lights*. A particularly impressive aurora occurred on March 20, 2001 in the Anchorage Alaska area, which allowed Leroy Zimmerman to take the rather amazing photograph shown below.

Aurora March 20, 2001 Fairbanks Alaska
Photograph by Leroy Zimmerman at http://www.photosymphony.com/.
More aurora pictures can be found at
http://spaceweather.com/aurora/gallery_20march01.html

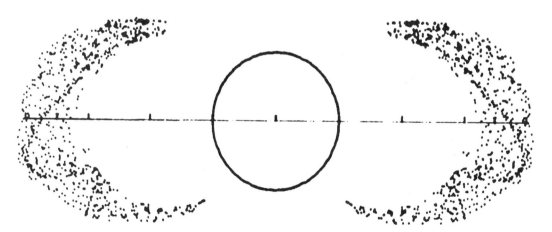

Figure 44
In this computer plot, all the data points from Figure 43 are plotted as dots in one plane. From this we see the shape of a Van Allen radiation belt emerge. (Figures (43) and (44) from a student project by Jeff Lelek.)

Chapter 29
Ampere's Law

In this chapter our main focus will be on Ampere's law, a general theorem that allows us to calculate the magnetic fields of simple current distributions in much the same way that Gauss' law allowed us to calculate the electric field of simple charge distributions.

As we use them, Gauss' and Ampere's laws are integral theorems. With Gauss' law we related the total flux out through a closed surface to $1/\varepsilon_0$ times the net charge inside the surface. In general, to calculate the total flux through a surface we have to perform what is called a **surface integral**. *Ampere's law will relate the integral of the magnetic field around a closed path to the total current flowing through that path. This integral around a closed path is called a* **line integral**.

Until now we have concentrated on examples that did not require us to say much about integration. But as we discuss Ampere's law in this chapter and the remaining Maxwell equations in the next few chapters, it will be convenient to draw upon the formalism of the surface and line integral. Therefore we will take a short break to discuss the mathematical concepts involved in these integrals.

THE SURFACE INTEGRAL

In our discussion of Gauss' law near the end of Chapter 24, we defined the flux Φ of a fluid in a flow tube as the *amount of water per second flowing past the cross-sectional area of the tube* as shown in Figure (1). This is equal to the velocity v times the cross-sectional area $A_\perp$ of the tube as given in Equation (24-46)

$$\Phi = vA_\perp \qquad \textit{flux in a flow tube} \qquad (24\text{-}46)$$

As seen in Figure (2), if we slice the flow tube by a plane that is not normal to the flow tube, the area A of the intersection of the tube and plane is larger than the cross-sectional area $A_\perp$. The relationship is $A_\perp = A \cos \theta$ where θ is the angle between $\vec{v}$ and $\vec{A}$ (see Equation 24-45). Defining the vector $\vec{A}$ as having a magnitude A and direction normal to the plane, we have

$$\vec{v} \cdot \vec{A} = vA \cos \theta = vA_\perp$$

and the formula for the flux in the flow tube is

$$\Phi = \vec{v} \cdot \vec{A} \qquad (24\text{-}47a)$$

In Chapter 24 we considered only problems where $A_\perp$ was something simple like a sphere around a point source, or a cylinder around a line source, and we could easily write a formula for the total flux. We now wish to consider how we should calculate, at least in principle, the flux in a more complex flow like the stream shown in Figure (3).

To give our flux calculation a sense of reality, suppose that we wish to catch all the salmon swimming up a stream to spawn. As shown in Figure (3), we place a net that goes completely across the stream, from bank to bank, from the surface to the bottom. The total flux Φ_T of water flowing through this net is therefore equal to the total current in the stream.

To calculate the total flux Φ_T, we break the stream flow up into a number of small flow tubes bounded by stream lines as shown in (3). Focusing our attention on the i th flow tube, we see that the tube intersects an area $d\vec{A}_i$ of the fish net. The flux through the fish net due to the i th tube is

$$d\Phi_i = \vec{v}_i \cdot d\vec{A}_i \qquad (1)$$

where $\vec{v}_i$ is the velocity of the water at the intersection of the tube and the net. The total flux or current of water in the stream is simply the sum of the fluxes in each flow tube, which can be written

$$\Phi_T = \sum_{\substack{\text{all flow} \\ \text{tubes}}} \Phi_i = \sum_i \vec{v}_i \cdot d\vec{A}_i \qquad (2)$$

where the $d\vec{A}_i$ are just those areas on the fish net marked out by the flow tubes.

If we go to infinitesimal sized flow tubes, the sum in Equation (2) becomes an integral which can be written as

$$\boxed{\Phi_T = \int_{\substack{\text{area of} \\ \text{fish net}}} \vec{v} \cdot d\vec{A}} \qquad \textit{surface integral} \qquad (3)$$

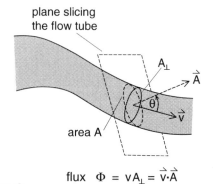

plane slicing
the flow tube

flux $\Phi = vA_\perp = \vec{v} \cdot \vec{A}$

Figure 2
If we have an area A that is not normal to the stream, then the cross-sectional area is $A_\perp = A \cos \theta$, and the flux is $\Phi = vA_\perp = vA \cos \theta$, which can also be written $\Phi = \vec{v} \cdot \vec{A}$.

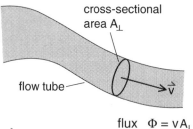

cross-sectional
area $A_\perp$

flow tube

flux $\Phi = vA_\perp$

Figure 1
The flux of water Φ through a flow tube is the amount of water per second flowing past a cross-sectional area $A_\perp$.

where the $d\vec{A}$'s are infinitesimal pieces of area on the fish net and our sum or integration extends over the entire submerged area of the net. Because we are integrating over an area or surface in Equation (3), this integral is called a *surface integral*.

Think of Equation (3), not as an integral you "do", like $\int x^2 dx = x^3/3$, but more as a formal statement of the steps we went through to calculate the total flux Φ_T.

Suppose, for example, someone came up to you and asked how you would calculate the total current in the stream. If you were a mathematician you might answer, "I would calculate the integral

$$\Phi_T = \int_S \vec{v} \cdot d\vec{A} \tag{3a}$$

where S is a surface cutting the stream."

If you were a physicist, you might answer, "Throw a fish net across the stream, making sure that there are no gaps that the fish can get through. (This defines the mathematician's surface S). Then measure the flux of water through each hole in the net (these are the $\vec{v} \cdot d\vec{A}$'s of Equation 3a), and then add them up to get the total flux (do the integral)."

Basically, the mathematician's statement in Equation (3a) is short hand notation for all the steps that the physicist would carry out.

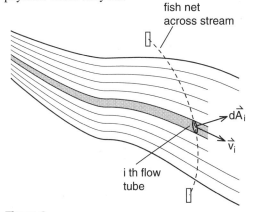

fish net
across stream

$d\vec{A}_i$

$\vec{v}_i$

i th flow
tube

Figure 3
To calculate the flux of water through a fish net, we can first calculate the flux of water through each hole in the net, and then add up the fluxes to get the total flux.

Gauss' Law

The statement of Gauss' law applied to electric fields in Chapter (24) was that the total electric flux Φ_T out through a closed surface was equal to $1/\varepsilon_0$ times the total charge Q_{in} inside the surface. Our surface integral of Equation (3) allows us to give a more formal (at least more mathematical sounding) statement of Gauss' law.

Suppose we have a collection of charged particles as shown in Figure (4), which are completely surrounded by a closed surface S. (*Think of the closed surface as being the surface of an inflated balloon. There cannot be any holes in the surface or air would escape.*) The total flux of the field $\vec{E}$ out through the surface S is formally given by the surface integral

$$\Phi_T = \int_S \vec{E} \cdot d\vec{A} \tag{4}$$

where the $d\vec{A}$ are small pieces of the surface, and $\vec{E}$ is the electric field vector at each $d\vec{A}$.

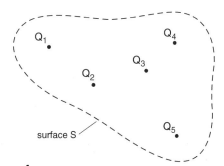

Figure 4
Closed surface S completely surrounding a collection of charges. The flux of $\vec{E}$ out through the closed surface is equal to $1/\varepsilon_0$ times the total charge inside.

The total charge Q_{in} inside the surface S is obtained by adding up all the charges we find inside. Any charges outside do not count. *(We have to have a completely closed surface so that we can decide whether a charge is inside or not.)* Then equating the total flux Φ_T to Q_{in}/ε_0 we get the **integral equation**

$$\oint_{\substack{\text{closed}\\\text{surface S}}} \vec{E}\cdot d\vec{A} = \frac{Q_{in}}{\varepsilon_0} \qquad \begin{array}{l}\textit{formal}\\\textit{statement of}\\\textit{Gauss' law}\end{array} \qquad (5)$$

There is nothing really new in Equation (5) that we did not say back in Chapter (24). What we now have is a convenient short hand notation for all the steps we discussed earlier.

We will now use Equation (5) to calculate the electric field of a point charge. Although we have done this same calculation before, we will do it again to remind us of the steps we actually go through to apply Equation (5). A formal equation like this becomes real or useful only when we have an explicit example to remind us how it is used. *When you memorize such an equation, also memorize an example to go with it.*

In Figure (5), we have a point charge +Q that produces a radial electric field $\vec{E}$ as shown. To apply Gauss' law we draw a spherical surface S of radius r around the charge. For this surface we have

$$\Phi_T = \int_S \vec{E}\cdot d\vec{A} = EA_\perp = E(r)4\pi r^2$$

Inside the surface the total charge is Q. Thus Gauss' law, Equation (5), gives

$$\int_S \vec{E}\cdot d\vec{A} = \frac{Q_{in}}{\varepsilon_0}$$

$$E(r)4\pi r^2 = \frac{Q}{\varepsilon_0}$$

$$E(r) = \frac{Q}{4\pi\varepsilon_0 r^2} \qquad (6)$$

The only thing that is new here is the use of the notation $\int_S \vec{E}\cdot d\vec{A}$ for total flux Φ_T. When we actually wish to calculate Φ_T, we look for a surface that is perpendicular to $\vec{E}$ so that we can use the simple formula $EA_\perp$.

If the charge distribution were complex, more like Figure (4), we could calculate $\int_S \vec{E}\cdot d\vec{A}$ by casting a fish net all around the charges and evaluating $\vec{E}_i\cdot d\vec{A}_i$ for each hole in the net. The formal expression of the surface integral at least gives us a procedure we can follow if we are desperate.

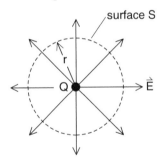

Figure 5
For the electric field of a point charge, we know immediately that the total flux Φ_T out through the spherical surface is the area $4\pi r^2$ times the strength $\vec{E}(r)$ of the field. Thus

$$\Phi_T \equiv \int_S \vec{E}\cdot d\vec{A} = \vec{E}(r)\times 4\pi r^2$$

THE LINE INTEGRAL

Another formal concept which we will use extensively in the remaining chapters on electromagnetic theory is the line integral. You have already been exposed to the idea in earlier discussions of the concept of work. If we exert a force $\vec{F}$ on a particle while the particle moves from Point (1) to Point (2) as shown in Figure (6), then the work we do is given by the integral

$$\text{Work } W = \int_1^2 \vec{F} \cdot d\vec{x} \tag{7}$$

where we are integrating along the path in Figure (6).

Equation (7) is short hand notation for many steps. What it really says is to draw the path taken by the particle in going from Point (1) to (2), break the path up into lots of little steps $d\vec{x}_i$, calculate the work dW_i we do during each step, $dW_i = \vec{F}_i \cdot d\vec{x}_i$, and then add up all the dW_i to get the total work W.

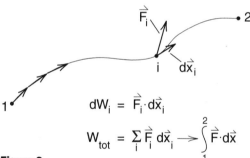

$$dW_i = \vec{F}_i \cdot d\vec{x}_i$$

$$W_{tot} = \sum_i \vec{F}_i \, d\vec{x}_i \longrightarrow \int_1^2 \vec{F} \cdot d\vec{x}$$

Figure 6
To calculate the total work done moving a particle from point (1) to point (2) along a path , first break the path up into many short displacements $d\vec{x}$. The work dW_i is $dW_i = \vec{F}_i \cdot d\vec{x}_i$. The total work W is the sum of all the dW_i.

The first thing we have to worry about in discussing Equation (7), is what path the particle takes in going from Point (1) to Point (2). If we are moving an eraser over a blackboard, the longer the path, the more work we do. In this case, we cannot do the line integral until the path has been specified.

On the other hand, if we are carrying the particle around the room, exerting a force $\vec{F} = -\vec{F}_g$ that just over-comes the gravitational force, then the work we do is stored as gravitational potential energy. The change in potential energy, and therefore the line integral of Equation (7), depends only on the end points (1) and (2) and not on the path we take. When the line integral of a force does not depend upon the path, we say that the force is **conservative**.

A formal statement that a force is conservative is that the line integrals are equal for any two paths -- for example, path (a) and path (b) in Figure (7).

$$\int_{1(\text{path a})}^2 \vec{F} \cdot d\vec{x} = \int_{1(\text{path b})}^2 \vec{F} \cdot d\vec{x} \tag{8}$$

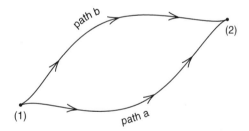

Figure 7
If the work done in carrying the particle from point (1) to point (2) does not depend upon which path we take, we say that the force is conservative.

Let us write Equation (8) in the form

$$\int_{1(\text{path a})}^{2} \vec{F} \cdot d\vec{x} - \int_{1(\text{path b})}^{2} \vec{F} \cdot d\vec{x} = 0$$

Now take the minus sign inside the integral over Path (b) so that we have a sum of $\vec{F}_i \cdot \left(-d\vec{x}_i\right)$

$$\int_{1(\text{path a})}^{2} \vec{F} \cdot d\vec{x} + \int_{1(\text{path b})}^{2} \vec{F} \cdot \left(-d\vec{x}\right) = 0 \qquad (9)$$

For the path (b) integral, we have reversed the direction of each step. The sum of the reversed steps is the same as going back, from point (2) to point (1) as illustrated in Figure (8). Thus Equation (9) becomes

$$\int_{1(\text{path a})}^{2} \vec{F} \cdot d\vec{x} + \int_{2(\text{path b})}^{1} \vec{F} \cdot d\vec{x} = 0 \qquad (10)$$

If Equation (10) applies for any Path (a) and (b), then the force $\vec{F}$ is conservative.

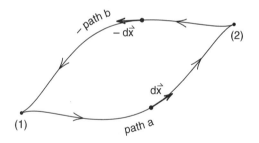

Figure 8
If we take path b backwards, i.e. go from point (2) to point (1), the $d\vec{x}_i$ on path b are reversed and the integral along path b changes sign. If the line integral from (1) to (2) does not depend upon the path, then the line integral for any return trip must be the negative of the integral for the trip out, and the sum of the two integrals must be zero.

Equation (10) does not really depend upon the Points (1) and (2). More generally, it says that if you go out and then come back to your starting point, and the sum of all your $\vec{F}_i \cdot d\vec{x}_i$ is zero, then the force is conservative. This special case of a line integral that comes back to the starting point as in Figure (9) is called ***the line integral around a closed path***, and is denoted by an integral sign with a circle in the center

$$\oint \vec{F} \cdot d\vec{x} \equiv \left\{ \begin{array}{l} \text{the line integral} \\ \text{around a closed path} \\ \text{as in Figure 9} \end{array} \right. \qquad (11)$$

With the notation of Equation (11), we can formally define a conservative force $\vec{F}$ as one for which

$$\oint_{\substack{\text{for any} \\ \text{closed path}}} \vec{F} \cdot d\vec{x} = 0 \qquad \begin{array}{l} \textbf{\textit{definition of a}} \\ \textbf{\textit{conservative force}} \end{array} \qquad (12)$$

This line integral around a closed path will turn out to be an extremely useful mathematical tool. We have already seen that it distinguishes a conservative force like gravity, where $\oint \vec{F} \cdot d\vec{x} = 0$, from a non conservative force like friction on a blackboard eraser, for which $\oint \vec{F} \cdot d\vec{x} \neq 0$. In another case, namely Ampere's law to be discussed next, the line integral of the magnetic field around a closed path tells us something about the currents that flow through the path.

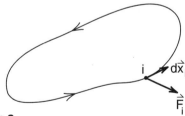

Figure 9
For a conservative force, the line integral $\oint \vec{F} \cdot d\vec{x}$, that goes completely around a closed path, must be zero. It is not necessary to specify where the calculation starts.

AMPERE'S LAW

Figure (10), which is similar to Figure (28-14), is a sketch of the magnetic field produced by a current in a straight wire. In this figure the current i is directed up and out of the paper, and the magnetic field lines travel in counter clockwise circles as shown. We saw from equation (28-18) that the strength of the magnetic field is given by

$$\left|\vec{B}\right| = \frac{\mu_0 i}{2\pi r} \tag{28-18}$$

In Figure (11) we have drawn a circular path of radius r around the wire and broken the path into a series of steps indicated by short vectors $d\vec{\ell}_i$. We drew the stick figure to emphasize the idea that this is really a path and that $d\vec{\ell}_i$ shows the length and direction of the i th step.

For each of the steps, calculate the dot product $\vec{B}_i \cdot d\vec{\ell}_i$ where $\vec{B}_i$ is the magnetic field at that step, and then add up the $\vec{B}_i \cdot d\vec{\ell}_i$ for all the steps around the path to get

$$\sum_{\substack{\text{all steps} \\ \text{around path}}} \vec{B}_i \cdot d\vec{\ell}_i \rightarrow \oint \vec{B} \cdot d\vec{\ell} \tag{13}$$

The result is the line integral of $\vec{B}$ around the closed path.

Why bother calculating this line integral? Let us put in the value for $\left|\vec{B}\right|$ given by Equation (28-18) and see why. We happen to have chosen a path where each step $d\vec{\ell}_i$ is parallel to $\vec{B}$ at that point, so that

$$\vec{B}_i \cdot d\vec{\ell}_i = B_i d\ell_i \tag{14}$$

and Equation (13) becomes

$$\oint \vec{B} \cdot d\vec{\ell} = \oint B d\ell \tag{15}$$

In addition, our path has a constant radius r, so that $B = \mu_0 i / 2\pi r$ is constant all around the path. We can take this constant outside the integral in Equation (15) to get

$$\oint \vec{B} \cdot d\vec{\ell} = B \oint d\ell \tag{16}$$

Next we note that $\oint d\ell$ is just the sum of the lengths of our steps around the circle; i.e., it is just the circumference $2\pi r$ of the circle, and we get

$$\oint \vec{B} \cdot d\vec{\ell} = B \oint d\ell = B \times 2\pi r \tag{17}$$

Finally substituting the value of B from Equation (28-18) we get

$$\oint \vec{B} \cdot d\vec{\ell} = \frac{\mu_0 i}{2\pi r} \times 2\pi r = \mu_0 i \tag{18}$$

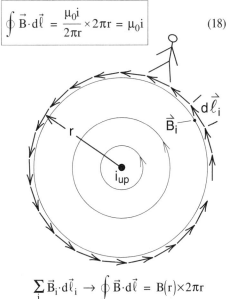

$$\sum_i \vec{B}_i \cdot d\vec{\ell}_i \rightarrow \oint \vec{B} \cdot d\vec{\ell} = B(r) \times 2\pi r$$

Figure 11
Circular path of radius r around the wire. As we walk around the path, each step represents a displacement $d\vec{\ell}$. To calculate the line integral $\oint \vec{B} \cdot d\vec{\ell}$, we take the dot product of $d\vec{\ell}$ with B at each interval and add them up as we go around the entire path. In this case the result is simply B(r) times the circumference $2\pi r$ of the path.

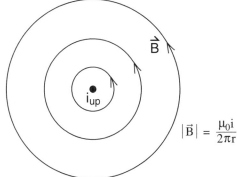

$$\left|\vec{B}\right| = \frac{\mu_0 i}{2\pi r}$$

Figure 10
Circular magnetic field of a wire.

There are several points we want to make about Equation (18). First we made the calculation easy by choosing a circular path that was parallel to $\vec{B}$ all the way around. This allowed us to replace the dot product $\vec{B} \cdot d\vec{\ell}$ by a numerical product $B d\ell$, pull the constant B outside the integral, and get an answer almost by inspection. This should be reminiscent of our work with Gauss' law where we chose surfaces that made it easy to solve the problem.

The second point is that we get an exceptionally simple answer for the line integral of $\vec{B}$ around the wire, namely

$$\boxed{\oint \vec{B} \cdot d\vec{\ell} = \mu_0 i} \qquad \text{Ampere's Law} \quad (18a)$$

The line integral depends only on the current i through the path and not on the radius r of the circular path.

What about more general paths that go around the wire? To find out, we have to do a slightly harder calculation, but the answer is interesting enough to justify the effort.

In Figure (12) we have constructed a closed path made up of three arc sections of lengths $r_1\theta_1$ $r_2\theta_2$, and $r_3\theta_3$ connected by radial sections as shown. These arcs are sections of circles of radii r_1, r_2 and r_3, respectively. We wish to calculate $\oint \vec{B} \cdot d\vec{\ell}$ for this path and see how the answer compares with what we got for the circular path.

The first thing to note as we go around our new path is that in all the *radial* sections, $\vec{B}$ and $d\vec{\ell}$ are perpendicular to each other, so that $\vec{B} \cdot d\vec{\ell} = 0$. The radial sections do not contribute to our line integral and all we have to do is add up the contributions from the three arc segments. These are easy to calculate because $\vec{B} \cdot d\vec{\ell} = B d\ell$ and B is constant over each arc, so that the integral of $\vec{B} \cdot d\vec{\ell}$ over an arc segment is just the value of B times the length $r\theta$ of the arc. We get

$$\int_{\text{arc }1} \vec{B} \cdot d\ell = B_1 r_1 \theta_1 = \frac{\mu_0 i}{2\pi r_1} r_1 \theta_1 = \frac{\mu_0 i \theta_1}{2\pi}$$

$$\int_{\text{arc }2} \vec{B} \cdot d\ell = B_2 r_2 \theta_2 = \frac{\mu_0 i}{2\pi r_2} r_2 \theta_2 = \frac{\mu_0 i \theta_2}{2\pi}$$

$$\int_{\text{arc }3} \vec{B} \cdot d\ell = B_3 r_3 \theta_3 = \frac{\mu_0 i}{2\pi r_3} r_3 \theta_3 = \frac{\mu_0 i \theta_3}{2\pi}$$

Adding the contribution from each arc segment we get the line integral around the closed path

$$\oint \vec{B} \cdot d\vec{\ell} = \frac{\mu_0 i \theta_1}{2\pi} + \frac{\mu_0 i \theta_2}{2\pi} + \frac{\mu_0 i \theta_3}{2\pi}$$

$$= \frac{\mu_0 i}{2\pi} \left(\theta_1 + \theta_2 + \theta_3 \right)$$

But $\left(\theta_1 + \theta_2 + \theta_3 \right)$ is the sum of the angles around the circle, and is therefore equal to 2π. Thus we get for the path of Figure (12)

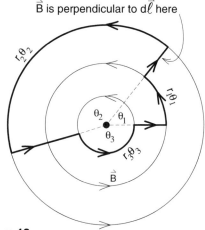

$\vec{B}$ is perpendicular to $d\vec{\ell}$ here

Figure 12
A somewhat arbitrary path around the wire is made of arc sections connected by radial sections. Since B is perpendicular to $d\vec{\ell}$ in the radial sections, the radial sections do not contribute to the $\oint \vec{B} \cdot d\vec{\ell}$ for this path. In the arcs, the length of the arc increases with r, but B decreases as 1/r, so that the contribution of the arc does not depend upon how far out it is. As a result the $\oint \vec{B} \cdot d\vec{\ell}$ is the same for this path as for a circular path centered on the wire.

$$\oint \vec{B} \cdot d\vec{\ell} = \frac{\mu_0 i}{2\pi}(2\pi) = \mu_0 i \qquad (19)$$

which is the same answer we got for the circular path.

The result in Equation (19) did not depend upon how many line segments we used, because each arc contributed an angle θ, and if the path goes all the way around, the angles always add up to 2π. In Figure (13) we have imitated a smooth path (the dotted line) by a path consisting of many arc sections. The more arcs we use the closer the imitation. We can come arbitrarily close to the desired path using paths whose integral $\oint \vec{B} \cdot d\vec{\ell}$ is $\mu_0 i$. In this sense we have proved that Equation (19) applies to any closed path around the wire.

It is another story if the path does not go around the wire. In Figure (14) we have such a path made up of two arc and two radial segments as shown. As before, we can ignore the radial segments because $\vec{B}$ and $d\vec{\ell}$ are perpendicular and $\vec{B} \cdot d\vec{\ell} = 0$.

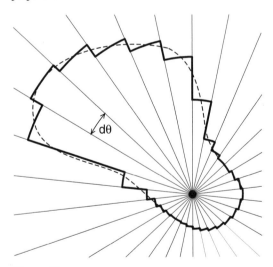

Figure 13
We can approximate an arbitrary path (dotted line) by a series of connected radial and arc sections. The smaller the angle $d\theta$ marking the arc sections, the better the approximation. In calculating $\vec{B} \cdot d\vec{\ell}$, the radial sections do not count, and we can bring all the arc sections back to a single circle centered on the wire. As a result, $\oint \vec{B} \cdot d\vec{\ell}$ does not depend upon the shape of the path, as long as the path goes around the wire.

On the outer segment we are going in the same direction as $\vec{B}$ so that $\vec{B} \cdot d\vec{\ell}$ is positive and we get

$$\int_{arc\,1} \vec{B} \cdot d\vec{\ell} = B_1\left(r_1\theta\right) = \frac{\mu_0 i}{2\pi r_1} r_1\theta = \frac{\mu_0 i}{2\pi}\theta$$

On the inner arc we are coming back around in a direction opposite to $\vec{B}$, the quantity $\vec{B} \cdot d\vec{\ell}$ is negative, and we get

$$\int_{arc\,2} \vec{B} \cdot d\vec{\ell} = -B_2\left(r_2\theta\right) = \frac{-\mu_0 i}{2\pi r_2} r_2\theta = \frac{-\mu_0 i}{2\pi}\theta$$

Adding up the two contributions from the two arcs, we get

$$\oint_{\substack{Path\,of \\ Figure\,14}} \vec{B} \cdot d\vec{\ell} = \left(\frac{\mu_0 i\theta}{2\pi}\right)_{arc\,1} + \left(\frac{-\mu_0 i\theta}{2\pi}\right)_{arc\,2} = 0 \quad (20)$$

For this closed path which does not go around the current, we get $\oint \vec{B} \cdot d\vec{\ell} = 0$. This result is not changed if we add more arcs and radial segments to the path. As long as the path does not go around the current, we get zero for $\oint \vec{B} \cdot d\vec{\ell}$.

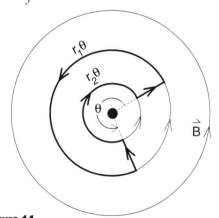

Figure 14
In this example, where the path does not go around the wire, the sections labeled $r_1\theta$ and $r_2\theta$ contribute equal and opposite amounts to the line integral $\oint \vec{B} \cdot d\vec{\ell}$. As a result $\oint \vec{B} \cdot d\vec{\ell}$ is zero for this, or any path that does not go around the wire.

Several Wires

It is relatively straightforward to generalize our results to the case where we have several wires as in Figures (15 a,b). Here we have three currents i_1, i_2, and i_3 each alone producing a magnetic field $\vec{B}_1$, $\vec{B}_2$ and $\vec{B}_3$ respectively.

The first step is to show that the net field $\vec{B}$ at any point is the vector sum of the fields of the individual wires. We can do this by considering the force on a test particle of charge q moving with a velocity $\vec{v}$ as shown in Figure (15a). Our earlier results tell us that the current i_1 exerts a force

$$\vec{F}_1 = q\vec{v} \times \vec{B}_1$$

Similarly i_2 and i_3 exert forces

$$\vec{F}_2 = q\vec{v} \times \vec{B}_2$$

$$\vec{F}_3 = q\vec{v} \times \vec{B}_3$$

Newton's second law required us to take the vector sum of the individual forces to get the total force $\vec{F}$ acting on an object

$$\vec{F} = \vec{F}_1 + \vec{F}_2 + \vec{F}_3$$
$$= q\vec{v} \times \left(\vec{B}_1 + \vec{B}_2 + \vec{B}_3\right) \tag{21}$$

If we write this total force in the form

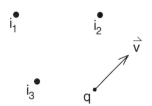

$$\vec{F}_T = q\vec{v} \times \vec{B} \tag{22}$$

where $\vec{B}$ is the effective field acting on the test particle, then Equations (21) and (22) give

$$\boxed{\vec{B} = \vec{B}_1 + \vec{B}_2 + \vec{B}_3} \tag{23}$$

The fact that magnetic fields add vectorially is a consequence of the vector addition of forces and our use of the magnetic force law to define $\vec{B}$.

With Equation (23) we can now calculate $\oint \vec{B} \cdot d\vec{\ell}$ for the field of several wires. Let us draw a path around two of the wires, as shown in Figure (15b). For this path, we get

$$\oint_{\substack{\text{Closed path} \\ \text{of Fig. 29-15}}} \vec{B} \cdot d\vec{\ell} = \oint \left(\vec{B}_1 + \vec{B}_2 + \vec{B}_3\right) \cdot d\vec{\ell}$$

$$= \oint \vec{B}_1 \cdot d\vec{\ell} + \oint \vec{B}_2 \cdot d\vec{\ell} + \oint \vec{B}_3 \cdot d\vec{\ell} \tag{24}$$

Since the closed path goes around currents i_1 and i_2, we get from Equation (19)

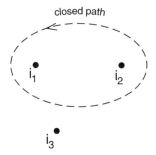

Figure 15a

A charge q moving in the vicinity of three currents i_1, i_2 and i_3. If the magnetic field $\vec{B}$ at the charge is the vector sum of the fields $\vec{B}_1$, $\vec{B}_2$ and $\vec{B}_3$ of the three wires, then the net magnetic force $\vec{F}_B$ on q is given by

$$\vec{F}_B = q\vec{v} \times \vec{B} = q\vec{v} \times \left(\vec{B}_1 + \vec{B}_2 + \vec{B}_3\right)$$
$$= q\vec{v} \times \vec{B}_1 + q\vec{v} \times \vec{B}_2 + q\vec{v} \times \vec{B}_3$$
$$= \vec{F}_{B1} + \vec{F}_{B2} + \vec{F}_{B3}$$

and we get the desired result that the net force on q is the vector sum of the forces exerted by each wire.

Figure 15b

Calculating $\oint \vec{B} \cdot d\vec{\ell}$ for a path that goes around two of the wires.

$$\oint \vec{B}_1 \cdot d\vec{\ell} = \mu_0 i_1$$

$$\oint \vec{B}_2 \cdot d\vec{\ell} = \mu_0 i_2$$

Since the path misses i_3, we get

$$\oint \vec{B}_3 \cdot d\vec{\ell} = 0$$

and Equation (24) gives

$$\underset{\substack{\text{Closed path}\\\text{of Fig. 15}}}{\oint} \vec{B} \cdot d\vec{\ell} = \mu_0 (i_1 + i_2) = \mu_0 \times \begin{pmatrix} \text{current} \\ \text{enclosed} \\ \text{by path} \end{pmatrix} \quad (25)$$

Equation (25) tells us that $\oint \vec{B} \cdot d\vec{\ell}$ around a closed path is equal to μ_0 times the total current $i = i_1 + i_2$ encircled by the path. This has the flavor of Gauss' law which said that the total flux or surface integral of $\vec{E}$ out through a closed surface was $1/\varepsilon_0$ times the total charge Q_{in} inside the surface. Just as charge outside the closed surface did not contribute to the surface integral of $\vec{E}$, currents outside the closed path do not contribute to the line integral of $\vec{B}$.

We derived Equation (25) for the case that all our currents were in parallel straight wires. It turns out that it does not matter if the wires are straight, bent, or form a hideous tangle. As a general rule, if we construct a closed path, then the line integral of $\vec{B}$ around the closed path is μ_0 times the net current $i_{enclosed}$ flowing through the path

$$\underset{\substack{\text{any closed}\\\text{path}}}{\oint} \vec{B} \cdot d\vec{\ell} = \mu_0 i_{enclosed} \qquad \begin{array}{c} (26) \\ \textit{Ampere's} \\ \textit{Law} \end{array}$$

This extremely powerful and general theorem is known as *Ampere's law*.

So far in this chapter we have focused on mathematical concepts. Let us now work out some practical applications of Ampere's law to get a feeling for how the law is used.

Field of a Straight Wire

Our first application of Ampere's law will be to calculate the magnetic field of a straight wire. We will use this trivial example to illustrate the steps used in applying Ampere's law.

First we sketch the situation as in Figure (16), and then write down Ampere's law to remind us of the law we are using

$$\oint \vec{B} \cdot d\vec{\ell} = \mu_0 i_{enclosed}$$

Next we choose a closed path that makes the line integral as simple as possible. Generally the path should either be along $\vec{B}$ so that $\vec{B} \cdot d\vec{\ell} = B d\ell$, or perpendicular so that $\vec{B} \cdot d\vec{\ell} = 0$. The circular path of Figure (16) gives $\vec{B} \cdot d\vec{\ell} = B d\ell$ with B constant, thus

$$\oint \vec{B} \cdot d\vec{\ell} = \oint B d\ell = B \oint d\ell = B * 2\pi r = \mu_0 i$$

The result is

$$\boxed{B = \frac{\mu_0 i}{2\pi r}}$$

which we expected. When you memorize Ampere's law, *memorize an example like this to go with it*.

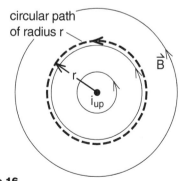

circular path of radius r

$\vec{B}$

r

i_{up}

Figure 16
Using Ampere's law to calculate the magnetic field of a wire. We have $\oint \vec{B} \cdot d\vec{\ell} = B \times 2\pi r$ around the path. Thus Ampere's law $\oint \vec{B} \cdot d\vec{\ell} = \mu_0 i$ gives $B = \mu_0 i / 2\pi r$.

Exercise 1

Each of the indicated eight conductors in Figure (17) carries 2.0A of current into (dark) or out of (white) the page. Two paths are indicated for the line integral $\oint \vec{B} \cdot d\vec{\ell}$. What is the value of the integral for (a) the dotted path? (b) the dashed path?

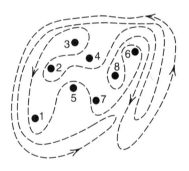

Figure 17

Exercise 2

Eight wires cut the page perpendicularly at the points shown in Figure (18). A wire labeled with the integer k (k = 1, 2..., 8) bears the current ki_0. For those with odd k, the current is up, out of the page; for those with even k it is down, into the page. Evaluate $\oint \vec{B} \cdot d\vec{\ell}$ along the closed path shown, in the direction shown.

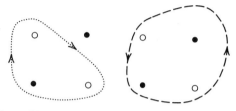

Figure 18

Exercise 3

Show that a uniform magnetic field B cannot drop abruptly to zero as one moves at right angles to it, as suggested by the horizontal arrow through point a in Figure (19). (Hint: Apply Ampere's law to the rectangular path shown by the dashed lines.) In actual magnets "fringing" of the lines of B always occurs, which means that B approaches zero gradually.

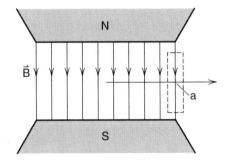

Figure 19

Exercise 4

Figure (20) shows a cross section of a long cylindrical conductor of radius a, carrying a uniformly distributed current i. Assume a = 2.0 cm, i = 100A, and sketch a plot of B(r) over the range 0 < r < 4 cm.

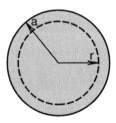

Figure 20

(The above are some choice problems from Halliday and Resnick.)

Exercise 5

Figure (21) shows a cross section of a hollow cylindrical conductor of radii a and b, carrying a uniformly distributed current i.

a) Show that B(r) for the range b < r < a is given by

$$B(r) = \frac{\mu_0 i}{2\pi r}\left(\frac{r^2 - b^2}{a^2 - b^2}\right)$$

b) Test this formula for the special cases of r = a, r = b, and r = 0.

c) Assume a = 2.0 cm, b = 1.8 cm, and i = 100 A. What is the value of B at r = a? (Give your answer in tesla and gauss.)

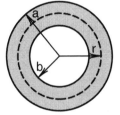

Figure 21

Exercise 6

Figure (22) shows a cross section of a long conductor of a type called a coaxial cable. Its radii (a, b, c) are shown in the figure. Equal but opposite currents i exist in the two conductors. Derive expressions for B(r) in the ranges

a) r < c,

b) c < r < b,

c) b < r < a, and

d) r > a.

e) Test these expressions for all the special cases that occur to you.

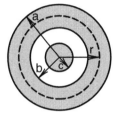

coaxial cable

Figure 22

Exercise 6 is a model of a ***coaxial cable***, where the current goes one way on the inner conductor and back the other way on the outside shield. If we draw any circuit outside the cable, there is no net current through the circuit, thus there is no magnetic field outside. As a result, coaxial cables confine all magnetic fields to the inside of the cable. This is important in many electronics applications where you do not want fields to radiate out from your wires. The cables we use in the lab, the ones with the so called BNC connectors, are coaxial cables, as are the cables that carry cable television.

FIELD OF A SOLENOID

As with Gauss' law, Ampere's law is most useful when we already know the field structure and wish to calculate the strength of the field. The classic example to which ampere's law is applied is the calculation of the magnetic field of a long straight solenoid.

A long solenoid is a coil of wire in which the length L of the coil is considerably larger than the diameter d of the individual turns. The shape of the field produced when a current i flows through the coil was illustrated in Figure (28-21) and is sketched here in Figure (23). Iron filings gave us the shape of the field and Ampere's law will tell us the strength.

The important and useful feature of a solenoid is that we have a nearly uniform magnetic field inside the coil and nearly zero field outside. The longer the solenoid, relative to the diameter d, the more uniform the field $\vec{B}$ inside and the more nearly it is zero outside.

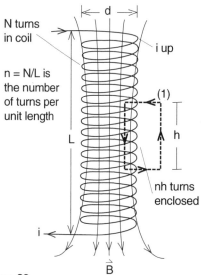

Figure 23
Calculating the magnetic field of a long solenoid. Around the path starting at point (1) we have $\oint \vec{B} \cdot d\vec{\ell} = 0 + Bh + 0 + 0$. The amount of current enclosed by the path is $i_{tot} = (nh)i$ where n is the number of turns per unit length. Thus Ampere's law $\oint \vec{B} \cdot d\vec{\ell} = \mu_0 i_{tot}$ gives $Bh = \mu_0 nhi$ or $B = \mu_0 ni$.

Right Hand Rule for Solenoids

The direction of the field inside the solenoid is a bit tricky to figure out. As shown in Figure (24), up near the wires and in between the turns, the field goes in a circle around the wire just as it does for a straight wire. As we go out from the wire the circular patterns merge to create the uniform field in the center of the solenoid.

We see, from Figure (24), that if the current goes around the coil in such a way that the current is up out of the paper on the right side and down into the paper on the left, then the field close to the wires will go in counterclockwise circles on the right and clockwise circles on the left. For both these sets of circles, the field inside the coil points down. As a result the uniform field inside the coil is down as shown.

There is a simple way to remember this result without having to look at the field close to the wires. Curl the fingers of your ***right*** hand in the direction of the flow of the current i in the solenoid, and your thumb will point in the direction of the magnetic field inside the solenoid. We will call this the right hand rule for solenoids.

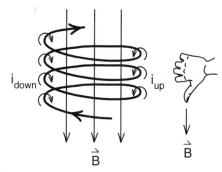

Figure 24
If you know the direction of the current in the wire, you can determine the direction of the magnetic field by looking very close to the wire where the field goes around the wire. You get the same answer if you curl the fingers of your right hand around in the direction the current in the coil is flowing. Your thumb then points in the direction of the field.

Evaluation of the Line Integral

Figure (25) is a detail showing the path we are going to use to evaluate $\oint \vec{B} \cdot d\vec{\ell}$ for the solenoid. This path goes down the solenoid in the direction of $\vec{B}$ (side 1), and out through the coil (side 2), up where $\vec{B} = 0$ (side 3) and back into the coil (side 4).

We can write $\oint \vec{B} \cdot d\vec{\ell}$ as the sum of four terms for the four sides

$$\oint \vec{B} \cdot d\vec{\ell} = \int_{\text{side 1}} \vec{B} \cdot d\vec{\ell} + \int_{\text{side 2}} \vec{B} \cdot d\vec{\ell}$$
$$+ \int_{\text{side 3}} \vec{B} \cdot d\vec{\ell} + \int_{\text{side 4}} \vec{B} \cdot d\vec{\ell}$$

On sides 2 and 4, when the path is inside the coil, $\vec{B}$ and $d\vec{\ell}$ are perpendicular and we get $\vec{B} \cdot d\vec{\ell} = 0$. Outside the coil it is still 0 because there is no field there. Likewise $\vec{B} \cdot d\vec{\ell} = 0$ for side 3 because there is no field there. The only contribution we get is from side 1 inside the coil. If h is the height of our path, then

$$\oint \vec{B} \cdot d\vec{\ell} = \int_{\text{side 1}} \vec{B} \cdot d\vec{\ell} = Bh \tag{27}$$

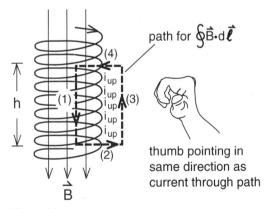

path for $\oint \vec{B} \cdot d\vec{\ell}$

thumb pointing in same direction as current through path

Figure 25
Right hand rule for using Ampere's law. We define the positive direction around the path as the direction you curl the fingers of your right hand when the thumb is pointing in the direction of the current through the path. (As you see, we can come up with a right hand rule for almost anything.)

Calculation of i_{enclosed}

From Figure (25) we see that we get a current i up through our path each time another turn comes up through the path. On the left side of the coil the current goes down into the paper, but these downward currents lie outside our path and therefore are not included in our evaluation of i_{enclosed}. Only the positive upward currents count, and i_{enclosed} is simply i times the number of turns that go up through the path.

To calculate the number of turns in a height h of the coil, we note that if the coil has a length L and a total of N turns, then the number of turns per unit length n is given by

$$\left. \begin{array}{c} \text{number of turns} \\ \text{per unit length} \end{array} \right\} \equiv n = \frac{N}{L} \tag{28}$$

and in a height h there must be nh turns

$$\left. \begin{array}{c} \text{number of turns} \\ \text{in a height h} \end{array} \right\} = nh \tag{29}$$

With nh turns, each carrying a current i, going up through our path, we see that i_{enclosed} must be

$$i_{\text{enclosed}} = inh \tag{30}$$

Using Ampere's law

We are now ready to apply Ampere's law to evaluate the strength B of the field inside the solenoid. Using Equation (27) for $\oint \vec{B} \cdot d\vec{\ell}$, and Equation (31) for i_{enclosed}, we get

$$\int \vec{B} \cdot d\vec{\ell} = \mu_0 i_{\text{enclosed}}$$

$$Bh = \mu_0 nih$$

$$\boxed{B = \mu_0 ni} \qquad \begin{array}{l} \textit{magnetic field} \\ \textit{inside a solenoid} \end{array} \tag{31}$$

The uniform magnetic field inside a long solenoid is proportional to the current i in the solenoid, and the number of turns per unit length, n.

Exercise 7

We will so often be using solenoids later in the course, that you should be able to derive the formula $B = \mu_0 ni$, starting from Ampere's law without looking at notes. This is a good time to practice. Take a blank sheet of paper, sketch a solenoid of length L with N turns. Then close the text and any notes, and derive the formula for B.

We have mentioned that equations like $\oint \vec{B}_1 \cdot d\vec{\ell} = \mu_0 i_{enclosed}$ are meaningless hen scratching until you know how to use them. The best way to do that is learn worked examples along with the equation. Two good examples for Ampere's law are to be able to calculate the magnetic field inside a wire (Exercise 4), and to be able to derive the magnetic field inside a solenoid. If you can do these two derivations without looking at notes, you should have a fairly good grasp of the law.

One More Right Hand Rule

If we really want to be careful about minus signs (and it is not always necessary), we have to say how the sign of $i_{enclosed}$ is evaluated in Figure (25). If, as in Figure (26) we reversed the direction of our path, then on side (1) $\vec{B} \cdot d\vec{\ell}$ is negative because our path is going in the opposite direction to $\vec{B}$. Thus for this path the complete integral $\oint \vec{B} \cdot d\vec{\ell}$ is negative, and somehow our $i_{enclosed}$ must also be negative, so that we get the same answer we got for Figure (25).

If we curl the fingers of our ***right hand*** in the direction that we go around the path, then in Figure (25) our thumb points up parallel to the current through the path, and in Figure (26) our thumb points down, opposite to the current. If we define the direction indicated by our right hand thumb as the positive direction ***through*** the path, as shown in Figure (27), then the current is going in a positive direction in Figure (25) but in a negative direction in Figure (26). This gives us a negative $i_{enclosed}$ for Figure (26) which goes along with the minus sign we got in the evaluation of $\oint \vec{B} \cdot d\vec{\ell}$.

By now you should be getting the idea of how we define directions in magnetic formula. ***Always use your right hand***. After a while you get so used to using your right hand that you do not have to remember the individual right hand rules.

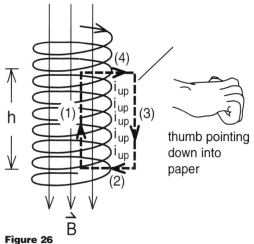

thumb pointing down into paper

$\vec{B}$

Figure 26
If we go around the wrong way, we just get two minus signs and all the results are the same. Here we went around the path so that our thumb pointed opposite to the direction of the current through the path. As a result the magnetic field in the solenoid points opposite to the direction of the path in the solenoid.

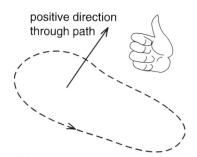

positive direction through path

Figure 27
In general, we use the right hand convention to associate a positive direction around a path to a positive direction through a path.

The Toroid

If we take a long solenoid, bend it in a circle and fit the ends together, we get what is called a toroid shown in Figure (28). The great advantage of a toroid is that there are no end effects. In the straight solenoid the magnetic field at the ends fanned out into space as seen in our iron filing map of Figure (28-23). With the toroid there are no ends. The field is completely confined to the region inside the toroid and there is essentially no field outside. For this reason a toroid is an ideal magnetic field storage device.

It is easy to use Ampere's law to calculate the magnetic field inside the toroid. In Figure (28) we have drawn a path of radius r inside the toroid. Going around this path in the same direction as $\vec{B}$, we immediately get

$$\oint \vec{B} \cdot d\vec{\ell} = B \times 2\pi r \tag{32}$$

because B is constant in magnitude and parallel to $d\vec{\ell}$.

If there are N turns of wire in the toroid, and the wire carries a current i, then all N turns come up through the path on the inside of the solenoid, and $i_{enclosed}$ is given by

$$i_{enclosed} = Ni \tag{33}$$

Using Equations (32) and (33) in Ampere's law gives

$$\oint \vec{B} \cdot d\vec{\ell} = \mu_0 i_{enclosed}$$

$$B \times 2\pi r = \mu_0 Ni$$

$$\boxed{B = \frac{\mu_0 Ni}{2\pi r}} \qquad \begin{array}{l}\textit{magnetic field} \\ \textit{of a toroid}\end{array} \tag{34}$$

Note that $N/2\pi r$ is the number of turns per unit length, n, so that Equation (34) can be written $B = \mu_0 ni$ which is the solenoid formula of Equation (31). To a good approximation the field in a toroid is the same as in the center of a straight solenoid.

The **derivation** of Equation (34) is so easy and such a good illustration of the use of Ampere's law that it should be remembered as an example of Ampere's law.

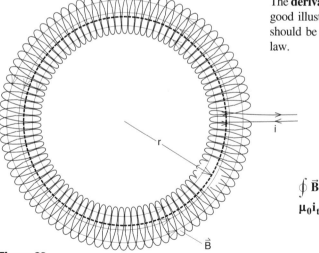

$$\left.\begin{array}{l}\oint \vec{B} \cdot d\vec{\ell} = B * 2\pi r \\ \mu_0 i_{tot} = \mu_0 Ni\end{array}\right\} \Rightarrow B = \frac{\mu_0 Ni}{2\pi r}$$

Figure 28
When the solenoid is bent into the shape of a toroid, there are no end effects. The magnetic field is confined to the region inside the toroid, and Ampere's law is easily applied. (You should remember this as an example of the use of Ampere's law.)

Exercise 8

Figure 29 shows a 400-turn solenoid that is 47.5 cm long and has a diameter of 2.54 cm. (The 10 turns of wire wrapped around the center are for a later experiment.) Calculate the magnitude of the magnetic field B near the center of the solenoid when the wire carries a current of 3 amperes. (Give your answer in tesla and gauss.)

Exercise 9

Figure 30 shows the toroidal solenoid that we use in several experiments later on. The coil has 696 turns wound on a 2.6 cm diameter plastic rod bent into a circle of radius 21.5 cm. What is the strength of the magnetic field inside the coil when a current of 1 amp is flowing through the wire? (Give your answer in tesla and gauss.)

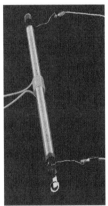

Figure 29
A 400 turn straight solenoid 47.5 cm long,
wound on a 2.54 cm diameter rod.

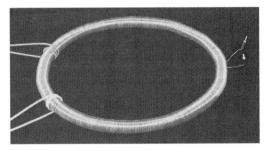

Figure 30
A 696 turn toroidal solenoid wound on
a 2.6 cm diameter plastic rod bent into
a circle of radius 21.5 cm.

Chapter 30
Faraday's Law

In this chapter we will discuss one of the more remarkable, and in terms of practical impact, important laws of physics – Faraday's law. This law explains the operation of the air cart speed detector we have used in air track experiments, the operation of AC voltage generators that supply most of the electrical power in the world, and transformers and inductors which are important components in the electronic circuits in radio and television sets.

In one form, Faraday's law deals with the line integral $\oint \vec{E} \cdot d\vec{\ell}$ of an electric field around a closed path. As an introduction we will begin with a discussion of this line integral for electric fields produced by static charges. (Nothing very interesting happens there.) Then we will analyze an experiment that is similar to our air cart speed detector to see why we get a voltage proportional to the speed of the air cart. Applying the principle of relativity to our speed detector, i.e., riding along with the air cart gives us an entirely new picture of the behavior of electric fields, a behavior that is best expressed in terms of the line integral $\oint \vec{E} \cdot d\vec{\ell}$. After a discussion of this behavior, we will go through some practical applications of Faraday's law.

ELECTRIC FIELD OF STATIC CHARGES

In this somewhat formal section, we show that $\oint \vec{E} \cdot d\vec{\ell} = 0$ for the electric field of static charges. With this as a background, we are in a better position to appreciate an experiment in which $\oint \vec{E} \cdot d\vec{\ell}$ is not zero.

In Figure (1), we have sketched a closed path through the electric field $\vec{E}$ of a point charge, and wish to calculate the line integral $\oint \vec{E} \cdot d\vec{\ell}$ for this path. To simplify the calculation, we have made the path out of arc and radial sections. But as in our discussion of Figure 29-13, we can get arbitrarily close to any path using arc and radial sections, thus what we learn from the path of Figure (1) should apply to a general path.

Because the electric field is radial, $\vec{E}$ is perpendicular to $d\vec{\ell}$ and $\vec{E} \cdot d\vec{\ell}$ is zero on the arc sections. On the radial sections, for every step out where $\vec{E} \cdot d\vec{r}$ is positive there is an exactly corresponding step back where $\vec{E} \cdot d\vec{r}$ is negative. Because we come back to the starting point, we take the same steps back as we took out, all the radial $\vec{E} \cdot d\vec{r}$ cancel and we are left with $\oint \vec{E} \cdot d\vec{\ell} = 0$ for the electric field of a point charge.

Now consider the distribution of fixed point charges shown in Figure (2). Let $\vec{E}_1$ be the field of Q_1, $\vec{E}_2$ of Q_2, etc. Because an electric field is the force on a unit test charge, and because forces add as vectors, the total electric field $\vec{E}$ at any point is the vector sum of the individual fields at that point

$$\vec{E} = \vec{E}_1 + \vec{E}_2 + \vec{E}_3 + \vec{E}_4 + \vec{E}_5 \qquad (1)$$

We can now use Equation (1) to calculate $\oint \vec{E} \cdot d\vec{\ell}$ around the closed path in Figure (2). The result is

$$\oint \vec{E} \cdot d\vec{\ell} = \oint \left(\vec{E}_1 + \vec{E}_2 + ... + \vec{E}_5 \right) \cdot d\vec{\ell}$$
$$= \oint \vec{E}_1 \cdot d\vec{\ell} + \cdots + \oint \vec{E}_5 \cdot d\vec{\ell} \qquad (2)$$

But $\oint \vec{E}_1 \cdot d\vec{\ell} = 0$ since $\vec{E}_1$ is the field of a point charge, and the same is true for $\vec{E}_2 \ldots \vec{E}_5$. Thus the right side of Equation (2) is zero and we have

$$\oint \vec{E} \cdot d\vec{\ell} = 0 \quad \left\{ \begin{array}{l} \text{for the field } \vec{E} \text{ of} \\ \text{any distribution of} \\ \text{static charges} \end{array} \right. \qquad (3)$$

Equation (3) applies to any distribution of static charges, a point charge, a line charge, and static charges on conductors and in capacitors.

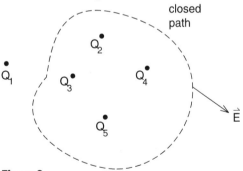

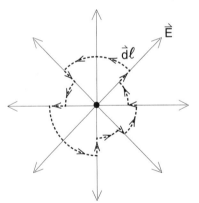

Figure 1
Closed path through the electric field of a point charge. The product $\vec{E} \cdot d\vec{\ell}$ is zero on the arc sections, and the path goes out just as much as it comes in on the radial sections. As a result $\oint \vec{E} \cdot d\vec{\ell} = 0$ when we integrate around the entire path.

Figure 2
Closed path in a region of a distribution of point charge. Since $\oint \vec{E} \cdot d\vec{\ell} = 0$ is zero for the field of each point charge alone, it must also be zero for the total field $\vec{E} = \vec{E}_1 + \vec{E}_2 + \vec{E}_3 + \vec{E}_4 + \vec{E}_5$

A MAGNETIC FORCE EXPERIMENT

Figures (3a,b) are two views of an experiment designed to test for the magnetic force on the conduction electrons in a moving copper wire. We have a wire loop with a gap and the loop is being pulled out of a magnet. At this instant only the end of the loop, the end opposite the gap, is in the magnetic field. It will soon leave the field since it is being pulled out at a velocity $\vec{v}$ as shown.

In our earlier discussions we saw that a copper atom has two loosely bound conduction electrons that are free to flow from one atom to another in a copper wire. These conduction electrons form a negatively charged electric fluid that flows in a wire much like water in a pipe.

Because of the gap we inserted in the wire loop of Figure (3), the conduction electrons in this loop cannot flow. If we move the loop, the conduction electrons must move with the wire. That means that the conduction electrons have a velocity $\vec{v}$ to the right as shown, perpendicular to the magnetic field which is directed into the page. Thus we expect that there should be a magnetic force

$$\vec{F}_{mag} = -e\vec{v} \times \vec{B} \qquad (4)$$

acting on the electrons. This force will be directed down as shown in Figure (3b).

Since the gap in the loop does not allow the conduction electrons to flow along the wire, how are we going to detect the magnetic force on them? There is no net force on the wire because the magnetic field exerts an equal and opposite force on the positive copper ions in the wire.

Our conjecture is that this magnetic force on the conduction electrons would act much like the gravitational force on the water molecules in a static column of water. The pressure at the bottom of the column is higher than the pressure at the top due to the gravitational force. Perhaps the pressure of the negatively

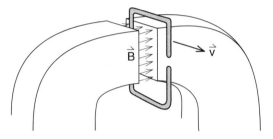

Figure 3a
Wire loop moving through magnetic field of iron magnet.

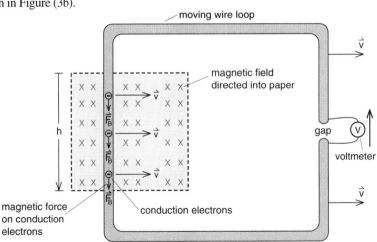

Figure 3b
When you pull a wire loop through a magnetic field, the electrons, moving at a velocity $\vec{v}$ with the wire, feel a magnetic force $\vec{F}_B = (-e)\vec{v} \times \vec{B}$ if they are in the field. This force raises the pressure of the electron fluid on the bottom of the loop and reduces it on the top, creating a voltage V across the gap. The arrow next to the voltmeter indicates a voltage rise for positive charge, which is a voltage drop for negative charge.

charged electric fluid is higher at the bottom of the loop than the top due to the magnetic force.

To find out if this is true, we use an electrical pressure gauge, which is a voltmeter. A correctly designed voltmeter measures an electrical pressure drop without allowing any current to flow. Thus we can place the voltmeter across the gap and still not let the conduction electrons flow in the loop.

If our conjecture is right, we should see a voltage reading while the magnetic force is acting. Explicitly there should be a voltage reading while the wire is moving and one end of the loop is in the magnetic field as shown. The voltage should go to zero as soon as the wire leaves the magnetic field. If we reverse the direction of motion of the loop, the velocity $\vec{v}$ of the conduction electrons is reversed, the magnetic force $-e\vec{v} \times \vec{B}$ should also be reversed, and thus the sign of the voltage on the voltmeter should reverse. If we oscillate the wire back and forth, keeping one end in the magnetic field, we should get an oscillating voltage reading on the meter.

The wonderful thing about this experiment is that all these predictions work precisely as described. There are further simple tests like moving the loop faster to get a stronger magnetic force and therefore a bigger voltage reading. Or stopping the wire in the middle of the magnetic field and getting no voltage reading. They all work!

The next step is to calculate the magnitude of the voltage reading we expect to see. As you follow this calculation, do not worry about the sign of the voltage V because many sign conventions (right hand rules, positive charge, etc.) are involved. Instead concentrate on the basic physical ideas. (In the laboratory, the sign of the voltage V you read on a voltmeter depends on how you attached the leads of the voltmeter to the apparatus. If you wish to change the sign of the voltage reading, you can reverse the leads.)

Since voltage has the dimensions of the potential energy of a unit test charge, the magnitude of the voltage in Figure (3) should be the strength of the force on a unit test charge, $(-e)\,\vec{v} \times \vec{B}$ with $(-e)$ replaced by 1, times the height h over which the force acts. This height h is the height of the magnetic field region in Figure (3). Since v and B are perpendicular, $\left|\vec{v} \times \vec{B}\right| = vB$ and we expect the voltage V to be given by

$$V = \begin{pmatrix} \text{force on unit} \\ \text{test charge} \end{pmatrix} \times \begin{pmatrix} \text{distance over} \\ \text{which force acts} \end{pmatrix}$$

$$\boxed{V = vB \times h}$$ *voltage V on loop moving at speed v through field B*   (5)

Figure 3c
Pulling the coil out of the magnet

AIR CART SPEED DETECTOR

The air cart velocity detector we have previously discussed, provides a direct verification of Equation (5). The only significant difference between the air cart speed detector and the loop in Figure (3) is that the speed detector coil has a number of turns (usually 10). In order to see the effect of having more than one turn in the coil, we show a two turn coil being pulled out of a magnetic field in Figure (4).

Figure (4) is beginning to look like a plumbing diagram for a house. To analyze the diagram, let us start at Position (1) at the top of the voltmeter and follow the wire all the way around until we get to Position (6) at the bottom end of the voltmeter. When we get to Position (2), we enter a region from (2) to (3) where the magnetic force is increasing the electron fluid pressure by an amount vBh, as in Figure (3).

Now instead of going directly to the voltmeter as in Figure (3), we go around until we get to Position (4) where we enter another region, from (4) to (5), where the magnetic force is increasing the fluid pressure. We get another increase of vBh, and then go to Position (6) at the bottom of the voltmeter. In Figure (4) we have two voltage rises as we go around the two loops, and we should get twice the reading on the voltmeter.

$$V = 2vBh \qquad \text{\textit{voltage reading for 2 loops}}$$

It is an easy abstraction to see that if our coil had N turns, the voltage rise would be N times as great, or

$$\boxed{V = NvBh} \qquad \text{\textit{voltage on an N turn coil being pulled out of a magnetic field}} \qquad (6)$$

Adding more turns is an easy way to increase or amplify the voltage.

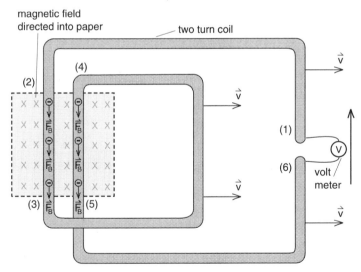

Figure 4
A two turn loop being pulled through a magnetic field. With two turns we have twice as much force pushing the electric fluid toward the bottom of the gap giving twice the voltage V.

The setup for the air cart speed detector is shown in Figure (6). A multi turn coil, etched on a circuit board as shown in Figure (5), is mounted as a sail on top of an air cart. Suspended over the air cart are two angle iron bars with magnets set across the top as shown. This produces a reasonably uniform magnetic field that goes across from one bar to the other as seen in the end view of Figure (6).

In Figure (7), we show the experiment of letting the cart travel at constant speed through the velocity detector. In the initial position (a), the coil has not yet reached the magnetic field and the voltage on the coil is zero, as indicated in the voltage curve at the bottom of the figure.

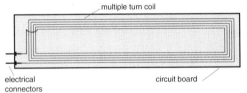

multiple turn coil

electrical connectors

circuit board

Figure 5
The multi turn coil that rides on the air cart. (Only 5 turns are shown.)

The situation most closely corresponding to Figure (4) is position (d) where the coil is leaving the magnet. According to Equation (6), the voltage at this point should be given by $V = N v B h$, where $N = 10$ for our 10 turn coil, v is the speed of the carts, B is the strength of the magnetic field between the angle iron bars, and h is the average height of the coils. (Since the coils are drawn on a circuit board the outer loop has the greatest height h and the inner loop the least.) The first time you use this apparatus, you can directly measure V, N, v and h and use Equation (6) to determine the magnetic field strength B. After that, you know the constants N, B and h, and Equation (6) written as

$$v = V \times \left(\frac{1}{NBh} \right) \qquad (6a)$$

gives you the cart's speed in terms of the measured voltage V. Equation (6a) explains why the apparatus acts as a speed detector.

Let us look at the voltage readings for the other cart positions. The zero readings at Positions (a) and (e) are easily understood. None of the coil is in the magnetic field and therefore there is no magnetic force or voltage.

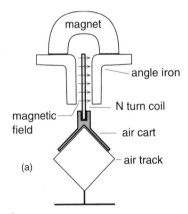

magnet

angle iron

N turn coil

magnetic field

air cart

air track

(a)

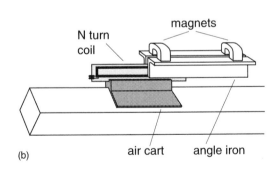

N turn coil

magnets

air cart   angle iron

(b)

Figure 6
The Faraday velocity detector. The apparatus is reasonably easy to build. We first constructed a 10 turn coil by etching the turns of the coil on a circuit board. This was much better than winding a coil, for a wound coil tends to have wrinkles that produce bumps in the data. Light electrical leads, not shown, go directly from the coil to the oscilloscope. The coil is mounted on top of an air cart and moves through a magnetic field produced by two pieces of angle iron with magnets on top as shown. Essentially we have reproduced the setup shown in Figures 3 and 4, but with the coil mounted on an air cart. As long as the coil remains with one end in the magnetic field and the other outside, as shown in (b), there will be a voltage on the leads to the coil that is proportional to the velocity of the cart.

Figure 6c
Velocity detector apparatus. The magnetic field goes across, between the two pieces of angle iron. The coil, mounted on a circuit board, is entering the magnetic field.

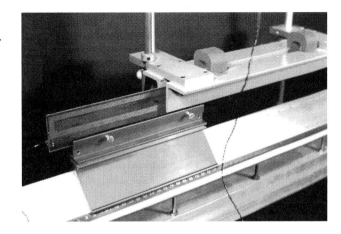

Figure 7
Voltage on the coil as it moves at constant speed through the magnetic field. At position (a) the coil has not yet reached the field and there is no voltage. At position (b) one end of the cart is in the field, the other outside, and we get a voltage proportional to the speed of the cart. At (c) there is no voltage because both ends of the cart are in the magnetic field and the magnetic force on the two ends cancel. (There is no change of magnetic flux at this point.) At (d), the other end alone is inside the field, and we get the opposite voltage from the one we had at (b). (Due to the thickness of the coil and fringing of the magnetic field, the voltage rises and falls will be somewhat rounded.)

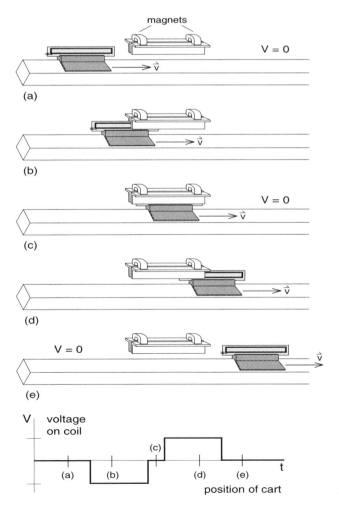

We need a closer look to understand the changes in voltage, when all or part of the coil is inside the magnetic field. This situation, for a one turn coil, is illustrated in Figure (8). For easier interpretation we have moved the gap and voltmeter to the bottom of the coil as shown. It turns out that it does not matter where the gap is located, we get the same voltage reading. We have also labeled the figures (b), (c), and (d) to correspond to the positions of the air cart in Figure (7).

In Figure (8c) where both ends of the coil are in the magnetic field, the conduction electrons are being pulled down in both ends and the fluid is balanced. The electron fluid would not flow in either direction if the gap were closed, thus there is no pressure across the gap and no voltage reading. In contrast, in Figure (8d) where only the left end of the coil is in the magnetic field, the magnetic force on the left side would cause the conduction electrons to flow counterclockwise around the loop if it were not for the gap. There must be an electric pressure or voltage drop across the gap to prevent the counterclockwise flow. This voltage drop is what we measure by the voltmeter.

In Figure (8b), where the coil is entering the magnetic field, the magnetic force on the right side of the coil would try to cause a clockwise flow of the conduction electrons. We should get a pressure or voltage opposite to Figure (8d) where the coil is leaving. This reversal in voltage is seen in the air cart experiment of Figure (7), as the cart travels from (b) to (d).

Note that in Figure (8), where the horizontal sections of the coil are also in the magnetic field, the magnetic force is across rather than along the wire in these sections. This is like the gravitational force on the fluid in a horizontal section of pipe. It does not produce any pressure drops.

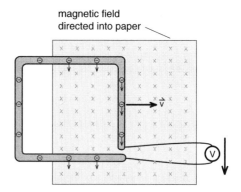

(b) coil entering magnetic field

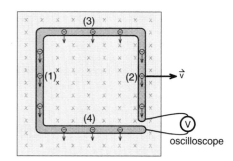

(c) coil completely in magnetic field

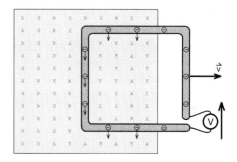

(d) coil leaving magnetic field

Figure 8
When the coil is completely in the magnetic field, the magnetic force on the electrons in the left hand leg (1) is balanced by the force on the electrons in the right hand leg (2), and there is no net pressure or voltage across the gap. When the coil is part way out, there is a voltage across the gap which balances the magnetic force on the electrons. The sign of the voltage depends upon which leg is in the magnetic field.

A RELATIVITY EXPERIMENT

Now that we have seen, from Figure (7), extensive experimental evidence for the magnetic force on the conduction electrons in a wire, let us go back to Figure (3) where we first considered these forces, and slightly modify the experiment. Instead of pulling the coil out of the magnet, let us pull the magnet away from the coil as shown in Figure (9b).

In Figure (9a) we have redrawn Figure (3), and added a stick figure to represent a student who happens to be walking by the apparatus at the same speed that we are pulling the coil out of the magnet. To this moving observer, the coil is at rest and she sees the magnet moving to the left as shown in (9b). In other words, pulling the magnet away from the coil is precisely the same experiment as pulling the coil from the magnet, except it is viewed by a moving observer.

The problem that the moving observer faces in Figure (9b) is that, to her, the electrons in the coil are at rest. For her the electron speed is $v = 0$ and the magnetic force $\vec{F}_B$, given by

$$\vec{F}_B = (-e)\vec{v} \times \vec{B} = 0 \quad (\text{for Figure 9b}) \quad (7)$$

is zero! Without a magnetic force to create the pressure in the electrical fluid in the wire, she might predict that there would be no voltage reading in the voltmeter.

But there is a voltage reading on the voltmeter! We have used this voltage to build our air cart velocity detector. If the voltmeter had a digital readout, for example, then it is clear that everyone would read the same number no matter how they were moving, whether they were like us moving with the magnet (9a), or like her moving with the coil (9b). In other words, she has to find some way to explain the voltage reading that she must see.

The answer she needs lies in the Lorentz force law that we discussed in Chapter 28. This law tells us the total electromagnetic force on a charge q due to either electric or magnetic fields, or both. We wrote the law in the form

$$\vec{F} = q\vec{E} + q\vec{v} \times \vec{B} \quad (28\text{-}20)$$

where $\vec{E}$ and $\vec{B}$ are the electric and magnetic fields acting on the charge.

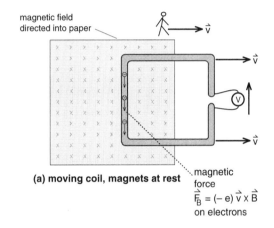

magnetic field
directed into paper

(a) moving coil, magnets at rest

magnetic force
$\vec{F}_B = (-e)\,\vec{v} \times \vec{B}$
on electrons

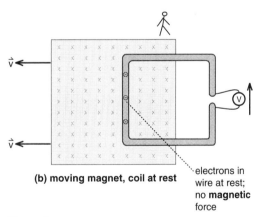

(b) moving magnet, coil at rest

electrons in wire at rest; no **magnetic** force

Figure 9
The only difference between (a) and (b) is the point of view of the observer. In (a) we see a magnetic force $\vec{F}_B = (-e)\,\vec{v} \times \vec{B}$ because the electrons are moving at a speed v through a magnetic field $\vec{B}$. To the observer in (b), the magnet is moving, not the electrons. Since the electrons are at rest, there is no magnetic force on them. Yet the voltmeter reading is the same from both points of view.

Let us propose that the Lorentz force law is generally correct even if we change coordinate systems. In Figure (9a) where we explained everything in terms of a magnetic force on the conduction electrons, there was apparently no electric field and the Lorentz force law gave

$$\vec{F} = q\vec{E} + q\vec{v} \times \vec{B}$$

$$= \left(-e\right)\vec{v} \times \vec{B} \qquad \left(\begin{array}{c} in\ Figure\ 9a, \\ \vec{E} = 0 \end{array}\right) \qquad (8a)$$

In Figure (9b), where $\vec{v} = 0$, we have

$$\vec{F} = q\vec{E} + q\vec{v} \times \vec{B}$$

$$= \left(-e\right)\vec{E} \qquad \left(\begin{array}{c} in\ Figure\ 9b, \\ \vec{v} = 0 \end{array}\right) \qquad (8b)$$

In other words, we will assume that the magnetic force of Figure (9a) has become an electric force in Figure (9b). Equating the two forces gives

$$\vec{E} \left(\begin{array}{c} That\ should\ be \\ in\ Figure\ 9b \end{array}\right) = \vec{v} \times \vec{B} \left(\begin{array}{c} From \\ Figure\ 9a \end{array}\right) \qquad (9)$$

In Figure (9c) we have redrawn Figure (9b) showing an electric field causing the force on the electrons. Because the electrons have a negative charge, the electric field must point up in order to cause a downward force.

That the magnetic force of Figure (9a) becomes an electric force in Figure (9c) should not be a completely surprising result. In our derivation of the magnetic force law, we also saw that an electric force from one point of view was a magnetic force from another point of view. The Lorentz force law, which includes both electric and magnetic forces, has the great advantage that it gives the correct electromagnetic force from any point of view.

Exercise 1

Equation (9) equates $\vec{E}$ in Figure (9c) with $\vec{v} \times \vec{B}$ in Figure (9a). Show that $\vec{E}$ and $\vec{v} \times \vec{B}$ point in the same direction.

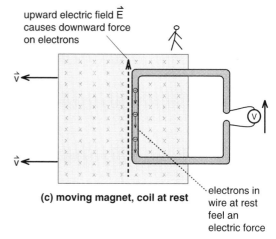

upward electric field $\vec{E}$ causes downward force on electrons

electrons in wire at rest feel an electric force

(c) moving magnet, coil at rest

Figure 9c
From the point of view that the coil is at rest, the downward force on the electrons in the coil must be produced by an upward directed electric field.

FARADAY'S LAW

An experiment whose results may be surprising, is shown in Figure (10). Here we have a magnetic field produced by an electromagnet so that we can turn $\vec{B}$ on and off. We have a wire loop that is large enough to surround but not lie in the magnetic field, so that $\vec{B} = 0$ all along the wire. Again we have a gap and a voltmeter to measure any forces that might be exerted on the conduction electrons in the wire.

We have seen that if we pull the wire out of the magnet, Figure (9a), we will get a voltage reading while the loop is leaving the magnetic field. We have also seen, Figure (9c), that we get a voltage reading if the magnetic field is pulled out of the loop. In both cases we started with a magnetic field through the loop, ended up with no magnetic field through the loop, and got a reading on the voltmeter while the amount of magnetic field through the loop was decreasing.

Now what we are going to do in Figure (10) is simply shut off the electromagnet. Initially we have a magnetic field through the loop, finally no field through the loop. It may or may not be a surprise, but *when we shut off the magnetic field, we also get a voltage reading*. We get a voltage reading if we pull the loop out of the field, the field out of the loop, or shut off the field. We are seeing that we *get a voltage reading whenever we change the amount of magnetic field, the flux of magnetic field, through the loop*.

Magnetic Flux

In our discussion of velocity fields and electric fields, we used the concept of the flux of a field. For the velocity field, the flux Φ_v of water was the volume of water flowing per second past some perpendicular area $A_\perp$. For a uniform stream moving at a speed v, the flux was $\Phi_v = v A_\perp$. For the electric field, the formula for flux was $\Phi_E = E A_\perp$.

In Figures (9 and 10), we have a magnetic field that "flows" through a wire loop. Following the same convention that we used for velocity and electric fields, we will define the magnetic flux Φ_B as the strength of the field $\vec{B}$ times the perpendicular area $A_\perp$ through which the field is flowing

$$\boxed{\Phi_B = BA_\perp} \quad \begin{array}{l}\text{Definition of}\\ \text{magnetic flux}\end{array} \quad (10)$$

In both figures (9) and (10), the flux Φ_B through the wire loop is decreasing. In Figure (9), Φ_B decreases because the perpendicular area $A_\perp$ is decreasing as the loop and the magnet move apart. In Figure (10), the flux Φ_B is decreasing because $\vec{B}$ is being shut off. The important observation is that whenever the flux Φ_B through the loop decreases, whatever the reason for the change may be, we get a voltage reading V on the voltmeter.

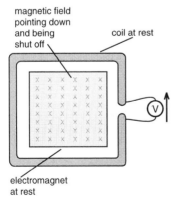

Figure 10
Here we have a large coil that lies completely outside the magnetic field. Thus there is no magnetic force on any of the electrons in the coil wire. Yet when we turn the magnet on or off, we get a reading in the volt meter.

One Form of Faraday's Law

The precise relationship between the ***voltage*** and the ***change in the magnetic flux*** through the loop is found from our analysis of Figure (9) where the loop and the magnet were pulled apart. We got a voltage given by Equation (5) as

$$V = vBh \tag{5}$$

Let us apply Equation (5) to the case where the magnet is being pulled out of the loop as shown in Figure (11). In a time dt, the magnet moves to the left a distance dx given by

$$dx = vdt \tag{11}$$

and the area of magnetic field that has left the loop, shown by the cross hatched band in Figure (11), is

$$dA = hdx = \begin{cases} \text{area of magnetic} \\ \text{field that has} \\ \text{left the loop} \end{cases} \tag{12}$$

This decrease in area causes a decrease in the magnetic flux $\Phi_B = BA_\perp$ through the loop. The change in flux $d\Phi_B$ is given by

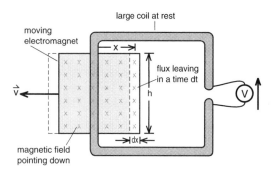

large coil at rest

moving electromagnet

flux leaving in a time dt

magnetic field pointing down

Figure 11
As the magnet and the coil move away from each other, the amount of magnetic flux through the coil decreases. When the magnet has moved a distance dx, the decrease in area is hdx, and the magnetic flux decreases by B×hdx.

$$d\Phi_B = -BdA = -Bhdx$$
$$= -Bhvdt \tag{13}$$

where the – sign indicates a reduction in flux, and we used Equation (11) to replace dx by vdt.

Dividing both sides of Equation (13) by dt gives

$$\frac{d\Phi_B}{dt} = -Bhv \tag{14}$$

But Bhv is just our voltmeter reading. Thus we get the surprisingly simple formula

$$\boxed{V = -\frac{d\Phi_B}{dt}} \quad \begin{matrix} \textit{One form of} \\ \textit{Faraday's law} \end{matrix} \tag{15}$$

Equation (15) is one form of Faraday's law.

Equation (15) has a generality that goes beyond our original analysis of the magnetic force on the conduction electrons. It makes no statement about what causes the magnetic flux to change. We can pull the loop out of the field as in Figure (9a), the field out of the loop as in Figure (9b), or shut the field off as in Figure (10). In all three cases Equation (15) predicts that we should see a voltage, and we do.

If we have a coil with more than one turn, as we had back in Figure (4), and put a voltmeter across the ends of the coil, then we get N times the voltage, and Equation (15) becomes

$$V = N\left(-\frac{d\Phi_B}{dt}\right) \quad \begin{matrix} \textit{for a coil} \\ \textit{with N turns} \end{matrix} \tag{15a}$$

provided $d\Phi_B/dt$ is the rate of change of magnetic flux in each loop of the coil.

Exercise 2

Go back to Figure (7) and explain the voltage plot in terms of the **rate of change of the flux of magnetic field** through the coil riding on top of the air cart.

A Circular Electric Field

In Figure (10), where we shut the magnet off and got a voltage reading on the voltmeter, there must have been some force on the electrons in the wire to produce the voltage. Since there was no magnetic field out at the wire, the force must have been produced by an electric field. We already have a hint of what that electric field looks like from Figure (9c). In that figure, we saw that the moving magnetic field created an upwardly directed electric field acting on the electrons on the left side of the wire loop.

To figure out the shape of the electric field produced when we shut off the magnet, consider Figure (12), where we have a circular magnet and a circular loop of wire . We chose this geometry so that the problem would have circular symmetry (except at the gap in the loop).

To produce the same kind of voltage V that we have seen in the previous experiments, the electric field at the wire must be directed up on the left hand side, as it was in Figure (9c). But because of the circular symmetry of the setup in Figure (12), the upwardly directed electric field on the left side, which is parallel to the wire, must remain parallel to the wire as we go around the wire loop. In other words, the only way we can have an upwardly directed electric field acting on the electrons on the left side of the loop, and maintain circular symmetry, is to have the electric field go in a circle all the way around the loop as shown in Figure (12).

We can determine the strength of this circular electric field, by figuring out how strong an electric field must act on the electrons in the wire, in order to produce the voltage V across the gap. We then use Equation (15) to relate this voltage to the rate of change of the magnetic flux through the loop.

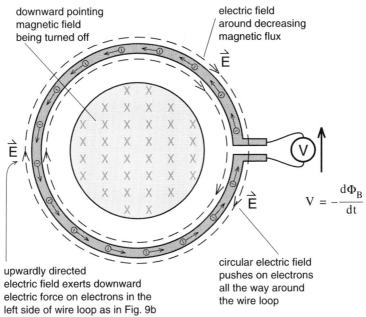

downward pointing
magnetic field
being turned off

electric field
around decreasing
magnetic flux

$\vec{E}$

$\vec{E}$

$\vec{E}$

$$V = -\frac{d\Phi_B}{dt}$$

upwardly directed
electric field exerts downward
electric force on electrons in the
left side of wire loop as in Fig. 9b

circular electric field
pushes on electrons
all the way around
the wire loop

Figure 12
When the magnetic field in the magnet is turned off, a circular electric field is generated. This electric field exerts a force on the electrons in the wire, creating a pressure in the electric fluid that is recorded as a voltage pulse by the voltmeter.

Recall that the definition of electric voltage used in deriving Equation (5) was

$$V = \begin{pmatrix} \text{force on unit} \\ \text{test charge} \end{pmatrix} \times \begin{pmatrix} \text{distance over} \\ \text{which force acts} \end{pmatrix}$$

For Figure (12), the force on a unit test charge is the electric field $\vec{E}$, and this force acts over the full circumference $2\pi r$ of the wire loop. Thus the voltage V across the gap is

$$V = E \times 2\pi r$$

Equating this voltage to the rate of change of magnetic flux through the wire loop gives

$$V = E \times 2\pi r = -\frac{d\Phi_B}{dt} \qquad (16)$$

Equation (16) tells us that the faster the magnetic field dies, i.e. the greater $d\Phi_B/dt$, the stronger the electric field $\vec{E}$ produced.

Line Integral of $\vec{E}$ around a Closed Path

In Figure (13) we have removed the wire loop and volt meter from Figure (12) so that we can focus our attention on the circular electric field produced by the decreasing magnetic flux. This is not the first time we have encountered a circular field. The velocity field of a vortex and the magnetic field of a straight current carrying wire are both circular. We have redrawn Figure (29-10) from the last chapter, showing the circular magnetic field around a wire.

The formula for the strength of the magnetic field in Figure (29-10) is

$$B \times 2\pi r = \mu_0 i \qquad (28\text{-}18)$$

a result we derived back in Equation 28-18. This should be compared with the formula for the strength of the electric field in Figure (13)

$$E \times 2\pi r = -\frac{d\Phi_B}{dt} \qquad (16)$$

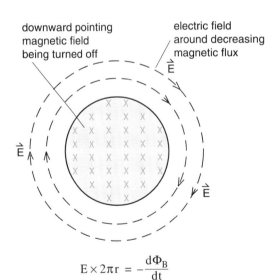

$$E \times 2\pi r = -\frac{d\Phi_B}{dt}$$

Figure 13
*Circular electric field around
a changing magnetic flux.*

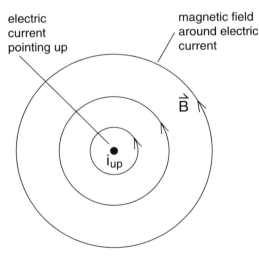

$$B \times 2\pi r = \mu_0 i$$

Figure 29-10
*Circular magnetic field
around an electric current.*

In our discussion of Ampere's law, we called $\mu_0 i$ the "source" of the circular magnetic field. By analogy, we should think of the rate of change of magnetic flux, $-d\Phi_B/dt$, as the "source" of the circular electric field.

In Chapter 29, we generalized Ampere's law by replacing $B * 2\pi r$ by the line integral $\oint \vec{B} \cdot d\vec{\ell}$ along a closed path around the wire. The result was

$$\boxed{\oint \vec{B} \cdot d\vec{\ell} = \mu_0 i} \quad \begin{array}{l} \textit{Ampere's law} \\ \textit{for} \\ \textit{magnetic fields} \end{array} \quad (29\text{-}18)$$

where the line integral can be carried out along any closed path surrounding the wire. Because of close analogy between the structure and magnitude of the magnetic field in Figure (29-10) and the electric field in Figure (13), we expect that the more general formula for the electric field produced by a changing magnetic flux is

$$\boxed{\oint \vec{E} \cdot d\vec{\ell} = -\frac{d\Phi_B}{dt}} \quad \begin{array}{l} \textit{Faraday's law} \\ \textit{for} \\ \textit{electric fields} \end{array} \quad (17)$$

Equation 17 is the most general form of Faraday's law. It says that the line integral of the electric field around any closed path is equal to (minus) the rate of change of magnetic flux through the path.

USING FARADAY'S LAW

Up until now we have been looking for arguments leading up to Faraday's law. Let us now reverse the procedure, treating Equation 17 as a basic law for electric fields, and see what the consequences are.

Electric Field of an Electromagnet

As a beginning exercise in the use of Faraday's law, let us use Equation (17) to calculate the electric field of the electromagnet in Figure (13). We first argue that because of the circular symmetry, the electric field should travel in circles around the decreasing magnetic field. Thus we choose a circular path, shown in Figure (13a), along which we will calculate $\oint \vec{E} \cdot d\vec{\ell}$. Then using the assumption (because of circular symmetry) that $\vec{E}$ is parallel to $d\vec{\ell}$ and has a constant magnitude all the way around the circular path, we can write

$$\oint \vec{E} \cdot d\vec{\ell} = \oint E\, d\ell = E \oint d\ell = E\, 2\pi r \quad (18)$$

Using this result in Equation (17) gives

$$\oint \vec{E} \cdot d\vec{\ell} = E\, 2\pi r = -\frac{d\Phi_B}{dt} \quad (19)$$

which is the result we had in Equation (16).

Right Hand Rule for Faraday's Law

We can get the correct direction for $\vec{E}$ with the following right hand rule. Point the thumb of your right hand in the direction of the magnetic field. If the magnetic flux is decreasing (if $-d\Phi_B/dt$ is positive), then the fingers of your right hand curl in the direction of $\vec{E}$. If the magnetic flux is increasing, then $\vec{E}$ points the other way. *Please practice this right hand rule on Figures (13a), (9c), and (15).*

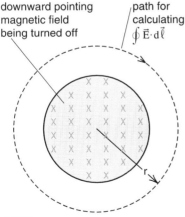

downward pointing magnetic field being turned off

path for calculating $\oint \vec{E} \cdot d\vec{\ell}$

r

Figure 13a
Using Faraday's law to calculate $\vec{E}$.

Electric Field of Static Charges

If all we have around are static electric charges, then there are no magnetic fields, no magnetic flux, and no changing magnetic flux. For this special case, $d\Phi_B/dt = 0$ and Faraday's law gives

$$\oint \vec{E} \cdot d\vec{\ell} = 0 \quad \begin{cases} \text{for electric fields} \\ \text{produced by} \\ \text{static charges} \end{cases} \quad (20)$$

When the line integral of a force is zero around any closed path, we say that the force is ***conservative***. (See Equation 29-12.) Thus we see that if we have only static electric charge (or constant magnetic fields), the electric field is a conservative field.

In contrast, if we have changing magnetic fields, if $d\Phi_B/dt$ is not zero, the electric field is not conservative. This can lead to some rather interesting results which we will see in our discussion of a device called the *betatron*.

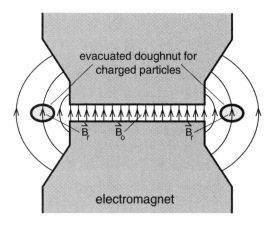

Figure 14a
Cross-sectional view of a betatron, showing the central field $\vec{B}_0$ and the field $\vec{B}_r$ out at the evacuated doughnut. The relative strength of $\vec{B}_0$ and $\vec{B}_r$ can be adjusted by changing the shape of the electromagnet pole pieces.

THE BETATRON

As we have mentioned before, when you encounter a new and strange equation like Faraday's law, it is essential to have an example that you know inside out that illustrates the equation. This transforms the equation from a collection of symbols into a set of instructions for solving problems and making predictions. One of the best examples to learn for the early form of Faraday's law, Equation (15a), was the air cart speed detector experiment shown in Figure (7). (You should have done Exercise 2 analyzing the experiment using Equation (15a).

The most direct example illustrating Faraday's law for electric fields, Equation (17), is the particle accelerator called the betatron. This device was used in the 1950s for study of elementary particles, and later for creating electron beams for medical research.

A cross-sectional view of the betatron is shown in Figure (14a). The device consists of a large electromagnet with a circular evacuated doughnut shaped chamber for the electrons. The circular shape of the electromagnet and the evacuated chamber are more clearly seen in the top view, Figure (14b). In that view we show the strong upward directed magnetic field $\vec{B}_0$ in the gap and the weaker upward directed magnetic field out at the evacuated doughnut.

The outer magnetic field B_r is required to keep the electrons moving along a circular orbit inside the evacuated chamber. This field exerts a force $\vec{F}_B = (-e) \vec{v} \times \vec{B}_r$ that points toward the center of the circle and has a magnitude mv^2/r in order to produce the required radial acceleration. Thus B_r is given by

$$B_r = \frac{mv}{er}$$

which is our familiar formula for electrons moving along a circular path in a magnetic field. *(As a quick review, derive the above equation.)*

Since a magnetic field does no work we need some means of accelerating the electrons. In a synchrotron, shown in Figure (28-27), a cavity which produces an electric accelerating field is inserted into the electron's path. As an electron gains energy and momentum (mv) each time it goes through the cavity, the magnetic field

B was increased so that the electron's orbital radius $r = mv/eB$ remains constant. (The synchronizing of B with the momentum mv leads to the name synchrotron.)

In the betatron of Figure (14), we have a magnetic field B_r to keep the electrons in a circular orbit, and as the electrons are accelerated, B_r is increased to keep the electrons in an orbit of constant radius r. But what accelerates the electrons? There is no cavity as in a synchrotron.

Suppose that both B_0 and B_r are increased simultaneously. In the design shown in Figure (14a), $\vec{B}_0$ and $\vec{B}_r$ are produced by the same electromagnet, so that we can increase both together by turning up the electromagnet. If the strong central field $\vec{B}_0$ is increased, we have a large change in the magnetic flux through the electron orbit, and therefore by Faraday's law $\oint \vec{E} \cdot d\vec{\ell} = -d\Phi_B/dt$ we must have a circular electric

field around the flux as shown in Figure (15), just as in Figure (13). This electric field is exactly parallel to the orbit of the electrons and accelerates them continuously as they go around.

What is elegant about the application of Faraday's law to the electrons in the betatron, is that $\oint \vec{E} \cdot d\vec{\ell}$, which has the dimensions of voltage, is the *voltage gained by an electron going once around the circular orbit*. *The energy gained is just this voltage in electron volts*

$$\left.\begin{array}{l}\text{energy gained} \\ \text{(in eV) by electron} \\ \text{going around once}\end{array}\right\} = \oint \vec{E} \cdot d\vec{\ell} \qquad (21)$$

This voltage is then related to $d\Phi_B/dt$ by Faraday's law.

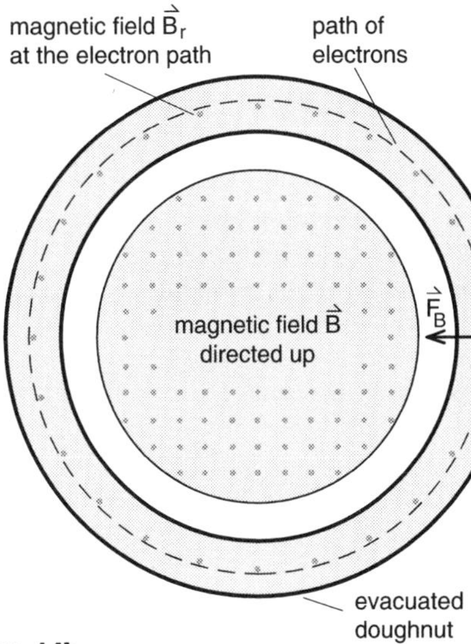

magnetic field $\vec{B}_r$ at the electron path   path of electrons

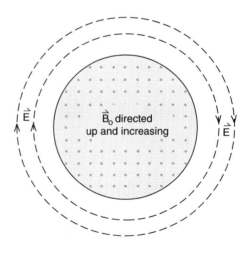

Figure 14b
Top view of the betatron showing the evacuated doughnut, the path of the electrons, and the magnetic fields $\vec{B}_0$ in the center and $\vec{B}_r$ out at the electron path. In order to keep the electrons moving on a circular path inside the doughnut, the magnetic force $\vec{F}_B = (-e)\vec{v} \times \vec{B}_r$ must have a magnitude $\vec{F}_B = mv^2/r$ where r is the radius of the evacuated doughnut.

Figure 15
When the strong central field $\vec{B}_0$ in the betatron is rapidly increased, it produces a circular electric field that is used to accelerate the electrons. The electric field $\vec{E}$ is related to the flux Φ_B of the central field $\vec{B}_0$ by Faraday's law $\oint \vec{E} \cdot d\vec{\ell} = -d\Phi_B/dt$.

Let us consider an explicit example to get a feeling for the kind of numbers involved. In the 100 MeV betatron built by General Electric, the electron orbital radius is 84 cm, and the magnetic field B_0 is cycled from 0 to .8 tesla in about 4 milliseconds. (The field B_0 is then dropped back to 0 and a new batch of electrons are accelerated. The cycle is repeated 60 times a second.)

The maximum flux Φ_m through the orbit is

$$\Phi_m = \left(B_0\right)_{max} \pi r^2 = .8 \text{ tesla} \times \pi \times (.84\text{m})^2$$

$$\Phi_m = 1.8 \text{ tesla m}^2$$

If this amount of flux is created in 4 milliseconds, then the average value of the rate of change of magnetic flux Φ_B is

$$\frac{d\Phi_B}{dt} = \frac{\Phi_m}{.004 \text{ sec}} = \frac{1.8}{.004} = 450 \text{ volts}$$

Thus each electron gains 450 electron volts of kinetic energy each time it goes once around its orbit.

Exercise 3

(a) How many times must the electron go around to reach its final voltage of 100 MeV advertised by the manufacturer?

(b) For a short while, until the electron's kinetic energy gets up to about the electron's rest energy m_0c^2, the electron is traveling at speeds noticeably less than c. After that the electron's speed remains very close to c. How many orbits does the electron have to make before its kinetic energy equals its rest energy? What fraction of the total is this?

(c) How long does it take the electron to go from the point that its kinetic energy equals its rest energy, up to the maximum of 100 MeV? Does this time fit within the 4 milliseconds that the magnetic flux is being increased?

TWO KINDS OF FIELDS

At the beginning of the chapter we showed that the line integral $\oint \vec{E} \cdot d\vec{\ell}$ around a closed path was zero for any electric field produced by static charges. Now we see that the line integral is not zero for the electric field produced by a changing magnetic flux. Instead it is given by Faraday's law $\oint \vec{E} \cdot d\vec{\ell} = -d\Phi_B/dt$. These results are shown schematically in Figure (16) where we are looking at the electric field of a charged rod in (16a) and a betatron in (16b).

In Figure (17), we have sketched a wire loop with a voltmeter, the arrangement we used in Figure (12) to measure the $\oint \vec{E} \cdot d\vec{\ell}$. We will call this device an "$\oint \vec{E} \cdot d\vec{\ell}$ *meter*". If you put the $\oint \vec{E} \cdot d\vec{\ell}$ meter over the changing magnetic flux in Figure (16b), the voltmeter will show a reading of magnitude $V = d\Phi_B/dt$. If we put the $\oint \vec{E} \cdot d\vec{\ell}$ meter over the charged rod in Figure (16a), the meter reads $V = 0$. Thus we have a simple physical device, our $\oint \vec{E} \cdot d\vec{\ell}$ meter, which can distinguish the radial field in Figure (16a) from the circular field in Figure (16b). In fact it can distinguish the circular field in (16b) from any electric field $\vec{E}$ whatsoever that we can construct from static charges. Our $\oint \vec{E} \cdot d\vec{\ell}$ meter allows us to separate all electric fields into two kinds, those like the one in (16b) that can give a **non zero reading**, and those, produced by static charges, which give a **zero reading**.

Fields which register on our $\oint \vec{E} \cdot d\vec{\ell}$ meter generally close on themselves like the circular fields in (16b). Since these fields do not appear to have sources, they are called sourceless or "**solenoidal**" fields. An $\oint \vec{E} \cdot d\vec{\ell}$ meter is the kind of device we need to detect solenoidal fields.

The conservative fields produced by static charges never close on themselves. They always start on positive charge, end on negative charge, or come from or go to infinity. These fields diverge from point charges and thus are sometimes called "**divergent**" fields. Our $\oint \vec{E} \cdot d\vec{\ell}$ meter does not work on the divergent fields because we always get a zero reading.

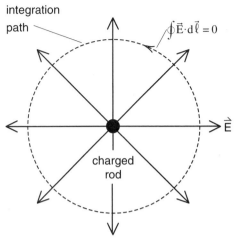

integration
path

$$\oint \vec{E} \cdot d\vec{\ell} = 0$$

charged
rod

$\vec{E}$

(a) Electric field of a static charge distribution
has the property $\oint \vec{E} \cdot d\vec{\ell} = 0$

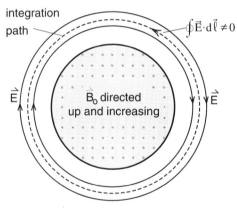

integration
path

$$\oint \vec{E} \cdot d\vec{\ell} \neq 0$$

B_o directed
up and increasing

$\vec{E}$ $\vec{E}$

(b) Electric field produced by a changing
magnetic flux has $\oint \vec{E} \cdot d\vec{\ell} = - d\vec{\Phi}_B/dt$

Figure 16
*Two kinds of electric field. Only the
field produced by the changing magnetic
flux has a non zero line integral.*

Although the $\oint \vec{E} \cdot d\vec{\ell}$ meter does not work on divergent fields, Gauss' law with the surface integral does. In a number of examples we used Gauss' law

$$\int_{\substack{closed \\ surface}} \vec{E} \cdot d\vec{A} = \frac{Q_{in}}{\varepsilon_0} \qquad (29\text{-}5)$$

to calculate the electric field of static charges. We are seeing now that we *use a surface integral to measure divergent fields, and a line integral to measure solenoidal fields*. There are two kinds of electric fields, and we have two kinds of integrals to detect them.

It turns out to be a general mathematical theorem that any vector field can be separated into a purely divergent part and a purely solenoidal part. The field can be uniquely specified if we have both an equation involving a Gauss' law type surface integral to tell us the divergent part, and an equation involving a Faraday's law type line integral to tell us the solenoidal part.

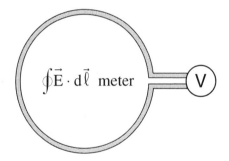

$$\oint \vec{E} \cdot d\vec{\ell} \text{ meter}$$

V

Figure 17
*Wire loop and a volt meter can be used directly to
measure $\oint \vec{E} \cdot d\vec{\ell}$ around the loop. We like to call
this apparatus an $\oint \vec{E} \cdot d\vec{\ell}$ meter.*

Exercise 4

a) Maxwell's equations are a set of equations that completely define the behavior of electric fields $\vec{E}$ and magnetic fields $\vec{B}$.   One of Maxwell's equations is Faraday's law

$$\oint \vec{E} \cdot d\vec{\ell} = -d\Phi_B/dt$$

which gives the line integral for the electric field. *How many Maxwell equations are there?* (How many equations will it take to completely define both $\vec{E}$ and $\vec{B}$?)

b) Are any of the other equations for electric and magnetic fields we have discussed earlier, candidates to be one of Maxwell's equations?

c) At least one of Maxwell's equations is missing – we have not discussed it. Can you guess what the equation is and write it down? Explain what you can about your guess.

d) Back in our early discussion of velocity fields and Gauss' law, we said that a *point source* for the velocity field of an incompressible fluid like water, was a small "magic" sphere in which water molecules were created. Suppose we do not believe in magic and assume that for real water there is no way that water molecules can be created or destroyed. Write down an integral equation for real water that expresses the fact that the $\vec{v}_{real\ water}$ has no sources (that create water molecules) or sinks (that destroy them).

Do the best you can on these exercises now. Keep a record of your work, and see how well you did when we discuss the answers later in chapter 32.

Note on our $\oint \vec{E} \cdot d\vec{\ell}$ meter

Back in Figure (17) we used a wire loop and a voltmeter as an $\oint \vec{E} \cdot d\vec{\ell}$ meter. I.e., we are saying that the voltage reading V on the voltmeter gives us the integral of E around the closed path defined by the wire loop. This is strictly true for a loop at rest, where the conduction electrons experience no magnetic force and all forces creating the electric pressure are caused by the electric field E.

*Earlier, in Figure (9), we had two views of an $\oint \vec{E} \cdot d\vec{\ell}$ meter. In the bottom view, (9b) the loop is at rest and the voltage must be caused by an electric force. The moving magnetic field must have an electric field associated with it. But in Figure (9a) where the magnet is at rest, there is no electric field and the voltage reading is caused by the magnetic force on the conduction electrons in the moving wire. Strictly speaking, in Figure (9a) the wire loop and voltmeter are measuring a pressure caused by magnetic forces and not an $\oint \vec{E} \cdot d\vec{\ell}$. The wire loop must be at rest, **the path for our line integral cannot move**, if we are measuring $\oint \vec{E} \cdot d\vec{\ell}$.*

In practice, however, it makes little difference whether we move the magnet or the loop, because the principle of relativity requires that we get the same voltage V.

APPLICATIONS OF FARADAY'S LAW

The last few sections have been somewhat heavy on theory. To end this chapter on a more practical note, we will consider some simple applications of Faraday's law, one that has immense practical applications and another that we can use in the laboratory. First we will discuss the AC voltage generator which is used by most power stations throughout the world. We will also describe a field mapping experiment in which we use our $\oint \vec{E} \cdot d\vec{\ell}$ meter to map the magnetic field of a pair of Helmholtz coils. In the next chapter Faraday's law is used to explain the operation of transformers and inductors that are common circuit elements in radio and television sets.

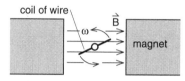

a) end view of a coil of wire rotating in a magnetic field

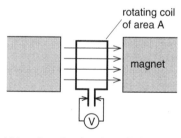

b) top view showing the coil of area A

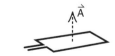

c) Vector $\vec{A}$ representing the area of the loop

Figure 18
An electric generator consists of a coil of wire rotating in a magnetic field.

The AC Voltage Generator

In Figure (18) we have inserted a wire loop of area $\vec{A}$ in the magnetic field $\vec{B}$ of a magnet. We then rotate the coil at a frequency ω about an axis of the coil as shown. We also attach a voltmeter to the coil, using sliding contacts so that the voltmeter leads do not twist as the coil spins.

As shown in Figure (19), as the loop turns, the magnetic flux changes sinusoidally from a maximum positive flux in (19a) to zero flux in (c) to a maximum negative flux in (d) to zero in (e). In (18c), we have shown the vector $\vec{A}$ representing the area of the coil ($\vec{A}$ points

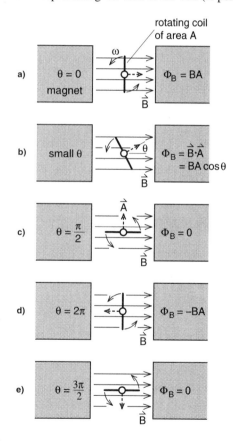

Figure 19
The changing magnetic flux through the rotating loop. The general formula for Φ_B is $BA\cos\theta$ where θ is the angle shown in (b), between the magnetic field and the normal to the loop. If the coil is rotating uniformly, then $\theta = \omega t$, and $\Phi_B = BA\cos\omega t$

perpendicular to the plane of the coil) and we can use our usual formula for magnetic flux to get

$$\Phi_B = \vec{B} \cdot \vec{A} = BA\cos\theta \qquad (22)$$

If the coil is rotating at a constant angular velocity ω, then $\theta = \omega t$ and we have

$$\Phi_B = BA\cos\omega t \qquad (23)$$

Differentiating Equation (23) with respect to time gives

$$\frac{d\Phi_B}{dt} = -\omega BA\sin\omega t \qquad (24)$$

Finally we use Faraday's law in the form

$$V = -\frac{d\Phi_B}{dt} \qquad (15)$$

to predict that the voltage V on the voltmeter will be

$$V = \omega BA\sin\omega t \qquad (25)$$

If we use a coil with N turns, we get a voltage N times as great, or

$$V = \omega NBA\sin\omega t = V_0\sin\omega t \qquad (26)$$

where V_0 is the amplitude of the sine wave as shown in Figure (20). Equation (26) shows that by rotating a coil in a magnetic field, we get an alternating or "AC" voltage. Power stations use this same principle to generate AC voltages.

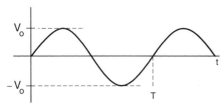

Figure 20
Amplitude and period of a sine wave.

Equation 26 predicts that the voltage amplitude V_0 produced by an N turn coil of area A rotating in a magnetic field B is

$$V_0 = \omega NBA \qquad (27)$$

where the angular frequency ω radians per second is related to the frequency f cycles per second and the period T seconds per cycle by

$$\omega \frac{\text{rad}}{\text{sec}} = 2\pi \frac{\text{rad}}{\text{cycle}} \times f \frac{\text{cycle}}{\text{sec}} = 2\pi \frac{\text{rad}}{\text{cycle}} \times \frac{1}{T \frac{\text{sec}}{\text{cycle}}}$$

Exercise 5
Suppose that you have a magnetic field B = 1 tesla, and you rotate the coil at 60 revolutions (cycles) per second. Design a generator that will produce a sine wave voltage whose amplitude is 120 volts.

Exercise 6
Figures (21a,b) show the voltage produced by a coil of wire rotating in a uniform magnetic field of a fairly large electromagnet. (The setup is similar to that shown in Figures 18 and 19.) The coil was square, 4 cm on a side, and had 10 turns. To go from the results shown in Figure (21a) to those shown in Figure (21b), we increased the rotational speed of the motor turning the coil. In both diagrams, we have selected one cycle of the output wave, and see that the frequency has increased from 10 cycles per second to nearly 31 cycles per second.

a) Explain why the amplitude of the voltage signal increased in going from Figure (21a) to (21b). Is the increase what you expected?

b) Calculate the strength of the magnetic field of the electromagnet used. Do you get the same answer using Figure (21a) and using Figure (21b)?

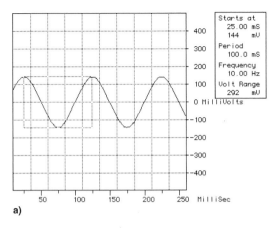

a)

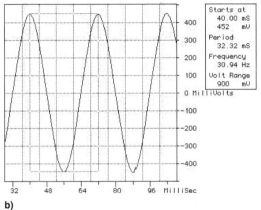

b)

Figure 21
Voltage output from a coil rotating in a uniform magnetic field. The coil was 4 cm on a side, and had 10 turns. In each figure we have selected one cycle of the output wave, and see that the frequency of rotation increased from 10 cycles per second in a) to nearly 31 cycles per second in b).

Gaussmeter

Exercise 6 demonstrates one way to measure the strength of the magnetic field of a magnet. By spinning a coil in a magnetic field, we produce a voltage amplitude given by Equation 27 as $V_0 = \omega NBA$. Thus by measuring V_0, ω, N, and A, we can solve for the magnetic field B.

A device designed to measure magnetic fields is called a *gaussmeter*. A commercial gaussmeter, used in our plasma physics lab, had a small coil mounted in the tip of a metal tube as shown in Figure (22). A small motor also in the tube spun the coil at high speed, and the amplitude V_0 of the coil voltage was displayed on a meter. The meter could have been calibrated using Equation (27), but more likely was calibrated by inserting the spinning coil into a known magnetic field.

In an attempt to measure the magnetic field in the Helmholtz coils used for our electron gun experiments, students have also built rotating coil gaussmeters. Despite excellent workmanship, the results were uniformly poor. The electrical noise generated by the sliding contacts and the motor swamped the desired signal except when B was strong. This approach turned out not to be the best way to measure $\vec{B}$ in the Helmholtz coils.

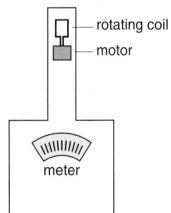

rotating coil
motor

meter

Figure 22
A commercial gauss meter, which measures the strength of a magnetic field, has a motor and a rotating coil like that shown in Figure 18. The amplitude V_0 of the voltage signal is displayed on a meter that is calibrated in gauss.

A Field Mapping Experiment

To measure the magnetic field in the Helmholtz coils, it is far easier to "rotate the field" than the detector loop. That is, use an alternating current in the Helmholtz coils, and you will get an alternating magnetic field in the form

$$B = B_0 \sin \omega t \tag{28}$$

where w is the frequency of the AC current in the coils. Simply place a stationary detector loop in the magnetic field as shown in Figure (23) and the magnetic flux through the detector loop will be

$$\Phi_B = \vec{B} \cdot \vec{A} = \vec{B}_0 \cdot \vec{A} \sin \omega t \tag{29}$$

where $\vec{A}$ is the area of the detector loop. By Faraday's law, the voltage in the voltmeter or oscilloscope attached to the detector loop is given by

$$V = -\frac{d\Phi_B}{dt} = -\left(\omega \vec{B}_0 \cdot \vec{A}\right) \cos \omega t \tag{30}$$

If our detector loop has N turns of wire, then the voltage will be N times as great, and the amplitude V_0 we see on the oscilloscope screen will be

$$\boxed{V_0 = N\omega \left(\vec{B}_0 \cdot \vec{A}\right)} \tag{31}$$

This is essentially the same formula we had for the rotating coil gaussmeter, Equation (27). The difference is that by "rotating the field" rather than the coil, we avoid sliding contacts, motors, electrical noise, and can make very precise measurements.

A feature of Equation (31) that we did not have when we rotated the coil is the dot product $\vec{B}_0 \cdot \vec{A}$. When the detector coil is aligned so that its area vector $\vec{A}$ (which is perpendicular to the plane of the detector coil) is parallel to $\vec{B}_0$, the dot product $\vec{B}_0 \cdot \vec{A}$ is a maximum. Thus we not only measure the magnitude of $\vec{B}_0$, we also get the direction by reorienting the detector coil until the V_0 is a maximum.

As a result, a small coil attached to an oscilloscope, which is our $\oint \vec{E} \cdot d\vec{\ell}$ meter, can be used to accurately map the magnitude and direction of the magnetic field of the Helmholtz coils, or of any coil of wire. Unlike our earlier electric field mapping experiments, there are no mysteries or unknown constants. Faraday's law, through Equation (31), gives us a precise relation between the observed voltage and the magnetic field. The experimental setup is seen in Figure (24).

Still another way to measure magnetic fields is illustrated in Exercise 7.

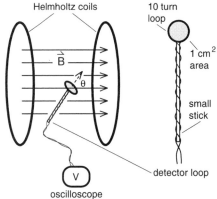

Figure 23
If you use an alternating current in the Helmholtz coils, then B has an alternating amplitude $B = B_0 \cos(\omega t)$. You can then easily map this field with the detector loop shown above. If you orient the loop so that the signal on the oscilloscope is a maximum, then you know that $\vec{B}$ is perpendicular to the detector loop and has a magnitude given by
$V = V_0 \sin(\omega t) = d\Phi_B/dt = d/dt\big(NAB\cos(\omega t)\big).$

Figure 24
Experimental setup for the magnetic field mapping experiment. A 60 cycle AC current is running through the Helmholtz coils, producing an alternating magnetic flux through the 10 turn search coil. The resulting induced voltage is seen on the oscilloscope screen.

Exercise 7

The point of this experiment is to determine the strength of the magnetic field produced by the small magnets that sat on the angle iron bars in the velocity detector apparatus. We placed a short piece of wood between two magnets so that there was a small gap between the ends as seen in the actual size computer scan of Figure (25). The pair of magnets were then suspended over the air track as shown in Figure (26).

On top of the air cart we mounted a **single turn** coil. When the air cart passes under the magnets, the single turn coil passes through the lower gap between the magnets as shown. The dimensions of the single turn coil are shown in Figure 27. We also show the dimensions and location of the lower end of one of the magnets at a time when the coil has passed part way through the gap. You can see that, at this point, all the magnetic flux across the lower gap is passing completely through the single turn coil. Figure 28 is a recording of the induced voltage in the single turn coil as the coil passes completely through the gap. The left hand blip was produced when the coil entered the gap, and the right hand blip when the coil left the gap. The air track was horizontal, so that the speed of the air cart was constant as the coil moved through the gap. Determine the strength of the magnetic field B in the gap. Show and explain your work.

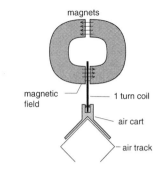

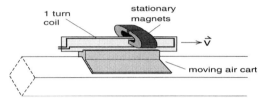

Figure 26
A single turn coil, mounted on an air cart, moves through the lower gap between the magnets.

Figure 25
Two C Magnets with wood spacer.

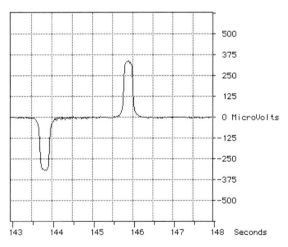

Figure 28
Voltage induced in the single turn coil.

Figure 27
Dimensions of the single turn coil. We also show the dimensions of the end of the magnets through which the coil is passing.

Exercise 8

As shown in Figure (29), we started with a solenoid with 219 turns wrapped in a 1" diameter plastic tube. The coil is 45.4 cm long. The current going through the coil first goes through a .1 Ω resistor. By measuring the voltage V_1 across that resistor, we can determine the current through the solenoid. V_1 is shown as the lower curve in Figure (30).

a) Using V_1 from Figure 30, calculate the magnitude B of the magnetic field in the solenoid.

We then wound 150 turns of wire around the center section of the solenoid, as indicated in Figure (29). You can see that the entire flux Φ_1 of the Magnetic field of the solenoid, goes through all the turns of the outer coil.

b) Use this fact to predict the voltage V_2 across the outer coil, and then compare your prediction with the experimental V_2 shown in the upper curve of Figure (30).

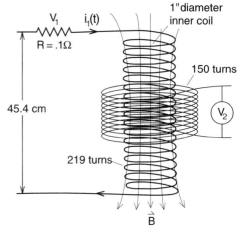

Figure 29
The inner (primary) coil 1 is 45.4 cm long, has 219 turns and is wound on a 2.54 cm (1") diameter tube. The outer (secondary) coil consists of 150 turns wound tightly around the center section of the primary coil. The current through the primary coil goes through a .1Ω resistor, and the voltage V_1 is measured across that resistor. V_2 is the voltage induced in the secondary coil.

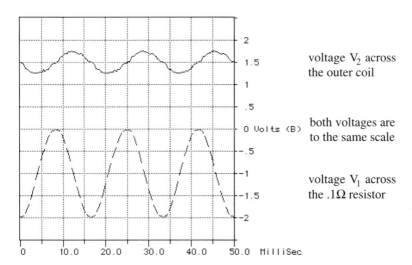

voltage V_2 across the outer coil

both voltages are to the same scale

voltage V_1 across the .1Ω resistor

Figure 30
The voltage V_1 across the .1Ω resistor measures the current in the primary (219 turn) coil. V_2 is the voltage induces in the secondary (outer 150 turn) coil.

Chapter 31
Induction and
Magnetic Moment

In this chapter we discuss several applications of
Faraday's law and the Lorentz force law. The first is
to the **inductor** which is a common electronic circuit
element. We will pay particular attention to a circuit
containing an inductor and a capacitor, in which an
electric current oscillates back and forth between the
two. Measurements of the period of the oscillation and
dimensions of the circuit elements allows us to predict
the speed of light without looking at light. Such a
prediction leads to one of the basic questions faced by
physicists around the beginning of the 20th century:
who got to measure this predicted speed? The answer
was provided by Einstein and his special theory of
relativity.

In the second part of this chapter we will discuss the
torque exerted by a magnetic field on a current loop,
and introduce the concept of a **magnetic moment**.
This discussion will provide some insight into how the
presence of iron greatly enhances the strength of the
magnetic field in an electromagnet. However the main
reason for developing the concept of magnetic moment
and the various magnetic moment equations is for our
later discussion of the behavior of atoms and elementary particles in a magnetic field. It is useful to clearly
separate the classical ideas discussed here from the
quantum mechanical concepts to be developed later.

THE INDUCTOR

In our discussion of Faraday's law and the betatron in Chapter 30, particularly in Figure (30-15), we saw that an increasing magnetic field in the core of the betatron creates a circular electric field around the core. This electric field was used to accelerate the electrons.

A more common and accessible way to produce the same circular electric field is by turning up the current in a solenoid as shown in Figure (1). As we saw in our discussion of Ampere's law in Chapter 29, a current i in a long coil of wire with n turns per unit length,

produces a nearly uniform magnetic field inside the coil whose strength is given by the formula

$$B = \mu_0 n i \tag{29-31}$$

and whose direction is given by the right hand rule as shown in the side view, Figure (1a).

If the coil has a cross-sectional area A, as seen in the top view Figure (1b), then the amount of magnetic flux Φ_B "flowing" up through the coil is given by

$$\Phi_B = BA = \mu_0 n A i \tag{1}$$

And if we are increasing the current i in the coil, then the rate of increase of this flux is (since μ_0, n and A are constants)

$$\frac{d\Phi_B}{dt} = \mu_0 n A \frac{di}{dt} \tag{2}$$

It is the changing magnetic flux that creates the circular electric field E shown in Figure (1b).

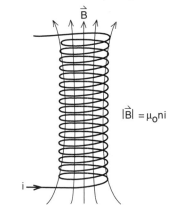

$$|\vec{B}| = \mu_0 n i$$

i ⟶

a) side view of coil and magnetic field

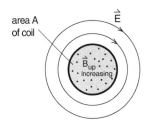

area A
of coil

$\vec{E}$

b) top view showing the electric field
surrounding the increasing magnetic flux

Figure 1
When we turn up the current in a solenoid, we increase the magnetic field and therefore the magnetic flux up through the coil. This increasing magnetic flux is the source of the circular electric field seen in the top view.

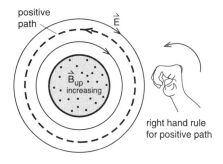

positive path

$\vec{E}$

$\vec{B}_{up}$ increasing

right hand rule
for positive path

$$\oint \vec{E} \cdot d\vec{\ell} = -\frac{d\Phi_B}{dt} \quad \text{(Faraday's Law)}$$

Figure 2 Sign conventions
We start by defining up, out of the paper, as the positive direction. Then use the right hand rule to define a positively oriented path. As a result, counter clockwise is positive, clockwise is negative. With these conventions, $d\Phi_B/dt$ is positive for an increasing upward directed magnetic flux. In calculating the line integral $\oint \vec{E} \cdot d\vec{\ell}$, we go around in a positive direction, counter clockwise. Everything is positive except the – sign in Faraday's law, thus the electric field goes around in a negative direction, clockwise as shown.

In Figure (2) we have shown the top view of the solenoid in Figure (1) and added in the circular electric field we would get if we had an increasing magnetic flux up through the solenoid. We have also drawn a circular path of radius r around the solenoid as shown. If we calculate the line integral $\oint \vec{E} \cdot d\vec{\ell}$ for this closed path, we get by Faraday's law

$$\oint \vec{E} \cdot d\vec{\ell} = -\frac{d\Phi_B}{dt} \qquad (30\text{-}17)$$

$$E \times 2\pi r = -\mu_0 n A \frac{di}{dt} \qquad (3)$$

where the integral $\oint \vec{E} \cdot d\vec{\ell}$ is simply E times the circumference of the circle, and we used Equation (2) for $d\Phi_B/dt$.

The minus sign in Equation (3) tells us that if we use a positive path as given by the right hand rule, and we are increasing the flux up through this path, then $\vec{E} \cdot d\vec{\ell}$ must be negative. I.e., the electric field must go clockwise, opposite to the positive path. (Do not worry too much about signs in this discussion. We will shortly find a simple, easily remembered, rule that tells us which way the electric field points.)

Equation (3) tells us that the strength E of the circular field is proportional to the rate of change of current i in the solenoid, and drops off as 1/r if we are outside the solenoid. In the following exercise, you are to show that we also have a circular field inside the solenoid, a field that decreases linearly to zero at the center.

Exercise 1

Use Faraday's law to calculate the electric field inside the solenoid. Note that for a circular path of radius r inside the solenoid, the flux Φ_B through the path is proportional to the area of the path and not the area A of the solenoid.

The calculation of the circular electric field inside and outside a solenoid, when i is changing, is a good example of the use of both Ampere's law to calculate $\vec{B}$ and Faraday's law to calculate $\vec{E}$. It should be saved in your collection of good examples.

Direction of the Electric Field

In Figure (2) and in the above exercise, we saw that an increasing magnetic flux in the coil created a clockwise circular electric field both inside and outside the wire as shown in Figure (3). In particular we have a circular electric field *at the wire*, and this circular electric field will act on the charges carrying the current in the wire.

To maintain our sign conventions, think of the current in the wire as being carried by the flow of positive charge. The up directed magnetic field of Figure (3) will be produced by a current flowing counterclockwise as shown (right hand rule). In order to have an increasing flux, this counterclockwise current must be increasing.

We saw that the electric field is clockwise, opposite to the direction of the current. We are turning up the current to increase the magnetic field, and the electric field is opposing the increase.

If we already have a current in a solenoid, already have an established $\vec{B}$ field and try to decrease it, di/dt is negative for this operation, and we get an extra minus sign in Equation (3) that reverses the direction of $\vec{E}$. As

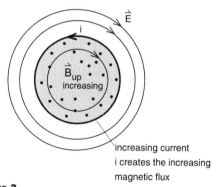

increasing current
i creates the increasing
magnetic flux

Figure 3
If the sign conventions described in Fig. 2 seemed too arbitrary, here is a physical way to determine the direction of $\vec{E}$. The rule is that <u>the electric field $\vec{E}$ opposes any change in the current i.</u> *In this case, to create an increasing upward directed magnetic flux, the current i must be flowing counter clockwise as shown, and be increasing. To oppose this increase, the electric field must be clockwise.*

a result we get a counterclockwise electric field that exerts a force in the direction of i. Thus when we try to decrease the current, the electric field tries to maintain it.

There is a general rule for determining the direction of the electric field. *The electric field produced by the changing magnetic flux always opposes the change.* If you have a counterclockwise current and increase it, you will get a clockwise electric field that opposes the increase. If you have a counterclockwise current and decrease it you get a counterclockwise electric field that opposes the decrease. If you have a clockwise current and try to increase it, you get a counter clockwise electric field that opposes the increase, etc. There are many possibilities, but one rule—the electric field always opposes the change.

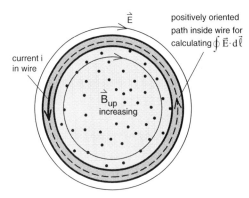

Figure 4
The electric field penetrates the wire, opposing the change in the current i. The line integral $\oint \vec{E} \cdot d\ell$ around the coil is just equal to the change in voltage ΔV_1 around each turn of the coil.

Induced Voltage

We have just seen that the changing magnetic flux in a solenoid creates an electric field that acts on the current in the solenoid to oppose the change in the current. From Equation (3), we see that the formula for the line integral of this electric field around one loop of the coil is given by

$$\oint \vec{E} \cdot d\vec{\ell} = -\mu_0 n A \frac{di}{dt} \qquad (4)$$

where the path is at the wire as shown in Figure (4). The n in Equation (4), which comes from the formula for the magnetic field of a solenoid, is the number of turns per unit length in the solenoid.

In our discussion of the betatron, we saw that the circular electric field accelerated electrons as they went around the evacuated donut. Each time the electrons went around once, they gained an amount of kinetic energy which, in electron volts, was equal to $\oint \vec{E} \cdot d\vec{\ell}$. In our discussion of the electron gun, we saw that using a battery of voltage V_{acc} to accelerate the electrons, produced electrons whose kinetic energy, in electron volts, was equal to V_{acc}. In other words, the circular electric field can act like a battery of voltage $V_{acc} = \oint \vec{E} \cdot d\vec{\ell}$.

When acting on the electrons in one loop of wire, the circular electric field produces a voltage change ΔV_1 given by

$$\Delta V_1 = \oint \vec{E} \cdot d\vec{\ell} \qquad \begin{array}{l} \textit{change in electric} \\ \textit{voltage in one} \\ \textit{turn of the coil} \end{array} \qquad (5a)$$

If we have a coil with N turns as shown in Figure (5), then the change in voltage ΔV_N across all N turns is N times as great, and we have

$$\Delta V_N = N \oint \vec{E} \cdot d\vec{\ell} \qquad \begin{array}{l} \textit{change in electric} \\ \textit{voltage in N} \\ \textit{turns of the coil} \end{array} \qquad (5b)$$

Using Equation 4 for the $\oint \vec{E} \cdot d\vec{\ell}$ for a solenoid, we see that the voltage change ΔV_N across the entire solenoid has a magnitude

$$|\Delta V_N| = \left| N \oint \vec{E} \cdot d\vec{\ell} \right| = (\mu_0 N n A)\frac{di}{dt} \qquad (6)$$

where N is the total number of turns, n = N/h is the number of turns per unit length, A is the cross-sectional area of the solenoid, and i the current through it.

To get the correct sign of ΔV_N, to see whether we have a voltage rise or a voltage drop, we will use the rule that the circular electric field opposes any change in the current. This rule is much easier to use than trying to keep track of all the minus signs in the equations.

In summary, Equation (6) is telling us that if you try to change the amount of current flowing in a solenoid, if di/dt is not zero, then a voltage will appear across the ends of the solenoid. The voltage has a magnitude proportional to the rate di/dt that we are trying to change the current, and a direction that opposes the change. It is traditional to call this voltage ΔV_N the *induced* voltage. One says that the changing magnetic flux in the coil *induces* a voltage. Such a coil of wire is often called an *inductor*.

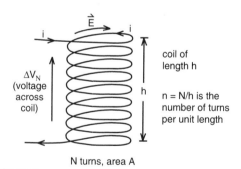

N turns, area A

Figure 5
Our standard coil with N turns, area A and length h. If you try to increase the current i in the coil, you get an opposing voltage.

Inductance

If you take a piece of insulated wire, tangle it up in any way you want, and run a current through it, you will get an induced voltage $V_{induced}$ that is proportional to the rate of change of current di/dt, and directed in a way such that it opposes the change in the current. If we designate the proportionality constant by the letter L, then the relationship between $V_{induced}$ and di/dt can be written

$$V_{induced} = L\frac{di}{dt} \qquad \begin{array}{l}\textit{induced voltages}\\ \textit{are proportional}\\ \textit{to di/dt}\end{array} \qquad (7)$$

The constant L is called the *inductance* of the coil or tangle of wire. In the MKS system, inductance has the dimension of volt seconds/ampere, which is called a *henry*.

Comparing Equations (6) and (7), we immediately obtain the formula for the inductance of a solenoid

$$L = \mu_0 N n A = \frac{\mu_0 N^2 A}{h} \qquad \begin{array}{l}\textit{inductance of}\\ \textit{a solenoid}\end{array} \qquad (8)$$

where N is the number of turns in the solenoid, A the cross-sectional area and h the length. In the middle term, n = N/h is the number of turns per unit length.

Example 1 The toroidal Inductor

The simplest solenoid we can use is a toroidal one, like that shown in Figure (6), where the magnetic field is completely confined to the region inside the coil. Essentially, the toroid is an ideal solenoid (no end effects) of length $h = 2\pi R$. To develop an intuitive feeling for inductance and the size of a henry, let us calculate the inductance of the toroidal solenoid shown in the photograph of Figure (6b). This solenoid has 696 turns and a radius of $R = 21.5$ cm. Each coil has a radius of $r = 1.3$ cm. Thus we have

$$N = 696 \text{ turns}$$
$$R = 21.5 \text{cm}$$
$$h = 2\pi R = 2\pi *.215 = 1.35 \text{m}$$
$$r = 1.3 \text{cm}$$
$$A = \pi r^2 = \pi \times .013^2 = 5.31 \times 10^{-4} \text{m}^2$$
$$\mu_0 = 1.26 \times 10^{-6} \text{ henry/m}$$

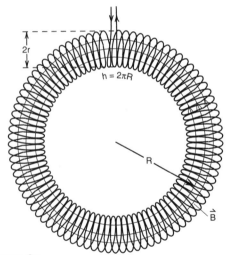

Figure 6a

A toroid is an ideal solenoid of length h = $2\pi R$.

With

$$L = \frac{\mu_0 N^2 A}{h}$$

we get

$$L = \frac{1.26 \times 10^{-6} \times (696)^2 \times 5.31 \times 10^{-4}}{1.35}$$

$$= 2.40 \times 10^{-4} \text{ henry}$$

We see that even a fairly big solenoid like the one shown in Figure (6) has a small inductance at least when measured in henrys. At the end of the chapter we will see that inserting an iron core into a solenoid greatly increases the inductance. Inductances as large or larger than one henry are easily obtained with iron core inductors.

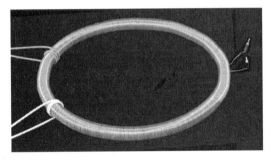

Figure 6b

Photograph of the toroidal solenoid used in various experiments. Although the coil looks big, the inductance is only 2.40×10^{-4} henry. (If you put iron inside the coil, you could greatly increase the inductance, but you would not be able to calculate its value.)

INDUCTOR AS A CIRCUIT ELEMENT

Because a changing electric current in a coil of wire produces a voltage rise, small coils are often used as circuit elements. Such a device is called an inductor, and the symbol used in circuit diagrams is a sketch of a solenoid —⦚⦚⦚⦚⦚— and usually designated by the symbol L. The voltage rise across the three circuit elements we have considered so far are

$$V_R = iR \qquad \text{resistor} \qquad (27\text{-}8)$$

$$V_C = \frac{Q}{C} \qquad \text{capacitor} \qquad (27\text{-}31)$$

$$V_L = L\frac{di}{dt} \qquad \text{inductor} \qquad (7)$$

As shown in Figure (7) the direction of the rise is opposite to the current in a resistor, toward the positive charge in a capacitor, and in a direction to oppose a change in the current i in an inductor. In (c), we are showing the direction of the voltage rise for an increasing current. The voltage in the inductor is opposing an increase in the current, just as the voltage in the resistor (a) opposes the current i itself.

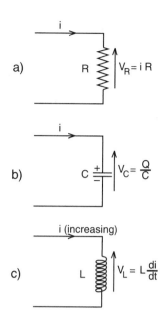

Figure 7
The resistor R, capacitor C and inductor L as circuit elements.

The LR Circuit

We will begin our discussion of the inductor as a circuit element with the LR circuit shown in Figure (8). Although this circuit is fairly easy to analyze, it is a bit tricky to get the current i started. One way to start the current is shown in Figure (9) where we have a battery and another resistor R_1 attached as shown.

When the switch of Figure (9) has been closed for a while, we have a constant current i_0 that flows down out of the battery, through the resistor R_1, around up through the inductor and back to the battery. Because the current is constant, $di_0/dt = 0$ and there is no voltage across the inductor. When we have constant "DC" currents, inductors act like short circuits. That is why the current, given the choice of going up through the inductor L or the resistor R, all goes up through L. (To say this another way, since the voltage across L is zero, the voltage V_R across R must also be zero, and the current $i_R = V_R/R = 0$.) Since R_1 is the only thing that limits the current in Figure (9), i_0 is given by

$$i_0 = \frac{V_B}{R_1} \tag{9}$$

Equation (9) tells us that we have a serious problem if we forget to include the current limiting resistor R_1.

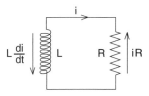

Figure 8
The LR circuit. If we have a decreasing current, the voltage in the inductor opposes the decrease and creates a voltage that continues to push the current through the resistor. But, to label the voltages for Kirchoff's law, it is easier to work with positive quantities. I.e. we label the circuit as if both i and di/dt were positive. With a positive di/dt, the voltage on the inductor opposes the current, as shown.

When we open the switch of Figure (9), the battery and resistor R_1 are immediately disconnected from the circuit, and we have the simple LR circuit shown in Figure (8). Everything changes instantly **except the current i in the inductor**. The inductor instantly sets up a voltage V_L to oppose any change in the current.

Figure (10) is a recording of the voltage V_L across the inductor, where the switch in Figure (9) is opened at time $t = 0$. Before $t = 0$, we have a constant current i_0 and no voltage V_L. When the switch is opened the voltage jumps up to $V_L = V_0$ and then decays exponentially just as in the RC circuit. What we want to do is apply Kirchoff's law to Figure (8) and see if we can determine the time constant for this exponential decay.

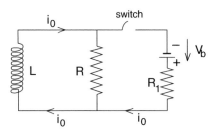

Figure 9
To get a current started in an LR circuit, we begin with the extra battery and resistor attached as shown. With the switch closed, in the steady state all the current i_0 flows up through the inductor because it (theoretically) has no resistance. When the switch is opened, the battery is disconnected, and we are left with the RL circuit starting with an initial current i_0.

If we walk around the circuit of Figure (8) in the direction of i, and add up the voltage rises we encounter, and set the sum equal to zero (Kirchoff's law), we get

$$-iR - L\frac{di}{dt} = 0$$

$$\frac{di}{dt} + \frac{R}{L}i = 0 \qquad (10)$$

Equation (10) is a simple first order differential equation for the current i. We guess from our experimental results in Figure (10) that i should be given by an exponential decay of the form

$$i = i_0 e^{-\alpha t} \qquad (11)$$

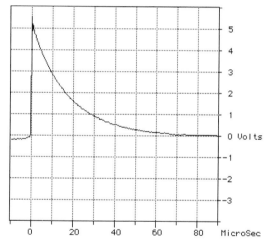

Figure 10
Experimental recording of the voltage in an RL circuit. We see that once the switch of Fig. 9 is opened, the voltage across the inductor jumps from zero to $V_0 = i_0 R$. This voltage on the inductor is trying to maintain the current now that the battery is disconnected. The voltage and the current then die with an exponential decay. (For this experiment, we used the toroidal inductor of Figure 6, with $R = 15\Omega$, $R1 = 4\Omega$, and $V_b = 2.5$ volts.)

Differentiating Equation (11) to get di/dt, we have

$$\frac{di}{dt} = -\alpha i_0 e^{-\alpha t} \qquad (12)$$

and substituting (11) and (12) in Equation (10) gives

$$-\alpha i_0 e^{-\alpha t} + \frac{R}{L} i_0 e^{-\alpha t} = 0 \qquad (13)$$

In Equation (13), i_0 and $e^{-\alpha t}$ cancel and we get

$$\alpha = R/L$$

Equation (11) for i becomes

$$i = i_0 e^{-(R/L)t} = i_0 e^{-t/(L/R)}$$

$$\equiv i_0 e^{-t/T} \qquad (14)$$

We see from Equation (14) that the time constant T for the decay is

$$\boxed{T = \frac{L}{R} = \begin{cases} \text{time constant} \\ \text{for the decay} \\ \text{of an LR circuit} \end{cases}} \qquad (15)$$

Everything we said about exponential decays and time constants for RC circuits at the end of Chapter (27) applies to the LR circuit, except that the time constant is now L/R rather than RC.

Exercise 2

The LR circuit that produced the experimental results shown in Figure (10) had a resistor whose resistance R was 15 ohms. Quickly estimate the inductance L. (You should be able to make this estimate accurate to within about 10% simply by sketching a straight line on the graph of Figure 10.) Compare your result with the inductance of the toroidal solenoid discussed in Figure (6) on page 6.

THE LC CIRCUIT

The next circuit we wish to look at is the LC circuit shown in Figure (11). All we have done is replace the resistor R in Figure (8) with a capacitor C as shown. It does not seem like much of a change, but the behavior of the circuit is very different. The exponential decays we saw in our LR and RC circuits occur because we are losing energy in the resistor R. In the LC circuit we have no resistor, no energy loss, and we will not get an exponential decay. To see what we should get, we will apply Kirchoff's law to the LC circuit and see if we can guess the solution to the resulting differential equation.

Walking clockwise around the circuit in Figure (11) and setting the sum of the voltage rises to zero, we get

$$-\frac{Q}{C} - L\frac{di}{dt} = 0$$

$$\frac{di}{dt} + \frac{Q}{LC} = 0 \tag{16}$$

The problem we have with Equation (16) is that we have two variables, i and Q, and one equation. But we had this problem before in our analysis of the RC circuit, and solved it by noticing that the charge Q on the capacitor is related to the current i flowing into the capacitor by

$$i = \frac{dQ}{dt} \tag{17}$$

If we differentiate Equation (16) once with respect to time to get

$$\frac{d^2i}{dt^2} + \frac{1}{LC}\frac{dQ}{dt} = 0$$

Finally use Equation (17) i = dQ/dt and we get the second order differential equation

$$\frac{d^2i}{dt^2} + \frac{1}{LC}i = 0 \tag{18}$$

The fact that we get a second order differential equation (with a second derivative of i) instead of the first order differential equations we got for LR and RC circuits, shows that we have a very different kind of problem. If we try an exponential decay in Equation (18), it will not work.

Exercise 3

Try the solution $i = i_0 e^{-\alpha t}$ in Equation (18) and see what goes wrong.

We have previously seen a second order differential equation in just the form of Equation (18) in our discussion of simple harmonic motion. We expect a sinusoidal solution of the form

$$i = i_0 \sin \omega t \tag{19}$$

In order to try this guess, Equation (19), we differentiate twice to get

$$\frac{di}{dt} = \omega i_0 \cos \omega t$$

$$\frac{d^2i}{dt^2} = -\omega^2 i_0 \sin \omega t \tag{20}$$

and substitute Equation (20) into (18) to get

$$-\omega^2 i_0 \sin \omega t + \frac{i_0}{LC} \sin \omega t = 0 \tag{21}$$

The quantity $i_0 \sin \omega t$ cancels from Equation (21) and we get

$$\omega^2 = \frac{1}{LC} \; ; \qquad \boxed{\omega = \frac{1}{\sqrt{LC}}} \tag{22}$$

We see that an oscillating current is a solution to Kirchoff's law, and that the frequency ω of oscillation is determined by the values of L and C.

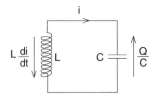

Figure 11
The LC circuit. This is the same as the LR circuit of Figure (8), except that the resistor has been replaced by a capacitor.

Exercise 4

In Figure (12) we have an LC circuit consisting of a toroidal coil shown in Figure (6) (on page 31-6), and the parallel plate capacitor made of two aluminum plates with small glass spacers. The voltage in Figure (12c) is oscillating at the natural frequency of the circuit.

a) What is the capacitance of the capacitor?

b) The aluminum plates have a radius of 11 cm. Assuming that we can use the parallel plate capacitor formula

$$C = \frac{\varepsilon_0 A_C}{d}$$

where A_C is the area of the plates, estimate the thickness d of the glass spacers used in this experiment.

(The measured value was 1.56 millimeters. You should get an answer closer to 1 mm. Errors could arise from fringing fields, effect of the glass, and non-uniformity of the surface of the plates.)

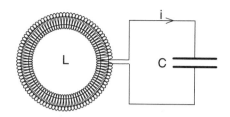

a) The LC circuit

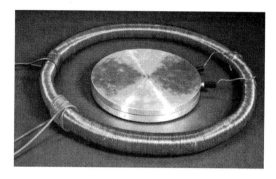

b) Inductor and capacitor used in the experiment

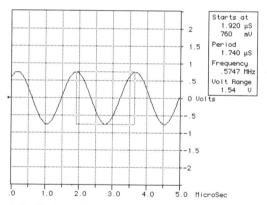

	Starts at
	1.920 µs
	760 mV
	Period
	1.740 µs
	Frequency
	.5747 MHz
	Volt Range
	1.54 V

c) Oscillating voltage at the resonant frequency.

Figure 12
Oscillating current in an LC circuit consisting of the toroidal inductor of Fig. 6 and a parallel plate capacitor. We will discuss shortly how we got the current oscillating and measured the voltage.

Intuitive Picture of the LC Oscillation

The rather striking behavior of the LC circuit deserves an attempt at an intuitive explanation. The key to understanding why the current oscillates lies in understanding the behavior of the inductor. As we have mentioned, the voltage rise on an inductor is always in a direction to oppose a change in current. The closest analogy is the concept of inertia. If you have a massive object, a large force is required to accelerate it. But once you have the massive object moving, a large force is required to stop it.

An inductor effectively supplies inertia to the current flowing through it. If you have a large inductor, a lot of work is required to get the current started. But once the current is established, a lot of effort is required to stop it. In our LR circuit of Figure (10), once we got a current going through the inductor L, the current continued to flow, even though there was no battery in the circuit, because of the inertia supplied to the current by the inductor.

Let us now see why an LC circuit oscillates. One cycle of an oscillation is shown in Figure (13) where we begin in (a) with a current flowing up through the inductor and over to the capacitor. The capacitor already has some positive charge on the upper plate and the current is supplying more. The capacitor voltage V_C is opposing the flow of the current, but the inertia supplied to the current by the inductor keeps the current flowing.

In the next stage, (b), so much charge has built up in the capacitor, V_C has become so large, that the current stops flowing. Now we have a charged up capacitor which in (c) begins to discharge. The current starts to flow back down through tin inductor. The current continues to flow out of the capacitor until we reach (d) where the capacitor is finally discharged.

The important point in (d) is that, although the capacitor is empty, we still have a current and the inductor gives the current inertia. The current will continue to flow even though it is no longer being pushed by the capacitor. Now in (e), the continuing current starts to charge the capacitor up the other way. The capacitor voltage is trying to slow the current down but the inductor voltage keeps it going.

Finally, in (f), enough positive charge has built up on the bottom of the capacitor to stop the flow of the current. In (g) the current reverses and the capacitor begins to discharge. The inductor supplies the inertia to keep this reversed current going until the capacitor is charged the other way in (i). But this is the same picture as (a), and the cycle begins again.

This intuitive picture allows you to make a rough estimate of how the frequency of the oscillation should depend upon the size of the inductance L and capacitance C. If the inductance L is large, the current has more inertia, it will charge up the capacitor more, and should take longer. If the capacitance C is larger, it should take longer to fill up. In other words, the period should be longer, the frequency ω lower, if either L or C are increased. This is consistent with the result $\omega = 1/\sqrt{LC}$ we saw in Equation (22).

Before leaving Figure (13) go back over the individual sketches and check two things. First, verify that Kirchoff's law works for each stage; i.e., that the sum of the voltage rises around the circuit is zero for each stage. Then note that whenever there is a voltage V_L on the inductor, the direction of V_L always opposes the *change* in current.

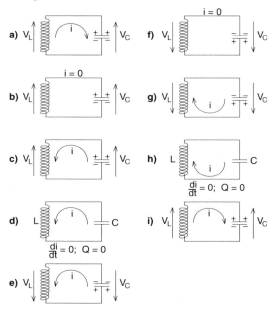

Figure 13
The various stages in the oscillation of the electric charge in an LC circuit.

The LC Circuit Experiment

The oscillation of the LC circuit in Figure (13) is a resonance phenomena and the frequency $\omega_0 = 1/\sqrt{LC}$ is the resonance frequency of the circuit. If we drive the circuit, force the current to oscillate, it will do so at any frequency, but the response is biggest when we drive the circuit at the resonant frequency ω_0.

There turns out to be a very close analogy between the LC circuit and a mass hanging on a spring as shown in Figure (14). The amplitude of the current in the circuit is analogous to the amplitude of the motion of the mass. If we oscillate the upper end of the spring at a low frequency ω much less than the resonant frequency ω_0, the mass just moves up and down with our hand. If ω is much higher than ω_0, the mass vibrates at a small amplitude and its motion is out of phase with the motion of our hand. I.e., when our hand comes down, the spring comes up, and vice versa. But when we oscillate our hand at the resonant frequency, the amplitude of vibration increases until either the mass jumps off the spring or some form of dampening or energy loss comes into play.

It is clear from Figure (14) how to drive the motion of a mass on a spring; just oscillate our hand up and down. But how do we drive the LC circuit? It turns out that for the parallel plate capacitor and air core toroidal inductor we are using, the resonance is so delicate that if we insert something into the circuit to drive it, we kill the resonance. We need a way to drive it from the outside, and an effective way to do that is shown in Figure (15).

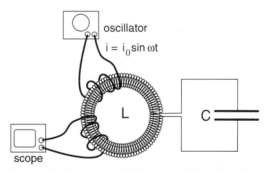

Run a wire from an oscillator around the coil and back to the oscillator. Do the same for the scope.

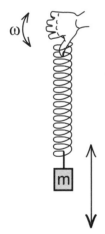

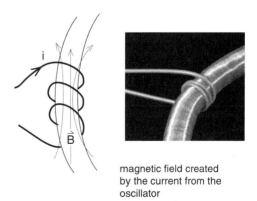

magnetic field created by the current from the oscillator

Figure 14
To get the mass on the end of a spring oscillating at some frequency ω you move your hand up and down at the frequency ω. If ω is the resonant frequency of the mass and spring system, the oscillations become quite large.

Figure 15
Driving the LC circuit. The turns of wire from the oscillator produce an oscillating magnetic field inside the coil. This in turn produces an electric field at the coil wires which also oscillates and drives the current in the coil. (A second wire wrapped around the coil is used to detect the voltage. The alternating magnetic field in the coil produces a voltage in the scope wire.)

In that figure we have taken a wire lead, wrapped it around the toroidal coil a couple of times, and plugged the ends into an oscillator as shown. (Some oscillators might not behave very well if you short them out this way. You may have to include a series resistor with the wire that goes to the oscillator.) When we turn on the oscillator, we get a current $i_{osc} = i_0 \sin \omega t$ in the wire, where the frequency ω is determined by the oscillator setting.

The important part of this setup is shown in Figure (15b) where the wire lead wraps a few times around the toroid. Since the wire lead itself forms a small coil and since it carries a current i_{osc}, it will create a magnetic field $\vec{B}_{osc}$ as shown. Part of the field B_{osc} will lie inside the toroid and create magnetic flux Φ_{osc} down the toroid. Since the current producing $\vec{B}_{osc}$ is oscillating at a frequency ω, the field and the flux will also oscillate at the same frequency. As a result we have an oscillating magnetic flux in the toroid, which by Faraday's law creates an electric field of magnitude $\oint \vec{E} \cdot d\vec{\ell} = -d\Phi_{osc}/dt$ around the turns of the solenoid. This electric field induces a voltage in the toroid

which drives the current in the LC circuit. We can change the driving frequency simply by adjusting the oscillator.

To detect the oscillating current in the coil, we wrap another wire around the coil, and plug that into an oscilloscope. The changing magnetic flux in the coil induces a voltage in the wire, a voltage that is detected by the scope.

In Figure (16a) we carried out the experiment shown in Figure (15), and recorded the amplitude V_C of the capacitor voltage as we changed the frequency ω on the oscillator. We see that the amplitude is very small until we get to a narrow band of frequencies centered on ω_0, in what is a typical resonance curve. The height of the peak at $\omega = \omega_0$ is limited by residual resistance in the LC circuit. Theory predicts that if there were no resistance, the amplitude at $\omega = \omega_0$ would go to infinity, but the wires in the toroid would melt first. In general, however, the less resistance in the circuit, the narrower the peak in Figure (16a), and the sharper the resonance.

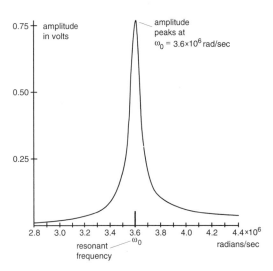

Figure 16a
As we tune the oscillator frequency through the resonant frequency, the amplitude of the LC voltage goes through a peak.

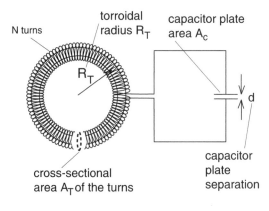

Figure 16b
The LC apparatus.

MEASURING THE SPEED OF LIGHT

The main reason we have focused on the LC resonance experiment shown in Figure (15) is that this apparatus can be used to measure the speed of light. We will first show how, and then discuss the philosophical implications of such a measurement.

The calculation is straightforward but a bit messy. We start with Equation (22) for the resonance frequency ω_0

$$\omega_0 = \frac{1}{\sqrt{LC}} \tag{22}$$

and then use Equation (8) for the inductance L of a solenoid

$$L = \mu_0 N^2 A/h \tag{8}$$

and Equation (27-32) for the capacitance of a parallel plate capacitor

$$C = \varepsilon_0 A_C/d \tag{27-32}$$

For the apparatus shown in Figure (16b), the length of the toroidal solenoid is $h = 2\pi R_T$, and the cross-sectional area is $A = A_T$, so that Equation (8) becomes

$$L_{toroid} = \mu_0 N^2 A_T/2\pi R_T \tag{23}$$

For the capacitor, A_C is the area of the plates, d their separation, and we can use Equation (27-32) as it stands. If we square Equation (22) to remove the square root

$$\omega_0{}^2 = \frac{1}{LC} \tag{22a}$$

and use Equations (23) and (27-32) for L and C, we get

$$\omega_0{}^2 = \frac{2\pi R_T}{\mu_0 N^2 A_T} \times \frac{d}{\varepsilon_0 A_C}$$

$$= \left(\frac{1}{\mu_0 \varepsilon_0}\right) \times \left(\frac{2\pi R_T d}{N^2 A_T A_C}\right) \tag{24}$$

The important point is that the product $\mu_0 \varepsilon_0$ appears in Equation (24), and we can solve for $1/\mu_0\varepsilon_0$ to get

$$\frac{1}{\mu_0 \varepsilon_0} = \frac{\omega_0^2 N^2 A_T A_C}{2\pi R_T d} \tag{25}$$

Finally, recall in our early discussion of magnetism, that $\mu_0 \varepsilon_0$ was related to the speed of light c by

$$c^2 = \frac{1}{\mu_0 \varepsilon_0} \tag{27-18}$$

Using Equation (25) in (27-18), and taking the square root gives

$$c = \frac{1}{\sqrt{\mu_0 \varepsilon_0}} = \omega_0 N \sqrt{\frac{A_T A_C}{2\pi R_T d}} \tag{26}$$

Exercise 5

Show that c in Equation (26) has the dimensions of a velocity. (Radians are really dimensionless.)

At first sight Equation (26) appears complex. But look at the quantities involved.

ω_0 = the measured resonant frequency

N = the number of turns in the solenoid

A_T = cross–sectional area of the toroid

A_C = area of capacitor plates

R_T = radius of toroid

d = separation of capacitor plates

Although it is a lot of stuff, everything can be counted, measured with a ruler, or in the case of ω_0, determined from the oscilloscope trace. And the result is the speed of light c. *We have determined the speed of light from a table top experiment that does not involve light.*

Exercise 6

The resonant curve in Figure (16a) was measured using the apparatus shown in Figure (16b). For an inductor, we used the toroid described in Figure (6). The parallel plates have a radius of 11 cm, and a separation d = 1.56mm. Use the experimental results of Figure (16a), along with the measured parameters of the toroid and parallel plates to predict the speed of light. (The result is about 20% low due to problems determining the capacitance, as we discussed in Exercise 3.)

In our initial discussion of the special theory of relativity in Chapter 1, we pointed out that according to Maxwell's theory of light, the speed of light c could be predicted from a table top experiment that did not involve light. This theory, developed in 1860, predicted that light should travel at a speed $c = 1/\sqrt{\mu_0\varepsilon_0}$, and Maxwell knew that the product $\mu_0\varepsilon_0$ could be determined from an experiment like the one we just described. (Different notation was used in 1860, but the ideas were the same.)

This raised the fundamental question: if you went out and actually measured the speed of a pulse of light as it passed by, would you get the predicted answer $1/\sqrt{\mu_0\varepsilon_0}$? If you did, that would be evidence that you were at rest. If you did not, then you could use the difference between the observed speed of the pulse and $1/\sqrt{\mu_0\varepsilon_0}$ as a measurement of your speed through space. This was the basis for the series of experiments performed by Michaelson and Morley to detect the motion of the earth. It was the basis for the rather firm conviction during the last half of the 19th century that the principle of relativity was wrong.

It was not until 1905 that Einstein resolved the problem by assuming that anyone who measured the speed of a pulse of light moving past them would get the answer $c = 1/\sqrt{\mu_0\varepsilon_0} = 3\times10^8$ m/s, no matter how they were moving. And if everyone always got the same answer for c, then a measurement of the speed of light could not be used as a way of detecting one's own motion and violating the principle of relativity. The importance of the LC resonance experiment, of the determination of the speed of light without looking at light, is that it focuses attention on the fundamental questions that lead to Einstein's special theory of relativity.

In the next chapter we will discuss Maxwell's equations which are the grand finale of electricity theory. It was the solution of these equations that led Maxwell to his theory of light and all the interesting problems that were raised concerning the principle of relativity.

Exercise 7

In Figure (17a) we have an LRC series circuit driven by a sinusoidal oscillator at a frequency ω radians/sec. The voltage V_R is given by the equation

$$V_R = V_{R0}\cos(\omega t)$$

as shown in the upper sketch of Figure (17b).

Knowing V_R, find the formulas and sketch the voltages for V_L and V_C. Determine the formulas for the amplitudes V_{L0} and V_{C0} in terms of V_{R0} and ω.

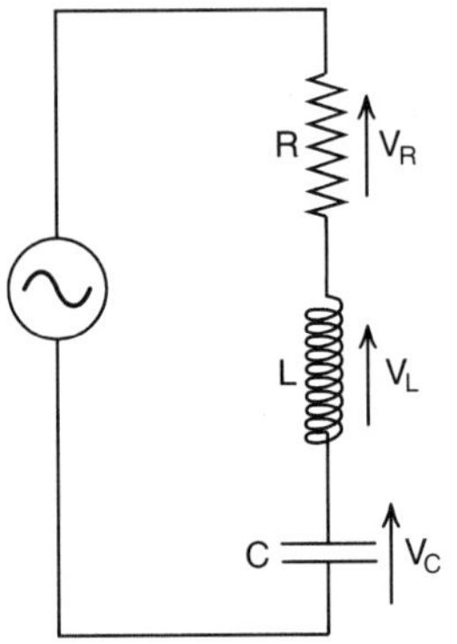

Figure 17a
An LRC circuit driven by a sinusoidal oscillator. The voltage V_R across the resistor is shown in Figure (17b).

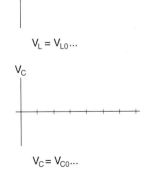

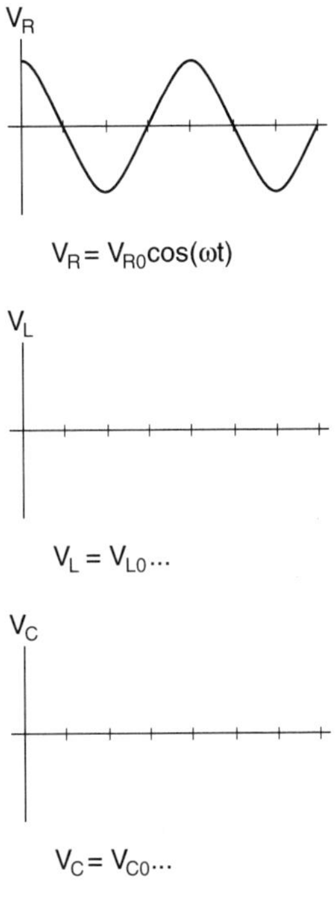

Figure 17b
Knowing V_R, find the formulas and sketch the voltages for V_L and V_C.

V_R

$$V_R = V_{R0}\cos(\omega t)$$

V_L

$$V_L = V_{L0}\cdots$$

V_C

$$V_C = V_{C0}\cdots$$

The second half of this chapter which discusses the concept of magnetic moment, provides additional laboratory oriented applications of Faraday's law and the Lorentz force law. This topic contains essential background material for our later discussion of the behavior of atoms and elementary particles in a magnetic field, but is not required for the discussion of Maxwell's equations in the next chapter. You may wish to read through the magnetic moment discussion to get the general idea now, and worry about the details when you need them later.

MAGNETIC MOMENT

We will see, using the Lorentz force law, that when a current loop (a loop of wire with a current flowing in it) is placed in a magnetic field, the field can exert a torque on the loop. This has an immediate practical application in the design of electric motors. But it also has an impact on an atomic scale. For example, iron atoms act like current loops that can be aligned by a magnetic field. This alignment itself produces a magnetic field and helps explain the magnetic properties of iron. On a still smaller scale elementary particles like the electron, proton, and neutron behave somewhat like a current loop in that a magnetic field can exert a torque on them. The phenomena related to this torque, although occurring on a subatomic scale, are surprisingly well described by the so called "classical" theory we will discuss here.

Magnetic Force on a Current

Before we consider a current loop, we will begin with a derivation of the force exerted by a straight wire carrying a current i as shown in Figure (18a). In that figure we have a positive current i flowing to the right and a uniform magnetic field $\vec{B}$ directed down into the paper.

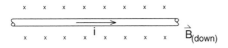

a) current in a magnetic field

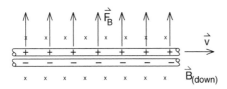

b) model of the neutral current as two rods of charge, the positive rod moving in the direction of the current i

Figure 18
Magnetic force on the moving charges when a current i is placed in a magnetic field.

In order to calculate the force exerted by $\vec{B}$ on i, we will use our model of a current as consisting of rods of charge moving past each other as shown in Figure (18b). The rods have equal and opposite charge densities Q/ℓ, and the positive rod is moving at a speed v to represent a positive current. The current i is the amount of charge per second carried past any cross-sectional area of the wire. This is the amount of charge per meter, Q/ℓ, times the number of meters per second, v, passing the cross-sectional area. Thus

$$i = \frac{Q}{\ell}\left(\frac{\text{coulombs}}{\text{meter}}\right) \times v\left(\frac{\text{meter}}{\text{second}}\right)$$

$$= \frac{Q}{\ell}v\left(\frac{\text{coulombs}}{\text{second}}\right) \tag{27}$$

In Figure (18b) we see that the downward magnetic field $\vec{B}$ acts on the moving positive charges to produce a force $\vec{F}_Q$ of magnitude

$$\left|\vec{F}_Q\right| = Q\left|\vec{v} \times \vec{B}\right| = QvB$$

which points toward the top of the page. The force $\vec{f}$ on a unit length of the wire is equal to the force on one charge Q times the number of charges per unit length, which is $1/\ell$. Thus

$$\left.\begin{array}{r}\text{force on a}\\ \text{unit length}\\ \text{of wire}\end{array}\right\} \equiv f = F_Q \times \left(\frac{1}{\ell}\right) = \frac{QvB}{\ell} \tag{28}$$

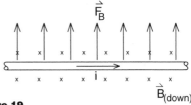

Figure 19
Sideways magnetic force on a current in a magnetic field. The force per unit length $\vec{f}$ is related to the charge per unit length λ by $\vec{f} = \lambda\vec{v} \times \vec{B}$. Since $\lambda\vec{v}$ is the current $\vec{i}$, we get $\vec{f} = \vec{i} \times \vec{B}$.

Using Equation (27) to replace Qv/ℓ by the current i in (28), and noting that $\vec{f}$ points in the direction of $\vec{i} \times \vec{B}$, as shown in Figure (19), we get

$$\boxed{\vec{f} = \vec{i} \times \vec{B}}$$ force per unit length exerted by a magnetic field $\vec{B}$ on a current $\vec{i}$ (29)

where $\vec{i}$ is a vector of magnitude i pointing in the direction of the positive current.

Example 2

Calculate the magnetic force between two straight parallel wires separated by a distance r, carrying parallel positive currents i_1 and i_2 as shown in Figure (21).

Solution

The current i_1 produces a magnetic field $\vec{B}_1$, which acts on i_2 as shown in Figure (20) (and vice versa). Since $\vec{B}_1$ is the field of a straight wire, it has a magnitude given by Equation (28-18) as

$$B_1 = \frac{\mu_0 i_1}{2\pi r}$$ (29-18)

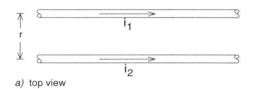

a) top view

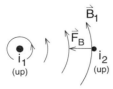

b) end view showing the magnetic field of current i_1 exerting a magnetic force on current i_2
The force between parallel currents is attractive.

Figure 20
Force between two currents.

The resulting force per unit length on $\vec{B}_2$ is

$$\vec{f} = \vec{i}_2 \times \vec{B}_1$$

which is directed in toward i_1 and has a magnitude

$$\left| \vec{f} \right| = i_2 B_1 = \frac{\mu_0 i_1 i_2}{2\pi r}$$ (30)

Equation (30) is used in the MKS definition of the ampere and the coulomb. In 1946 the following definition of the ampere was adopted:

The ampere is the constant current which, if maintained in two straight parallel conductors of infinite length, of negligible circular cross section, and placed 1 meter apart in a vacuum, would produce on each of these conductors a force equal to 2×10^{-7} newtons per meter of length.

Applying this definition to Equation (30), we set $i_1 = i_2 = 1$ to represent one ampere currents, $r = 1$ to represent the one meter separation, and $f = 2 \times 10^{-7}$ as the force per meter of length. We get

$$2 \times 10^{-7} = \frac{\mu_0}{2\pi}$$

From this we see that μ_0 is now a defined constant with the exact value

$$\boxed{\mu_0 = 4\pi \times 10^{-7}} \qquad \textit{by definition}$$ (31)

With the above definition of the ampere, the coulomb is officially defined by the amount of charge carried by a one ampere current, per second, past a cross-sectional area of a wire .

Looking back over our derivation of the formula $\vec{f} = \vec{i} \times \vec{B}$, and then the above MKS definitions, we see that it is the magnetic force law $\vec{F} = Q \vec{v} \times \vec{B}$ which now underlies the official definitions of charge and current.

Torque on a Current Loop

In an easily performed experiment, we place a square loop of wire of sides (ℓ) and (w) as shown in Figure (21a), into a uniform magnetic field as shown in (21b). The loop is allowed to rotate around the axis and is now orientated at an angle θ as seen in (21c).

If we now turn on a current i, we get an upward magnetic force proportional to $\vec{i} \times \vec{B}$ in the section from point (1) to point (2), and a downward magnetic force proportional to $\vec{i} \times \vec{B}$ in the section from point (3) to point (4). These two forces exert a torque about the axis of the loop, a torque that is trying to increase the angle θ. (This torque is what turns the armature of an electric motor.)

Following our earlier right hand conventions, we will define the area $\vec{a}$ of the loop as a vector whose magnitude is the area ($a = \ell$ w) of the loop, and whose direction is given by a right hand rule for the current in the loop. Curl the fingers of your right hand in the direction of the positive current i and your thumb points in the direction of $\vec{a}$ as shown in Figure (22).

With this convention, the loop area $\vec{A}$ points toward the upper left part of the page in Figure (21b) as shown. And we see that **the torque caused by the magnetic forces, is trying to orient the loop so that the loop area $\vec{a}$ is parallel to $\vec{B}$**. This is a key result we will use often.

To calculate the magnitude of the magnetic torque, we note that the magnitude of the force on side (1)-(2) or side (3)-(4) is the force per unit length $\vec{f} = \vec{i} \times \vec{B}$ times the length ℓ of the side

$$\left| \vec{F}_{1,2} \right| = \left| \vec{F}_{3,4} \right| = f\ell = iB\ell$$

When the loop is orientated at an angle θ as shown, then the lever arm for these forces is

$$\text{lever arm} = \frac{w}{2} \sin \theta$$

Since both forces are trying to turn the same way, the total torque is twice the torque produced by one force, and we have

$$\text{torque} = 2 \times \frac{w}{2}\sin\theta \times iB\ell \quad \begin{array}{l} \textit{2 times} \\ \textit{lever arm} \\ \textit{times force} \end{array}$$

$$\text{torque} = iB\left(w\ell\right) \sin \theta \tag{32}$$

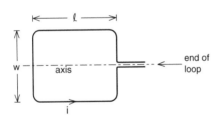

a) a wire loop carrying a current i, free to turn on the axis

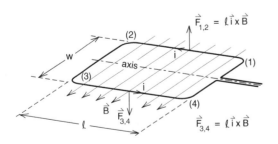

b) magnetic force acting on horizontal loop

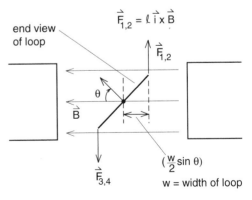

c) magnetic force acting on tilted loop

Figure 21
Analysis of the forces on a current loop in a magnetic field.

The final step is to convert Equation (32) into a vector equation. First recall that the vector torque $\vec{\tau}$ is defined as

$$\vec{\tau} = \vec{r} \times \vec{F}$$

where $\vec{r}$ and $\vec{F}_{1,2}$ are shown in Figure (23). In the figure we see that $\vec{r} \times \vec{F}$ and therefore $\vec{\tau}$ points up out of the paper.

Next we note that in Equation (32), θ is the angle between the magnetic field $\vec{B}$ and the loop area $\vec{a}$. In addition, the vector cross product $\vec{a} \times \vec{B}$ has a magnitude

$$\left| \vec{a} \times \vec{B} \right| = aB\sin\theta = (w\ell)B\sin\theta$$

and points up, in the same direction as the torque $\vec{\tau}$. Thus Equation (32) immediately converts to the vector equation

$$\vec{\tau} = i\,\vec{a} \times \vec{B} \tag{33}$$

where i is the current in the loop, and $\vec{a}$ is the vector area defined by the right hand convention of Figure (22).

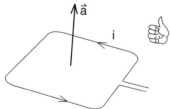

Figure 22
Right hand convention for the loop area A.

Magnetic Moment

When you put a current loop in a magnetic field, there is no net force on the loop ($\vec{F}_{1,2} = -\vec{F}_{3,4}$ in Figure 19b), but we do get a torque. Thus magnetic fields do not accelerate current loops, but they do turn them. In the study of the behavior of current loops, it is the torque that is important, and the torque is given by the simple formula of Equation (33).

This result can be written in an even more compact form if we define the ***magnetic moment*** $\vec{\mu}$ of a current loop as the current i times the vector area $\vec{a}$ of the loop

$$\boxed{\vec{\mu} \equiv i\vec{a}} \quad \begin{array}{l} \textit{definition of} \\ \textit{magnetic moment} \end{array} \tag{34}$$

With this definition, the formula for the torque on a current loop reduces to

$$\boxed{\vec{\tau} = \vec{\mu} \times \vec{B}} \tag{35}$$

Although we derived Equations (34) and (35) for a square loop, they also apply to other shapes such as round loops.

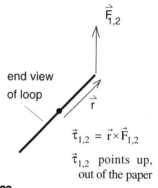

end view of loop

$$\vec{\tau}_{1,2} = \vec{r} \times \vec{F}_{1,2}$$

$\vec{\tau}_{1,2}$ points up, out of the paper

Figure 23
The torque $\vec{\tau}_{1,2}$ exerted by the force $\vec{F}_{1,2}$ acting on the side of the current loop. The vector $\vec{r}$ is the lever arm of $\vec{F}_{1,2}$ about the axis of the coil. You can see that $\vec{r} \times \vec{F}_{1,2}$ points up out of the paper.

Magnetic Energy

In Figure (24) we start with a current loop with its magnetic moment μ aligned with the magnetic field as shown in (24a). We saw in Figure (21b) that this is the orientation towards which the magnetic force is trying to turn the loop.

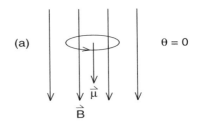

If we grab the loop and rotate it around as shown in (24b) until $\vec{\mu}$ is finally orientated opposite $\vec{B}$ as in (24c) we have to do work on the loop. We can calculate the amount of work we do rotating the loop from an angle $\theta = 0$ to $\theta = \pi$ using the angular analogy for the formula for work. The linear formula for work is

$$W = \int_{x_1}^{x_2} F_x dx \qquad (10\text{-}19)$$

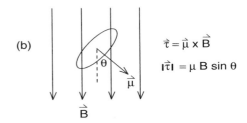

Replacing the linear force F_x by the angular force τ, and the linear distances dx, x_1, and x_2 by the angular distances $d\theta$, 0 and π, we get

$$W = \int_0^{\pi} \tau d\theta \qquad (36)$$

If we let go of the loop, the magnetic force will try to reorient the loop back in the $\theta = 0$ position shown in Figure (24a). We can think of the loop as falling back down to the $\theta = 0$ position releasing all the energy we stored in it by the work we did. In the $\theta = \pi$ position of Figure (24c) the current loop has a potential energy equal to the work we did in rotating the loop from $\theta = 0$ to $\theta = \pi$.

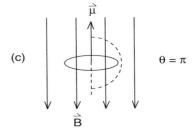

Figure 24
The resting, low energy position of a current loop is with $\vec{\mu}$ parallel to $\vec{B}$ as shown in (a). To turn the loop the other way, we have to do work against the restoring torque $\vec{\tau} = \vec{\mu} \times \vec{B}$ as shown in (b). The total work we do to get the loop into its high energy position (c) is $2\mu B$. We can think of this as magnetic potential energy that would be released if we let the loop flip back down again. We choose the zero of this potential energy half way between so that the magnetic potential energy ranges from $+\mu B$ in (c) to $-\mu B$ in (a).

We can calculate the magnetic potential energy by evaluating the integral in Equation (36). From Equation (35) we have

$$\left| \vec{\tau} \right| = \left| \vec{\mu} \times \vec{B} \right| = \mu B \sin \theta$$

so that

$$W = \int_0^{\pi} \mu B \sin \theta d\theta$$

$$= -\mu B \cos \theta \Big|_0^{\pi} = 2\mu B \qquad (36a)$$

Thus the current loop in the $\theta = \pi$ position of (24c) has an energy $2\mu B$ greater than the energy in the $\theta = 0$ position of (24a).

It is very reasonable to define the zero of magnetic potential energy for the position $\theta = \pi/2$, half way between the low and high energy positions. Then the magnetic potential energy is $+\mu B$ in the high energy position and $-\mu B$ in the low energy position. We immediately guess that a more general formula for magnetic potential energy of the current loop is

$$\left.\begin{array}{l}\text{magnetic potential} \\ \text{energy of a} \\ \text{current loop}\end{array}\right\} \quad E_{mag} = -\vec{\mu} \cdot \vec{B}$$
$$(37)$$

This gives $E_{mag} = +\mu B$ when the loop is in the high energy position with $\vec{\mu}$ opposite $\vec{B}$, and $E_{mag} = -\mu B$ in the low energy position where $\vec{\mu}$ and $\vec{B}$ are parallel. At an arbitrary angle θ, Equation (37) gives $E_{mag} = -\mu B \cos \theta$, a result you can obtain from equation (36a) if you integrate from $\theta = 0$ to $\theta = \theta$, and adjust the zero of potential energy to be at $\theta = \pi/2$.

Summary of Magnetic Moment Equations

Since we will be using the magnetic moment equations in later discussions, it will be convenient to summarize them in one place. They are a short set of surprisingly compact equations.

$\vec{A}$ = area of current loop

$$\vec{\mu} \equiv i\vec{A} \qquad \qquad (34)$$

$$\vec{\tau} = \vec{\mu} \times \vec{B} \qquad \qquad (35)$$

$$E_{mag} = -\vec{\mu} \cdot \vec{B} \qquad \qquad (37)$$

Charge q in a Circular Orbit

Most applications of the concept of magnetic moment are to atoms and elementary particles. In the case of atoms, we can often picture the magnetic moment as resulting from an electron traveling in a circular orbit like that shown in Figure (25). In that figure we show a charge q traveling at a speed v in a circular orbit of radius r. Since charge is being carried around this loop, this is a current loop, where the current i is the amount of charge per second being carried past a point on the orbit. In one second the charge q goes around $v/2\pi r$ times, therefore

$$i = q\,coulombs\left(\frac{v\,meter/\sec}{2\pi r\,meter}\right)$$

$$i = q\left(\frac{v}{2\pi r}\right)$$

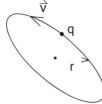

Figure 25
A charged particle in a circular orbit acts like a current loop. Its magnetic moment turns out to be $\mu = qvr/2$.

Since the area of the loop is πr^2, we get as the formula for the magnetic moment

$$\mu = iA = q\left(\frac{v}{2\pi r}\right) * \left(\pi r^2\right)$$

$$\boxed{\mu = \frac{qvr}{2}} \qquad \begin{array}{l}\textit{magnetic moment of} \\ \textit{a charge q traveling} \\ \textit{at a speed v in a} \\ \textit{circular orbit of radius r}\end{array} \qquad (38)$$

We can make a further refinement of Equation (38) by noting that the angular momentum J (we have already used L for inductance) of a particle of mass m traveling at a speed v in a circle of radius r has a magnitude

$$J = mvr$$

More importantly, $\vec{J}$ points perpendicular to the plane of the orbit in a right handed sense as shown in Figure (26a). This is the same direction as the magnetic moment $\vec{\mu}$ seen in (26b), thus if we write Equation (38) in the form

$$\mu = \frac{q}{2m}(mvr) \qquad \qquad (38a)$$

and use J for mvr, we can write (38a) as the vector equation

$$\boxed{\vec{\mu} = \frac{q}{2m}\,\vec{J}} \qquad \begin{array}{l}\textit{relation between the angular} \\ \textit{momentum } \vec{J} \textit{ and magnetic} \\ \textit{moment } \vec{\mu} \textit{ for a particle} \\ \textit{traveling in a circular orbit}\end{array} \qquad (39)$$

Equation (39) is as far as we want to go in developing magnetic moment formulas using strictly classical physics. We will come back to these equations when we study the behavior of atoms in a magnetic field.

(angular momentum)

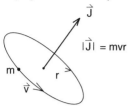

$\vec{J}$

$|\vec{J}| = mvr$

m

$\vec{v}$

r

(a) angular momentum of a particle in
   a circular orbit

(magnetic moment)

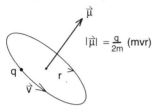

$\vec{\mu}$

$|\vec{\mu}| = \frac{q}{2m}\ (mvr)$

q

$\vec{v}$

r

(b) magnetic moment of a charged particle
   in a circular orbit

Figure 26
*Comparing the magnetic moment $\vec{\mu}$ and angular
momentum $\vec{J}$ of a particle in a circular orbit, we see
that*

$$\vec{\mu} = \frac{q}{2m}\ \vec{J}$$

IRON MAGNETS

In iron and many other elements, the atoms have a net magnetic moment due to the motion of the electrons about the nucleus. The classical picture is a small current loop consisting of a charged particle moving in a circular orbit as shown previously in Figures (25) and (26).

If a material where the atoms have a net magnetic moment is placed in an external magnetic field $\vec{B}_{ext}$ the torque exerted by the magnetic field tends to line up the magnetic moments parallel to $\vec{B}_{ext}$ as illustrated in Figure (27). This picture, where we show all the atomic magnetic moments aligned with $\vec{B}_{ext}$ is an exaggeration. In most cases the thermal motion of the atoms seriously disrupts the alignment. Only at temperatures of the order of one degree above absolute zero and in external fields of the order of one tesla do we get a nearly complete alignment.

Iron and a few other elements are an exception. A small external field, of the order of 10 gauss (.001 tesla) or less, can align the magnetic moments at room temperature. This happens because neighboring atoms interact with each other to preserve the alignment in an effect called *ferromagnetism*. The theory of how this interaction takes place, and why it suddenly disappears at a certain temperature (at 1043 K for iron) has been and still is one of the challenging problems of theoretical physics. (The problem was solved by Lars Onsager for a two-dimensional array of iron atoms, but no one has yet succeeded in working out the theory for a three-dimensional array.)

The behavior of iron or other ferromagnetic materials depends very much on how the substance was physically prepared, i.e., on how it was cooled from the molten mixture, what impurities are present, etc. In one extreme, it takes a fairly strong external field to align the magnetic moments, but once aligned they stay there. This preparation, called magnetically hard iron, is used for permanent magnets. In the other extreme, a small external field of a few gauss causes a major alignment which disappears when the external field is removed. This preparation called *magnetically soft* iron is used for electromagnets.

Our purpose in this discussion of iron magnets is not to go over the details of how magnetic moments are aligned, what keeps them aligned or what disrupts the alignment. We will consider only the more fundamental question – what is the effect of lining up the magnetic moments in a sample of matter. What happens if we line them all up as shown in Figure (27)?

A current loop has its own magnetic field which we saw in our original discussion of magnetic field patterns and which we have reproduced here in Figure (28). This is a fairly complex field shape. (Out from the loop at distances of several loop radii, the field has the shape of what is called a "dipole" magnetic field. In certain regions earth's magnetic field has this dipole magnetic field shape.)

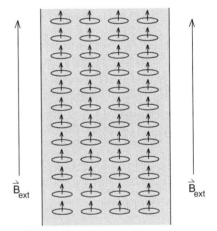

Figure 27
Ferromagnetism. When you apply an external magnetic field to a piece of "magnetically soft" iron (like a nail), the external magnetic field aligns all of the magnetic moments of the iron atoms inside the iron. The magnetic field of the current loops can be enormous compared to the external field lining them up. As a result a small external field produced say by a coil of wire, can create a strong field in the iron and we have an electromagnet. This phenomena is called ferromagnetism.

When you have a large collection of aligned current loops as shown in Figure (27), the magnetic fields of each of the current loops add together to produce the magnetic field of the magnet. The magnetic field of a single current loop, shown in Figure (28), is bad enough. What kind of a mess do we get if we add up the fields of thousands, billions, 6×10^{23} of these current loops? The calculation seems impossibly difficult.

Ampere discovered a simple, elegant way to solve the problem. Instead of adding up the magnetic fields of each current loop, he first added up the currents using a diagram like that shown in Figure (29).

We can think of Figure (29) as the top view of the aligned current loops of Figure (27). If you look at Figure (29) for a while, you see that all the currents inside the large circle lie next to, or very close to, an equal current flowing in the opposite direction. We can say that these currents inside the big circle cancel each other. They do not carry a net charge, and therefore do not produce a net magnetic field.

The cancellation is complete everywhere except at the outside surface. At this surface we essentially have a single large current loop with a current i equal to the current i in each of the little loops. It was in this way that Ampere saw that the magnetic field produced by all the small current loops packed together must be the same as the magnetic field of one big loop. What an enormous simplification!

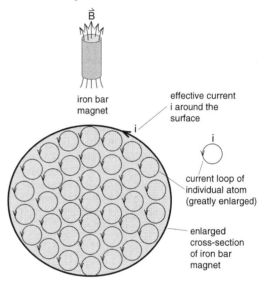

Figure 29
In an iron bar magnet, the iron atoms are permanently aligned. In the cross-sectional view we are looking down on the aligned current loops of the atoms. Inside the iron, we picture the currents as cancelling, leaving a net current i (the same as the current in each loop) going around the surface of the nail. This picture of a surface current replacing the actual current loops was proposed by Ampere, and the surface current is known as an Amperian current.

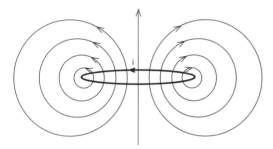

Figure 28
Magnetic field of a current loop.

Now let us return to Figure (27), redrawn in (30a), where we had a collection of aligned current loops. We can think of this as a model of a magnetized iron rod where all the iron atom magnetic moments are aligned. From Figure (29) we see that one horizontal layer of these current loops can be replaced by one large loop carrying a current i that goes all the way around the iron rod. This is shown again at the top of Figure (30).

Now our rod consists of a number of layers of small loops shown in (30a). If each of these layers is replaced by a single large loop, we end up with the stack of large loops shown in (30b). But this stack of large loops is just the **same current distribution we get in a solenoid!** Thus we get the remarkable result that the vector sum of the magnetic fields of all the current loops in Figure (30a) is just the simple field of a solenoid. This is why a bar magnet and a solenoid of the same size have the same field shape, as seen in Figure (31).

Although a bar magnet and a solenoid have the same field shape, the strength of the field in a bar magnet is usually far stronger. If the majority of the magnetic moments in an iron bar are aligned, we get a field of the order of one tesla inside the bar. To obtain comparable field strengths inside a solenoid made using copper wire, we would have to use currents so strong that the copper wire would soon heat up and perhaps melt due to electrical resistance in the copper.

The Electromagnet

If we insert a magnetically soft iron rod into the core of a solenoid as shown in Figure (32), we have an electromagnet. It only takes a small external field to align a majority of the magnetic moments in magnetically soft iron. And when the moments are aligned, we an get fields approaching one tesla, 10^4 gauss, as a result. This is the principle of an electromagnet where a weak field produced by a small current in the windings produces a strong field in the iron.

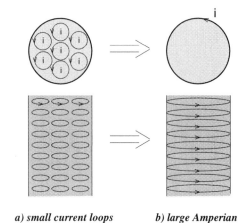

a) small current loops   *b) large Amperian currents around surface*

Figure 30
Ampere's picture of replacing small current loop throughout the substance by large ones on the surface.

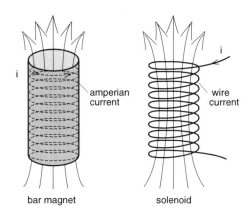

bar magnet   solenoid

Figure 31
Comparison of the magnetic fields of an iron magnet and a solenoid. The fields are essentially the same because the Amperian currents in a bar magnet are essentially the same as the current in the coils of a solenoid.

Figure (33) is a graph showing the strength of the magnetic field inside the iron core of an electromagnet as a function of the strength of the external magnetic field produced by the windings of the solenoid. In this case a toroidal solenoid was used, the iron core is an iron ring inside the toroid, and the results in Figure (33) are for one particular sample of iron. We can get different results for different samples of iron prepared in different ways.

The vertical axis in Figure (33) shows the percentage of the maximum field B_{max} we can get in the iron. B_{max} is the "saturated" field we get when all the iron atoms magnetic moments are aligned and has a typical value of about 1 tesla. We see that a very small external field of 2 gauss brings the magnetic field up to 50% of its saturated value. Getting the other 50% is much harder. We can more or less turn on the electromagnet using a 2 gauss external field, and that not much is to be gained by using a stronger external field.

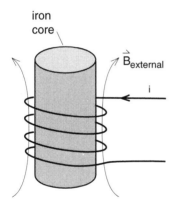

iron
core

$\vec{B}_{external}$

i

Figure 32
In an electromagnet, some turns of wire are wrapped around an iron bar. When a current i is turned on, the magnetic field of the turns of wire provide the external field to align the iron atoms. When the current i is shut off, and the external field disappears, the iron atoms return to a random alignment and the electromagnet shuts off. Whether the iron atoms remain aligned or not, whether we have a permanent magnet or not, depends upon the alloys (impurities) in the iron and the way the iron was cooled after casting.

The Iron Core Inductor

When the external field is less than 2 gauss in Figure (33), we have a more or less linear relationship shown by the dotted line between the external field and the field in the iron. In this region of the curve, for $B_{ext} <$ 2 gauss, the iron is essentially acting as a magnetic field amplifier. For this sample, a 2 gauss external field produces a 50,000 gauss magnetic field in the iron, an amplification by a factor of 25,000.

If we amplify the magnetic field in our solenoid 25,000 times, we are also amplifying the magnetic flux Φ_B by the same factor. If we have a varying current in the solenoid, but keep B_{ext} under 2 gauss, we will get a varying magnetic field in the iron and a varying magnetic flux Φ_B that is roughly proportional to the current i in the solenoid. The difference that the iron makes is that the flux Φ_B, and the rate of change of flux $d\Phi_B/dt$ will be 25,000 times larger. And so will the induced voltage in the turns of the solenoid. This means that the inductance of the solenoid is also increased by 25,000 times. If we inserted an iron ring into our air core solenoid shown in Figure (6), and the iron had the same magnetic properties as the iron sample studied in Figure (33), the inductance of our toroidal solenoid would increase 25,000 times from 1.8×10^{-4} henry up to about 4 henrys.

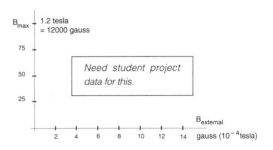

Figure 33
Example of a magnetization curve for magnetically soft iron. The impressive feature is that an external field of only a few gauss can produce fields in excess of xxxxx gauss inside the iron.

We can easily get large inductances from iron core inductors, but there are certain disadvantages. The curve in Figure (33) is not strictly linear, therefore the inductance has some dependence on the strength of the current in the coil. When we use an AC current in the solenoid, the iron atoms have to flip back and forth to keep their magnetic moments aligned with the AC external field. There is always some energy dissipated in the process and the iron can get hot. And if we try to go to too high a frequency, the iron atoms may not be able to flip fast enough, the magnetic field in the iron will no longer be able to follow the external field, and the amplification is lost. None of these problems is present with a air core inductor that has no iron.

Superconducting Magnets

The fact that iron saturates, the fact that we can do no better than aligning all the iron atoms current loops, places a fundamental limit on the usefulness of electromagnets for producing strong magnetic fields. Instead it is necessary to return to air core solenoids or other arrangements of coils of wire, and simply use huge currents.

The problem with using copper wire for coils that produce magnetic fields stronger than 1 tesla is that such strong currents are required that even the small resistance in copper produces enough heat to melt the wire. The only solutions for copper are to use an elaborate cooling system to keep the copper from heating, or do the experiment so fast that either the copper does not have time to heat, or you do not mind if it melts.

The introduction in the early 1970s of superconducting wire that could carry huge currents yet had zero electrical resistance revolutionized the design of strong field magnets. Magnets made from superconducting wire, called superconducting magnets are routinely designed to create magnetic fields of strengths up to 5 tesla. Such magnets will be used in the superconducting supercollider discussed earlier, and are now found in the magnetic resonance imaging devices in most large hospitals. The major problem with the superconducting magnets is that the superconducting wire has to be cooled by liquid helium to keep the wire in its superconducting state. And helium is a rare substance (at least on earth) that is difficult to liquefy and hard to maintain as a liquid.

In the late 1980s substances were discovered that are superconducting when immersed in liquid nitrogen, an inexpensive substance to create and maintain. So far we have not been able to make wires out of these "high temperature" superconductors that can carry the huge currents needed for big superconducting magnets. But this seems to be an engineering problem that when solved, may have a revolutionary effect not just on the design and use of superconducting magnets, but on technology in general.

APPENDIX

THE LC CIRCUIT
AND FOURIER ANALYSIS

The special feature of an LC circuit, like the one shown in Figure (A1), is it's resonance at an angular frequency $\omega_0 = 1/\sqrt{LC}$. If you drive the circuit with an oscillator that puts out a sine wave voltage $V = V_0 \sin \omega t$, the circuit will respond with a large voltage output when the driving frequency ω equals the circuit's resonant frequency ω_0. We saw this resonance in our discussion of the LC circuit shown in Figure (12) (p 31-11).

In Chapter 16, in our discussion of Fourier analysis, we saw that a square wave of frequency ω_0 can be constructed by adding up a series of harmonic sine waves. The first harmonic, of the form $A_1 \sin \omega_1 t$, has the same frequency as the square wave. For a square wave the second and all even harmonics are missing. The third harmonic is of the form $A_3 \sin \omega_3 t$. That is, the third harmonic's frequency is three times the frequency ω_1 of the first harmonic. The fifth harmonic is of the form $A_5 \sin \omega_5 t$. The amplitudes A_1, A_3, $A_5, \ldots$ of the harmonic sine waves present in the square wave, which are shown by the vertical bars in Figure (A2), were determined by Fourier analysis. (For a square wave, there is the simple relationship $A_3 = A_1/3$, $A_5 = A_1/5$, etc.)

The point of this lab is to demonstrate the physical reality of the harmonics in a square wave. We have seen that an LC circuit can be driven to a large amplitude resonance only when the driving frequency is equal to or close to the resonant frequency $\omega_0 = 1/\sqrt{LC}$. To put it another way, we can use the LC circuit to detect the presence of a sine wave of frequency ω_0 in the driving signal. If that frequency is present in the driving signal, the circuit will resonate. If it is not present, the circuit will not resonate.

Our experiment is to drive an LC circuit with a square wave, and see if the various harmonics in the square wave can each cause a resonance in the circuit. For example, if we adjust the frequency ω_1 of the square wave to equal ω_0, then we expect the first harmonic $A_1 \sin \omega_1 t$ to drive the circuit in resonance.

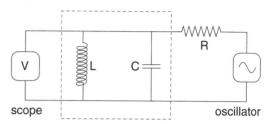

scope   oscillator

Figure A1
The LC circuit. For this experiment we used a fairly large commercial 9.1×10^{-3} henry inductor. This large inductance made the circuit more stable and less noisy than when we tried to do the experiment with the toroidal inductor of Figure (12). This inductor turned out to have an internal capacitance of 9.9×10^{-10} farads (990 picofarads) due to the coil windings themselves. We used this internal capacitance for the capacitor C of the circuit. (The internal capacitance was determined by measuring the resonant frequency $\omega_0 = 1/\sqrt{LC}$ and solving for C.)

By using a large inductor L, we can attach the scope directly across the inductor, as shown, without the scope having a serious effect on the circuit. The resistance R, with $R = 150 K\Omega$, partially isolated the LC circuit from the oscillator. This allowed the oscillator to gently drive the circuit without putting out much current and without distorting the shape of the square wave. (The oscillator could not maintain a square wave when we used the LC circuit shown in Figure (12).)

If one wants to try values of C other than the internal capacitance of the coil, one can add an external capacitor in parallel with L and C of Figure (A1). If you use an external capacitor more than about 10 times the internal capacitance, the internal capacitance of the inductor can be neglected.

If we then lower the frequency ω_1 of the square wave so that $3\omega_1 = \omega_0$, then we expect that the third harmonic $A_3 \sin \omega_3 t$ should drive the circuit in resonance. We should get another resonance when $5\omega_1 = \omega_0$, and another at $7\omega_1 = \omega_0$, etc. When the LC circuit is driven by a square wave, there should be a whole series of resonances, where in each case one of the harmonics has the right frequency to drive the circuit. These resonances provide direct experimental evidence that the various harmonics are physically present in the square wave, that they have energy that can drive the resonance.

In Figures A3 and A4 we look at the shape of the first few harmonics in the square wave of Figure (A2), and then watch as a square wave emerges as the harmonics are added together. (This is mostly a review of what we did back in Chapter 16.) After that, we study the resonances that occur when $\omega_1 = \omega_0$, $3\omega_1 = \omega_0$, $5\omega_1 = \omega_0$, etc. Finally we drop the square wave frequency to about $23\omega_1 = \omega_0$, and watch the LC circuit ring like a bell repeatedly struck by a hammer. Figure A1

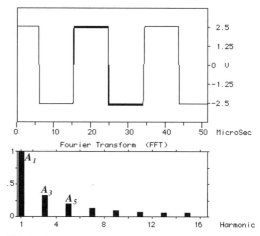

Figure A2

Fourier analysis of a square wave. The top part of the MacScope output shows an experimental square wave. We selected one cycle of the wave, chose Fourier analysis, and see that the wave consists of a series of odd harmonics. You can see the progression of amplitudes with $A_3 = A_1/3$, $A_5 = A_1/5$, etc. In Figure A3 we show the harmonics $A_1 \sin \omega_1 t$, $A_3 \sin \omega_3 t$, and $A_5 \sin \omega_5 t$. In Figure A4 we add together the harmonics to create a square wave.

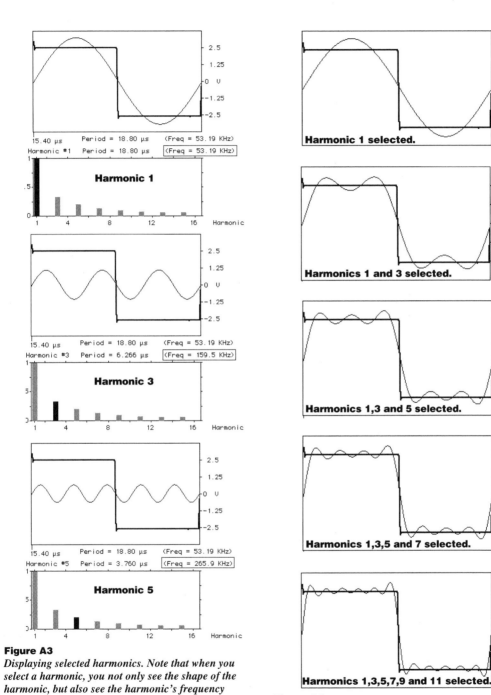

Figure A3
Displaying selected harmonics. Note that when you select a harmonic, you not only see the shape of the harmonic, but also see the harmonic's frequency displayed. (We highlighted this display with small rectangles.) You can see, for example, that the third harmonic is 159.5 kHz, 3 times the fundamental frequency of 53.19 kHz.

Figure A4
Adding up harmonics to create a square wave. The more harmonics we add, the closer we get.

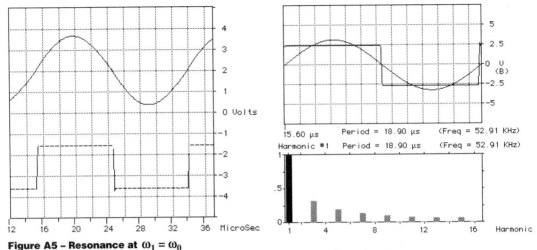

Figure A5 – Resonance at $\omega_1 = \omega_0$

We get the biggest resonance when the frequency ω_1 of the square wave is equal to the resonant frequency ω_0 of the circuit. We displayed the first harmonic by clicking on the bar over the 1 in the Fourier analysis window, and see that the frequency of the first harmonic is 52.91 kilohertz.

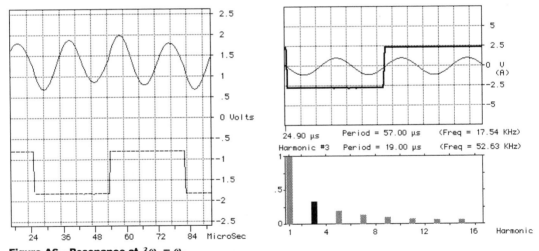

Figure A6 – Resonance at $3\omega_1 = \omega_0$

Lowering the frequency of the square wave to 17.54 kilohertz, we get another resonance shown above. In the Fourier analysis window, we selected the third harmonic, and see that the frequency of this harmonic is 52.63 kilohertz. To within experimental accuracy, this is equal to the LC circuit's resonant frequency of 52.91 kilohertz.

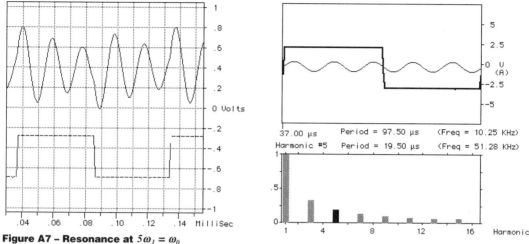

Figure A7 – Resonance at $5\omega_1 = \omega_0$

Lowering the frequency of the square wave to 10.25 kilohertz, we get another resonance. In the Fourier analysis window, we selected the fifth harmonic, and see that the frequency of this harmonic is 52.63 kilohertz. To within experimental accuracy, this is again equal to the LC circuit's resonant frequency of 52.91 kilohertz.

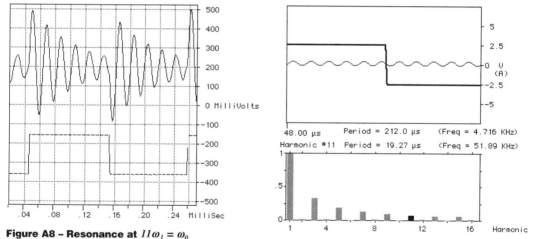

Figure A8 – Resonance at $11\omega_1 = \omega_0$

Skipping two resonances and lowering the frequency of the square wave to 4.712 kilohertz, we get a sixth resonance. In the Fourier analysis window, we selected the eleventh harmonic, and see that the frequency of this harmonic is 52.63 kilohertz. To within experimental accuracy, this is again equal to the LC circuit's resonant frequency of 52.91 kilohertz.

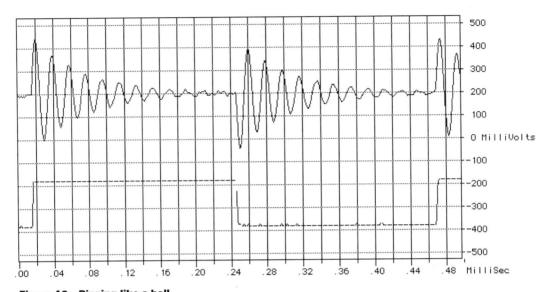

Figure A9 – Ringing like a bell

Dropping the square wave frequency even further, we see that every time the voltage of the square wave changes, the circuit responds like a bell struck by a hammer. This setup can be used as the starting point for the study of damped resonant (LRC) circuits.

Chapter 32

Maxwell's Equations

In 1860 James Clerk Maxwell summarized the entire content of the theory of electricity and magnetism in a few short equations. In this chapter we will review these equations and investigate some of the predictions one can make when the entire theory is available.

What does a complete theory of electricity and magnetism involve? We have to fully specify the electric field $\vec{E}$, the magnetic field $\vec{B}$, and describe what effect the fields have when they interact with matter.

The interaction is described by the Lorentz force law

$$\vec{F} = q\vec{E} + q\,\vec{v} \times \vec{B} \qquad (28\text{-}18)$$

which tells us the force exerted on a charge q by the $\vec{E}$ and $\vec{B}$ fields. As long as we stay away from the atomic world where quantum mechanics dominates, then the Lorentz force law combined with Newton's second law fully explains the behavior of charges in the presence of electric and magnetic fields, whatever the origin of the fields may be.

To handle the electric and magnetic fields, recall our discussion in Chapter 30 (on two kinds of fields) where we saw that any vector field can be separated into two parts; a divergent part like the electric field of static charges, and a solenoidal part like the electric field in a betatron or inductor. To completely specify a vector field, we need two equations – one involving a surface integral or its equivalent to define the divergent part of the field, and another involving a line integral or its equivalent defining the solenoidal part.

In electricity theory we have two vector fields $\vec{E}$ and $\vec{B}$, and two equations are needed to define each field. Therefore the total number of equations required must be four.

How many of the required equations have we discussed so far? We have Gauss' law for the divergent part of $\vec{E}$, and Faraday's law for the solenoidal part. It appears that we already have a complete theory of the electric field, and we do. Gauss' law and Faraday's law are two of the four equations needed.

For magnetism, we have Ampere's law that defines the solenoidal part of $\vec{B}$. But we have not written an equation involving the surface integral of $\vec{B}$. We are missing a Gauss' law type equation for the magnetic field. It would appear that the missing Gauss' law for $\vec{B}$, plus Ampere's law make up the remaining two equations.

This is not quite correct. The missing Gauss' law is one of the needed equations for $\vec{B}$, and it is easily written down because there are no known sources for a divergent $\vec{B}$ field. But Ampere's law, in the form we have been using

$$\oint \vec{B} \cdot d\vec{l} = \mu_0 i \qquad (29\text{-}18)$$

has a logical flaw that was discovered by Maxwell. When Maxwell corrected this flaw by adding another source term to the right side of Equation (29-18), he then had the complete, correct set of four equations for $\vec{E}$ and $\vec{B}$.

All Maxwell did was to add one term to the four equations for $\vec{E}$ and $\vec{B}$, and yet the entire set of equations are named after him. The reason for this is that with the correct set of equations, Maxwell was able to obtain solutions of the four equations, predictions of these equations that could not be obtained until Ampere's law had been corrected. The most famous of these predictions was that a certain structure of electric and magnetic fields could travel through empty space at a speed $v = 1/\sqrt{\mu_0 \varepsilon_0}$. Since Maxwell knew that $1/\sqrt{\mu_0 \varepsilon_0}$ was close to the observed speed 3×10^8 m/s for light, he proposed that this structure of electric and magnetic fields was light itself.

In this chapter, we will first describe the missing Gauss' law for magnetic fields, then correct Ampere's law to get the complete set of Maxwell's four equations. We will then solve these equations for a structure of electric and magnetic fields that moves through empty space at a speed $v = 1/\sqrt{\mu_0 \varepsilon_0}$. We will see that this structure explains various properties of light waves, radio waves, and other components of the electromagnetic spectrum. We will find, for example, that we can detect radio waves by using the same equipment and procedures we have used in earlier chapters to detect and map electric and magnetic fields.

GAUSS' LAW FOR MAGNETIC FIELDS

Let us review a calculation we have done several times now—the use of Gauss' law to calculate the electric field of a point particle. Our latest form of the law is

$$\int_{\substack{\text{closed} \\ \text{surface}}} \vec{E} \cdot d\vec{A} = \frac{Q_{in}}{\varepsilon_0} \tag{29-5}$$

where Q_{in} is the total amount of electric charge inside the surface.

In Figure (1), we have a point charge Q and have constructed a closed spherical surface of radius r centered on the charge. For this surface, $\vec{E}$ is everywhere perpendicular to the surface or parallel to every surface element $d\vec{A}$, thus $\vec{E} \cdot d\vec{A} = EdA$. Since E is of constant magnitude, we get

$$\int_{\substack{\text{closed} \\ \text{surface}}} \vec{E} \cdot d\vec{A} = E \int_{\substack{\text{closed} \\ \text{surface}}} dA$$

$$= E 4\pi r^2 \tag{1}$$

$$= \frac{Q_{in}}{\varepsilon_0}$$

where $Q_{in} = Q$.

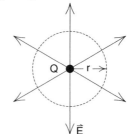

Figure 1
Field of point charge.

The solution of Equation (1) gives $E = Q/4\pi\varepsilon_0 r^2$ as the strength of the electric field of a point charge. A similar calculation using a cylindrical surface gave us the electric field of a charged rod. By being clever, or working very hard, one can use Gauss' law in the form of Equation (29-5) to solve for the electric field of any static distribution of electric charge.

But the simple example of the field of a point charge illustrates the point we wish to make. Gauss' law determines the diverging kind of field we get from a point source. Electric fields have point sources, namely electric charges, and it is these sources in the form of Q_{in} that appear on the right hand side of Equation (29-5).

Figure (2) shows a magnetic field emerging from a point source of magnetism. Such a point source of magnetism is given the name *magnetic monopole* and magnetic monopoles are predicted to exist by various recent theories of elementary particles. These theories are designed to unify three of the four basic interactions – the electrical, the weak, and the nuclear interactions. (They are called "Grand Unified Theories" or "GUT" theories. Gravity raises problems that are not handled by GUT theories.) These theories also predict that the proton should decay with a half life of 10^{32} years.

In the last 20 years there has been an extensive search for evidence for the decay of protons or the existence of magnetic monopoles. So far we have found no evidence for either. (You do not have to wait 10^{32} years to see if protons decay; instead you can see if one out of 10^{32} protons decays in one year.)

The failure to find the magnetic monopole, the fact that no one has yet seen a magnetic field with the shape shown in Figure (2), can be stated mathematically by writing a form of Gauss' law for magnetic fields with the magnetic charge Q_{in} set to zero

$$\int_{\substack{closed \\ surface}} \vec{B} \cdot d\vec{A} = 0 \qquad (2)$$

When reading Equation (2), interpret the zero on the right side of Equation (2) as a statement that the divergent part of the magnetic field has no *source term*. This is in contrast to Gauss' law for electric fields, where Q_{in}/ε_0 is the source term.

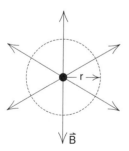

Figure 2
Magnetic field produced by a point source.

MAXWELL'S CORRECTION TO AMPERE'S LAW

As we mentioned in the introduction, Maxwell detected a logical flaw in Ampere's law which, when corrected, gave him the complete set of equations for the electric and magnetic fields. With the complete set of equations, Maxwell was able to obtain a theory of light. No theory of light could be obtained without the correction.

Ampere's law, Equation (29-18), uses the line integral to detect the solenoidal component of the magnetic field. We had

$$\oint \vec{B} \cdot d\vec{\ell} = \mu_0 i_{enclosed} \qquad \begin{matrix} Ampere's \\ Law \end{matrix} \qquad (29\text{-}18)$$

where $i_{enclosed}$ is the total current encircled by the closed path used to evaluate $\vec{B} \cdot d\vec{\ell}$. We can say that $\mu_0 i_{enclosed}$ is the source term for this equation, in analogy to Q_{in}/ε_0 being the source term for Gauss' law.

Before we discuss Maxwell's correction, let us review the use of Equation (29-18) to calculate the magnetic field of a straight current i as shown in Figure (3). In (3a) we see the wire carrying the current, and in (3b) we show the circular magnetic field produced by the current. To apply Equation (29-18) we draw a closed circular path of radius r around the wire, centering the

path on the wire so that $\vec{B}$ and sections $d\vec{\ell}$ of the path are everywhere parallel. Thus $\vec{B} \cdot d\vec{\ell} = B\,d\ell$, and since B is constant along the path, we have

$$\oint \vec{B} \cdot d\vec{\ell} = B \oint d\ell = B \times 2\pi r = \mu_0 i$$

which gives our old formula $B = \mu_0 i/2\pi r$ for the magnetic field of a wire.

To see the flaw with Ampere's law, consider a circuit where a capacitor is being charged up by a current i as shown in Figure (4). When a capacitor becomes charged, one plate becomes positively charged and the other negatively charged as shown. We can think of the capacitor being charged because a positive current is flowing into the left plate, making that plate positive, and a positive current is flowing out of the right plate, making that plate negative.

Figure (4) looks somewhat peculiar in that the current i almost appears to be flowing through the capacitor. We have a current i on the left, which continues on the right, with a break between the capacitor plates. To emphasize the peculiar nature of this discontinuity in the current, imagine that the wires leading to the capacitor are huge wires, and that the capacitor plates are just the ends of the wires as shown in Figure (5).

Now let us apply Ampere's law to the situation shown in Figure (5). We have drawn three paths, Path (1) around the wire leading into the positive plate of the capacitor, Path (2) around the wire leading out of the negative plate, and Path (3) around the gap between the plates. Applying Ampere's law we have

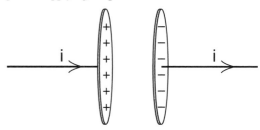

Figure 4
Charging up a capacitor.

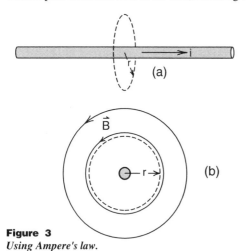

Figure 3
Using Ampere's law.

$$\oint_{path\,1} \vec{B} \cdot d\vec{\ell} = \mu_0 i \qquad \begin{array}{l} \textit{path 1 goes} \\ \textit{around a current i} \end{array} \qquad (3)$$

$$\oint_{path\,2} \vec{B} \cdot d\vec{\ell} = \mu_0 i \qquad \begin{array}{l} \textit{path 2 goes} \\ \textit{around a current i} \end{array} \qquad (4)$$

$$\oint_{path\,3} \vec{B} \cdot d\vec{\ell} = 0 \qquad \begin{array}{l} \textit{path 3 does not go} \\ \textit{around any current} \end{array} \qquad (5)$$

When we write out Ampere's law this way, the discontinuity in the current at the capacitor plates looks a bit more disturbing.

For greater emphasis of the problem, imagine that the gap in Figure (5) is very narrow, like Figure (5a) only worse. Assume we have a 1 mm diameter wire and the gap is only 10 atomic diameters. Then according to Ampere's law, $\oint \vec{B} \cdot d\vec{\ell}$ should still be zero if it is correctly centered on the gap. But can we possibly center a path on a gap that is only 10 atomic diameters wide? And even if we could, would $\oint \vec{B} \cdot d\vec{\ell}$ be zero for this path, and have the full value $\mu_0 i$ for the path 10 atomic diameters away? No, we simply cannot have such a discontinuity in the magnetic field and there must be something wrong with Ampere's law. This was the problem recognized by Maxwell.

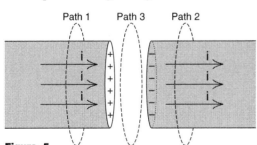

Figure 5
*Current flows through Paths (1)
& (2), but not through Path (3).*

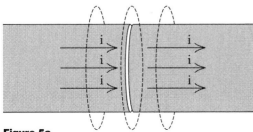

Figure 5a
Very narrow gap

Maxwell's solution was that even inside the gap at the capacitor there was a source for $\oint \vec{B} \cdot d\vec{\ell}$, and that the strength of the source was still $\mu_0 i$. What actually exists inside the gap is the electric field $\vec{E}$ due to the + and – charge accumulating on the capacitor plates as shown in Figure (6). Perhaps this electric field can somehow replace the missing current in the gap.

The capacitor plates or rod ends in Figure (6) have a charge density $\sigma = Q/A$ where Q is the present charge on the capacitor and A is the area of the plates. In one of our early Gauss' law calculations we saw that a charge density on a conducting surface produces an electric field of strength $E = \sigma/\varepsilon_0$, thus E between the plates is related to the charge Q on them by

$$E = \frac{\sigma}{\varepsilon_0} = \frac{Q}{\varepsilon_0 A} \;\; ; \quad Q = \varepsilon_0 E A \qquad (6)$$

The current flowing into the capacitor plates is related to the charge Q that has accumulated by

$$i = \frac{dQ}{dt} \qquad (7)$$

Using Equation (6) in Equation (7), we get

$$i = \varepsilon_0 \frac{d}{dt}\left(EA\right) \qquad (8)$$

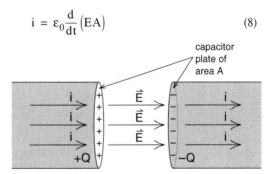

Figure 6
An electric field $\vec{E}$ exists between the plates.

Noting that the flux Φ_E of electric field between the plates is

$$\Phi_E = \vec{E} \cdot d\vec{A} = E A_\perp = E A \qquad (9)$$

and multiplying through by μ_0, we can write Equation (8) in the form

$$\mu_0 i = \mu_0 \varepsilon_0 \frac{d\Phi_E}{dt} \qquad (10)$$

We get the somewhat surprising result that $\mu_0 \varepsilon_0$ times the rate of change of electric flux inside the capacitor has the same magnitude as $\mu_0 i$, where i is the current in the wire leading to the capacitor. Maxwell proposed that $\mu_0 \varepsilon_0 d\Phi_E/dt$ played the same role, inside the capacitor, as a source term for $\oint \vec{B} \cdot d\vec{l}$, that $\mu_0 i$ did outside in the wire.

As a result, Maxwell proposed that Ampere's law be corrected to read

$$\oint \vec{B} \cdot d\vec{l} = \mu_0 i + \mu_0 \varepsilon_0 \frac{d\Phi_E}{dt} \quad \begin{array}{l} corrected \\ Ampere's \\ law \end{array} \quad (11)$$

Applying Equation (11) to the three paths shown in Figure (7), we have

$$\oint_{\text{paths 1\&2}} \vec{B} \cdot d\vec{l} = \mu_0 i \qquad (\Phi_E = 0) \qquad (12)$$

$$\oint_{\text{paths 3}} \vec{B} \cdot d\vec{l} = \mu_0 \varepsilon_0 \frac{d\Phi_E}{dt} \qquad (i = 0) \qquad (13)$$

I.e., for Paths (1) and (2), there is no electric flux through the path and the source of $\oint \vec{B} \cdot d\vec{l}$ is the current. For Path (3), no current flows through the path and the source of $\oint \vec{B} \cdot d\vec{l}$ is the changing electric flux. But, because $\mu_0 i$ in Equation (12) has the same magnitude $\mu_0 \varepsilon_0 d\Phi_E/dt$ in Equation (13), the term $\oint \vec{B} \cdot d\vec{l}$ has the same value for Path (3) as (1) and (2), and there is no discontinuity in the magnetic field.

Example: Magnetic Field between the Capacitor Plates

As an example of the use of the new term in the corrected Ampere's law, let us calculate the magnetic field in the region between the capacitor plates. To do this we draw a centered circular path of radius r smaller than the capacitor radius R as shown in Figure (8). There is no current through this path, but there is an electric flux $\Phi_E(r) = EA(r) = E\pi r^2$ through the path. Thus we set $i = 0$ in Ampere's corrected law, and replace Φ_E by the flux $\Phi_E(r)$ through our path to get

$$\oint \vec{B} \cdot d\vec{l} = \mu_0 \varepsilon_0 \frac{d\Phi_E(r)}{dt} \qquad (14)$$

Equation (14) tells us that because we have an increasing electric field between the plates, and thus an increasing electric flux through our path, there must be a magnetic field around the path.

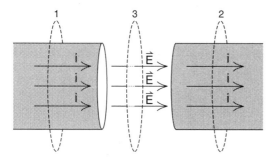

Figure 7
Path (2) surrounds a changing electric flux. Inside the gap, $\mu_0 \varepsilon_0 (d\Phi_E/dt)$ replaces $\mu_0 i$ as the source of $\vec{B}$.

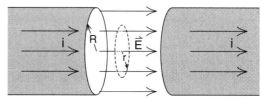

Figure 8
Calculating $\vec{B}$ in the region between the plates.

Due to the cylindrical symmetry of the problem, the only possible shape for the magnetic field inside the capacitor is circular, just like the field outside. This circular field and our path are shown in the end view, Figure (9). Since $\vec{B}$ and $d\vec{\ell}$ are parallel for all the steps around the circular path, we have $\vec{B} \cdot d\vec{\ell} = B\,d\ell$. And since B is constant in magnitude along the path, we get

$$\oint \vec{B} \cdot d\vec{\ell} = B\oint d\ell = B \times 2\pi r \qquad (15)$$

To evaluate the right hand side of Equation (14), note that the flux through our path $\Phi_E(r)$ is equal to the total flux $\Phi_E(\text{total})$ times the ratio of the area πr^2 of our path to the total area πR^2 of the capacitor plates

$$\Phi_E(r) = \Phi_E(\text{total}) \frac{\pi r^2}{\pi R^2} \qquad (16)$$

so that the right hand side becomes

$$\mu_0\varepsilon_0 \frac{d\Phi_E(r)}{dt} = \mu_0\varepsilon_0 \frac{d\Phi_E(\text{total})}{dt} \frac{r^2}{R^2}$$

$$= \mu_0 i \frac{r^2}{R^2} \qquad (17)$$

where in the last step we used Equation (10) to replace $\mu_0\varepsilon_0 d\Phi_E(\text{total})/dt$ by a term of the same magnitude, namely $\mu_0 i$.

Finally using Equation (15) and (17) in (14) we get

$$B \times 2\pi r = \mu_0 i \frac{r^2}{R^2}$$

$$B = \frac{\mu_0 i}{2\pi}\left(\frac{r}{R^2}\right) \quad \begin{array}{l}\textit{magnetic}\\ \textit{field between} \\ \textit{capacitor plates}\end{array} \qquad (18)$$

Figure (10) is a graph of the magnitude of B both inside and outside the plates. They match up at r = R, and the field strength decreases linearly to zero inside the plates.

Exercise 1

Calculate the magnetic field inside the copper wires that lead to the capacitor plates of Figure (5). Use Ampere's law and a circular path of radius r inside the copper as shown in Figure (11). Assuming that there is a uniform current density in the wire, you should get Equation (18) as an answer. Thus the magnetic field is continuous as we go out from the copper to between the capacitor plates.

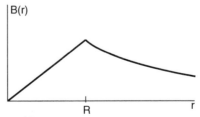

Figure 11

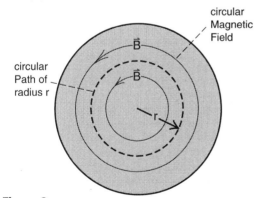

Figure 9
End view of capacitor plate.

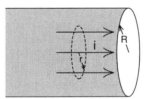

Figure 10
Magnetic fields inside and outside the gap.

MAXWELL'S EQUATIONS

Now that we have corrected Ampere's law, we are ready to write the four equations that completely govern the behavior of classical electric and magnetic fields. They are

$$
\begin{array}{lll}
\text{(a)} \quad \displaystyle\int_{\substack{\text{closed} \\ \text{surface}}} \vec{E} \cdot d\vec{A} \;=\; \frac{Q_{in}}{\varepsilon_0} & \quad \textit{Gauss'} \\[2.5em]
& \quad \textit{Law} \\[0.5em]
\text{(b)} \quad \displaystyle\int_{\substack{\text{closed} \\ \text{surface}}} \vec{B} \cdot d\vec{A} \;=\; 0 & \quad \textit{No} \\[0.5em]
& \quad \textit{Monopole} \\[1em]
\text{(c)} \quad \displaystyle\oint \vec{B} \cdot d\vec{\ell} \;=\; \mu_0 i + \mu_0 \varepsilon_0 \frac{d\Phi_E}{dt} & \quad \textit{Ampere's} \\[0.5em]
& \quad \textit{Law} \\[1em]
\text{(d)} \quad \displaystyle\oint \vec{E} \cdot d\vec{\ell} \;=\; -\frac{d\Phi_B}{dt} & \quad \textit{Faraday's} \\[0.5em]
& \quad \textit{Law} \\[0.5em]
& \hspace{4em} (19)
\end{array}
$$

The only other thing you need for the classical theory of electromagnetism is the Lorentz force law and Newton's second law to calculate the effect of electric and magnetic fields on charged particles.

$$
\vec{F} \;=\; q\vec{E} + q\vec{v} \times \vec{B} \qquad \begin{array}{l} \textit{Lorentz} \\ \textit{Force} \\ \textit{Law} \end{array}
$$
$$
(20)
$$

This is a complete formal summary of everything we have learned in the past ten chapters.

Exercise 2

This is one of the most important exercises in the text. The four Maxwell's equations and the Lorentz force law represent an elegant summary of many ideas. But these equations are nothing but hen scratchings on a piece of paper if you do not have a clear idea of how each term is used.

The best way to give these equations meaning is to know inside out at least one specific example that illustrates the use of each term in the equations. For Gauss' law, we have emphasized the calculation of the electric field of a point and a line charge. We have the nonexistence of the divergent magnetic field in Figure (2) to illustrate Gauss' law for magnetic fields. We have used Ampere's law to calculate the magnetic field of a wire and a solenoid. The new term in Ampere's law was used to calculate the magnetic field inside a parallel plate capacitor that is being charged up.

Faraday's law has numerous applications including the air cart speed meter, the betatron, the AC voltage generator, and the inductance of a solenoid. Perhaps the most important concept with Faraday's law is that $\oint \vec{E} \cdot \vec{d\ell}$ is the voltage rise created by solenoidal electric fields, which for circuits can be read directly by a voltmeter. This lead to the interpretation of a loop of wire with a voltmeter attached as an $\oint \vec{E} \cdot \vec{d\ell}$ meter. We used an $\oint \vec{E} \cdot \vec{d\ell}$ meter in the design of the air cart speed detector and experiment where we mapped the magnetic field of a Helmholtz coil.

Then there is the Lorentz force law with the formulas for the electric and magnetic force on a charged particle. As an example of an electric force we calculated the trajectory of an electron beam between charged plates, and for a magnetic force we studied the circular motion of electrons in a uniform magnetic field.

The assignment of this exercise is to write out Maxwell's equations one by one, and with each equation write down a fully worked out example of the use of each term. Do this neatly, and save it for later reference. This is what turns the hen scratchings shown on the previous page into a meaningful theory. When you buy a T-shirt with Maxwell's equations on it, you will be able to wear it with confidence.

We have just crossed what you might call a continental divide in our study of the theory of electricity and magnetism. We spent the last ten chapters building up to Maxwell's equations. Now we descend into applications of the theory. We will focus on applications and discussions that would not have made sense until we had the complete set of equations—discussions on the symmetry of the equations and applications like Maxwell's theory of light.

SYMMETRY OF MAXWELL'S EQUATIONS

Maxwell's Equations (19 a, b, c, and d), display considerable symmetry, and a special lack of symmetry. But the symmetry or lack of it is clouded by our choice of the MKS units with its historical constants μ_0 and ε_0 that appear, somewhat randomly, either in the numerator or denominator at various places.

For this section, let us use a special set of units where the constants μ_0 and ε_0 have the value 1

$$\mu_0 = 1 \; ; \quad \varepsilon_0 = 1 \quad \begin{pmatrix} \text{in a special} \\ \text{set of units} \end{pmatrix} \qquad (21)$$

Because the speed of light c is related to μ_0 and ε_0 by $c = 1/\sqrt{\mu_0 \varepsilon_0}$, we are now using a set of units where the speed of light is 1. If we set $\mu_0 = \varepsilon_0 = 1$ in Equations (19) we get

$$\int_{\substack{\text{closed} \\ \text{surface}}} \vec{E} \cdot d\vec{A} = Q_{in} \qquad (22a)$$

$$\int_{\substack{\text{closed} \\ \text{surface}}} \vec{B} \cdot d\vec{A} = 0 \qquad (22b)$$

$$\oint \vec{B} \cdot d\vec{\ell} = i + \frac{d\Phi_E}{dt} \qquad (22c)$$

$$\oint \vec{E} \cdot d\vec{\ell} = -\frac{d\Phi_B}{dt} \qquad (22d)$$

Stripping out μ_0 and ε_0 gives a clearer picture of what Maxwell's equations are trying to say. Equation (22a) tells us that electric charge is the source of divergent electric fields. Equation (22b) says that we haven't found any source for divergent magnetic fields. Equation (22c) tells us that an electric current or a changing electric flux is a source for solenoidal magnetic fields, and (22d) tells us that a changing magnetic flux creates a solenoidal electric field.

Equations (22) immediately demonstrate the lack of symmetry caused by the absence of magnetic monopoles, and so does the Lorentz force law of Equation (20). If the magnetic monopole is discovered, and we assign to it the "magnetic charge" Q_B, then for example Equation (22b) would become

$$\int_{\substack{\text{closed} \\ \text{surface}}} \vec{B} \cdot d\vec{A} = Q_B \qquad (22b')$$

If we have magnetic monopoles, a magnetic field should exert a force $\vec{F}_B = Q_B\vec{B}$ and perhaps an electric field should exert a force something like $\vec{F}_E = Q_B\vec{v} \times \vec{E}$.

Aside from Equation (22b), the other glaring asymmetry is the presence of an electric current i in Ampere's law (22c) but no current term in Faraday's law (22d). If, however, we have magnetic monopoles we can also have a current i_B of magnetic monopoles, and this asymmetry can be removed.

Exercise 3

Assume that the magnetic monopole has been discovered, and that we now have magnetic charge Q_B and a current i_B of magnetic charge. Correct Maxwell's Equations (22) and the Lorentz force law (20) to include the magnetic monopole. For each new term you add to these equations, provide a worked-out example of its use.

In this exercise, use symmetry to guess what terms should be added. If you want to go beyond what we are asking for in this exercise, you can start with the formula $\vec{F}_B = Q_B\vec{B}$ for the magnetic force on a magnetic charge, and with the kind of thought experiments we used in the chapter on magnetism, derive the formula for the electric force on a magnetic charge Q_B. You will also end up with a derivation of the correction to Faraday's law caused by a current of magnetic charge. (This is more of a project than an exercise.)

MAXWELL'S EQUATIONS IN EMPTY SPACE

In the remainder of this chapter we will discuss the behavior of electric and magnetic fields in empty space where there are no charges or currents. A few chapters ago, there would not have been much point in such a discussion, for electric fields were produced by charges, magnetic fields by currents, and without charges and currents, we had no fields. But with Faraday's law, we see that a changing magnetic flux $d\Phi_B/dt$ acts as the source of a solenoidal electric field. And with the correction to Ampere's law, we see that a changing electric flux is a source of solenoidal magnetic fields. Even without charges and currents we have sources for both electric and magnetic fields.

First note that if we have no electric charge (or magnetic monopoles), then we have no sources for either a divergent electric or divergent magnetic field. In empty space diverging fields do not play an important role and we can focus our attention on the equations for the solenoidal magnetic and solenoidal electric field, namely Ampere's and Faraday's laws.

Setting $i = 0$ in Equation (19c), the Equations (19c) and (19d) for the solenoidal fields in empty space become

$$\oint \vec{B} \cdot d\vec{\ell} = \mu_0 \varepsilon_0 \frac{d\Phi_E}{dt} \tag{23a}$$

$$\oint \vec{E} \cdot d\vec{\ell} = -\frac{d\Phi_B}{dt} \tag{23b}$$

We can make these equations look better if we write $\mu_0 \varepsilon_0$ as $1/c^2$, where $c = 3 \times 10^8$ m/s as determined in our LC circuit experiment. Then Equations (23) become

$$\oint \vec{B} \cdot d\vec{\ell} = \frac{1}{c^2} \frac{d\Phi_E}{dt} \tag{24a}$$

$$\oint \vec{E} \cdot d\vec{\ell} = -\frac{d\Phi_B}{dt} \tag{24b}$$

Maxwell's equations in empty space

Equations (24a, b) suggest a coupling between electric and magnetic fields. Let us first discuss this coupling in a qualitative, somewhat sloppy way, and then work out explicit examples to see precisely what is happening.

Roughly speaking, Equation (24a) tells us that a changing electric flux or field creates a magnetic field, and (24b) tells us that a changing magnetic field creates an electric field. These fields interact, and in some sense support each other.

If we were experts in integral and differential equations, we would look at Equations (24) and say, "Oh, yes, this is just one form of the standard wave equation. The solution is a wave of electric and magnetic fields traveling through space." Maxwell was able to do this, and solve Equations (24) for both the structure and the speed of the wave. The speed turns out to be c, and he guessed that the wave was light.

Because the reader is not expected to be an expert in integral and differential equations, we will go slower, working out specific examples to see what kind of structures and behavior we do get from Equations (24). We are just beginning to touch upon the enormous subject of electromagnetic radiation.

A Radiated Electromagnetic Pulse

We will solve Equations (24) the same way we have been solving all equations involving derivatives or integrals—by guessing and checking. The rules of the game are as follows. Guess a solution, then apply Equations (24) to your guess in every possible way you can think of. If you cannot find an inconsistency, your guess may be correct.

In order to guess a solution, we want to pick an example that we know as much as possible about and use every insight we can to improve our chances of getting the right answer. Since we are already familiar with the fields associated with a current in a wire, we will focus on that situation. Explicitly, we will consider what happens, what kind of fields we get, when we first turn on a current in a wire. We will see that a structure of magnetic and electric fields travels out from the wire, in what will be an example of a radiated electromagnetic pulse or wave.

A Thought Experiment

Let us picture a very long, straight, copper wire with no current in it. At time t = 0 we start an upward directed current i everywhere in the wire as shown in Figure (12). This is the tricky part of the experiment, having the current i start everywhere at the same time. If we closed a switch, the motion of charge would begin at the switch and advance down the wire. To avoid this, imagine that we have many observers with synchronized watches, and they all reach into the wire and start the positive charge moving at t = 0. However you want to picture it, just make sure that there is no current in the wire before t = 0, and that we have a uniform current i afterward.

In our previous discussions, we saw that a current i in a straight wire produced a circular magnetic field of magnitude $B = \mu_0 i / 2\pi r$ everywhere outside the wire. This cannot be the solution we need because it implies that as soon as the current is turned on, we have a magnetic field throughout all of space. The existence of the magnetic field carries the information that we have turned on the current. Thus the instantaneous spread of the field throughout space carries this information faster than the speed of light and violates the principle of causality. As we saw in Chapter 1, we could get answers to questions that have not yet been asked.

Using our knowledge of special relativity as a guide, we suspect that the solution $B = \mu_0 i / 2\pi r$ everywhere in space, instantaneously, is not a good guess. A more reasonable guess is that the magnetic field grows at some speed v out from the wire. Inside the growing front, the field may be somewhat like its final form $B = \mu_0 i / 2\pi r$, but outside we will assume B = 0.

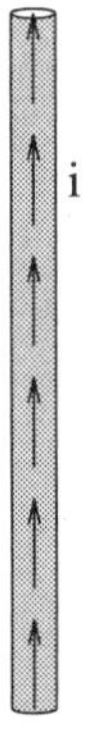

Figure 12
A current i is started all along the wire at time t = 0.

The pure, expanding magnetic field shown in Figure (13) seems like a good guess. But it is wrong, as we can see if we apply Ampere's law to Path (a) which has not yet been reached by the growing magnetic field. For this path that lies outside the magnetic field, $\oint \vec{B} \cdot d\vec{\ell} = 0$, and the corrected Ampere's law, Equation (19c), gives

$$\oint \vec{B} \cdot d\vec{\ell} = \mu_0 i + \mu_0 \varepsilon_0 \frac{d\Phi_E}{dt} = 0 \qquad (25)$$

In our picture of Figure (13) we have no electric field, therefore $\Phi_E = 0$ and Equation (25) implies that $\mu_0 i$ is zero, or the current i through Path (a) is zero. But **the current is not zero** and we thus have an inconsistency. The growing magnetic field of Figure (13) is not a solution of Maxwell's equations. (This is how we play the game. Guess and try, and this time we failed.)

Equation (25) gives us a hint of what is wrong with our guess. It says that

$$\frac{d\Phi_E}{dt} = -\frac{i}{\varepsilon_0} \qquad (25a)$$

thus if we have a current i and have the growing magnetic field shown in Figure (13) we must also have a changing electric flux Φ_E through Path (a). Somewhere there must be an electric field $\vec{E}$ to produce the changing flux Φ_E, a field that points either up or down, passing through the circular path of Figure (13).

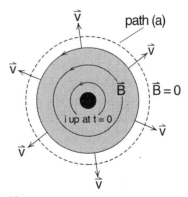

Figure 13
As a guess, we will assume that the magnetic field expands at a speed v out from the wire, when the current is turned on.

In our earlier discussion of inductance and induced voltage, we saw that a changing current creates an electric field that opposed the change. This is what gives an effective inertia to the current in an inductor. Thus when we suddenly turn on the upward directed current as shown in Figure (12), we expect that we should have a downward directed electric field as indicated in Figure (14), opposing our trying to start the current.

Initially the downward directed electric field should be inside the wire where it can act on the current carrying charges. But our growing circular magnetic field shown in Figure (13) must also have started inside the wire. Since a growing magnetic field alone is not a solution of Maxwell's equations and since there must be an associated electric field, let us propose that both the circular magnetic field of Figure (13), and the downward electric field of Figure (14) grow together as shown in Figure (15).

In Figure (15), we have sketched a field structure consisting of a circular magnetic field and a downward electric field that started out at the wire and is expanding radially outward at a speed v as shown. This structure has not yet expanded out to our Path (1), so that the line integral $\oint \vec{B} \cdot d\vec{\ell}$ is still zero and Ampere's law still requires that

$$0 = \mu_0 i + \mu_0 \varepsilon_0 \frac{d\Phi_E}{dt} \qquad \begin{array}{l} \textit{Path 1 of} \\ \textit{Figure 15} \end{array}$$

$$\frac{d\Phi_E}{dt} = -\frac{i}{\varepsilon_0} \qquad (26)$$

which is the same as Equation (25).

Looking at Figure (15), we see that the downward electric field gives us a negative flux Φ_E through our path. (We chose the direction of the path so that by the right hand convention, the current i is positive.) And as the field structure expands, we have more negative flux through the path. This increasing negative flux is just what is required by Equation (26).

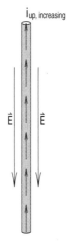

i up, increasing

Figure 14
When a current starts up, it is opposed by an electric field.

$\vec{E} \quad \vec{E}$

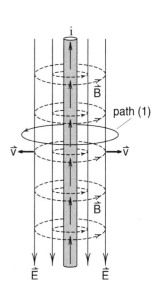

Figure 15
As a second guess, we will assume that there is a downward directed electric field associated with the expanding magnetic field. Again, Path (1) is out where the fields have not yet arrived.

i

$\vec{B}$

path (1)

$\vec{v}$ ← → $\vec{v}$

$\vec{B}$

$\vec{E} \qquad \vec{E}$

What happens when the field structure gets to and passes our path? The situation suddenly changes. Now we have a magnetic field at the path, so that $\oint \vec{B} \cdot d\vec{\ell}$ is no longer zero. And now the expanding front is outside our path so that the expansion no longer contributes to $-d\Phi_E/dt$. The sudden appearance of $\oint \vec{B} \cdot d\vec{\ell}$ is precisely compensated by the sudden loss of the $d\Phi_E/dt$ due to expansion of the field structure.

The alert student, who calculates $\oint \vec{E} \cdot d\vec{\ell}$ for some paths inside the field structure of Figure (15) will discover that we have not yet found a completely satisfactory solution to Maxwell's equations. The electric fields in close to the wire eventually die away, and only when they have gone do we get a static magnetic field given by $\oint \vec{B} \cdot d\vec{\ell} = \mu_0 i + 0$.

The problems associated with the electric field dying away can be avoided if we turn on the current at time $t = 0$, and then shut it off a very short time later. In that case we should expect to see an expanding cylindrical shell of electric and magnetic fields as shown in Figure (16). The front of the shell started out when the current was turned on, and the back should start out when the current is shut off. We will guess that the front and back should both travel radially outward at a speed v as shown.

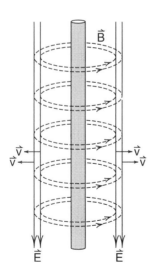

Figure 16
Electromagnetic pulse produced by turning the current on and then quickly off. We will see that this structure agrees with Maxwell's equations.

Speed of an Electromagnetic Pulse

Let us use Figure (16), redrawn as Figure (17a), as our best guess for the structure of an electromagnetic pulse. The first step is to check that this field structure obeys Maxwell's equations. If it does, then we will see if we can solve for the speed v of the wave front.

In Figure (17a), where we have shut the current off, there is no net charge or current and all we need to consider is the expanding shell of electric and magnetic fields moving through space. We have no divergent fields, no current, and the equations for $\vec{E}$ and $\vec{B}$ become

$$\oint \vec{B} \cdot d\vec{\ell} = \mu_0 \varepsilon_0 \frac{d\Phi_E}{dt} \qquad (23a)$$

$$\oint \vec{E} \cdot d\vec{\ell} = -\frac{d\Phi_B}{dt} \qquad (23b)$$

which we wrote down earlier as Maxwell's equation for empty space.

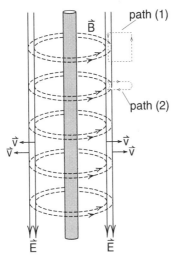

Figure 17a
In order to analyze the electromagnetic pulse produced by turning the current on and off, we introduce the two paths shown above. Path (1) has one side parallel to the electric field, while Path (2) has a side parallel to the magnetic field.

In order to apply Maxwell's equations to the fields in Figure (17a), we will focus our attention on a small piece of the shell on the right side that is moving to the right at a speed v. For this analysis, we will use the two paths labeled Path (1) and Path (2). Path (1) has a side parallel to the electric field, and will be used for Equation 23b. Path (2) has a side parallel to the magnetic field, and will be used for Equation 23a.

Analysis of Path 1

In Figure (17b), we have a close up view of Path (1). The path was chosen so that only the left edge of length h was in the electric field, so that

$$\oint \vec{E} \cdot d\vec{\ell} = Eh \qquad (27)$$

In order to make $\vec{E} \cdot d\vec{\ell}$ positive on this left edge, we went around Path (1) in a counterclockwise direction. By the right hand convention, any vector up through this path is positive, therefore the downward directed magnetic field is going through Path (1) in a negative direction. (We will be very careful about signs in this discussion.)

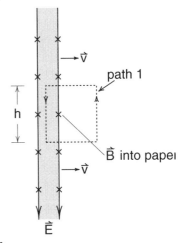

Figure 17b
Side view showing path (1). An increasing (negative) magnetic flux flows down through Path (1).

In Figure (18), we are looking at Path (1), first at a time t (18a) where the expanding front has reached a position x as shown, then at a time $t + \Delta t$ where the front has reached $x + \Delta x$. Since the front is moving at an assumed speed v, we have

$$v = \frac{\Delta x}{\Delta t}$$

At time $t + \Delta t$, there is additional magnetic flux through Path (1). The amount of additional magnetic flux $\Delta \Phi_B$ is equal to the strength B of the field times the additional area $(h\Delta x)$. Since $\vec{B}$ points down through Path (1), in a negative direction, the additional flux is negative and we have

$$\Delta \Phi_B = -B(\Delta A_\perp) = -B(h\Delta x) \tag{28}$$

Dividing Equation (28) through by Δt, and taking the limit that Δt goes to dt, gives

$$\frac{\Delta \Phi_B}{\Delta t} = -Bh\frac{\Delta x}{\Delta t}$$

$$\frac{d\Phi_B}{dt} = -Bh\frac{dx}{dt} = -Bhv \tag{29}$$

We now have a formula for $\vec{E} \cdot d\vec{\ell}$ (Equation 27) and for $d\Phi_E/dt$ (Equation 29) which we can substitute into Faraday's law (23b) to get

$$\oint \vec{E} \cdot d\vec{\ell} = -\frac{d\Phi_B}{dt}$$

$$Eh = -(-Bvh) = +Bhv$$

The factor of h cancels and we are left with

$$\boxed{E = Bv} \qquad \begin{array}{l}\textit{from} \\ \textit{Faraday's law}\end{array} \tag{30}$$

which is a surprisingly simple relationship between the strengths of the electric and magnetic fields.

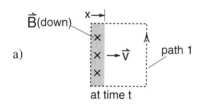

a)

at time t

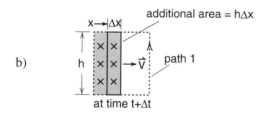

b)

at time $t + \Delta t$

Figure 18
As the front expands, there is more magnetic flux down through Path (1).

Analysis of Path 2

Path (2), shown in Figure (17c), is chosen to have one side in and parallel to the magnetic field. We have gone around clockwise so that $\vec{B}$ and $d\vec{\ell}$ point in the same direction. Integrating $\vec{B}$ around the path gives

$$\oint \vec{B} \cdot d\vec{\ell} = Bh \tag{31}$$

Combining Equation 31 with Ampere's law

$$\oint \vec{B} \cdot d\vec{\ell} = \mu_0 \varepsilon_0 \frac{d\Phi_E}{dt}$$

gives

$$Bh = \mu_0 \varepsilon_0 \frac{d\Phi_E}{dt} \tag{32}$$

To evaluate $d\Phi_E/dt$, we first note that for a clockwise path, the positive direction is down into the paper in Figure (17c). This is the same direction as the electric field, thus we have a positive electric flux through path (2).

In Figure (19), we show the expanding front at time t (19a) and at time $t + \Delta t$ (19b). The increase in electric flux $\Delta\Phi_E$ is (E) times the increased area ($h\Delta x$)

$$\Delta\Phi_E = E\left(h\Delta x\right)$$

Dividing through by Δt, and taking the limit that Δt goes to dt, gives

$$\frac{\Delta\Phi_E}{\Delta t} = Eh\frac{\Delta x}{\Delta t}$$

$$\frac{d\Phi_E}{dt} = Eh\frac{dx}{dt} = Ehv \tag{33}$$

Using Equation 33 in 32, and then cancelling h, gives

$$Bh = \mu_0 \varepsilon_0 Ehv \tag{34}$$

$$\boxed{B = \mu_0 \varepsilon_0 Ev} \qquad \begin{array}{l} \textit{From} \\ \textit{Ampere's law} \end{array}$$

which is another simple relationship between E and B.

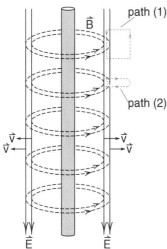

Figure 17a (repeated)
We will now turn our attention to path 2 which has one side parallel to the magnetic field.

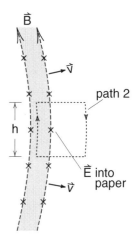

Figure 17c
An increasing (negative) electric flux flows down through Path (2).

If we divide Equation (30) $B\,v = E$, by Equation (34) $B = \mu_0\varepsilon_0\,E\,v$, both E and B cancel giving

$$\frac{B\,v}{B} = \frac{E}{\mu_0\varepsilon_0 E\,v}$$

$$v^2 = \frac{1}{\mu_0\varepsilon_0}$$

$$\boxed{v = \frac{1}{\sqrt{\mu_0\varepsilon_0}}} \qquad \text{\textit{speed of light!!!}} \qquad (35)$$

Thus the electromagnetic pulse of Figures (16) and (17) expands outward at the speed $1/\sqrt{\mu_0\varepsilon_0}$ which we have seen is 3×10^8 meters per second. *Maxwell recognized that this was the speed of light and recognized that the electromagnetic pulse must be closely related to light itself.*

Using $v = 1/\sqrt{\mu_0\varepsilon_0} = c$ in Equation (34) we get

$$\boxed{B = \frac{E}{c}} \qquad (36)$$

as the relative strength of the electric and magnetic fields in an electromagnetic pulse, or as we shall see, any light wave. If we had used a reasonable set of units where c = 1 (like feet and nanoseconds), then E and B would have equal strengths in a light wave.

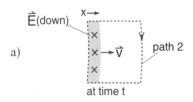
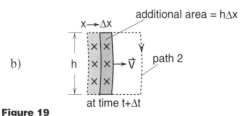

Figure 19
As the front expands, there is more electric flux down through Path (2).

Exercise 4

Construct paths like (1) and (2) of Figure (17), but which include the back side, rather than the front side, of the electromagnetic pulse. Repeat the kind of steps used to derive Equation (35) to show that the back of the pulse also travels outward at a speed $v = 1/\sqrt{\mu_0\varepsilon_0}$. As a result the pulse maintains its thickness as it expands out through space.

Exercise 5

After a class in which we discussed the electromagnetic pulse shown in Figure (20a), a student said she thought that the electric field would get ahead of the magnetic field as shown in Figure (20b). Use Maxwell's equations to show that this does not happen.

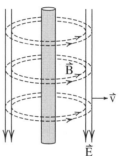

Figure 20a
The radiated electromagnetic pulse we saw in Figures (16) and (17).

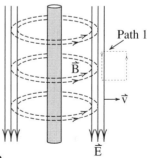

Figure 20b
The student guessed that the electric fields would get out ahead of the magnetic field. Use Path (1) to show that this does not happen.

ELECTROMAGNETIC WAVES

The single electromagnetic pulse shown in Figure (17) is an example of an electromagnetic wave. We usually think of a wave as some kind of oscillating sinusoidal thing, but as we saw in our discussion of waves on a Slinky in Chapter 1, the simplest form of a wave is a single pulse like that shown in Figure (21). The basic feature of the Slinky wave pulse was that it maintained its shape while it moved down the Slinky at the wave speed v . Now we see that the electromagnetic pulse maintains its structure of $\vec{E}$ and $\vec{B}$ fields while it moves at a speed v = c through space.

We made a more or less sinusoidal wave on the Slinky by shaking one end up and down to produce a series of alternate up and down pulses that traveled together down the Slinky. Similarly, if we use an alternating current in the wire of Figure (17), we will get a series of electromagnetic pulses that travel out from the wire. This series of pulses will more closely resemble what we usually think of as an electromagnetic wave.

Figure (22a) is a graph of a rather jerky alternating current where we turn on an upward directed current of magnitude i_0, then shut off the current for a while, then turn on a downward directed current i_0, etc. This series of current pulse produces the series of electromagnetic pulses shown in Figure (22b). Far out from the wire where we can neglect the curvature of the magnetic field, we see a series of pulses shown in the close-up view, Figure (23a). This series of flat or non-curved pulses is called a *plane wave* of electromagnetic radiation.

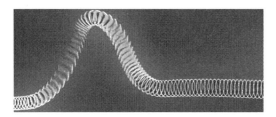

Figure 21
Slinky wave pulse.

If we used a sinusoidally oscillating current in the wire of Figure (22), then the series of electromagnetic pulses would blend together to form the sinusoidally varying electric and magnetic fields structure shown in Figure (23b). This is the wave structure one usually associates with an electromagnetic wave.

When you think of an electromagnetic wave, picture the fields shown in Figure (23), moving more or less as a rigid object past you at a speed c. The distance λ between crests is called the *wavelength* of the wave. The time T it takes one wavelength or cycle to pass you is

$$T \frac{\text{second}}{\text{cycle}} = \frac{\lambda \dfrac{\text{meter}}{\text{cycle}}}{c \dfrac{\text{meter}}{\text{second}}} = \frac{\lambda}{c} \frac{\text{second}}{\text{cycle}} \qquad (37)$$

T is called the *period* of the wave. The *frequency* of the wave, the number of wavelengths or full cycles of the wave that pass you per second is

$$f \frac{\text{cycle}}{\text{second}} = \frac{c \dfrac{\text{meter}}{\text{second}}}{\lambda \dfrac{\text{meter}}{\text{cycle}}} = \frac{\lambda}{c} \frac{\text{cycle}}{\text{second}} \qquad (38)$$

In Equations (37) and (38) we gave λ the dimensions meters/cycle, T of seconds/cycle and f of cycles/second so that we can use the dimensions to remember the

a) Graph of current pulses in wire

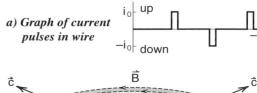

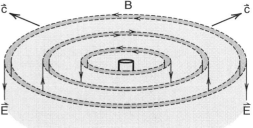

b) Resulting electric and magnetic fields

Figure 22
Fields produced by a series of current pulses.

formulas $T = \lambda/c$, $f = c/\lambda$. (It is now common to use "hertz" or "Hz" for the dimensions of frequency. This is a classic example of ruining simple dimensional analysis by using people's names.) Finally, the angular frequency ω radians per second is defined as

$$\omega \frac{\text{radians}}{\text{second}} = 2\pi \frac{\text{radians}}{\text{cycle}} \times f \frac{\text{cycles}}{\text{second}}$$

$$= 2\pi f \frac{\text{radians}}{\text{second}} \qquad (39)$$

You can remember where the 2π goes by giving it the dimensions 2π radians/cycle. (Think of a full circle or full cycle as having 2π radians.) We will indiscriminately use the word *frequency* to describe either f cycles/second or ω radians/second, whichever is more appropriate. If, however, we say that something has a frequency of so many hertz, as in 60 Hz, we will always mean cycles/second.

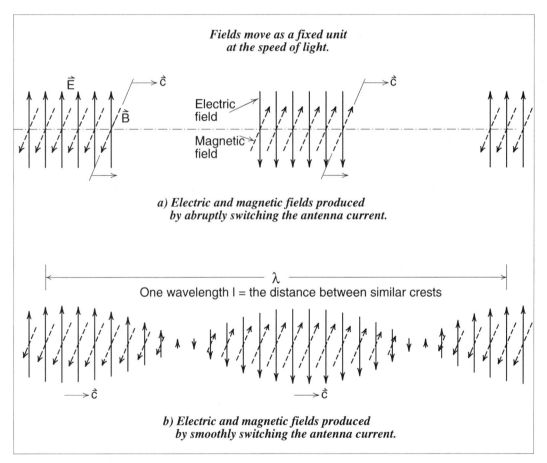

Fields move as a fixed unit at the speed of light.

a) *Electric and magnetic fields produced by abruptly switching the antenna current.*

b) *Electric and magnetic fields produced by smoothly switching the antenna current.*

Figure 23
Structure of electric and magnetic fields in light and radio waves.

ELECTROMAGNETIC SPECTRUM

We have seen by direct calculation that the electromagnetic pulse of Figure (17), and the series of pulses in Figure (22) are a solution of Maxwell's equations. It is not much of an extension of our work to show that the sinusoidal wave structure of Figure (23b) is also a solution. The fact that all of these structures move at a speed $c = 1/\sqrt{\mu_0\varepsilon_0} = 3 \times 10^8$ m/s is what suggested to Maxwell that these electromagnetic waves were light, that he had discovered the theory of light.

But there is nothing in Maxwell's equations that restricts our sinusoidal solution in Figure (23b) to certain values or ranges of frequency or wavelength. One hundred years before Maxwell it was known from interference experiments (which we will discuss in the next chapter) that light had a wave nature and that the wavelengths of light ranged from about 6×10^{-5} cm in the red part of the spectrum down to 4×10^{-5} cm in the blue part. With the discovery of Maxwell's theory of light, it became clear that there must be a complete spectrum of electromagnetic radiation, from very long down to very short wavelengths, and that visible light was just a tiny piece of this spectrum.

More importantly, Maxwell's theory provided the clue as to how you might be able to create electromagnetic waves at other frequencies. We have seen that an oscillating current in a wire produces an electromagnetic wave whose frequency is the same as that of the current. If, for example, the frequency of the current is 1030 kc (1030 kilocycles) $= 1.03 \times 10^6$ cycles/sec, then the electromagnetic wave produced should have a wavelength

$$\lambda \frac{\text{meters}}{\text{cycle}} = \frac{c \dfrac{\text{meters}}{\text{second}}}{f \dfrac{\text{cycles}}{\text{sec ond}}} = \frac{3 \times 10^8 \text{m/s}}{103 \times 10^6 \text{c/s}}$$

$$= 297 \text{ meters}$$

Such waves were discovered within 10 years of Maxwell's theory, and were called radio waves. The frequency 1030 kc is the frequency of radio station WBZ in Boston, Mass.

Components of the Electromagnetic Spectrum

Figure (24) shows the complete electromagnetic spectrum as we know it today. We have labeled various components that may be familiar to the reader. These components, and the corresponding range of wavelengths are as follows:

Radio Waves	10^6 m to .05 mm
AM Band	500 m to 190 m
Short Wave	60 m to 15 m
TV VHF Band	10 m to 1 m
TV UHF Band	1 m to 10 cm
Microwaves	10 cm to .05 mm
Infrared Light	.05 mm to 6×10^{-5} cm
Visible Light	6×10^{-5} to 4×10^{-5} cm
Ultraviolet Light	4×10^{-5} cm to 10^{-6} cm
X Rays	10^{-6} cm to 10^{-9} cm
γ Rays	10^{-9} cm and shorter

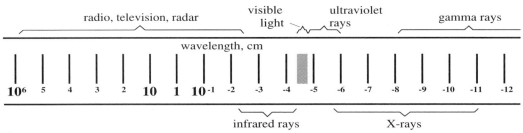

Figure 24
The electromagnetic spectrum extends from long wavelength radio waves down to short wavelength X rays and gamma rays. The visible part of the spectrum is indicated by the small box.

In each of these ranges, the most efficient way to emit or detect the radiation is to use antennas whose size is comparable to the wavelength of the radiation. For radio waves the antennas are generally some kind of a structure made from wire. In the infrared and the visible region, radiation is generally emitted by molecules and atoms. The short wavelength x rays and γ rays generally come from atomic nuclei or subatomic particles.

The longest wavelength radio waves that have been studied are the so-called "whistlers", radio waves with an audio frequency, that are produced by lightening bolts and reflected back and forth around the earth by charged particles trapped in the earth's magnetic field. On a shorter scale of distance are the long wavelength radio waves which penetrate the ocean and are used for communications with submarines. The radio station in Cutler, Maine, shown in Figure (25), has twenty-six towers over 1000 feet tall to support the antenna to produce such waves. This station, operated by the United States Navy, is the world's most powerful.

As we go to shorter wavelengths and smaller antennas, we get to the broadcast band, short wave radio, then to the VHF and UHF television frequencies. (FM radio is tucked into the VHF band next to Channel 6). The wavelengths for VHF television are of the order of meters, while those for UHF are of the order of a foot. Those with separate VHF and UHF television antennas will be familiar with the fact that the UHF antenna, which detects the shorter wavelengths, is smaller in size.

Adjusting the rabbit ears antenna on a television set provides practical experience with the problems of detecting an electromagnetic wave. As the TV signal strikes the antenna, the electric field in the wave acts on the electrons in the TV antenna wire. If the wire is parallel to the electric field, the electrons are pushed along in the wire producing a voltage that is detected by the television set. If the wire is perpendicular, the electrons will not be pushed up and down and no voltage will be produced.

The length of the wire is also important. If the antenna were one half wavelength, then the electric field at one end would be pushing in the opposite direction from the field at the other end, the integral $\oint \vec{E} \cdot \vec{dl}$ down the antenna would be zero, and you would get no net voltage or signal. You want the antenna long enough to get a big voltage, but not so long that the electric field in one part of the antenna works against the field in another part. One quarter wavelength is generally the optimum antenna length.

Figure 25
The worlds largest radio station at Cutler, Maine. This structure, with 75 miles of antenna wire and 26 towers over 1000 ft high, generates long wavelength low frequency, radio waves for communications with submarines.

The microwave region, now familiar from microwave communications and particularly microwave ovens, lies between the television frequencies and infrared radiation. The fact that you heat food in a microwave oven emphasizes the fact that electromagnetic radiation carries energy. One can derive that the energy density in an electromagnetic wave is given by the formula

$$\left.\begin{array}{l} \text{energy density in an} \\ \text{electromagnetic} \\ \text{wave} \end{array}\right\} = \frac{\varepsilon_0 E^2}{2} + \frac{B^2}{2\mu_0} \quad (37)$$

We have already seen the first term $\varepsilon_0 E^2 / 2$, when we calculated the energy stored in a capacitor (see Equation 27-36 on page 27-19). If we had calculated the energy to start a current in an inductor, we would have gotten the formula $B^2 / 2\mu_0$ for the energy density in that device. Equation (37) tells us the amount of energy is associated with electric and magnetic fields whenever we find them.

Blackbody Radiation

Atoms and molecules emit radiation in the infrared, visible and ultraviolet part of the spectrum. One of the main sources of radiation in this part of the spectrum is the so-called *blackbody* radiation emitted by objects due to the thermal motion of their atoms and molecules.

If you heat an iron poker in a fire, the poker first gets warm, then begins to glow a dull red, then a bright red or even, orange. At higher temperatures the poker becomes white, like the filaments in an electric light bulb. At still higher temperatures, if the poker did not melt, it would become bluish. The name *blackbody radiation* is related to the fact that an initially cold, black object emits these colors of light when heated.

There is a well studied relationship between the temperature of an object and the predominant frequency of the blackbody radiation it emits. Basically, the higher the temperature, the higher the frequency. Astronomers use this relationship to determine the temperature of stars from their color. The infrared stars are quite cool, our yellow sun has about the same temperature as the yellow filament in an incandescent lamp, and the blue stars are the hottest.

All objects emit blackbody radiation. You, yourself, are like a small star emitting infrared radiation at a wavelength corresponding to a temperature of 300K. In an infrared photograph taken at night, you would show up distinctly due to this radiation. Infrared photographs are now taken of houses at night to show up hot spots and heat leaks in the house.

Perhaps the most famous example of blackbody radiation is the 3K cosmic background radiation which is the remnant of the big bang which created the universe. We will say much more about this radiation in Chapter 34.

UV, X Rays, and Gamma Rays

When we get to wavelengths shorter than the visible spectrum, and even in the visible spectrum, we begin to run into problems with Maxwell's theory of light. These problems were first clearly displayed by Max Planck who in 1900 developed a theory that explained the blackbody spectrum of radiation. The problem with Planck's theory of blackbody radiation is that it could not be derived from Maxwell's theory of light and Newtonian mechanics. His theory involved arbitrary assumptions that would not be understood for another 23 years, until after the development of quantum mechanics.

Despite the failure of Newton's and Maxwell's theories to explain all the details, the electromagnetic spectrum continues right on up into the shorter wavelengths of ultraviolet (UV) light, then to x rays and finally to γ rays. Ultraviolet light is most familiar from the effect it has on us, causing tanning, sunburns, and skin cancer depending on the intensity and duration of the dose. The ozone layer in the upper atmosphere, as long as it lasts, is important because it filters out much of the ultraviolet light emitted by the sun.

X rays are famous for their ability to penetrate flesh and produce photographs of bones. These rays are usually emitted by the tightly bound electrons on the inside of large atoms, and also by nuclear reactions. The highest frequency radiation, γ rays, are emitted by the smallest objects—nuclei and elementary particles.

POLARIZATION

One of the immediate tests of our picture of a light or radio wave, shown in Figure (23), is the phenomena of *polarization*. We mentioned that the reason that you had to adjust the angle of the wires on a rabbit ears antenna was that the electric field of the television signal had to have a significant component parallel to the wires in order to push the electrons up and down the wire. Or, in the terminology of the last few chapters, we needed the parallel component of $\vec{E}$ so that the voltage $V = \oint \vec{E} \cdot \vec{dl}$ would be large enough to be detected by the television circuitry. (In this case, the line integral $\oint \vec{E} \cdot \vec{dl}$ is along the antenna wire.) Polarization is a phenomena that results from the fact that the electric field $\vec{E}$ in an electromagnetic wave can have various orientations as the wave moves through space.

Although we have derived the structure of an electromagnetic wave for the specific case of a wave produced by an alternating current in a long, straight wire, some of the general features of electromagnetic waves are clearly present in our solution. The general features that are present in all electromagnetic waves are:

1) *All electromagnetic waves are a structure consisting of an electric field $\vec{E}$ and a magnetic field $\vec{B}$.*

2) *$\vec{E}$ and $\vec{B}$ are at right angles to each other as shown in Figure (23).*

3) *The wave travels in a direction perpendicular to the plane of $\vec{E}$ and $\vec{B}$.*

4) *The speed of the wave is $c = 3 \times 10^8 m/s$.*

5) *The relative strengths of $\vec{E}$ and $\vec{B}$ are given by Equation (36) as $B = E/c$.*

Even with these restrictions, and even if we consider only flat or plane electromagnetic waves, there are still various possible orientations of the electric field as shown in Figure (26). In Figure (26a) we see a plane wave with a vertical electrical field. This would be called a vertically polarized wave. In Figure (26b), where the electric field is horizontal, we have a horizontally polarized wave. By convention we say that the direction of polarization is the direction of the electric field in an electromagnetic wave.

Because $\vec{E}$ must lie in the plane perpendicular to the direction of motion of an electromagnetic wave, **$\vec{E}$ has only two independent components**, which we can call the vertical and horizontal polarizations, or the x and y polarizations as shown in Figures (27a) and (27b) respectively. If we happen to encounter an electromagnetic wave where $\vec{E}$ is neither vertical or horizontal, but at some angle θ, we can decompose $\vec{E}$ into its x and y components as shown in (27c). Thus we can consider a wave polarized at an arbitrary angle θ as a mixture of the two independent polarizations.

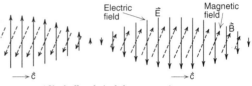

a) Vertically polarized electromagnetic wave.

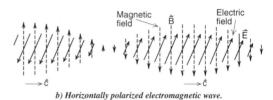

b) Horizontally polarized electromagnetic wave.

Figure 26
Two possible polarizations of an electromagnetic wave.

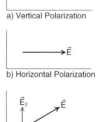

Figure 27
We define the direction of polarization of an electromagnetic wave as the direction of the electric field.

a) Vertical Polarization

b) Horizontal Polarization

c) Mixture

Polarizers

A polarizer is a device that lets only one of the two possible polarizations of an electromagnetic wave pass through. If we are working with microwaves whose wavelength is of the order of a few centimeters, a frame strung with parallel copper wires, as seen in Figure (28), makes an excellent polarizer. If a vertically polarized wave strikes this vertical array of wires, the electric field $\vec{E}$ in the wave will be parallel to the wires. This parallel $\vec{E}$ field will cause electrons to move up and down in the wires, taking energy out of the incident wave. As a result the vertically polarized wave cannot get through. (One can observe that the wave is actually reflected by the parallel wires.)

If you then rotate the wires 90°, so that the $\vec{E}$ field in the wave is perpendicular to the wires, the electric field can no longer move electrons along the wires and the wires have no effect. The wave passes through without attenuation.

If you do not happen to know the direction of polarization of the microwave, put the polarizer in the beam and rotate it. For one orientation the microwave beam will be completely blocked. Rotate the polarizer by 90° and you will get a maximum transmission.

Figure 28
Microwave polarizer, made from an array of copper wires. The microwave transmitter is seen on the other side of the wires, the detector is on this side. When the wires are parallel to the transmitted electric field, no signal is detected. Rotate the wires 90 degrees, and the full signal is detected.

Light Polarizers

We can picture light from the sun as a mixture of light waves with randomly oriented polarizations. (The $\vec{E}$ fields are, of course, always in the plane perpendicular to the direction of motion of the light wave. Only the angle in that plane is random.) A polarizer made of an array of copper wires like that shown in Figure (28), will not work for light because the wavelength of light is so short $\left(\lambda \approx 5 \times 10^{-5}\,\text{cm}\right)$ that the light passes right between the wires. For such a polarizer to be effective, the spacing between the wires would have to be of the order of a wavelength of light or less.

A polarizer for light can be constructed by imbedding long-chain molecules in a flexible plastic sheet, and then stretching the sheet so that the molecules are aligned parallel to each other. The molecules act like the wires in our copper wire array, but have a spacing of the order of the wavelength of light. As a result the molecules block light waves whose electric field is parallel to them, while allowing waves with a perpendicular electric field to pass. (The commercial name for such a sheet of plastic is *Polaroid*.)

Since light from the sun or from standard electric light bulbs consists of many randomly polarized waves, a single sheet of Polaroid removes half of the waves no matter how we orient the Polaroid (as long as the sheet of Polaroid is perpendicular to the direction of motion of the light beam). But once the light has gone through one sheet of Polaroid, all the surviving light waves have the same polarization. If we place a second sheet of Polaroid over the first, all the light will be absorbed if the long molecules in the second sheet are perpendicular to the long molecules in the first sheet. If the long molecules in the second sheet are parallel to those in the first, most of the waves that make it through the first, make it through the second also. This effect is seen clearly in Figure (29).

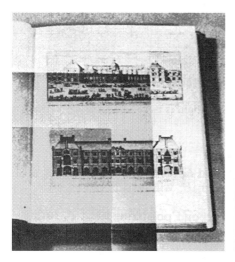

Figure 29
Light polarizers. Two sheets of polaroid are placed on top of a drawing. On the left, the axes of the sheets are parallel, so that nearly half the light passes through. On the right, the axes are perpendicular, so that no light passes through. (Photo from Halliday & Resnick)

Magnetic Field Detector

So far, our discussion of electromagnetic radiation has focused primarily on detecting the *electric field* in the wave. The rabbit ear antenna wire had to be partially parallel to the electric field so that $\oint \vec{E} \cdot d\vec{\ell}$ and therefore the voltage on the antenna would not be zero. In our discussion of polarization, we aligned the parallel array of wires or molecules parallel to the electric field when we wanted the radiation to be reflected or absorbed.

It is also fairly easy to detect the **magnetic field** in a radio wave by using one of our $\oint \vec{E} \cdot d\vec{\ell}$ meters to detect a changing magnetic flux (an application of Faraday's law). This is the principle behind the radio direction finders featured in a few World War II spy pictures.

In a typical scene we see a car with a metal loop mounted on top as shown in Figure (30a). It is chasing another car with a hidden transmitter, or looking for a clandestine enemy transmitter.

If the transmitter is a radio antenna with a vertical transmitting wire as shown in Figure (30b), the magnetic field of the radiated wave will be concentric circles as shown. Objects on the ground, the ground itself, and nearby buildings and hills can distort this picture, but for now we will neglect the distortions.

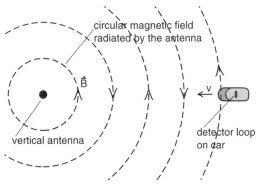

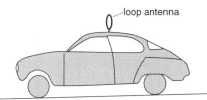

Figure 30a
Car with radio direction finder loop antenna mounted on top.

Figure 30b
Car driving toward radio transmitter.

Figure 31
In a January 1998 National Geographic article on Amelia Earhardt, there appeared a picture of a vintage Electra airplane similar to the one flown by Earhardt on her last trip in 1938. On the top of the plane, you can see the kind of radio direction finder we have been discussing. (The plane is being flown by Linda Finch.)

In Figures (32a) and (32b), we show the magnetic field of the radio wave as it passes the detector loop mounted on the car. A voltmeter is attached to the loop as shown in Figure (33). In (32a), the plane of the loop is parallel to $\vec{B}$, the magnetic flux Φ_B through the loop is zero, and Faraday's law gives

$$V = \oint \vec{E} \cdot d\vec{\ell} = \frac{d\Phi_B}{dt} = 0$$

In this orientation there is no voltage reading on the voltmeter attached to the loop.

In the orientation of Figure (32b), the magnetic field passes through the loop and we get a maximum amount of magnetic flux Φ_B. As the radio wave passes by the loop, this flux alternates signs at the frequency of the wave, therefore the rate of change of flux $d\Phi_B/dt$ is at a maximum. In this orientation we get a maximum voltmeter reading.

The most sensitive way to use this radio direction finder is to get a zero or "null" reading on the voltmeter. Only when the loop is oriented as in Figure (32a), with its plane perpendicular to the direction of motion of the radio wave, will we get a null reading. At any other orientation some magnetic flux will pass through the loop and we get some voltage.

Spy pictures, set in more modern times, do not show antenna loops like that in Figure (30) because modern radio direction finders use so-called "ferrite" antennas that detect the electric field in the radio wave. We get a voltage on a ferrite antenna when the electric field in the radio wave has a component along the ferrite rod, just as it needed a component along the wires of a rabbit ears antenna. Again these direction finders are most accurate when detecting a null or zero voltage. This occurs only when the rod is parallel to the direction of motion of the radio wave, i.e. points toward the station. (This effect is very obvious in a small portable radio. You will notice that the reception disappears and you get a null detection, for some orientations of the radio.)

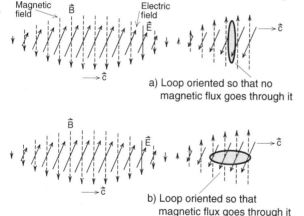

a) Loop oriented so that no magnetic flux goes through it

b) Loop oriented so that magnetic flux goes through it

Figure 32

Electromagnetic field impinging upon a loop antenna. In (a), the magnetic field is parallel to the plane of the loop, and therefore no magnetic flux goes through the loop. In (b), the magnetic flux goes through the loop. As the wave passes by, the amount of flux changes, inducing a voltage in the loop antenna.

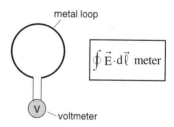

Figure 33

We can think of a wire loop connected to a voltmeter as an $\oint \vec{E} \cdot d\ell$ meter. Any changing magnetic flux through the loop induces a voltage around the loop. This voltage is read by the voltmeter.

RADIATED ELECTRIC FIELDS

One of the best computer simulations of physical phenomena is the series of short films about the electric fields produced by moving and accelerated charges. We will describe a few of the frames from these films, but nothing replaces watching them.

Two basic ideas underlie these films. One is Gauss' law which requires that *electric field lines not break*, do not end, in empty space. The other is that disturbances on an *electric field line travel outward at the speed of light*. No disturbance, no change in the electric field structure, can travel faster than the speed of light without violating causality. (You could get answers to questions that have not yet been asked.)

As an introduction to the computer simulations of radiation, let us see how a simple application of these two basic ideas leads to the picture of the electromagnetic pulse shown back in Figure (16). In Figure (34a) we show the electric field of a stationary, positively charged rod. The electric field lines go radially outward to infinity. (It's a long rod, and it has been at rest for a long time.)

At time $t = 0$ we start moving the entire rod upward at a speed v. By Gauss' law the electric field lines must stay attached to the charges Q in the rod, so that the ends of the electric field lines have to start moving up with the rod.

No information about our moving the rod can travel outward from the rod faster than the speed of light. If the time is now $t > 0$, then beyond a distance ct, the electric field lines must still be radially outward as in Figure (34b). To keep the field lines radial beyond $r = ct$, and keep them attached to the charges +Q in the rod, there must be some kind of expanding kink in the lines as indicated.

At time $t = t_1$, we stop moving the positively charged rod. The information that the charged rod has stopped moving cannot travel faster than the speed of light, thus the displaced radial field next to the rod cannot be any farther out than a distance $c(t-t_1)$ as shown in Figure (34c). The effect of starting, then stopping the positive rod is an outward traveling kink in the electric field lines. It is as if we had ropes attached to the positive rod, and jerking the rod produced an outward traveling kink or wave on the ropes.

In Figure (34d), we have added in a stationary negatively charged rod and the inward directed electric field produced by that rod. The charge density on the negative rod is opposite that of the positive rod, so that there is no net charge on the two rods. When we combine these rods, all we have left is a positive upward directed current during the time interval $t = 0$ to $t = t_1$. We have a short current pulse, and the electric field produced by the current pulse must be the vector sum of the electric fields of the two rods.

In Figure (34e), we add up the two electric fields. In the region $r > ct$ beyond the kink, the positive and negative fields must cancel exactly. In the region $r < (t - t_1)$ we should also have nearly complete cancellation. Thus all we are left with are the fields E_+ and E_- inside the kink as shown in Figure (34f). Since electric field lines cannot end in empty space, E_+ and E_- must add up to produce the downward directed E_{net} shown in Figure (34g). Note that this downward directed electric field pulse was produced by an upward directed current pulse. As we have seen before, this induced electric field opposes the change in current.

In Figure (34h) we added the expanding magnetic field pulse that should be associated with the current pulse. What we see is an expanding electromagnetic pulse that has the structure shown in Figure (16). Simple arguments based on Gauss' law and causality gave us most of the results we worked so hard to get earlier. What we did get earlier, however, when we applied Ampere's and Faraday's law to this field structure, was the explicit prediction that the pulse expands at the speed $1/\sqrt{\mu_0\varepsilon_0} = 3 \times 10^8 \, m/s$.

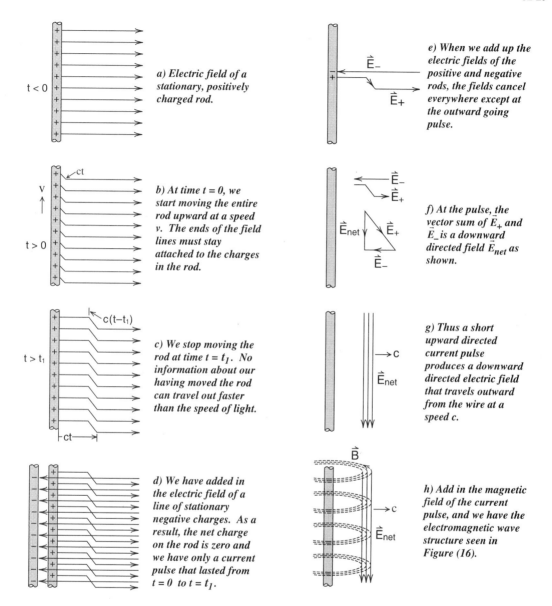

a) Electric field of a stationary, positively charged rod.

b) At time t = 0, we start moving the entire rod upward at a speed v. The ends of the field lines must stay attached to the charges in the rod.

c) We stop moving the rod at time t = t_1. No information about our having moved the rod can travel out faster than the speed of light.

d) We have added in the electric field of a line of stationary negative charges. As a result, the net charge on the rod is zero and we have only a current pulse that lasted from t = 0 to t = t_1.

e) When we add up the electric fields of the positive and negative rods, the fields cancel everywhere except at the outward going pulse.

f) At the pulse, the vector sum of $\vec{E}_+$ and $\vec{E}_-$ is a downward directed field $\vec{E}_{net}$ as shown.

g) Thus a short upward directed current pulse produces a downward directed electric field that travels outward from the wire at a speed c.

h) Add in the magnetic field of the current pulse, and we have the electromagnetic wave structure seen in Figure (16).

Figure 34
Using the fact that electric field lines cannot break in empty space (Gauss' Law), and the idea that kinks in the field lines travel at the speed of light, we can guess the structure of an electromagnetic pulse.

Field of a Point Charge

The computer simulations show the electric field of a point charge under varying situations. In the first, we see the electric field of a point charge at rest, as shown in Figure (35a). Then we see a charge moving at constant velocity $\vec{v}$. As the speed of the charge approaches c, the electric field scrunches up as shown in Figure (35b).

The next film segment shows what happens when we have a moving charge that stops. If the charge stopped at time $t = 0$, then at a distance $r = ct$ or greater, we must have the electric field of a moving charge, because no information that the charge has stopped can reach beyond this distance. In close we have the electric field of a static charge. The expanding kink that connects the two regions is the electromagnetic wave. The result is shown in Figure (35c). The final film segment shows the electric field of an oscillating charge. Figure (36) shows one frame of the film. This still picture does a serious injustice to the animated film. There is no substitute, or words to explain, what you see and feel when you watch this film.

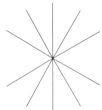

Figure 35a
Electric field of a stationary charge.

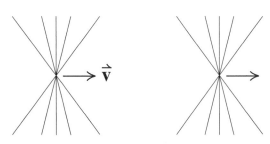

Figure 35b
Electric field of a moving charge. If the charge has been moving at constant speed for a long time, the field is radial, but squeezed up at the top and bottom.

Figure 35c
Field of a charge that stopped. Assume that the charge stopped t seconds ago. Inside a circle of radius ct, we have the field of a stationary charge. Outside, where there is no information that the charge has stopped, we still have the field of a moving charge. The kink that connects the two fields is the electromagnetic radiation.

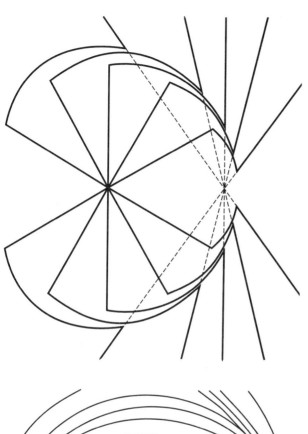

Figure 35c (enlarged)
Electric field of a charge that stopped. The dotted lines show the field structure we would have seen had the charge not stopped.

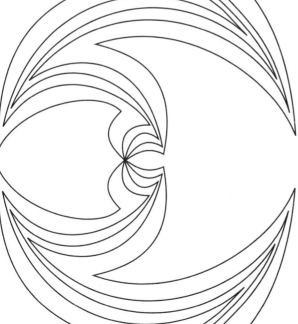

Figure 36
Electric field of an oscillating charge.

Exercise 6

Assume that we have a supply of ping pong balls and cardboard tubes shown in Figures (37). By looking at the fields outside these objects decide what could be inside producing the fields. Explicitly do the following for each case.

i) Write down the Maxwell equation which you used to decide what is inside the ball or tube, and explain how you used the equation.

ii)   If more than one kind of source could produce the field shown, describe both (or all) sources and show the appropriate Maxwell equations.

iii)   If the field is impossible, explain why, using a Maxwell equation to back up your explanation.

In each case, we have indicated whether the source is in a ball or tube. Magnetic fields are dashed lines, electric fields are solid lines, and the balls and tubes are surrounded by empty space.

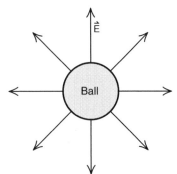

Figure 37a
Electric field emerging from ping pong ball.

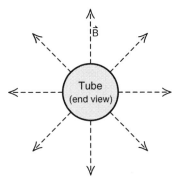

Figure 37b
Magnetic field emerging from ping pong ball.

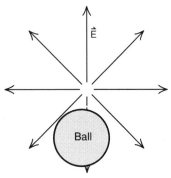

Figure 37c
Electric field emerging above ping pong ball.

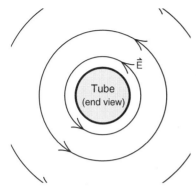

Figure 37d
Electric field around tube.

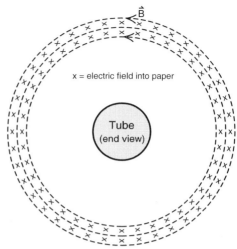

x = electric field into paper

Figure 37f
For this example, explain what is happening to the fields, what is in the tube, and what happened inside.

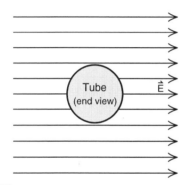

Figure 37e
Electric field passing through tube.

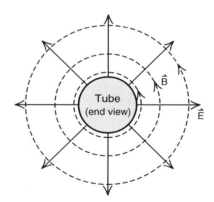

Figure 37g
There is only ONE object inside this tube. What is it? What is it doing?

Chapter 33
Light Waves

Ripples produced by rain drops. (Bill Jack Rodgers, Los Alamos Scientific Laboratory)

In the examples of wave motion we studied back in Chapter 15, like waves on a rope and sound in a gas, we could picture the wave motion as a consequence of the mechanical behavior particles in the rope or molecules in the gas. We used Newton's laws to predict the speed of a rope wave and could have done the same for a sound. When we discuss light waves, we go beyond the Newtonian behavior. Waves on a rope, on water or in a gas are mechanical undulations of an explicit medium. Light waves travel through empty space; there is nothing to undulate, nothing to which we can apply Newton's laws. Yet, in many ways, the behavior of light waves, water waves, sound waves, and even the waves of quantum theory, are remarkably similar.

There are general rules of wave motion that transcend the nature of the medium or type of wave. One is the principle of superposition that we used extensively in Chapter 15. It is the idea that as waves move through each other, they produce an overall wave whose amplitude is the sum of the amplitudes of the individual waves. The other is a concept we will use extensively in this chapter called the **Huygens principle**, named after its discoverer Christian Huygens, a contemporary of Isaac Newton.

We will see that a straightforward application of the principle of superposition and the Huygens principle allows us to make detailed predictions that even can be used as a test of the wave nature of the phenomena we are studying.

SUPERPOSITION OF CIRCULAR WAVE PATTERNS

When we studied the interaction of waves on a rope, it was a relatively simple process of adding up the individual waves to see what the resultant wave would be. For example, in Figure (15-6) reproduced here, we see that when a crest and a trough run into each other, for an instant they add up to produce a flat rope. At this instant the crest and the trough cancel each other. In contrast two crests add to produce a big crest, and two troughs add to produce a deeper trough.

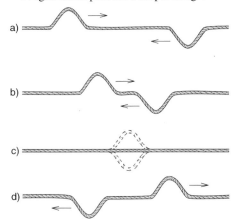

Figure 15-6
When a crest meets a trough, there is a short time when the waves cancel.

When we extend our study of wave motion to two and three dimensions, the principle of superposition works the same way, but now we have to add patterns rather than just heights along a line. If, for example, we are studying wave motion on the surface of water, and two wave patterns move through each other, the resulting wave is the sum of the heights of the individual waves at every point on the surface. We do the same addition as we did for one dimensional waves, but at many more points.

A relatively simple, but important example of the superposition of wave patterns is the pattern we get when concentric circular waves from two nearby sources run into each other. The pattern is easy to set up in a ripple tank using two oscillating plungers.

Figure (1a) shows the circular wave pattern produced by a single oscillating plunger. From this picture we can easily see the circular waves emerging from the plunger. The only difficulty is distinguishing crests from troughs. We will handle this by using a solid line to represent the crest of a wave and a dashed line for a trough, as illustrated in Figure (1b).

Figure 1a,b
Circular wave pattern produced in a ripple tank by a plunger. The pattern consists of alternate crests and troughs. To diagram the circular wave pattern, we will use solid lines for crests and dashed lines for troughs.

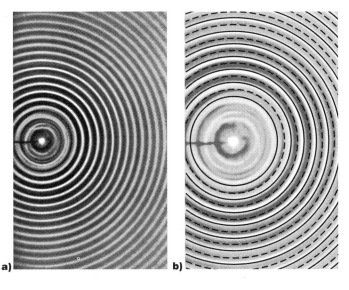

a)   b)

In Figure (2a), we see the wave pattern produced by two plungers oscillating side by side. Each plunger sends out a circular set of waves like that seen in Figure (1). When the two sets of circular waves cross each other, we get cancellation where crests from one set meet troughs from the other set (where a solid line from one set of circles meets a dashed line from the other set of circles in Figure (2b). This cancellation occurs along lines called *lines of nodes* which are clearly seen in Figure (2a).

Between the lines of nodes we get beams of waves. In each beam, crests from one plunger meet crests from the other producing a higher crest. And troughs from one set meet troughs from the other producing deeper troughs. In our drawing of circles, Figures (2b) and (2c), we get beams of waves along the lines where solid circles cross solid circles and dashed circles cross dashed circles.

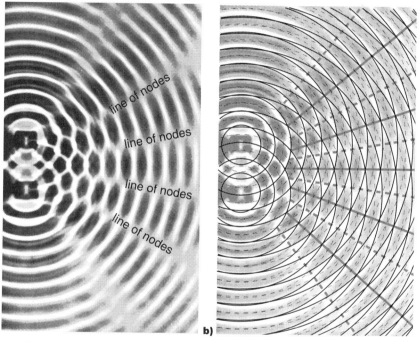

a) b)

Figure 2a,b
Ripple tank photograph of an interference pattern. When two sets of circular waves move through each other, there are lines along which crests from one set always meet troughs from the other set. These are called lines of nodes. *Between the lines of nodes, we get beams of waves. The resulting pattern is called an* interference pattern.

Figure 2c
We get beams of waves where crests meet crests and troughs meet troughs. The lines of nodes are where crests meet troughs and the waves cancel.

HUYGENS PRINCIPLE

When sunlight streams in through an open kitchen door, we see a distinct shadow on the floor. The shadow can be explained by assuming that the light beams travel in straight lines from the sun through the doorway. The whole subject of geometrical optics and lens design is based on the assumption that light travels in straight lines (except at the interface of two media of different indices of refraction).

In Figure (3) we see what happens when a wave impinges upon a slit whose width is comparable to the wavelength of the waves. Instead of there being a shadow of the slit, we see that the emerging wave comes out in all directions. The wave pattern on the right side of the slit is essentially identical to the wave pattern produced by the oscillating plunger in Figure (1a). We can explain Figure (3) by saying that the small piece of wave front that gets through the slit acts as a source of waves in much the same way that the oscillating plunger acted as a source of waves.

Christian Huygens noted this phenomena and from it developed a general principle of wave motion. His idea was that as a wave pattern evolved, each point of a wave front acts as the source of a new circular or spherical wave. To see how this principle can be applied,

consider the relatively smooth wave front shown in Figure (4). To predict the position of the wave front a short time later, we treat each point on the front of the wave as a source of circular waves. We can see the effect by drawing a series of circles at closely spaced points along the wave. The circular waves add up to produce a new wave front farther out. While you can use the same construction to figure out what is happening throughout the wave, it is much easier to see what is happening at the front.

Exercise 1

At some instant of time, the front of a wave has a sharp, right angle corner. Use Huygens principle to find the shape of the wave front at some later instant of time. (Draw a right angle corner and use the kind of construction shown in Figure (4).)

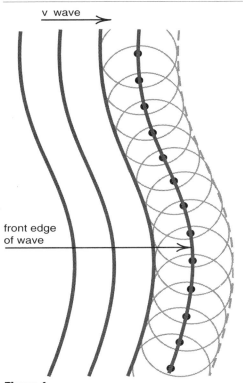

Figure 4
Huygen's construction. The future position of a wavefront can accurately be predicted by assuming that each point on the wavefront is a source of a new wave.

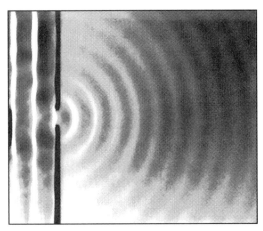

Figure 3
A wave emerging from a narrow slit spreads out in all directions, just as if the wave in the slit were a plunger.

By using the construction of Figure (4) to predict the future shape of a wave front, we see that if we use a slit to block all but a small section of the wave front, as illustrated in Figure (5), then the remaining piece of wave front will act as a source of circular waves emerging from the other side. This is what we saw in Figure (3). Thus the Huygens construction allows us to see not only how a smooth wave travels forward intact, but also why circular waves emerge from a narrow slit as we saw in Figure (3).

The Huygens construction also provides a picture of what happens as waves go through progressively wider slits. If the slit is wider than a wavelength then we have more sources in the slit and the waves from the sources begin to interfere with each other. In Figures (6, 7, 8) we see the wave patterns for increasingly wide slits and the corresponding Huygens constructions. For the wider slits, more of the wave goes through the center intact, but there is always a circular wave coming out at the edges. For the slit of Figure (8), the circular waves at the edges are relatively unimportant, and the edges of the slit cast a shadow. This is beginning to resemble our example of sunlight coming through the kitchen door-way. The name *diffraction* is used to describe the spreading of the waves that we see at the edges of the slits in Figures (5) through (8).

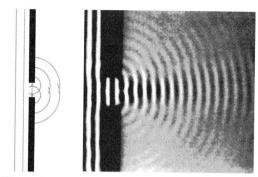

Figure 6
When the slit is about 2 wavelengths wide, the wave in the slit acts as 2 point sources.

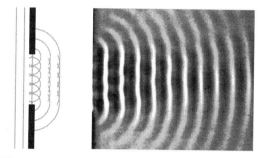

Figure 7
As the slit is widened, more of the wave comes through intact. In the center we are beginning to get a beam of waves, yet at the edges, the wavefront continues to act as a source of circular waves.

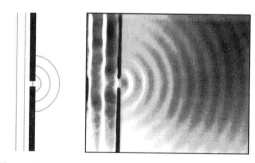

Figure 5
The small piece of wave in a narrow slit acting as a single point source.

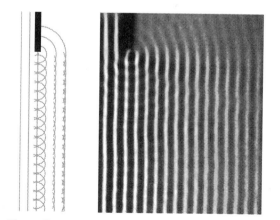

Figure 8
When the slit is wide compared to a wavelength, we get a distinct beam of waves. Yet no matter how wide the slit, there are still circular waves at the edges.

TWO SLIT
INTERFERENCE PATTERN

If a single narrow slit can produce the same wave pattern as an oscillating plunger, as we saw in Figure (3), then we should expect that two slits next to each other should produce an interference pattern similar to the one produced by two oscillating plungers seen in Figure (2). That this is indeed correct is demonstrated in Figure (9). On the left we have repeated the wave pattern of 2 plungers. On the right we have a wave impinging upon two narrow slits. We see that both have the same structure of lines of nodes, with beam of waves coming out between the lines of nodes. Because the patterns are the same, we can use the same analysis for both situations.

Sending a wave through two slits and observing the resulting wave pattern is a convenient way to analyze various kinds of wave motion. But in most cases we do not see the full interference pattern as we do for these ripple tank photographs. Instead, we observe only where the waves strike some object, and from this deduce the nature of the waves.

To illustrate what we mean, imagine a harbor with a sea wall and two narrow entrances in the wall as shown in Figure (10). Waves coming in from the ocean emerge as circular waves from each entrance and produce a two slit interference pattern in the harbor. Opposite the sea wall is a beach as shown.

If we are at point A on the beach directly across from the center of the two entrances, we are standing in the center beam of waves in the interference pattern. Here

Figure 9
The wave pattern emerging from 2 slits is similar to the wave pattern produced by two plungers.

large waves wash up on the beach. Walking north along the beach we cross the first line of nodes at point B. Here the water is calm. Going farther up to point C we are again in the center of a beam of waves. We will call this the *first maximum* above the *central maximum*. Farther up we cross the second line of nodes at point D and encounter the second minimum in the height of waves striking the shore.

Going south from point A we encounter the same alternate series of maxima and minima at points B', C', D', etc. If we graphed the amplitude of the waves striking the shore, we would get the pattern shown at the right side of Figure (10).

Now suppose that we walk along the beach on a calm day where there are no waves, but on the previous day there had been a storm. During the storm, the waves

striking the shore eroded the beach. As you walk along the beach you notice a series of indentations, at points A, C, C', etc. where the beach was eroded. The sand was not eroded at points B, B', D and D'. If someone asked what the ocean waves were like during the storm, could you tell them?

By measuring the distance between the maximum erosions and knowing the geometry of the harbor, you can determine the wavelength of the ocean waves that struck the sea wall during the storm. Similar calculations can be made to determine the wavelength of any kind of wave striking two narrow slits producing an interference pattern on the other side. We do not have to see the actual wave pattern, we only have to note the location of the maxima and minima of the waves striking an object like the shore in Figure (10).

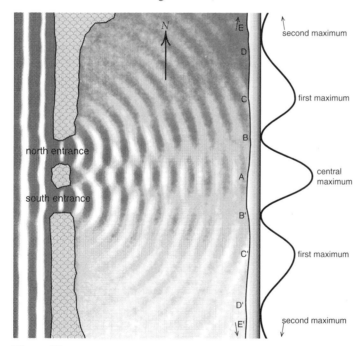

Figure 10

Hypothetical harbor with two entrances through the sea wall. If ocean waves are coming straight in toward the sea wall, there will be a 2 slit interference pattern inside the harbor, with a series of maxima and minima along the beach.

We begin our analysis of the two slit wave pattern by drawing a series of circles to represent the wave crests and troughs emerging from the two slits. The results, which are shown in Figure (11), are essentially the same as our analysis of the two plunger interference pattern in Figure (2). The maxima occur where crests meet crests and troughs meet troughs. The minima or lines of nodes are where crests meet troughs.

Exercise 2

On Figure (11), sketch the lines along which crests meet troughs, i.e., where solid and dashed circles intersect. This should be where the lines of nodes are located.

The First Maxima

The central maximum is straight across from the center of the two slits (if the incoming waves are parallel to the slits as in Figure 11). To figure out where the first maximum is located, consider the sketch in Figure (12). We have reduced the complexity of the sketch by drawing only the solid circles representing wave crests. In addition we have numbered the crests emerging

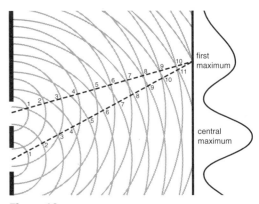

Figure 12
One more wave fits in the path from the bottom slit to the first maxima, than in the path from the top slit.

from each slit. We see that at first maximum, the 12th crest from the lower slit has run into the 11th crest from the upper slit, producing a maximum crest. *The distance from the lower slit to the first maximum is exactly one wavelength longer than the distance from the upper slit to the first maximum.* This is what determines the location of the first maximum.

In Figure (13) we have repeated the sketch of Figure (12), but now focus our attention on the difference in the length of the two paths from the slits to the first maximum. Since an extra wavelength λ fits into the lower path, the path length difference is λ as shown. The bottom path, with λ removed, and the upper path, both shown as dashed lines in Figure (13), are thus the same length and therefore form 2 sides of an isosceles triangle.

Let us denote by θ_1 the angle from the center of the two slits up to the first maximum. Since this line bisects the isosceles triangle formed by the two dashed lines, it is perpendicular to the base of the isosceles triangle which is the line from the center of the upper slit down to the point (a) on the lower path. As a result, the base of the isosceles triangle makes the same angle θ_1 with the plane of the slits as the line to the first maximum does with the horizontal line to the central maximum. (Picture rotating

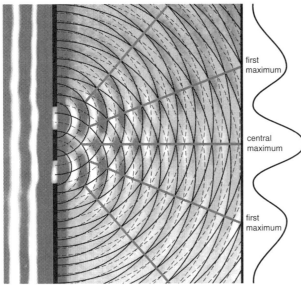

Figure 11
Analysis of the two slit wave pattern, assuming that circular waves emerge from each slit and interfere with each other. The maxima are where crests from one slit meet crests from the other. Cancellation occurs where crests meet troughs.

the isosceles triangle up around its base. If you rotate the isosceles triangle by an angle θ_1, its base will rotate by the same angle θ_1, thus the 2 angles labeled θ_1 in Figure (13) are the same.)

Our approximation in this analysis is that the separation d between the slits is very small compared to the distance over to where we are viewing the first maximum. If this is true, then the two paths to the first maximum are essentially parallel and the small bold triangle in Figure (13) is very nearly a right triangle. Assuming that this is a right triangle, we immediately get

$$\sin \theta_1 = \frac{\lambda}{d} \qquad \text{\textit{angle to first maximum}} \qquad (1)$$

In Figure (14) we have another right triangle involving the angle θ_1. If the distance from the slits to where we are viewing the maxima is D, and if we designate by Y_{max} the distance from the central to the first maximum, then the hypotenuse of this right triangle is given by the Pythagorean theorem as $\sqrt{D^2 + Y_{max}^2}$. From this triangle we have

$$\sin \theta_1 = \frac{Y_{max}}{\sqrt{D^2 + Y_{max}^2}} \qquad (2)$$

Equating the two formulas for $\sin \theta_1$ and solving for λ gives

$$\lambda = Y_{max} \frac{d}{\sqrt{D^2 + Y_{max}^2}} \qquad (3)$$

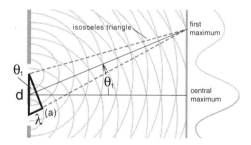

Figure 13
The path length difference to the first maximum is one wavelength λ.

An easy way to remember this derivation is to note that the two triangles in Figures (13) and (14), drawn separately in Figure (15), are similar triangles. Thus the ratios of the small sides to the hypotenuses must be equal, giving

$$\frac{\lambda}{d} = \frac{Y_{max}}{\sqrt{D^2 + Y_{max}^2}} \qquad (3')$$

The importance of Equation 3 is that it allows us to calculate the wavelength of a wave by observing the distance Y_{max} between maxima of the interference pattern. For example, in our problem of determining the character of the waves eroding the beach in Figure (10), we could use a map to determine the distance D from the breakwater to the shore and the distance d between entrances through the breakwater. Then pacing off the distance Y_{max} between erosions on the beach, we could use Equation 3 to determine what the wavelength of the waves were during the storm.

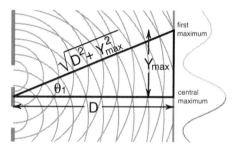

Figure 14
The angle θ_1 up to the first maxima is the same as the angle in the small triangle of Figure (13).

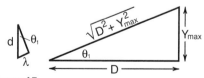

Figure 15
Since the two triangles are similar, we have $\lambda/d = y_{max} / \sqrt{D^2 + Y_{max}^2}$

Exercise 3

Repeat the derivation that led to Equation 3 except do the calculation in terms of the distance Y_{min} from the central maximum to the first minimum. (Now the path length difference is $\lambda/2$.)

Exercise 4

We will see that Equation 3 has an applicability that goes far beyond the analysis of two slit interference patterns. You will need this formula several times later in this course, and quite likely in other research work. Rather than memorizing the formula, it is much better to memorize the derivation. The best way to do this is to treat the derivation as a clean desk problem. Some time, a day or so after you have read this section, clean off your desk, take out a blank sheet of paper, and derive Equation 3. The first time you try it, you may have forgotten some steps. If that happens, review the derivation and try to do the clean desk problem a day or so later. It is worth the effort because the derivation summarizes all the formulas used in this chapter.

TWO SLIT PATTERN FOR LIGHT

Christian Huygens discovered his principle of wave motion in 1678, and developed a wave theory of light that competed with Newton's particle theory of light. It was not until 1801, over 120 years later, that Thomas Young first demonstrated the wave nature of light using a two slit interference experiment. Why did it take so long to do this demonstration?

Two major problems arise when you try to test for the wave nature of light. One is the fact that the wavelength of light is very short, on the order of one hundred thousand times shorter than the wavelengths of the water waves we observe in the ripple tank photographs. A more serious problem is that individual atoms in the sun or a light bulb emit short bursts of light that are not coordinated with each other. The result is a chopped up, incoherent beam of light that may also include a mixture of frequencies.

In our analogy of a sea wall with two entrances, it is likely that a real storm would produce a mixture of waves of different wavelengths heading in different directions. Many different interference patterns would be superimposed on the inside of the sea wall, different maxima and minima would overlap at the beach and the beach would be more or less uniformly eroded. Walking along the beach the next day, you would not find enough evidence to prove that the damage was done by ocean waves, let alone trying to determine the wavelength of the waves.

The invention of the laser by Charles Townes in 1960 eliminated the experimental problems. The laser emits a continuous coherent beam of light that more closely resembles the orderly ripple tank waves approaching the slits in Figure (10) than the confused wave motion seen in a storm. If you send a laser beam through two closely spaced slits, you cannot help but see a two slit interference pattern.

Even in a demonstration lecture, the two slit pattern produced by a laser beam can be used to measure the wavelength of the light in the beam. In Figure (16) we placed a two slit mask next to a millimeter scale on the top of an overhead projector and projected the image on a large screen. You can see that the spacing between the two slits is about 1/3 of a millimeter. In Figure (17) we aimed the red beam of a common helium neon laser through the two slits of Figure (16), onto a screen 10 meters from the slits. The resulting two slit pattern consisting of the alternate maxima and minima are easily seen by the class. Marking the separation of two maxima on a piece of paper and measuring the distance we found that the separation Y_{max} between maxima was about 2.3 cm.

In using Equation 3, $\lambda = Y_{max}d/\sqrt{D^2 + Y_{max}^2}$ to calculate the wavelength λ, we note that the 10 meter distance D is much greater than the 2.3 cm Y_{max}. Thus we can neglect the Y_{max}^2 in the square root and we get the simpler formula

$$\lambda \approx Y_{max}\frac{d}{D} \quad \left(if\ D >> Y_{max}\right) \tag{3a}$$

Putting in the numbers obtained from Figures (16) and (17), we get

$$\lambda = 2.3\ cm \times \frac{.3 \times 10^{-3}m}{10m} = 7 \times 10^{-5}cm \tag{4}$$

While this demonstration experiment gives fairly approximate results, accurate to about one significant figure, it may be somewhat surprising that a piece of apparatus as crude as the two slits seen in Figure (16) even allows us to measure something as small as $7 \times 10^{-5}cm$.

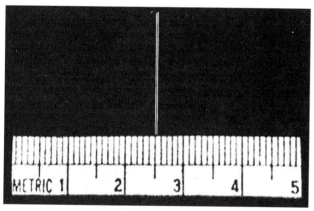

Shows that d = .3mm

Figure 16
The two slits and a plastic ruler are placed on an overhead projector and projected onto a screen 10 meters away. This is a photograph of the screen.

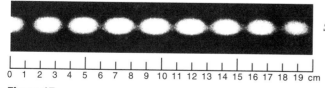

Shows that Y_{max} = 2.3 mm for D = 10 m.

Figure 17
The 2 slit laser pattern is then projected on the screen. Below is a centimeter scale, showing that the maxima are about 2.3 centimeters apart.

THE DIFFRACTION GRATING

The crudeness of our measurement of the wavelength of the laser light in our two slit experiment could be improved somewhat by a more accurate measurement of the separation of the two slits, but the improvement would not be great. There is, however, a simple way to make far more accurate measurements of the wavelength of a beam of light. The trick is simply add more slits.

To see why adding more slits gives more accurate results, we show in Figure (18) the wave patterns we get when the laser beam is sent through two slits, three slits, four slits, five slits, and seven slits. We created the slits using a Macintosh computer using the Adobe Photoshop program and a Linatronic printer to produce the film images of the slits. The Linatronic printer can draw precise lines one micron wide (10^{-6} meters); thus we had excellent control over the slit width and spacing. For these images, the slits are 50 microns (50μ) wide and spaced 150 microns apart on centers.

The photographs of the interference patterns produced by the slits of Figure (18) are all enlarged to the same scale. The important point to notice is that while the maxima become sharper as we increase the number of slits, the spacing between maxima remains the same. *Adding more identical slits sharpens the maxima but does not change their spacing!* As a result the two slit formula, Equation 3, can be applied to any number of slits as long as the spacing d between slits remains constant.

If there are many slits, the device is called a *diffraction grating* and Equation 3, which we repeat below, is known as the diffraction grating formula.

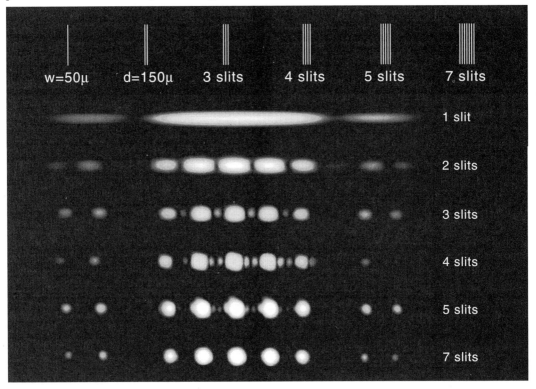

Figure 18
Interference patterns for various slit structures. If we keep the spacing between slits the same, then there is no change in the location of the maxima, no matter how many slits the laser beam passes through. Thus an analysis of the location of the maxima for 2 slits applies to any number of slits. Also note that the single slit pattern acts as an envelope for the multiple slit patterns.

$$\lambda = Y_{max}\frac{d}{\sqrt{D^2 + Y_{max}^2}} \quad \begin{array}{l} \textit{diffraction} \\ \textit{grating} \\ \textit{formula} \end{array} \quad (3\,repeat)$$

Exercise 5

In Figure (18) the separation of the slits is 150 microns and the separation of 10 maxima is 26.4 cm. The screen is a distance of 6.00 meters from the slits. From this determine the wavelength of the light in the laser beam

(a) using the exact formula, Equation 3.

(b) using the approximate formula, Equation 3a.

How many significant figures are meaningful in your result? To this accuracy, did it make any difference whether you used the exact Equation 3 or the approximate Equation 3a.

Figure (18) demonstrates that the more slits you use, the sharper the maxima and the more accurately you can determine the wavelength of the light passing through the slits. In the latter part of the 1800s, the diffraction grating was recognized as an excellent tool for scientific research, and a great effort was put into producing gratings with as many closely spaced lines as possible. Fine ruling machines were developed that produced gratings on the order of 6000 lines or slits per centimeter. With so many lines, very sharp maxima are produced and very precise wavelength measurements can be made. It is possible to make inexpensive plastic replicas of fine diffraction gratings for use in all kinds of laboratory work, or even for making jewelry. It turns out that compact disks (CDs) also make superb diffraction gratings. We will not tell you the spacing of the lines on a CD for it is a nice project to figure that out for yourself. (All you need is a common helium neon laser. The wavelength of the laser beam can be gotten from Exercise 6.)

Exercise 6

In Figure (19), a laser beam is sent through a smoke filled box with a diffraction grating at the center of the box as shown in the sketch (19a). The smoke allows you to see and photograph the central laser beam and two maxima on each side. You also see maxima reflected from the back side of the grating. (When you shine a laser beam on a CD you get only the reflected maxima, no light goes through the record.)

The grating used in Figure (19) had 15,000 lines per inch (1 inch = 2.54 cm). From this information and the photograph of Figure (19b), determine the wavelength of the laser beam used. Try both the exact Equation 3 and the approximate Equation 3a. Explain why Equation 3a does not work well for this case.

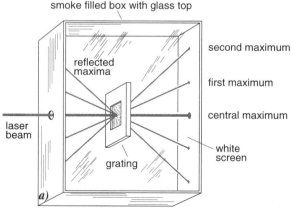

Figure 19
Laser beam passing through a diffraction grating. The beam is made visible by placing the grating in a smoke filled box. Because the lines are so close together, the maxima are widely separated. You can also see reflected maxima on the back side.

More About Diffraction Gratings

The results of Figure (18) demonstrated that the maxima got sharper but remained in the same place as we added slits. Let us now see why this happens.

The maxima of a diffraction grating occur at those points on the screen where the waves from every slit add up constructively. This can happen only when the path length difference between neighboring maxima is 0 (central maximum), λ (first maxima), 2λ (second maxima), etc. In Figure (20) we are looking at a small section of a diffraction grating where we have drawn in the paths to the first maxima. The path length differences between neighboring slits are all λ and the angle θ_1 to the first maxima is given by $\theta_1 = \lambda/d$, the same results we had for the two slit problem in Figure (13). This angle does not depend upon the number of slits, thus the position of the maxima do not change when we add slits as in Figure (18).

To see why the maxima become narrower as we add slits, let us consider the example of a 1000 slit grating illustrated in Figure (21). We have numbered the slits from 1 to 1000, and are showing the paths to a point just below the first maximum where the path length difference between neighboring slits is $(\lambda - \lambda/1000)$ instead of λ.

On the figure we are indicating, not the path length difference between neighboring slits, but instead, the path length difference between the first slit and the others. This difference is $(\lambda - \lambda/1000)$ for slit #2, $(2\lambda - 2\lambda/1000)$ for slit #3, $(3\lambda - 3\lambda/1000)$ for slit #4, etc.

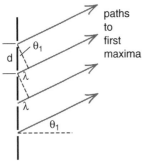

paths to first maxima

Figure 20
When the path length difference between neighboring paths is λ, then the waves from all slits add constructively and we get the first maxima.

When we get down to slit # 501, just over half way down, the path length difference is $500\lambda - 500\lambda/1000 = 500\lambda - \lambda/2$. In other words, the waves from slit 1 and slit 501 are precisely one half a wavelength out of phase, crests exactly meet troughs, and there is precise cancellation. A similar argument shows that waves from slits #2 and #502 are $\lambda/2$ out of phase and cancel exactly. The same goes for the pairs #3 and #503, #4 and #504, all the way down to 500 and 1000. In other words, the waves all cancel in pairs and we have a minimum, complete cancellation at the point just below the first maximum where the path length difference is $\lambda - \lambda/1000$ instead of λ. With two slits we got complete cancellation half way between maxima. With 1000 slits, we only have to go approximately 1/1000 the way toward the next maxima before we get complete cancellation. The maxima are roughly 500 times sharper. You can see that with n slits, the maxima will be about n/2 times sharper than for the two slit example.

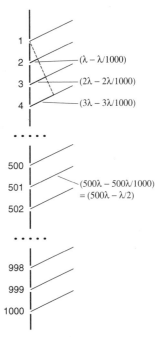

Figure 21
In a thousand slit grating, we get complete cancellation when the path length difference between neighboring slits is reduced from λ to $\lambda - \lambda/1000$.

The maxima will also be much more intense because the light is coming in from more slits. If we have n slits, the amplitude of the wave at the center of a maxima will be n times as great as the amplitude from a single slit. It turns out that the amount of energy in a wave, the intensity, or, for light, the brightness, is proportional to the square of the amplitude of the wave. Thus the brightness at the center of the maxima for an n slit grating is n^2 times as bright at the brightness we would have for a single slit. The maxima for the 1000 slit grating illustrated in Figure (21) would be one million times brighter than if we let light go through only one of the slits. (To see how the total energy works out, consider the following argument. Compared to one slit, when you have n slits, you have n times as much light energy that is compressed into a maxima that is only 1/n as wide. You get one factor of n in brightness due to the compression, and the other factor of n due to there being n slits.)

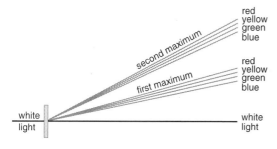

Figure 22
When white light passes through a diffraction grating, the maxima for different colors emerge at different angles. Since red light has the longest wavelength of the visible colors, it emerges at the greatest angle.

The Visible Spectrum

Thus far we have been using a laser beam to study the operation of a diffraction grating. Now we will reverse the process and use diffraction gratings to study the nature of beams of light.

If you send a beam of white light through a diffraction grating, you get a series of maxima. In all but the central maxima light is spread out into a rainbow of colors illustrated in Figure (22). In each maxima the red light is bent the most, and blue the least. As we saw from Equation 1, $\sin \theta_1 = \lambda/d$, the longer the wavelength the greater the angle the wave is bent or diffracted. Thus red light has the longest wavelength and blue the shortest in the mixture of wavelengths that make up white light.

The longest wavelength that the human eye can see is about 7.0×10^{-5} cm, a deep red light, and the shortest is about 4.0×10^{-5} cm, a deep purple. All other visible wavelengths, the entire spectrum of visible light, lies in the range between 4.0×10^{-5} cm to 7.0×10^{-5} cm. Yellow light, for example, has a wavelength around 5.7×10^{-5} cm, and green light is near 5.0×10^{-5} cm.

As we saw in Chapter 32, visible light is just a part of the complete electromagnetic spectrum. A surprisingly small part. As radio, television, microwave ovens, infra red sensors, ultraviolet sunscreens, x ray photographs, and γ ray bursts in the sky have entered our experience of the world, we have become familiar with a much greater range of the electromagnetic spectrum. As indicated in Figure (23), AM radio waves have wavelengths in the range of 10 to 100 meters, VHF television a few meters, VHF from around 10 cm to a meter, microwaves from around a millimeter to 10 cm, infra red from less than a millimeter down to visible red light at 7.0×10^{-5} cm. At shorter wavelengths

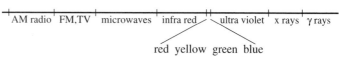

Figure 23
Visible light is a tiny piece of the electromagnetic spectrum.

than deep blue we have ultraviolet, then x rays, and the very shortest wavelengths are called γ (gamma) rays.

To study the electromagnetic spectrum, different devices are used at different wavelengths. In Chapter 32 we used a loop of wire and Faraday's law to detect the magnetic fields of a radio wave. This required the use of an oscilloscope that could display radio wave frequencies, typically of the order of a megacycle for AM radio. For visible light the frequencies are too high, the wavelengths too short for light to be studied by similar techniques. Instead the diffraction grating will be our main tool for studying the electromagnetic waves in the visible spectrum.

Exercise 7

What are the lowest and highest frequencies of the waves in the visible spectrum? What is the color of the lowest frequency? What is the color of the highest? What is the frequency of yellow light?

Atomic Spectra

Our main application of the diffraction grating will be to study the spectrum of light emitted by atoms. It has long been known that if you have a gas of a particular kind of atom, like nitrogen, oxygen, helium, or hydrogen, a special kind of light is emitted. You do not get the continuous blend of wavelengths seen in white light. Instead the light consists of a mixture of distinct wavelengths. Which wavelengths are involved depends upon the kind of atom emitting the light. The mixture of wavelengths provide a unique signature of that atom, better than a fingerprint, for identifying the presence of an atom in a gas. In fact, the element helium (named after the Greek word *helios* for sun) was first identified in the sun by a study of the spectrum of light from the sun. Only later was helium found here on earth.

The subject of modern astronomy is based on the study of the spectrum of light emitted by stars. Some stars consist mostly of hydrogen gas, others a mixture of hydrogen and helium, while still others contain various amounts of heavier elements. We learn the composition of the star by studying the spectrum of light emitted, and from the composition we can deduce

something about the age of the star and the environment in which it was formed.

Our main reason for studying the spectrum of light emitted by atoms will be to learn something about the atoms themselves. Since Rutherford's discovery of the atomic nucleus in 1912, it has been known that atoms consist of a positively charged nucleus surrounded by negatively charged electrons. If we apply Newtonian mechanics to predict the motion of the electrons, and Maxwell's equations to predict the kind of electromagnetic radiation the moving electrons should radiate, we get the wrong answer. There is no way that we can explain the spectrum of light emitted by atoms from Maxwell's equations and Newtonian mechanics. The existence of detailed atomic spectra is a clue that something is wrong with this classical picture of the atom. It is also the evidence upon which to test new theories.

We do not have to study many kinds of atoms to find something wrong with the predictions of classical theory. The simplest of all atoms, the hydrogen atom consisting of one proton for a nucleus, surrounded by one electron, is all we need. Heated hydrogen gas emits a distinct, orderly, spectrum of light that provides the essential clues of what is going on inside a hydrogen atom. In this chapter we will focus on using a diffraction grating to learn what the spectrum of hydrogen is. In the following chapters we use the hydrogen spectrum to study the atom itself.

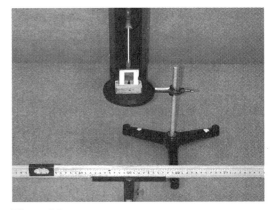

Figure 24
Apparatus to measure hydrogen spectrum.

THE HYDROGEN SPECTRUM

The apparatus required for studying the hydrogen spectrum can be as simple as the hydrogen source, meter stick and diffraction grating shown in the photograph of Figure (24). The hydrogen source consists of a narrow glass tube filled with hydrogen gas, with metal electrodes at the ends of the tube. When a high voltage is applied to the electrodes, an electric current flows through the gas, heating it and causing it to emit light. The diffraction grating is placed in front of the hydrogen tube, and the meter stick is used to measure the location of the maxima.

The setup of the apparatus is illustrated in Figure (25) and the resulting spectrum in Figure (26). In this spectrum we are looking at the first maxima on the left side of the meter stick as shown in Figure (25). The leftmost line, the one bent the farthest is a deep red line which is called the **hydrogen α** line, and labeled by α in the photograph. The next line is a spurious line caused by impurities in the hydrogen tube. More to the right is a bright, swimming-pool blue line called **hydrogen β**. Much harder to see is the third line called **hydrogen γ**, a deep violet line near the short wavelength end of the visible spectrum. The three lines α,

β and γ are the only lines emitted by pure hydrogen gas in the visible part of the electromagnetic spectrum. Their wavelengths are

$$\lambda_\alpha = 6.56 \times 10^{-5} \text{cm}$$

$$\lambda_\beta = 4.86 \times 10^{-5} \text{cm}$$

$$\lambda_\gamma = 4.34 \times 10^{-5} \text{cm} \qquad (5)$$

When actually performing the experiment shown in Figure (24), there are some steps one should take to improve the accuracy of the results. As shown in Figure (27), a small arrowhead is placed on the grating itself. You then place your eye behind the meter stick and move your head and the slider on the meter stick until the point on the slider lines up with the arrowhead on the grating and with the spectral line you are trying to measure.

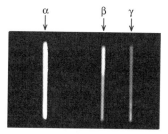

Figure 26
Photograph of the α, β and γ lines in the hydrogen spectrum.

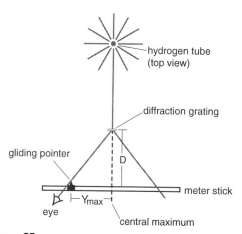

Figure 25
To determine the wavelength of light using a diffraction grating, you need to measure the distance Y_{max} to the first maximum, and the distance D shown. To measure Y_{max}, slide the pointer along the meter stick until it lines up with the first maximum.

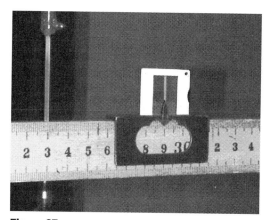

Figure 27
Looking through the grating, move your eye so that the spectral line is centered over the pointer as shown.

Rather than trying to measure the distance Y_{max} from the central maximum to the spectral line, it is more accurate to measure the distance $2Y_{max}$ from the first maximum on the left to the first maximum on the right, and then divide by 2. The wavelength of the line under study is then given by the diffraction grating formula, Equation 3

$$\lambda = Y_{max}\frac{d}{\sqrt{D^2 + Y_{max}^2}} \qquad \text{(3 repeated)}$$

where d is the separation of the slits in the grating and D the distance from the grating to the meter stick.

(When you first perform this experiment you may be confused by where the central maximum is. If you look straight through the grating at the tube, all you see is the tube. But that is the central maximum. It looks like the tube because all the colors go straight through the grating. There is no separation of colors or distortion of the image. To see a spectrum you have to look through the grating but far off to the side from the tube.)

Exercise 8

Derive a formula for the wavelength λ of a spectral line in terms of the distance $Y_{2\,max}$ from the central maximum to the second maxima of the line. The second maxima of the bright lines of an atomic spectra are quite easily seen using the apparatus of Figure (24).

The Experiment on Hydrogen Spectra

You should carry out the following steps when doing the hydrogen spectrum experiment shown in Figure (25).

(1) Determine the wavelength of all the spectral lines you can see, and compare your results with those given in Equation 5. Measure distances between first maxima, not to the central maxima.

(2) Measure the distances to the second maxima for the lines you can see out there and compute the corresponding wavelengths using our results from Exercise 8. Compare these wavelengths with those you get using the first maxima.

The Balmer Series

There are many spectral lines emitted by the hydrogen atom. Only three, however, are in the visible part of the spectrum. The complete spectrum consists of a number of series of lines, and the three visible lines belong to the series called the **Balmer series**. The red line, hydrogen α, is the longest wavelength line in the Balmer series, next comes the blue hydrogen β, then the violet hydrogen γ. Then there are many lines of the Balmer series out in the ultraviolet, which we cannot see by eye, but which we can record on photographic film.

Figure (28) shows part of the spectrum of light from a hydrogen star. These lines are in the ultraviolet and are all part of the Balmer series. Slightly different naming is used here. In the notation of Figure (28), we should call the red hydrogen α line H3, the blue β line H4, and the violet γ line H5. In Figure (28), the first 6 Balmer lines are missing. Here we see lines H9 through H40. As the lines increase in number they get closer and closer together. The whole series ends with very many, very closely spaced lines near 3.65×10^{-5} cm. It is called a *series* because the lines converge to a final wavelength in much the same way that many mathematical series converge to a final value.

It was the Swiss school teacher Johann Balmer who in 1885 discovered a formula for the wavelengths of the spectral lines seen in Figure (28). The wavelength of the m th line (m=3 for H3, m=4 for H4, etc.) is given by the formula

$$\lambda_m = 3.6456 \times 10^{-5}\,\text{cm} \times \left(\frac{m^2}{m^2 - 4}\right) \quad (6)$$

Equation 6 is known as the *Balmer formula*.

For m=3 we get from the Balmer formula

$$\begin{aligned} \lambda(H3) &= 3.6456 \times 10^{-5}\,\text{cm} \times \left(\frac{9}{9 - 4}\right) \\ &= 6.56 \times 10^{-5}\,\text{cm} \end{aligned} \quad (6a)$$

which agrees with Equation 5 for hydrogen α. Each higher value of m gives us the wavelength of a new line. At large values of m the factor $m^2/(m^2 - 4)$ approaches 1, and the lines get closer and closer together as seen in Figure (28). The end is at 3.65×10^{-5} cm where m is very large.

Exercise 9

(a) Use Equation 6 to calculate the wavelengths of the β and γ lines of the hydrogen spectrum and compare the results with Equation 5.

(b) Calculate the wavelength of H40 and compare your results with Figure (28).

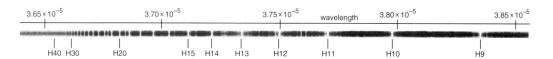

Figure 28
Spectrum of the star HD193182, showing ultra violet hydrogen lines near the limit of the Balmer series. This series of lines begins in the visible part of the spectrum with the lines we have called α, β, and γ, (which would be called H3, H4, and H5 in this diagram), and goes on to the ultra violet. The lines get closer and closer together, until the end just beyond the point labeled H40. The Swiss school teacher Johann Balmer discovered a formula for the wavelengths of these lines.

THE DOPPLER EFFECT

One phenomena of wave motion that is particularly easy to visualize is the ***Doppler effect***. As you can see in Figure (29), if the wave source is moving, the wavelength of the waves is compressed in front of the source and stretched out behind. This result, which is obvious for water waves, also applies to sound waves in air and to light waves moving through space.

To analyze the effect, we first note that if the source is at rest, then the waves all travel out from the source at a speed v_{wave}, have a wavelength λ_0 and a period T_0 given by

$$T_0 \frac{sec}{cycle} = \frac{\lambda_0 \; cm/cycle}{v_{wave} \; cm/sec} = \frac{\lambda_0}{v_{wave}} \frac{sec}{cycle} \quad (7)$$

If the source is moving forward at a speed v_{source}, then during one period T_0 the source will move forward a distance $x = v_{source}T_0$. But this is just the amount $\Delta\lambda$ by which the wavelength is shortened in front and stretched out in back. Thus

$$\Delta\lambda = v_{source}T_0 = v_{source}\frac{\lambda_0}{v_{wave}} \quad (8)$$

where we used Equation 7 to replace T_0 by λ_0/v_{wave}.

As a result the wavelengths in front and back of the source are

$$\lambda_{front} = \lambda_0 - \Delta\lambda = \lambda_0\left(1 - \frac{v_{source}}{v_{wave}}\right) \quad (9a)$$

$$\lambda_{back} = \lambda_0 + \Delta\lambda = \lambda_0\left(1 + \frac{v_{source}}{v_{wave}}\right) \quad (9b)$$

If we are in front of the moving source, the wave period T_{front} we observe is the time it takes the shortened wavelength λ_{front} to pass us at a speed v_{wave}, which is

$$T_{front} = \frac{\lambda_{front}}{v_{wave}} = \frac{\lambda_0}{v_{wave}}\left(1 - \frac{v_{source}}{v_{wave}}\right)$$

$$\boxed{T_{front} = T_0\left(1 - \frac{v_{source}}{v_{wave}}\right)} \quad (10a)$$

where we now replaced λ_0/v_{wave} by T_0. In the back, the period is extended to

$$\boxed{T_{back} = T_0\left(1 + \frac{v_{source}}{v_{wave}}\right)} \quad (10b)$$

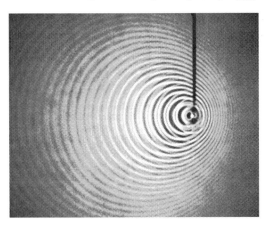

Figure 29
When the source of the wave is moving, the wavelengths are compressed in front and stretched out behind.

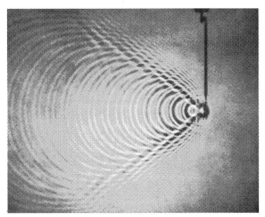

Figure 30
When the source is moving faster than the waves, the waves build up on the front edge to create a shock wave. For supersonic flight, this shock wave produces the sonic boom.

If the speed of the source approaches the speed of the wave, as in the case of a jet airplane approaching the speed of sound, the wavelength in front goes to zero. At speeds greater than the speed of the wave, as in supersonic flight, there are no waves ahead of the source; instead, the leading edge of the waves pile up as shown in Figure (30) to create what is called a *shock wave*. This shock wave is responsible for the sonic boom we hear when a jet passes overhead at supersonic speeds.

Exercise 10

There is a simple experiment you can perform to observe the Doppler effect. Stand beside a road and have a friend drive by at about 40 mi/hr while blowing the car horn. As the car passes, the pitch of the horn will suddenly drop because the wavelength of the sound waves, which was shortened as the car approached, is lengthened after it passes. The shorter, higher-pitched sound waves change to longer, lower-pitched waves.

For this exercise, assume the car is owned by a musician, and the car horn plays the musical note A at a frequency of 440 cycles per second.

a) What is the wavelength of a 440 cycle/sec note, if the speed of sound is 1000 ft/sec?

b) What is the wavelength of the note we hear if the car is approaching at a speed of 40 miles/hr?

c) What is the frequency we hear if the car is approaching at 40 miles/hr?

d) What is the frequency we hear when the car is going away from us at 40 miles/hr?

Stationary Source and Moving Observer

If the source is at rest but we, the observer, are moving, there is also a Doppler effect. In the case of water or sound waves, if we are moving through the medium toward the source, then the wave crests pass by us at an increased relative speed $v_{rel} = v_{wave} + v_{us}$. Even though the wavelength is unchanged, the increased speed of the wave will carry the crests by faster, giving us an apparently shorter period and higher frequency.

If our velocity through the medium is small compared to the wave speed, then we observe essentially the same decrease in period and increase in frequency as in the case when the source was moving. In particular, Equation 10 is approximately correct.

On the other hand, when the waves are in water or air and the relative speed of the source and observer approaches or exceeds the wave speed, there can be a considerable difference between a moving source and a moving observer. As illustrated in Figure (31a), if the source is moving faster than the wave speed, there is a shock wave and the observer detects no waves until the source passes. But if the source is at rest as in Figure (31b), there is no shock front and the observer moves through waves before getting to the source, even if the observer is moving faster than the wave speed.

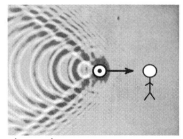 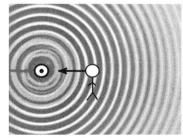

a) moving source   b) moving observer

Figure 31
For waves in water or air, there can be a significant difference between a moving source with a stationary observer, and a moving observer with a stationary source, even though the source and observer have the same relative velocity in the two cases. For light, the principle of relativity requires that the two cases be identical.

Doppler Effect for Light

When a source of light waves is moving toward or away from us, there is also a Doppler effect. If the source is moving toward us, the wavelengths we see are shortened. This means that the color of the light is *shifted toward the blue*. If the source is moving away, the wavelengths are stretched out, become longer, and the color *shifts toward the red*. When the speed of the source is considerably less than the speed of light, Equations 9 and 10 correctly give the observed wavelength λ and period T in terms of the source's wavelength λ_0 and period and T_0.

Principle of Relativity

There is one fundamental difference, however, between the Doppler effect for water and sound waves, and the Doppler effect for light waves. For water and sound waves we could distinguish between a source at rest with a moving observer and an observer at rest with a moving source. If the source were at rest, it was at rest *relative to the medium through which the wave moves*. We got different results depending on whether it was the source or the observer that was at rest.

In the case of light, the medium through which light moves is *space*. According to the principle of relativity, one cannot detect uniform motion relative to space. Since it is not possible to determine which one is at rest and which one is moving, we must have exactly the same Doppler effect formula for the case of a stationary source and a moving observer, or vice versa. The Doppler effect formula can depend only on the *relative velocity* of the source and observer.

One way to use the principle of relativity is to always assume that you yourself are at rest relative to space. (No one can prove you are wrong.) This suggests that we should start from Equations 9 and 10, which were derived for a stationary observer, and replace v_{wave} by the speed of light c, and interpret v_{source} as the relative velocity between the source and the observer.

Equations 9 and 10, modified this way, are correct as long as the source is not moving too fast. However if the source is moving relative to us at a speed approaching the speed of light, there is one more relativistic effect that we have to take into account . Remember that a moving clock runs slow by a factor of $\sqrt{1 - v^2/c^2}$. If the source is radiating a light wave of period T_0, then that period can be used in the construction of a clock. If we observe the source go by at a speed v_{source} , the period T_0 must appear to us to increase to $T_0{}'$ given by

$$T_0{}' = \frac{T_0}{\sqrt{1 - v_{source}^2/c^2}} \qquad \text{(see Eq 1-11)}$$

From our point of view, the source is radiating light of period $T_0{}'$. This is the light whose wavelength is stretched or compressed, depending on whether the source is moving away from or towards us. Thus we should use $T_0{}'$ instead of T_0 in Equation 10.

Replacing T_0 by $T_0{}'$ in Equation 10 gives

$$T_{front} = \frac{T_0}{\sqrt{1 - v_{source}^2/c^2}} \left(1 - \frac{v_{source}}{c}\right) \qquad \text{(11a)}$$

$$T_{back} = \frac{T_0}{\sqrt{1 - v_{source}^2/c^2}} \left(1 + \frac{v_{source}}{c}\right) \qquad \text{(11b)}$$

where v_{source} is the speed of the source relative to us, and we have set $v_{wave} = c$. Equations (11) are the relativistic Doppler effect equations for light. They are applicable for any source speed, even if the source is moving relative to us at speeds approaching the speed of light. The corresponding wavelengths are

$$\lambda_{front} = c \frac{cm}{sec} \times T_{front} \frac{sec}{cycle} = cT_{front}\frac{cm}{cycle}$$

$$\begin{aligned} \lambda_{front} &= cT_{front} \\ \lambda_{back} &= cT_{back} \end{aligned} \qquad \text{(12)}$$

Exercise 11

Using Equations (12), express λ_{front} and λ_{back} in terms of λ_0, v_{source} and c.

Exercise 12

In Figure 31b, where we picture a stationary source and a moving observer, the waves pass by the observer at a speed $v_{wave} + v_{observer}$. Why can't this picture be applied to light, simply replacing v_{wave} by c and letting $v_{observer}$ be the relative velocity between the source and observer.

Doppler Effect in Astronomy

The Doppler effect has become one of the most powerful tools astronomers use in the study of the universe. Assuming that distant stars and galaxies are made up of the same matter as nearby stars, we can compare the spectral lines emitted by distant galaxies with the corresponding spectral lines radiated by elements here on earth. A general shift in the wavelengths to the blue or the red, indicates that the source of the waves is moving either toward or away from us. Using Equations 12 we can then quite accurately determine how fast this motion toward or away from us is.

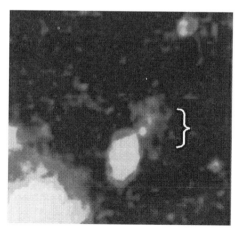

Figure 32
The most distant galaxy observed as of January 1995. This galaxy, given the romantic name 8C 1435+63, was photographed by the Keck telescope in Hawaii. The two halves of the distant galaxy are indicated by the white bracket. The galaxy is moving away from us at 95% the speed of light.

Until the 1960s astronomers did not have much need for the relativistic Doppler shift equations. The non relativistic Equations 10 were generally adequate because we did not observe stars or galaxies moving relative to us at speeds greater than 10 to 20% the speed of light. But that changed dramatically with the discovery of quasars in 1963. Quasars are now thought to be brilliant galaxies in the early stages of formation. They can be seen from great distances and are observed to move away from us at speeds as great as 95% the speed of light. To analyze such motion, the relativistic formulas, Equations 11 and 12 are clearly needed.

Exercise 13

The most rapidly receding galaxy observed by the spring of 1995 is the galaxy named 8C 1435 + 63 shown in the photograph of Figure (32) taken by the Keck telescope in Hawaii. Much of the light from this galaxy is radiated by hydrogen gas. This galaxy is moving away from us at a speed $v_{source} = .95c$.

(a) Assuming that the hydrogen in this galaxy radiates the same spectrum of light as the hydrogen gas in our discharge tube of Figure (24), what are the wavelengths of the first three Balmer series lines λ_α, λ_β, and λ_γ, by the time these waves reach us. (They will be greatly stretched out by the motion of the galaxy.)

(b) Astronomers use the letter z to denote the relative shift of the wavelength of light due to the Doppler effect. I.e.,

$$z = \frac{\Delta\lambda}{\lambda_0} = \frac{\lambda - \lambda_0}{\lambda_0} \quad \begin{array}{l} \textit{astronomers} \\ \textit{notation for} \\ \textit{the red shift} \end{array} \qquad (13)$$

where λ_0 is the wavelength of the unshifted spectral line, and λ is the Doppler shifted wavelength we see. What is z for galaxy 8C 1435 + 63?

The Red Shift and the Expanding Universe

In 1917 Albert Einstein published his relativistic theory of gravity, known as General Relativity. In applying his theory of gravity to the behavior of the stars and galaxies in the universe, he encountered what he thought was a serious problem with the theory. Any model of the universe he constructed was unstable. The galaxies tended either to collapse in upon themselves or fly apart. He could not find a solution to his equations that represented the stable unchanging universe everyone knew was out there.

Einstein then discovered that he could add a new term to his gravitational equations. By properly adjusting the value of this term, he could construct a model of the universe that neither collapsed or blew up. This term, that allowed Einstein to create a static model of the universe, became known as the *cosmological constant*.

In later life, Einstein said that his introduction of the cosmological constant was the greatest mistake he ever made. The reason is that the universe is not static. Instead it is *expanding*. The galaxies are all flying apart like the debris from some gigantic explosion. The expansion, or at least instability of the universe, could have been considered one of the predictions of Einstein's theory of gravity, had Einstein not found his cosmological constant. (Later analysis showed that the static model, obtained using the cosmological constant, was not stable. The slightest perturbation would cause it to either expand or contract.)

That the universe is not static was discovered by Doppler shift measurements. In the 1920s, the astronomer Edwin Hubble observed that spectral lines from distant galaxies were all shifted toward the red, and that the farther away the galaxy was, the greater the red shift. Interpreting the red shift as being due to the Doppler effect meant that the distant galaxies were moving away from us, and the farther away a galaxy was, the faster it was moving.

Hubble was the first astronomer to develop a way to measure the distance out to other galaxies. Thus he could compare the red shift or recessional velocity to the distance the galaxy is away from us. He found a simple rule known as Hubble's law. *If you look at galaxies twice as far away, they will be receding from us twice as fast*. Roughly speaking, he found that a galaxy .1 billion light years away would be receding at 1% the speed of light; a galaxy .2 billion light years away at 2% the speed of light, etc. In the 1930s, construction of the 200 inch Mt. Palomar telescope was started. It was hoped that this telescope (completed in 1946) would be able to observe galaxies as far away as 2 billion light years. Such galaxies should be receding at the enormous speeds of approximately 20% the speed of light.

With the discovery of quasars, we have been able to observe much more distant galaxies, with far greater recessional velocities. As we have just seen, the galaxy 8C 1435 + 63, photographed by the 10 meter (400 inch) telescope in Hawaii, is receding from us at a speed of 95% the speed of light. To analyze the Doppler effect for such a galaxy, the fully relativistic Doppler effect formula, Equation 12 is needed; non relativistic approximations will not do.

Hubble's law raises several interesting questions. First, it sounds as if we must be at the center of everything, since the galaxies in the universe appear to all be moving away from us. But this is simply a consequence of a uniform expansion. Someone in a distant galaxy will also observe the same Hubble law.

To see how a uniform expansion works, mark a number of equally spaced dots on a partially blown up balloon. Select any one of the dots to represent our galaxy, and then start blowing up the balloon to represent the expansion of the universe. You will notice that dots twice as far away move away twice as fast, no matter which dot you selected. Hubble's law is obeyed from the point of view of any of the dots on the balloon. (You can see this expansion in Figure (33), where we started with an array of light colored dots, and uniformly expanded the array to get the black dots.)

Another interesting question is related to nature's speed limit c. We cannot keep looking out twice as far to see galaxies receding twice as fast, because we cannot have galaxies receding faster than the speed of light. Something special has to happen when the recessional speeds approach the speed of light, as they have in the case of 8C 1435 + 63. This appears to place a limit on the size of the universe we can observe.

One of the things to remember when we look at distant galaxies is that we are not only looking far away, but we are also looking back in time. When we look at a galaxy 10 billion light years away, we are looking at light emitted 10 billion years ago, when the universe was 10 billion years younger. Recent studies have clearly shown that galaxies 10 billion years away look different than nearby galaxies. Over the past 10 billion years the universe has evolved; galaxies have aged, becoming more symmetric and less violent.

To predict what we will find as we look back in time, look at ever more distant galaxies, imagine that we take a moving picture of the universe and run the moving picture backwards.

If we reverse the moving picture of expanding galaxies, we see contracting galaxies. They are all contracting back to one point in space and time. Go back to that point and run the movie forward, and we see all of the universe rushing out of that point, apparently the consequence of a gigantic explosion. This explosion has become known as the **_Big Bang_**. (The name Big Bang was a derisive expression coined by the astronomer Fred Hoyle who had a competing theory of the origin of the universe.)

The idea that the universe started in a big bang, provides a simple picture of the Hubble law. From our point of view, galaxies emerged at various speeds in all directions from the Big Bang. Those that were moving away from us the fastest just after the explosion are now the farthest away from us. Galaxies moving away twice as fast are now twice as far away.

In the next chapter we will have more to say about the origin of the universe and evidence for the Big Bang. We will also introduce another way to interpret the Doppler effect and its relationship to the expansion of the universe.

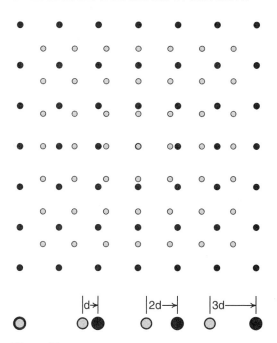

Figure 33
During a uniform expansion, neighboring galaxies move a certain distance away (d), galaxies twice as far away move twice as far (2d), etc. This is Hubble's law for the expanding universe.

A CLOSER LOOK AT INTERFERENCE PATTERNS

Our focus so far in this chapter has been on the application of the wave properties of light to the study of physical phenomena such as atomic spectra and the expansion of the universe. We now wish to turn our attention to a more detailed study of the wave phenomena itself. We will first take a closer look at the single slit diffraction pattern that serves as the envelope of the multiple slit patterns we saw back in Figure (18). We will then discuss an experimental technique for accurately recording various interference patterns produced by laser beams. We will then end the chapter with a demonstration of how Fourier analysis can be used to predict the structure of the interference patterns we observe.

The reason for these studies is to strengthen intuition about the behavior of waves. The remainder of the text deals with the inherent wave nature of matter, and here we wish to develop the conceptual and experimental tools to study this wave nature.

Single Slit Diffraction Pattern

In the 1600s, Francesco Maria Grinaldi discovered that light going through a fine slit cannot be prevented from spreading on the other side. He named this phenomenon *diffraction*. Independently Robert Hook, of Hook's law fame, made the same observation and provided a wave like explanation. The clearest explanation comes from the Huygens construction illustrated in Figures (3) through (8).

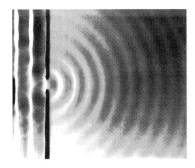

Figure 3 (repeated)
The simple diffraction pattern you get when the slit is narrow compared to a wavelength.

In Figure (3) reproduced here, we have a deceptively simple picture of the single slit diffraction pattern. In the photograph a wave is impinging upon a slit whose width is less than one wavelength, with the result that we get a simple circular wave emerging on the other side. In Figures (6), (7) and (8) we look at what happens when the slit becomes wider than a wavelength. These are all views of the wave pattern close to the slit. We see that as the slit becomes wider, more of the wave passes through undisturbed, creating a more or less distinct shadow effect. Nevertheless we always see circular waves at the edge of the shadow. If you carefully look at Figure (6), reproduced below, you can see lines of nodes coming out of the slit that is about 2 wavelengths wide.

In Figure (34), we have the diffraction pattern produced by a laser beam passing through a 50 micron wide single slit and striking a screen 10 meters away. This is a reproduction of the single slit pattern that acts as an envelope for the multiple slit patterns seen in Figure (18). A 50 micron slit is 50×10^{-6} meters or 500×10^{-5} cm wide. This is nearly a hundred times greater than the 6.4×10^{-5} cm wavelength of the laser light passing through the slit. Thus we are dealing with slits that are about 100 wavelengths wide.

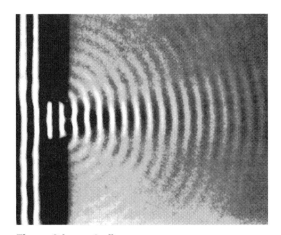

Figure 6 (repeated)
The pattern becomes more complex when the slit is wider than a wavelength. Here you can begin to see lines of nodes emerging from the slit.

The fact that the diffraction pattern was photographed 10 meters from the slits means that we are looking at the pattern nearly 20 million wavelengths away from the slits. Thus the ripple tank photographs of Figures (5) through (8), showing diffraction patterns within a few wavelengths of the slits, are not a particularly relevant guide as to what we could expect to see 20 million wavelengths away.

The general features of the diffraction pattern in Figure (34) is that we have a relatively broad central maximum, with nodes on either side. Then there are dimmer and narrower maxima on either side. There is a series of these side maxima that extend out beyond the photograph of Figure (34).

If the slit were narrow compared to a wavelength, if the wave spread out as in Figures (3), then we would get just one broad central maxima. Only when the slit is wider than a wavelength do we get the minima we see in Figure (34). These minima result from the interference and cancellation of waves from different parts of the slit. What we wish to do now is to show how this cancellation occurs and predicts where the minima will be located.

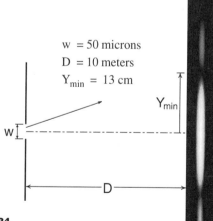

$$w = 50 \text{ microns}$$
$$D = 10 \text{ meters}$$
$$Y_{min} = 13 \text{ cm}$$

Figure 34
For this single slit laser beam diffraction pattern, the slit was about 100 wavelengths wide (w), and the screen was about 20 million wavelengths away (D).

Analysis of the Single Slit Pattern

In our discussion of diffraction gratings, we estimated the width of the maxima by determining how far from the center of the maxima the intensity first went to zero, where we first got complete cancellation. This occurred where light from pairs of slits cancelled. In our example of Figure (21), light from slit 1 cancelled that from slit 501, from slit 2 with slit 502, etc., all the way down to slits 500 and 1000.

We can use a similar analysis for the single slit pattern, except the one big slit is broken up, ***conceptually***, into many narrow slits, as illustrated in Figure (35). Suppose, for example, we think of the one wide slit of width w as being broken up into 1000 neighboring individual slits. The individual slits are so narrow that each piece of wave front in them should act as a source of a pure circular wave as shown back in Figure (3).

Now consider the light heading out in such a direction that the wave from the first conceptual slit is half a wavelength $\lambda/2$, in front of the wave from the middle slit, number 501. When the waves from these two "slits" strike the screen they will cancel. Similarly waves from slits 2 and 502 will cancel, as will those from 3 and 503, etc., down to 500 and 1000. Thus is the direction where the path length difference from the edge to the center of the opening is half a wavelength $\lambda/2$. Between the two edges of the opening, the path length difference to this minimum is λ, as shown.

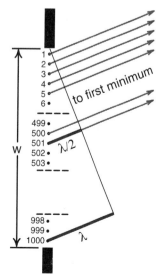

Figure 35
Conceptually break the single slit up into many individual slits. We get a minimum when light from the conceptual slits cancels in pairs.

From Figure (36), we can calculate the height of the first minimum using the familiar similar triangles we have seen in previous analysis. The small right triangle near the slit has a short side of length λ and a hypotenuse equal to the slit width w. The big triangle has a short side equal to Y_{min} and a hypotenuse given by the Pythagorean theorem as $\sqrt{D^2 + Y_{min}^2}$. Usually Y_{min} will be much smaller than the distance D, so that we can replace $\sqrt{D^2 + Y_{min}^2}$ by D, to get

$$\frac{\lambda}{w} = \frac{Y_{min}}{\sqrt{D^2 + Y_{min}^2}} \approx \frac{Y_{min}}{D}$$

or

$$\boxed{Y_{min} \approx \frac{\lambda D}{w}} \qquad \begin{array}{l} \textit{distance to the} \\ \textit{first minima of} \\ \textit{a single slit} \\ \textit{diffraction pattern} \end{array} \qquad (14)$$

Exercise 14

To obtain the single slit diffraction pattern seen in Figure (34), we used a slit 50 microns wide located 10 meters from the screen. The distance Y_{min} to the first minimum was about 13 cm. Use this result to determine the wavelength of the laser light used. Compare your answer with your results from Exercises 5 and 6, where the same wavelength light was used.

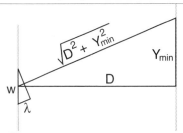

Figure 36
Similar triangles for calculating the distance to the first minimum of a single slit diffraction pattern.

RECORDING DIFFRACTION GRATING PATTERNS

Another way to record the diffraction pattern is to use a device called a ***photoresistor***. A photoresistor is an inexpensive resistor whose resistance R_p varies depending upon the intensity of the light striking the resistor. If you place the photoresistor in the circuit shown in Figure (37), along with a fixed resistance R_2, a battery of voltage V_b, and an oscilloscope, you can measure with the oscilloscope the intensity of light striking the photoresistor.

The analysis of the circuit in Figure (37) is as follows. The resistors R_p and R_1 are in series and thus have an effective resistance $R = R_p + R_1$. The current i in the circuit is thus $i = V_b/R = V_b/(R_p + R_1)$. Thus as the photoresistor's resistance R_p changes with changes in light intensity, the current i will also change. Finally the voltage V_1 that the oscilloscope sees across the fixed resistor R_1 is given by Ohm's law as $V_1 = iR_1$. Thus as i changes, V_1 changes and we see the change on the oscilloscope.

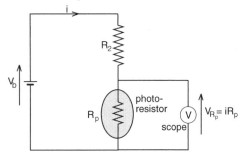

Figure 37
The photoresistor circuit. By making R_2 considerably bigger than the photoresistor resistance R_p, the current i stays relatively constant. As a result, the voltage $V_p = iR_p$ is nearly proportional to R_p. (We used $R_p = 6.8K\Omega$ and the EG&G opto VT30N4 photoresistor.)

For a number of years we tried various ways of moving the photoresistor through the diffraction pattern in order to record the intensity of the light in the diffraction pattern. We tried mounting the photoresistor on xy recorders and various home-built devices, but there was always some jitter and the results were only fair. The solution, as it turns out, is not to move the photoresistor, but move the diffraction pattern across a fixed photoresistor instead. This is easily done using a rotating mirror, a mirror attached to a clock motor as shown in Figure (38). We have found that if you use a motor with a speed of 1/2 revolutions per minute, you have plenty of time to make a stable noise-free recording of the bottom. Using the recording oscilloscope MacScope, we recorded the single slit diffraction pattern seen in Figure (39).

A photoresistor is sensitive to the intensity or energy density of the light striking it. And the intensity is proportional to the square of the amplitude of the waves in the beam. Thus in Figure (39) we are looking at a graph of the square of the wave amplitude in a single slit diffraction pattern. It is reasonable that the intensity should be proportional to the square of the amplitude, because amplitudes can be positive or negative, but intensities are always positive. You cannot have a negative intensity, and you do not get one if you square the amplitude since squares of real numbers are always positive.

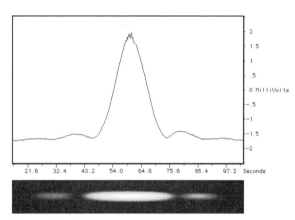

Figure 38
The rotating mirror. (We were careful to make sure that the axis of rotation was accurately perpendicular to the base. If it isn't, the laser beam wobbles up and down.)

Figure 39 a,b
Single slit diffraction pattern. Data from the project by Cham, Cole, and Layang.

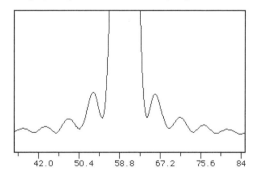

Figure 39c
Single slit diffraction pattern with the amplitude of the voltage amplified so that we can see the side lobes. Data from the project by Cham, Cole, and Layang.

Exercise 15

In Figure (40) we study the interference pattern produced by a laser beam passing through three equally spaced slits. Figure (40a) shows the experimental setup and (40b) the shape of the slits through which the laser beam went. Figure (40c) is a photograph of the interference pattern, and Figure (40d) is a recording in which the voltage is proportional to the intensity of the light striking a stationary photoresistor. The beam rotated at a rate of 1 revolution per hour, sweeping the beam past the photo resistor. (The mirror only turned at .5 revolutions per hour, but the reflected beam rotates twice as fast as the mirror. You can see this by the fact that when the mirror turns $45°$ the beam rotates $90°$.)

To Calculate the wavelength of the laser light from the experimental data in Figure 40, first note that the beam is sweeping past the photoresistor at a speed

$$V_{beam} = \frac{2\pi r \,(cm/revolution)}{3600 \,(sec/revolution)} = \frac{2\pi r}{3600} \frac{cm}{sec}$$

where r is the distance from the axis of the mirror to the photoresistor. If it takes a time T for two maxima to sweep past the photoresistor, then the distance Y_{max} between the maxima is

$$Y_{max} = V_{beam} \times T = \frac{2\pi r}{3600} \times T$$

If the slits are close to the mirror, then r is also equal to the distance D from the slits to the photoresistor (screen). The wavelength is then given by Equation 3a as

$$\lambda = Y_{max} \times \frac{d}{D} \approx Y_{max} \times \frac{d}{r}$$

$$= \left(\frac{2\pi r}{3600} \times T\right) \times \frac{d}{r}$$

The factors of r cancel, and we are left with

$$\lambda = \frac{2\pi d}{3600} \times T \qquad (15)$$

Thus if the slits are close to the mirror, we do not need to know the distance to the photoresistor. (You can see that if the rotating beam is twice as long, the end travels twice as fast. But the maxima are twice as far apart, thus it takes the same length of time for the maxima to pass the photoresistor.)

Use the results of Figure 40d to determine the wavelength of the laser beam. (In this experiment, the slits were close to the mirror.)

(More to come on the use of Fourier analysis to predict diffraction patterns.)

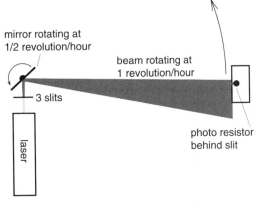

a) Experimental setup

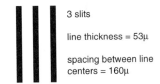

b) The slits. ($1\mu = 10^{-6}m = 10^{-4}cm$)

c) 3 slit diffraction pattern

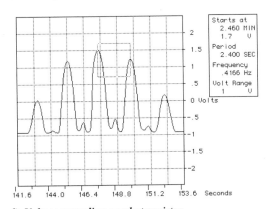

d) Voltage recording on photoresistor

Figure 40
Recording the 3 slit interference pattern

Chapter 34
Photons

The effort to determine the true nature of light has been a fitful process in the history of physics. Newton and Huygens did not agree on whether light was a wave or consisted of beams of particles. That issue was apparently settled by Thomas Young's two-slit experiment performed in 1801, nearly three quarters of a century after Newton's death. Young's experiment still did not indicate what light was a wave of. That insight had to come from Maxwell's theory of 1860 which showed that light was a wave of electric and magnetic fields.

In the late 1800s there were dramatic confirmations of Maxwell's theory. In 1888 Heinrich Hertz observed radio waves, the expected low frequency component of the electromagnetic spectrum. As we have seen from our own experiments, the electric and magnetic fields in a radio wave can be measured directly.

But as the nineteenth century was ending, not all predictions of Maxwell's theory were as successful. Applications of Maxwell's equations to explain the light radiated by matter were not working well. No one understood why a heated gas emitted sharp spectral lines, and scientists like Boltzman were unable to explain important features of light radiated by hot solid objects. The fact that Boltzman could get some features right, but not others, made the problem more vexing. Even harder to understand was the way beams of light could eject electrons from the surface of a piece of metal, a phenomenon discovered in 1897 by Hertz.

Many of these problems were cleared up by a picture developed by Max Planck and Einstein, a picture in which light consisted of beams of particles which became known as photons. The photon picture immediately explained the ejection of electrons from a metal surface and the spectrum of radiation from a heated solid object. In the past few years the observation of photons coming in uniformly from all directions in space has led to a new and surprisingly well confirmed picture of the origin of the universe.

In this chapter we will discuss the properties of photons and how discovering the particle nature of light solved some outstanding problems of the late nineteenth century. We will finish with a discussion of what photons have told us about the early universe.

What we will not discuss in this chapter is how to reconcile the two points of view about light. How could light behave as a wave in Thomas Young's experiment, and as a particle in experiments explained by Einstein. How could Maxwell's theory work so well in some cases and fail completely in others? These questions, which puzzled physicists for over a quarter of a century, will be the topic of discussion in the chapter on quantum mechanics.

BLACKBODY RADIATION

When we studied the spectrum of hydrogen, we saw that heated hydrogen gas emits definite spectral lines, the red hydrogen α, the blue hydrogen β and violet hydrogen γ. Other gases emit definite but different spectral lines. But when we look through a diffraction grating at the heated tungsten filament of a light bulb, we see something quite different. Instead of sharp spectral lines we see a continuous rainbow of all the colors of the visible spectrum. Another difference is that the color of the light emitted by the filament changes as you change the temperature of the filament. If you turn on the light bulb slowly, you first see a dull red, then a brighter red, and finally the filament becomes white hot, emitting the full spectrum seen in white light. In contrast, if you heat hydrogen gas, you see either no light, or you see all three spectral lines at definite unchanging wave lengths.

Some complications have to be dealt with when studying light from solid objects. The heated burner on an electric stove and a ripe McIntosh apple both look red, but for obviously different reasons. The skin of the McIntosh apple absorbs all frequencies of visible light except red, which it reflects. A stove burner, when it is cool, looks black because it absorbs all wavelengths of light equally. When the black stove burner is heated, the spectrum of light is not complicated by selective absorption or emission properties of the surface that might enhance the radiation at some frequencies. The light emitted by a heated black object has universal characteristic properties that do not depend upon what kind of black substance is doing the radiating. The light from such objects is *blackbody radiation*.

One reason for studying blackbody radiation is that you can determine the temperature of an object from the light it emits. For example, Figure (1) shows the intensity of light radiated at different wavelengths by a tungsten filament at a temperature of 5800 kelvins. The greatest intensity is at a wavelength of 5×10^{-5}cm, the middle of the visible spectrum at the color yellow. If we plot intensities of the various wavelengths radiated by the sun, you get essentially the same curve. As a result we can conclude that the temperature of the surface of the sun is 5800 kelvins. It would be hard to make this measurement any other way.

There are a few simple rules governing blackbody radiation. One is that the wavelength of the most intense radiation, indicated by λ_{max} in Figure (1), is inversely proportional to the temperature. The explicit formula, known as *Wein's displacement law* turns out to be

$$\lambda_{max} = \frac{2.898}{T}\text{mmK} \qquad (1)$$

where λ_{max} is in millimeters and the temperature T is in kelvins. For T= 5800K, Equation 1 gives

$$\lambda_{max}(5800K) = \frac{2.898\text{mmK}}{5800K} = 5 \times 10^{-4}\text{mm}$$
$$= 5.0 \times 10^{-5}\text{cm}$$

which is the expected result.

While λ_{max} changes with temperature, the relative shape of the spectrum of radiated intensities does not. Figure (1) is a general sketch of the blackbody radiation spectrum. To determine the blackbody spectrum for another temperature, first calculate the new value of λ_{max} using Equation 1 then shift the horizontal scale in Figure (1) so that λ_{max} has this new value.

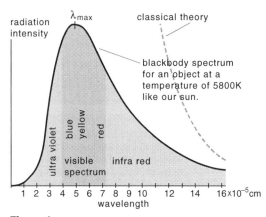

Figure 1
Blackbody spectrum at 5800 degrees on the kelvin scale. The solid line is the experimental curve, the dotted line represents the prediction of Newtonian mechanics combined with Maxwell's equations. The classical theory agrees with the experimental curve only at long wavelengths.

Knowledge of the blackbody spectrum is particularly useful in astronomy. Most stars radiate a blackbody spectrum of radiation. Thus a measurement of the value of λ_{max} determines the temperature of the surface of the star. There happens to be quite a variation in the surface temperature and color of stars. This may seem surprising at first, because most stars look white. But this is due to the fact that our eyes are not color sensitive in dim light. The variation in the color of the stars can show up much better in a color photograph.

As an example of the use of Equation 1, suppose you observe a red star that is radiating a blackbody spectrum with $\lambda_{max} = 7.0 \times 10^{-5}$cm . The surface temperature should then be given by

$$T = \frac{2.898\,mmK}{7.0 \times 10^{-4}\,mm} = 4140\,kelvin$$

Exercise 1

(a) What is the surface temperature of a blue star whose most intense wavelength is $\lambda_{max} = 4 \times 10^{-5}$cm?

(b) What is the wavelength λ_{max} of the most intense radiation emitted by an electric stove burner that is at a temperature of 600° C (873K)?

Another feature of blackbody radiation is that the intensity of the radiation increases rapidly with temperature. You see this when you turn up the voltage on the filament of a light bulb. Not only does the color change from red to white, the bulb also becomes much brighter.

The net amount of radiation you get from a hot object is the difference between the amount of radiation emitted and the amount absorbed from the surroundings. If the object is at the same temperature as its surroundings, it absorbs just as much radiation as it emits, with the result that there is no net radiation. This is why you cannot feel any heat from an electric stove burner before it is turned on. But after the burner is turned on and its temperature rises above the room temperature, you begin to feel heat. Even if you do not touch the burner you feel infrared radiation which is being emitted faster than it is being absorbed. By the time the burner becomes red hot, the amount of radiation it emits greatly exceeds the amount being absorbed.

In 1879, Joseph Stefan discovered that the total intensity, the total energy emitted per second in blackbody radiation was proportional to the fourth power of the temperature, to T^4 where T is in kelvins. Five years later Ludwig Boltzman explained the result theoretically. This result is thus known as the Stefan-Boltzman law.

As an example of the use of the Stefan-Boltzman law, suppose that two stars are of the same size, the same surface area, but one is a red star at a temperature of 4,000K while the other is a blue star at a temperature of 10,000K. How much more rapidly is the hot blue star radiating energy than the cool red star?

The ratio of the rates of energy radiation is equal to the ratio of the fourth power of the temperatures. Thus

$$\frac{\text{energy radiated by blue star}}{\text{energy radiated by red star}} = \frac{T_{blue}^4}{T_{red}^4} = \left(\frac{10,000K}{4,000K}\right)^4$$

$$= 2.5^4 \approx 40$$

We see that the blue star must be burning its nuclear fuel 40 times faster than the red star.

Planck Blackbody Radiation Law

Boltzman used a combination of Maxwell's equations, Newtonian mechanics, and the theory of statistics to show that the intensity of blackbody radiation increased as the fourth power of intensity. But neither he nor anyone else was able to derive the blackbody radiation spectrum shown in Figure (1). There was some success in predicting the long wavelength side of the curve, but no one could explain why the intensity curve dropped off again at short wavelengths.

In 1900 Max Planck tried a different approach. He first found an empirical formula for a curve that matched the blackbody spectrum. Then he searched for a derivation that would lead to his formula. The idea was to see if the laws of physics, as they were then known, could be modified in some way to explain his empirical blackbody radiation curve.

Planck succeeded in the following way. According to Maxwell's theory of light, the amount of radiation emitted or absorbed by a charged particle was related to the acceleration of the particle, and that could vary continuously. Planck found that he could get his empirical formula if he assumed that the electrons in a solid emitted or absorbed radiation only in discrete packets. The energy in each packet had to be proportional to the frequency of the radiation being emitted and absorbed. Planck wrote the formula for the energy of the packets in the form

$$E = hf \qquad (2)$$

where f is the frequency of the radiation. The proportionality constant h became known as ***Planck's constant***.

For over two decades physicists had suspected that something was wrong either with Newtonian mechanics, Maxwell's equations, or both. Maxwell was unable to derive a formula that explained the specific heat of gases (except the monatomic noble gases), and no one had the slightest idea why heated gases emitted sharp spectral lines. Planck's derivation of the blackbody radiation formula was the first successful derivation of a phenomena that could not be explained by Newtonian mechanics and Maxwell's equations.

But what did it mean that radiation could be emitted or absorbed only in discrete packets or ***quanta*** as Planck called them? What peculiar mechanism lead to this ***quantization*** of the emission and absorption process? Planck did not know.

THE PHOTOELECTRIC EFFECT

1905 was the year in which Einstein cleared up several outstanding problems in physics. We have seen how his focus on the basic idea of the principle of relativity lead to his theory of special relativity and a new understanding of the structure of space and time. Another clear picture allowed Einstein to explain why light was emitted and absorbed in discrete quanta in blackbody radiation. The same idea also explained a process called the *photoelectric effect*, a phenomenon first encountered in 1887 by Heinrich Hertz.

In the photoelectric effect, a beam of light ejects electrons from the surface of a piece of metal. This phenomenon can be easily demonstrated in a lecture, using the kind of equipment that was available to Hertz. You start with a gold leaf electrometer like that shown in Figure (2), an old but effective device for measuring the presence of electric charge. (This is the apparatus we used in our initial discussion of capacitors.) If a charged object is placed upon the platform at the top of the electrometer, some of the charge will flow down to the gold leaves that are protected from air currents by a glass sided container. The gold leaves, each receiving the same sign of charge, repel each other and spread apart as shown. Very small amounts of charge can be detected by the spreading of the gold leaves.

To perform the photoelectric effect experiment, clean the surface of a piece of zinc metal by scrubbing it with steel wool, and charge the zinc with a negative charge. We can be sure that the charge is negative by going back to Ben Franklin's definition. If you rub a rubber rod with cat fur, a negative charge will remain on the rubber rod. Then touch the rubber rod to the piece of zinc, and the zinc will become negatively charged. The presence of charge will be detected by the spreading of the gold leaves.

Now shine a beam of light at the charged piece of zinc. For a source of light use a carbon arc that is generated when an electric current jumps the narrow gap between two carbon electrodes. The arc is so bright that you do not need to use a lens to focus the light on the zinc. The setup is shown in Figure (3).

When the light is shining on the zinc, the gold leaves start to fall toward each other. Shut off or block the light and the leaves stop falling. You can turn on and off the light several times and observe that the gold leaves fall only when the light is shining on the zinc. Clearly it is the light from the carbon arc that is discharging the zinc.

Figure 2
The gold leaf electrometer. This is the same apparatus we used back in Figure 26-28 in our study of capacitors.

Figure 3
Photoelectric effect experiment.

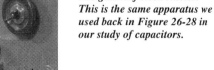
carbon arc
light source

A simple extension to the experiment is to see what happens if the zinc is given a positive charge. Following Ben Franklin's prescription, we can obtain a positive charge by rubbing a glass rod with a silk cloth. Then touch the positively charged glass rod to the zinc and again you see the gold leaves separate indicating the presence of charge. Now shine the light from the carbon arc on the zinc and nothing happens. The leaves stay spread apart, and the zinc is not discharged by the light.

When we charge the zinc with a negative charge, we are placing an excess of electrons on the zinc. From Gauss's law we know that there cannot be any net charge inside a conductor, thus the excess negative charge, the extra electrons, must be residing in the surface of the metal. The light from the carbon arc, which discharges the zinc, must therefore be knocking these extra electrons out of the metal surface. When we charged the zinc positively, we created a deficiency of electrons in the surface, and no electrons were knocked out.

In the context of Maxwell's equations, it is not particularly surprising that a beam of light should be able to knock electrons out of the surface of a piece of metal. According to Maxwell's theory, light consists of a wave of electric and magnetic fields. An electron, residing on the surface of the zinc, should experience an oscillating electric force when the light shines on the zinc. The frequency of oscillation should be equal to the frequency of the light wave, and the strength of the electric field should be directly related to the intensity of the light. (We saw earlier that the intensity of the light should be proportional to the square of the magnitude of the electric field.)

The question is whether the electric force is capable of ejecting an electron from the metal surface. A certain amount of energy is required to do this. For example, in our electron gun experiment we had to heat the filament in order to get an electron beam. It was the thermal energy that allowed electrons to escape from the filament. We now want to know whether the oscillating electric force of the light wave can supply enough energy to an electron for the electron to escape.

There are two obvious conclusions we should reach. One is that we do not want the frequency of oscillation to be too high, because the direction of the electric field reverses on each half cycle of the oscillation. The electron is pushed one way, and then back again. The longer the time it is pushed in one direction, the lower the frequency of the oscillation, the more time the electron has to pick up speed and gain kinetic energy. If the frequency is too high, just as the electron starts to move one way, it is pushed back the other way, and it does not have time to gain much kinetic energy.

The second obvious conclusion is that we have a better chance of ejecting electrons if we use a more intense beam of light. With a more intense beam, we have a stronger electric field which should exert a stronger force on the electron, producing a greater acceleration and giving the electron more kinetic energy. An intense enough beam might supply enough kinetic energy for the electrons to escape.

In summary, we expect that light might be able to eject electrons from the surface of a piece of metal if we use a low enough frequency and an intense enough beam of light. An intense beam of red light should give the best results.

These predictions, based on Maxwell's equations and Newtonian mechanics, are completely wrong!

Let us return to our photoelectric effect demonstration. During a lecture, a student suggested that we make the light from the carbon arc more intense by using a magnifying glass to focus more of the arc light onto the zinc. The more intense beam of light should discharge the zinc faster.

When you use a magnifying glass, you can make the light striking the zinc look brighter. But something surprising happens. The zinc stops discharging. The gold leaves stop falling. Remove the magnifying glass and the leaves start to fall again. *The magnifying glass prevents the discharge.*

You do not have to use a magnifying glass to stop the discharge. A pane of window glass will do just as well. Insert the window glass and the discharge stops. Remove it, and the gold leaves start to fall again.

How could the window glass stop the discharge? The window glass appears to have no effect on the light striking the zinc. The light appears just as bright. It was brighter when we used the magnifying glass, but still no electrons were ejected. The prediction from Maxwell's theory that we should use a more intense beam of light does not work for this experiment.

What the window glass does is block *ultraviolet radiation*. It is ultraviolet radiation that tans your skin (and can lead to skin cancer). It is difficult to get a tan indoors from sunlight that has gone through a window because the glass has blocked the ultraviolet component of the sun's radiation. Similarly the pane of window glass, or the glass in the magnifying lens, used in the photoelectric effect experiment, prevents ultraviolet radiation from the carbon arc from reaching the zinc. It is the high frequency ultraviolet radiation that is ejecting electrons from the zinc, not the lower frequency visible light. This is in direct contradiction to the prediction of Maxwell's theory and Newton's laws.

Einstein's explanation of the photoelectric effect is simple. He assumed that Newton was right after all, in that light actually consisted of beams of particles. The photoelectric effect occurred when a particle of light, a *photon*, struck an electron in the surface of the metal. All the energy of the photon would be completely absorbed by the electron. If this were enough energy the electron could escape, if not, it could not.

The idea that light actually consisted of particles explains why Planck had to assume that in blackbody radiation, light could only be emitted or absorbed in quantum units. What was happening in blackbody radiation, photons, particles of light, were being emitted or absorbed. As a result, Planck's formula for the energy of the quanta of emitted and absorbed radiation, must also be the formula for the energy of a photon. Thus Einstein concluded that a photon's energy is given by the equation

$$E_{photon} = hf \qquad \text{Einstein's photoelectric effect formula} \qquad (3)$$

where again f is the frequency of the light and h is Planck's constant. Equation 3 is known as Einstein's *photoelectric effect formula*.

With Equation 3, we can begin to understand our photoelectric effect demonstration. It turns out that visible photons do not have enough energy to knock an electron out of the surface of zinc. There are other metals that require less energy and for these metals visible light will produce a photoelectric effect. But for zinc, visible photons do not have enough energy. Even making the visible light more intense using a magnifying glass does not help. It is only the higher frequency, more energetic, ultraviolet photons that have enough energy to kick an electron out of the surface of zinc. We blocked these energetic photons with the window glass and the magnifying glass.

In 1921, Einstein received the Nobel prize, not for the special theory of relativity which was still controversial, nor for general relativity, but for his explanation of the photoelectric effect.

PLANCK'S CONSTANT h

Planck's constant h, the proportionality constant in Einstein's photoelectric effect formula, appears nowhere in Newtonian mechanics or Maxwell's theory of electricity and magnetism. As physicists were to discover in the early part of the twentieth century, Planck's constant appears just when Newtonian mechanics and Maxwell's equations began to fail. Something was wrong with the nineteenth century physics, and Planck's constant seemed to be a sign of this failure.

The value of Planck's constant is

$$ h = 6.63 \times 10^{-34} \text{joule sec} \tag{4} $$

where the dimensions of h have to be an energy times a time, as we can see from the photoelectric formula

$$ E_{joules} = h\left(\text{joule sec}\right) \times f\left(\frac{\text{cycles}}{\text{sec}}\right) \tag{3a} $$

The dimensions check because cycles are dimensionless.

It is not hard to see that Planck's constant also has the dimensions of angular momentum. Recall that the angular momentum L of an object is equal to the object's linear momentum p = mv times its lever arm $r_\perp$ about some point. Thus the formula for angular momentum is

$$ L = pr_\perp = m\left(kg\right)v\left(\frac{\text{meter}}{\text{sec}}\right) \times r_\perp\left(\text{meter}\right) $$
$$ = mvr_\perp\left(kg\frac{m^2}{\text{sec}}\right) $$

We get the same dimensions if we write Planck's constant in the form

$$ h\left(\text{joule sec}\right) = h\left(kg\frac{m^2}{\text{sec}^2}\right) \times \text{sec} $$
$$ = h\left(kg\frac{m^2}{\text{sec}}\right) \tag{5} $$

where we used the fact that the dimensions of energy are a mass times a velocity squared.

A fundamental constant of nature with the dimensions of angular momentum is not something to be expected in Newtonian mechanics. It suggests that there is something special about this amount of angular momentum, $6.63 \times 10^{-34}\left(kg\ m^2/\text{sec}\right)$ of it, and nowhere in Newtonian mechanics is there any reason for any special amount. It would be Neils Bohr in 1913 who first appreciated the significance of this amount of angular momentum.

PHOTON ENERGIES

Up to a point we have been describing the electromagnetic spectrum in terms of the frequency or the wavelength of the light. Now with Einstein's photoelectric formula, we can also describe the radiation in terms of the energy of the photons in the radiation. This can be convenient, for we often want to know how much energy photons have. For example, do the photons in a particular beam of light have enough energy to kick an electron out of the surface of a given piece of metal, or to break a certain chemical bond?

For visible light and nearby infrared light, the frequencies are so high that describing the light in terms of frequency is not particularly convenient. We are more likely to work in terms of the light's wavelength and the photon's energy, and want to go back and forth between the two. Using the formula

$$f\frac{cycles}{sec} = \frac{c \text{ meters/sec}}{\lambda \text{ meters/cycle}}$$

which we can get from dimensions, we can write the photoelectric formula in the form

$$E = hf = \frac{hc}{\lambda} \qquad (6)$$

Using MKS units in Equation 6 for h, c, and λ, we end up with the photon energy expressed in joules. But a joule, a huge unit of energy compared to the energy of a visible photon, is also inconvenient to use. A far more convenient unit is the electron volt. To see why, let us calculate the energy of the photons in the red hydrogen α line, whose wavelength was 6.56×10^{-5}cm or 6.56×10^{-7}m. First calculating the energy in joules, we have

$$E(H_{\alpha \text{ line}}) = \frac{hc}{\lambda_\alpha}$$

$$= \frac{6.63 \times 10^{-34} \text{joule sec} \times 3 \times 10^8 \text{m / sec}}{6.56^{-7} \times 10^{-7} \text{m}}$$

$$= 3.03 \times 10^{-19} \text{joules}$$

Converting this to electron volts, we get

$$E(H_{\alpha \text{ line}}) = \frac{3.03 \times 10^{-19} \text{joules}}{1.6 \times 10^{-19} \text{joules/eV}}$$

$$E(H_{\alpha \text{ line}}) = 1.89 \text{ eV} \qquad (7)$$

That is a convenient result. It turns out that the visible spectrum ranges from about 1.8 eV for the long wavelength red light to about 3.1 eV for the shortest wavelength blue photons we can see. It requires 3.1 eV to remove an electron from the surface of zinc. You can see immediately that visible photons do not quite have enough energy. You need ultraviolet photons with an energy greater than 3.1 eV.

Exercise 2

The blackbody spectrum of the sun corresponds to an object whose temperature is 5800 kelvin. The predominant wavelength λ_{max} for this temperature is 5.0×10^{-5}cm as we saw in the calculation following Equation 1. What is the energy, in electron volts, of the photons of this wavelength?

Exercise 3

The rest energy of an electron is .51MeV = 5.1×10^5 eV. What is the wavelength, in centimeters, of a photon whose energy is equal to the rest energy of an electron?

We will often want to convert directly from a photon's wavelength λ in centimeters to its energy E in electron volts. This is most easily done by starting with the formula $E = hc/\lambda$ and using conversion factors until E is in electron volts when λ is in centimeters. We get

$$E = \frac{hc}{\lambda}$$

$$= \frac{6.63 \times 10^{-34} \text{joule sec} \times 3 \times 10^{10} \frac{\text{cm}}{\text{sec}}}{\lambda \text{ cm}}$$

$$= \frac{1.989}{\lambda \text{ cm}} \times 10^{-23} \text{joule cm} \times \frac{1}{1.6 \times 10^{-19} \frac{\text{joule}}{\text{eV}}}$$

The desired formula is thus

$$\boxed{E_{photon}(\text{in eV}) = \frac{12.4 \times 10^{-5} \text{eV} \cdot \text{cm}}{\lambda(\text{in cm})}} \quad (8)$$

As an example in the use of Equation 8, let us recalculate the energy of the H_α photons whose wavelength is $6.56 \times 10^{-5} \text{cm}$. We get immediately

$$E_{H\alpha} = \frac{12.4 \times 10^{-5} \text{eV cm}}{6.56 \times 10^{-5} \text{cm}} = 1.89 \text{eV}$$

which is our previous result.

Exercise 4

The range of wavelengths of light in the visible spectrum is from $7 \times 10^{-5}\text{cm}$ in the red down to $4 \times 10^{-5}\text{cm}$ in the blue. What is the corresponding range of photon energies?

Exercise 5

(a) It requires 2.20 eV to eject an electron from the surface of potassium. What is the longest wavelength light that can eject electrons from potassium?

(b) You shine blue light of wavelength $4 \times 10^{-5}\text{cm}$ at potassium. What is the maximum kinetic energy of the ejected electrons?

Exercise 6

The human skin radiates blackbody radiation corresponding to a temperature of 32°C. (Skin temperature is slightly lower than the 37°C internal temperature.) What is the predominant energy, in eV of the photons radiated by a human? (This is the energy corresponding to λ_{max} for this temperature.)

Exercise 7

A 100 watt bulb uses 100 joules of energy per second. For this problem, assume that all this energy went into emitting yellow photons at a wavelength of $\lambda = 5.88 \times 10^{-5}\text{cm}$.

(a) What is the energy, in eV and joules, of one of these photons?

(b) How many of these photons would the bulb radiate in one second?

(c) From the results of part (b), explain why it is difficult to detect individual photons in a beam of light.

Exercise 8

Radio station WBZ in Boston broadcasts at a frequency of 1050 kilocycles at a power of 50,000 watts.

(a) How many photons per second does this radio station emit?

(b) Should these photons be hard to detect individually?

Exercise 9

In what part of the electromagnetic spectrum will photons of the following energies be found?

(a) 1 eV   (e) 5 eV

(b) 2.1 eV   (f) 1000 eV

(c) 2.5 eV   (g) $.51 \times 10^6 \text{eV}$ (.51 MeV)

(d) 3 eV   (h) $4.34 \times 10^{-9}\text{eV}$

(The rest energy of the electron is .51 MeV.)

Exercise 10

(a) Calculate the energy, in eV, of the photons in the three visible spectral lines in hydrogen

$$\lambda_\alpha(\text{red}) = 6.56 \times 10^{-5}\text{cm}$$

$$\lambda_\beta(\text{blue}) = 4.86 \times 10^{-5}\text{cm}$$

$$\lambda_\gamma(\text{violet}) = 4.34 \times 10^{-5}\text{cm}$$

It requires 2.28 eV to eject electrons from sodium.

(b) The red H_α light does not eject electrons from sodium. Explain why.

(c) The H_β and H_γ lines do eject electrons. What is the maximum kinetic energy of the ejected electrons for these two spectral lines?

PARTICLES AND WAVES

We gain two different perspectives when we think of the electromagnetic spectrum in terms of wavelengths and in terms of photon energies. The wavelength picture brings to mind Young's two slit experiment and Maxwell's theory of electromagnetic radiation. In the photon picture we think of electrons being knocked out of metals and chemical bonds being broken. These pictures are so different that it seems nearly impossible to reconcile them. Reconciling these two pictures will, in fact, be the main focus of the remainder of the text.

For now we seek to answer a more modest question. How can the two pictures coexist? How could some experiments, like our demonstration of the photoelectric effect exhibit only the particle nature and completely violate the predictions of Maxwell's equations, while other experiments, like our measurements of the magnetic field of a radio wave, support Maxwell's equations and give no hint of a particle nature?

In Figure (4) we show the electromagnetic spectrum both in terms of wavelengths and photon energies. It is in the low energy, long wavelength region, from radio waves to light waves, that the wave nature of the radiation tends to dominate. At shorter wavelengths and higher photon energies, from visible light through γ rays, the particle nature tends to dominate. The reason for this was well illustrated in Exercise 8.

In Exercise 8 you were asked to calculate how many photons were radiated per second by radio station WBZ in Boston. The station radiates 50,000 watts of power at a frequency of 1.05 megacycles. To solve the problem, you first had to calculate the energy of a 1.05 megacycle photon using Einstein's formula $E_{photon} = hf$. This turns out to be about 7×10^{-28} joules. The radio station is radiating 50,000 joules of energy every second, and thus emitting 7×10^{31} photons per second. It is hard to imagine an experiment in which we can detect individual photons when so many are being radiated at once. Any experiments should detect some kind of average effect, and that average effect is given by Maxwell's equations.

When we get up to visible photons, whose energies are in the 2—3 eV range and wavelengths of the order of 5×10^{-5} cm , it is reasonably easy to find experiments that can detect either the particle or the wave nature of light. With a diffraction grating we have no problem measuring wavelengths in the range of 10^{-5} cm . With the photoelectric effect, we can easily detect individual photons in the 2-3 eV range.

As we go to shorter wavelengths, individual photons have more energy and the particle nature begins to dominate. To detect the wave nature of X rays, we need something like a diffraction grating with line spacing of the order of the X ray wavelength. It turns out that the regular lines and planes of atoms in crystalline materials act as diffraction gratings allowing us to observe the wave nature of X ray photons. But when we get up into the γ ray region, where photons have energies comparable to the rest energies of electrons and protons, all we observe experimentally are particle reactions. At these high energies, the wave nature of the photon is basically a theoretical concept used to understand the particle reactions.

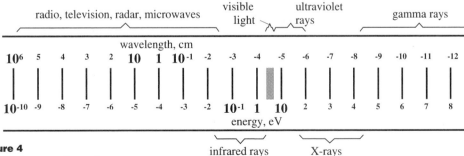

Figure 4
The electromagnetic spectrum.

While it is a rule of thumb that at wavelengths longer than visible light, the wave nature of electromagnetic radiation dominates, there are important exceptions. The individual photons in the WBZ radio wave can be detected! You might ask, what kind of experiment can detect an object whose energy is only 7×10^{-28} joules. This, however, happens to be the amount of energy required to flip the spin of an electron or a nucleus in a reasonably sized magnetic field. This spin flip process for electrons is called *electron spin resonance* and for nuclei, *nuclear spin resonance*. In Chapter 38 we will discuss an electron spin resonance experiment that is easily performed in the lab. Nuclear spin resonance, as you may be aware, is the basis of magnetic resonance imaging, an increasingly important medical diagnostic tool.

The truly amazing feature of the magnetic resonance experiments is that Maxwell's equations and Einstein's photoelectric effect formula make the same predictions! Einstein's photoelectric effect formula is easier to use and will be the way we analyze the electron spin resonance experiment. Maxwell's equations and the classical pictures of angular momentum and gyroscopes facilitate the more detailed analysis needed for the imaging apparatus. Texts describing the imaging apparatus use the classical approach. For this discussion the important point is that the two points of view come together in this low energy, long wavelength limit.

PHOTON MASS

The basic idea behind Einstein's famous formula $E = mc^2$ is that energy is mass. The factor c^2 is a conversion factor to go between energy measured in grams and energy measured in ergs. If we had used a different set of units, for example, measuring distances in feet, and time in nanoseconds, then the numerical value of c would be 1, and Einstein's equation would be $E = m$, the more revealing statement.

Photons have energy, thus they have mass. If we combine the photoelectric formula $E = hf$ with $E = mc^2$, we can solve for the mass m of a photon of frequency f. The result is

$$E = hf = m_{photon}c^2$$

$$m_{photon} = \frac{hf}{c^2} \qquad (10)$$

We can also express the photon mass in terms of the wavelength λ, using $f/c = 1/\lambda$

$$m_{photon} = \frac{h}{c}\left(\frac{f}{c}\right) = \frac{h}{c\lambda} \qquad (11)$$

The idea that photons have mass presents a certain problem. In our earliest discussions of mass in Chapter 6, we saw that the mass increased with velocity, increasing without bounds as the speed of the object approached the speed of light. The formula that described this increase in mass was

$$m = \frac{m_0}{\sqrt{1 - v^2/c^2}} \qquad (6\text{-}14)$$

where m_0 is the mass of the particle at rest and m its mass when traveling at a speed v.

The obvious problems with photons is that they are light—and therefore travel *at* the speed of light. Applying Equation 6-14 to photons gives

$$m_{photon} = \frac{m_0}{\sqrt{1 - c^2/c^2}} = \frac{m_0}{\sqrt{1-1}} = \frac{m_0}{0} \qquad (12)$$

a rather embarrassing result. The divisor in Equation 12 is exactly zero, not approximately zero. Usually division by 0 is a mathematical disaster.

There is only one way Equation 12 can be salvaged. The numerator m_0 must also be identically zero. Then Equation 12 gives m = 0/0, an undefined, but not disastrous result. The numerical value of 0/0 can be anything -1, 5, 10^{-17}, anything you want. In other words if the rest mass m_0 of a photon is zero, Equation 12 says nothing about what the actual mass m_{photon} is. Equation 12 only tells us that the rest mass of a photon must be zero.

Stop a photon and what do you have left? Heat! In the daytime many billions of photons strike your skin every second. But after they hit nothing is left except the warmth of the sunlight. When a photon is stopped it no longer exists—only its energy is left behind. That is what is remarkable about photons. Only if they are moving *at* the speed of light do they exist, carry energy and have mass. This distinguishes them from all the particles that have rest mass and cannot get up to the speed of light.

An interesting particle is the neutrino. We are not sure whether a neutrino (there are actually 3 different kinds of neutrinos) has a rest mass or not. If neutrinos have no rest mass, then they must travel at the speed of light, and obey the same mechanics as a photon. The evidence is highly suggestive of this interpretation. We saw, for example, that neutrinos from the 1987 supernova explosion raced photons for some 100,000 years, and took within an hour of the same amount of time to get here. That is very close to the speed of light. If the neutrinos took a tiny bit longer to reach us, if they moved at slightly less than the speed of light, then they would have to have some rest mass. The rest mass of an individual neutrino would have to be extremely small, but there are so many neutrinos in the universe that their total mass could make up a significant fraction of the mass in the universe. This might help explain some of the missing mass in the universe that astronomers are worrying about. At the present time, however, all experiments are consistent with the idea that a neutrino's rest mass is exactly zero.

For particles with rest mass, we used the formulas $E = mc^2$, $m = m_0/\sqrt{1 - v^2/c^2}$ to get the formulas for the rest energy and the kinetic energy of the particle. In particular we got the approximate formula $1/2\, m_0 v^2$ for the kinetic energy of a slowly moving particle. For photons, the formula $m_0 / \sqrt{1 - v^2/c^2}$ does not apply, there is no such thing as a slowly moving photon, and the kinetic energy formula $1/2\, mv^2$ is **completely wrong**! For photons, all the energy is kinetic energy, and the formula for the photon's kinetic energy is given by Einstein's photoelectric effect formula $E = hf = hc/\lambda$. The energy of a photon is determined by its frequency, not its speed.

Photon Momentum

While photons have no rest mass, and do not obey Newton's second law, they do obey what turns out to be a quite simple set of rules of mechanics. Like their massive counterparts, photons carry energy, linear momentum, and angular momentum all of which are conserved in interactions between particles. The formulas for these quantities can all be obtained straightforwardly from Einstein's photoelectric formula $E = hf$ and energy formula $E = mc^2$.

We have already combined these two equations to obtain Equation 11 for the mass of a photon

$$m_{photon} = \frac{h}{\lambda c} \tag{11a}$$

To find the momentum of the photon, we multiply its mass by its velocity. Since all photons move at the same speed c, the photon momentum p_{photon} is given by

$$\boxed{p_{photon} = m_{photon}c = \frac{h}{\lambda}} \tag{13}$$

In the next few chapters, we will find that Equation 13 applies to more than just photons. It turns out to be one of the most important equations in physics.

In our discussion of systems of particles in Chapter 11, we had an exercise where a boy washing a car, was squirting the hose at the door of the car. The water striking the door carried a certain amount of momentum per second, and as a result exerted a force $\vec{F} = d\vec{p}/dt$ on the door. The exercise was to calculate this force.

When you shine a beam of light at an object, if the photons in the beam actually carry momentum $p = h/\lambda$ then the beam should exert a force equal to the rate at which momentum is being absorbed by the object. If the object absorbs the photon, like a black surface would, the momentum delivered is just the momentum of the photons. If it is a reflecting surface, then we have to include the photon recoil, and the momentum transferred is twice as great.

There is a common toy called a radiometer that has 4 vane structures balanced on the tip of a needle as shown in Figure (5). One side of each vane is painted black, while the other side is reflecting. If you shine a beam of light at the vanes, they start to rotate. If, however, you look at the apparatus for a while, you will notice that the vanes rotate the wrong way. They move as if the black side were being pushed harder by the beam of light than the reflecting side.

In the toy radiometers, it is not the force exerted by the light, but the fact that there are some air molecules remaining inside the radiometer, that causes the vanes to rotate. When the light strikes the vanes, it heats the

Figure 5
The radiometer.

black side more than the reflecting side. Air molecules striking the black side are heated, gain thermal energy, and bounce off or recoil from the vane with more speed than molecules bouncing off the cooler reflecting side. It is the extra speed of the recoil of the air molecules from the black side that turns the vane. This thermal effect is stronger than the force exerted by the light beam itself.

We can see from the example of the radiometer that the measurement of the force exerted by a beam of light, measuring the so-called pressure of light, must be done in a good vacuum during a carefully controlled experiment. That measurement was first made by Nichols and Hull at Dartmouth College in 1901. While Maxwell's theory of light also predicts that a beam of light should exert a force, we can now interpret the Nichols and Hull experiment as the first experimental measurement of the momentum carried by photons.

The first experiments to demonstrate that individual photons carried momentum were carried out by Arthur Compton in 1923. In what is now known as the **Compton scattering** or the **Compton effect**, X ray photons are aimed at a thin foil of metal. In many cases the X ray photons collide with and scatter an electron rather than being absorbed as in the photoelectric effect. Both the struck electron and the scattered photon emerge from the back side of the foil as illustrated in Figure (6).

The collision of the photon with the electron in the metal foil is in many ways similar to the collision of the two steel balls studied in Chapter 7, Figures (1) and (2). The energy of the X ray photons used by Compton were of the order of 10,000 eV while the energy of the electron in the metal is of the order of 1 or 2 eV. Thus the X ray photon is essentially striking an electron at rest, much as the moving steel ball struck a steel ball at rest in Figure (7-2).

In both the collision of the steel balls and in the Compton scattering, both energy and linear momentum are conserved. In particular the momentum carried in by the incoming X ray photon is shared between the scattered X ray and the excited electron. This means that the X ray photon loses momentum in the scattering process. Since the photon's momentum is related to its wavelength by $p = h/\lambda$, a loss in momentum means an increase in wavelength. Thus, if the photon mechanics we have developed applies to X ray photons, then the scattered X rays should have a slightly longer wavelength than the incident X rays, a result which Compton observed.

According to Maxwell's theory, if a light wave impinges on a metal, it should start the electrons oscillating at the frequency of the incident wave. The oscillating electrons should then radiate light at the same frequency. This radiated light would appear as the scattered light in Compton's experiments. Thus Maxwell's theory predicts that the scattered X rays should have the same wavelength as the incident wave, a result which is not in agreement with experiment.

While the experiments we have just discussed involved delicate measurements in order to detect the photon momentum, in astronomy the momentum of photons and the pressure of light can have dramatic effects. In about 5 billion years our sun will finish burning the hydrogen in its core. The core will then cool and start to collapse. In one of the contradictory features of stellar evolution, the contracting core releases gravitational potential energy at a greater rate than energy was released by burning hydrogen. As a result the core becomes hotter and much brighter than it was before.

The core will become so bright, emit so much light, that the pressure of the escaping light will lift the surface of the sun out into space. As a result the sun will expand until it engulfs the orbit of the earth. At this point the sun will have become what astronomers call a *red giant* star. Because of its huge surface area it will become thousands of times brighter than it is now.

The red giant phase does not last long, only a few million years. If the sun were bigger than it is, the released gravitational potential energy would be enough to ignite helium and nuclear fusion would continue. But the red giant phase for the sun will be near the end of the road. The sun will gradually cool and shrink, becoming a white dwarf star about the size of the earth, and finally a black ember of about the same size.

The pressure of light played an even more important role in the evolution of the early universe. The light from the big bang explosion that created the universe was so intense that for the first 1/3 of a million years, it knocked the particles of matter around and prevented the formation of stars, and galaxies. But a dramatic event occurred when the universe reached an age of 1/3 of a million years. That was the point where the universe had cooled enough to become transparent. At that point the light from the big bang decoupled from matter and stars and galaxies began to form. We will discuss this event in more detail shortly.

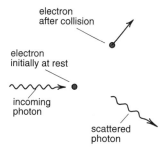

X rays which have collided with electrons in the slab are scattered out of the main beam. These X rays lose momentum, with the result that their wavelength is longer than those that were not scattered.

(a) observation of Compton scattering

(b) collision of photon and electron resulting in Compton scattering

Figure 6

Compton scattering.

ANTIMATTER

The fact that photons have no rest mass and travel only at the speed of light makes them seem quite different from particles like an electron or proton that have rest mass and make up the atoms and molecules. The distinction fades somewhat when we consider a process in which a photon is transformed into two particles with rest mass. The two particles can be any particle-antiparticle pair. Figure (7) is a bubble chamber photograph of the creation of an electron-positron pair by a photon.

In 1926 Erwin Schrödinger developed a wave equation to describe the behavior of electrons in atoms. The first equation he tried had a serious problem; it was a relativistic wave equation that appeared to have two solutions. One solution represented the ordinary electrons he was trying to describe, but the other solution appeared to represent a particle with a negative rest mass. Schrödinger found that if he went to the non relativistic limit, and developed an equation that applied only to particles moving at speeds much less than the speed of light, then the negative rest mass solutions did not appear. The non relativistic equation was adequate to describe most chemical phenomena, and is the famous Schrödinger equation.

A year later, Paul Dirac developed another relativistic wave equation for electrons. The equation was specifically designed to avoid the negative mass solutions, but the techniques used did not work. Dirac's equation correctly predicted some important relativistic phenomena, but as Dirac soon found out, the negative mass solutions were still present.

Usually one ignores undesirable solutions to mathematical equations. For example, if you want to solve for the hypotenuse of a triangle, the Pythagorean theorem tells you that $c^2 = a^2 + b^2$. This equation has two solutions, $c = \sqrt{a^2 + b^2}$ and $c = -\sqrt{a^2 + b^2}$. Clearly you want the positive solution, the negative solution in this case is irrelevant.

The problem Dirac faced was that he could not ignore the negative mass solution. If he started with a collection of positive mass particles and let them interact, the equation predicted that negative mass particles would appear, would be created. He could not avoid them.

Through a rather incredible trick, Dirac was able to reinterpret the negative mass solutions as positive mass solutions of another kind of matter—antimatter. In this interpretation, every elementary particle has a corresponding antiparticle. The antiparticle had the same rest mass but opposite charge from its corresponding particle. Thus a particle-antiparticle pair could be created or annihilated without violating the law of conservation of electric charge.

In 1927 when Dirac proposed his theory, no one had seen any form of antimatter, and no one was sure of exactly what to look for. The proton had the opposite charge from the electron, but its mass was much greater, and therefore it could not be the electron's antiparticle. If the electron antiparticle existed, it would have to have the same positive charge as the proton, but the same mass as an electron. In 1932 Carl Anderson at Caltech found just such a particle among the cosmic rays that rain down through the earth's atmosphere. That particle is the ***positron*** which is shown being created in the bubble chamber photograph of Figure (7).

(In the muon lifetime moving picture, discussed in Chapter 1, positively charged muons were stopped in the block of plastic, emitting the first pulse of light. When a positive muon decays, it decays into a positron and a neutrino. It was the positron that made the second flash of light that was used to measure the muon's lifetime.)

In the early 1950s, the synchrotron at Berkeley, the one shown in Figure (28-27b) was built just large enough to create antiprotons, and succeeded in doing so. Since then we have created antineutrons, and have observed antiparticles corresponding to all the known elementary particles. Nature really has two solutions—matter and antimatter.

The main question we have now concerning antimatter is why there is so little of it around at the present time. In the very early universe, temperatures were so high that there was a continual creation and annihilation of particles and antiparticles, with roughly equal but not exactly equal, numbers of particles and antiparticles. There probably was an excess of particles over antiparticles in the order of about one part in 10 billion. In a short while the universe cooled to the point where annihilation became more likely than creation, and the particle-antiparticle pairs annihilated. What was left behind was the slight excess of matter particles, the particles that now form the stars and galaxies of the current universe.

In 1964, James Cronin and Val Fitch, while working on particle accelerator experiments, discovered interactions that lead to an excess of particles over antiparticles. It could be that these interactions were active in the very early universe, creating the slight excess of matter over antimatter. But on the other hand, there may not have been time for known processes to create the observed imbalance. We do not yet have a clear picture of how the excess of matter over antimatter came about.

Exercise 11

Since an electron and a positron have opposite charge, they attract each other via the Coulomb electric force. They can go into orbit forming a small atom like object called *positronium*. It is like a hydrogen atom except that the two particles have equal mass and thus move about each other rather than having one particle sit at the center. The positronium atom lasts for about a microsecond, whereupon the positron and electron annihilate each other, giving off their rest mass energy in the form of photons. The rest mass energy of the electron and positron is so much greater than their orbital kinetic energy, that one can assume that the positron and electron were essentially at rest when they annihilated. In the annihilation both momentum and energy are conserved.

(a) Explain why the positron and electron cannot annihilate, forming only one photon. (What conservation law would be violated by a one photon annihilation?)

(b) Suppose the positronium annihilated forming two photons. What must be the energy of each photon in eV? What must be the relative direction of motion of the two photons?

The answer to part (b) is that each photon must have an energy of .51 MeV and the photons must come out in exactly opposite directions. By detecting the emerging photons you can tell precisely where the positronium annihilated. This phenomenon is used in the medical imaging process called *positron emission tomography* or *PET* scans.

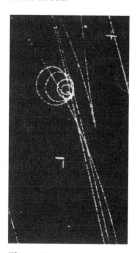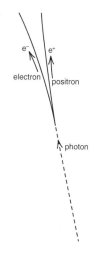

Figure 7
Creation of positron-electron pair. A photon enters from the bottom of the chamber and collides with a hydrogen nucleus. The nucleus absorbs some of the photon's momentum, allowing the photon's energy to be converted into a positron-electron pair. Since a photon is uncharged, it leaves no track in the bubble chamber; the photon's path is shown by a dotted line. (Photograph copyright The Ealing Corporation, Cambridge, Mass.)

INTERACTION OF PHOTONS AND GRAVITY

Because photons have mass, we should expect that photons should interact with gravity. But we should be careful about applying the laws of Newtonian gravity to photons, because Newtonian gravity is a non relativistic theory, while photons are completely relativistic particles.

If we apply the ideas of Newtonian gravity to photons, which we will do shortly, we will find that we get agreement with experiment if the photons are moving parallel to the gravitational force, for example, falling toward the earth. But if we do a Newtonian type of calculation of the deflection of a photon as it passes a star, we get half the deflection predicted by Einstein's general theory of relativity. It was in Eddington's famous eclipse expedition of 1917 where the full deflection predicted by Einstein's theory was observed. This observation, along with measurements of the precession of Mercury's orbit, were the first experimental evidence that Newton's theory of gravity was not exactly right.

In 1960, R. V. Pound and G.A. Rebka performed an experiment at Harvard that consisted essentially of dropping photons down a well. What they did was to aim a beam of light of precisely known frequency down a vertical shaft about 22 meters long, and observed that the photons at the bottom of the shaft had a slightly higher frequency, i.e., had slightly more energy than when they were emitted at the top of the shaft.

The way you can use Newtonian gravity to explain their results is the following. If you drop a rock of mass m down a shaft of height h, the rock's gravitational potential energy mgh at the top of the shaft is converted to kinetic energy at the bottom. For a rock, the kinetic energy shows up in the form of increased velocity, and is given by the formula $1/2 \, mv^2$. For a photon, all of whose energy is kinetic energy anyway, the kinetic energy gained from the fall shows up as an increased frequency of the photon.

Using Einstein's formula $E = hf$ for the kinetic energy of a photon, we predict that the photon energy at the bottom is given by

$$E_{bottom} = hf_{bottom} = hf_{top} + m_{photon}gy \quad (14)$$

where we are assuming that the same formula mgy for gravitational potential energy applies to both rocks and photons.

Since $m_{photon} = hf/c^2$, the mass of the photon changes slightly as the photon falls. But for a 22 meter deep shaft, the change in frequency is very small and we can quite accurately use hf_{top}/c^2 for the mass of the photon in Equation 14. This gives

$$hf_{bottom} = hf_{top} + \left(\frac{hf_{top}}{c^2}\right)gy$$

Cancelling the h's, we get

$$f_{bottom} = f_{top}\left(1 + \frac{gy}{c^2}\right) \quad (15)$$

as the formula for the increase in the frequency of the photon. This is in agreement with the results found by Pound and Rebka.

Exercise 12

(a) Show that the quantity gy/c^2 is dimensionless.

(b) What is the percentage increase in the frequency of the photons in the Pound-Rebka experiment? (Answer: $2.4 \times 10^{-13}\%$. This indicates how extremely precise the experiment had to be.)

To calculate the sideways deflection of a photon passing a star, we could use Newton's second law in the form $\vec{F} = d\vec{p}/dt$ to calculate the rate at which a sideways gravitational force added a sideways component to the momentum of the photon. The gravitational force would be $\vec{F}_g = m_{photon}\vec{g}$, with $m_{photon} = hf/c^2$. The result, as we have mentioned, is half the deflection predicted by Einstein's theory of gravity and half that observed during Eddington's eclipse expedition.

The gravitational deflection of photons, while difficult to detect in 1917, has recently become a useful tool in astronomy. In 1961, Allen Sandage at Mt. Palomar Observatory discovered a peculiar kind of object that seemed to be about the size of a star but which emitted radio waves like a radio galaxy. In 1963 Maarten Schmidt photographed the spectral lines of a second radio star and discovered that the spectral lines were all shifted far to the red. If this red shift were caused by the Doppler effect, then the radio star would be moving away from the earth at a speed of 16% the speed of light.

If the motion were due to the expansion of the universe, then the radio star would have to be between one and two billion light years away. An object that far away, and still visible from the earth, would have to be as bright as an entire galaxy.

The problem was the size of the object. The intensity of the radiation emitted by these radio stars was observed to vary significantly over times as short as weeks to months. This virtually guarantees that the object is no bigger than light weeks or light months across, because the information required to coordinate a major change in intensity cannot travel faster than the speed of light. Thus Schmidt had found an object, not much bigger than a star, radiating as much energy as the billions of stars in a galaxy. These rather dramatic objects, many more of which were soon found, became known as *quasars*, which is an abbreviation for *quasi stellar objects*.

It was hard to believe that something not much bigger than a star could be as bright as a galaxy. There were suggestions that the red shift detected by Maarten Schmidt was due to something other than the expansion of the universe. Perhaps quasars were close by objects that just happened to be moving away from us at incredible speeds. Perhaps they were very massive objects so massive that the photons escaping from the object lost a lot of their energy and emerged with lower frequencies and longer wavelengths. (This would be the opposite effect than that seen in the Pound-Rebka experiment where photons falling toward the earth gained kinetic energy and increased in frequency.)

Over the years, no explanation other than the expansion of the universe satisfactorily explained the huge red shifts seen in quasars, but there was this nagging doubt about whether the quasars were really that far away. Everything seemed to fit with the model that red shifts were caused by an expanding universe, but it would be nice to have direct proof.

The direct proof was supplied by gravitational lensing, a consequence of the sideways deflection of photons as they pass a massive object. In 1979, a photograph revealed two quasars that were unusually close to each other. Further investigation showed that the two quasars had identical red shifts and emitted identical spectral lines. This was too much of a coincidence. The two quasars had to be two images of the same quasar.

How could two images of a single quasar appear side by side on a photographic plate? The answer is illustrated in Figure (8). Suppose the quasar were directly behind a massive galaxy, so that the light from the quasar to the earth is deflected sideways as shown. Here on earth we could see light coming from the quasar from 2 or more different directions. The telescope forms images as if the light came in a straight line. Thus in Figure (8), light that came around the top side of the galaxy would look like it came from a quasar located above the actual quasar, while light that came around the bottom side would look as if it came from another quasar located below the actual quasar.

This gravitational lensing turned out to be a more common phenomena than one might have expected. More than a dozen examples of gravitational lensing have been discovered in the past decade. Figure (9), an image produced by the repaired Hubble telescope, shows a quasar surrounded by four images of itself. The four images were formed by the gravitational lensing of an intermediate galaxy.

The importance of gravitational lensing is that it provides definite proof that the imaged objects are more distant than the objects doing the imaging. The quasar in Figure (9) must be farther away from us than the galaxy that is deflecting the quasar's light. This proved that the quasars are distant objects and that the red shift is definitely due to the expansion of the universe.

Evidence over the years has indicted that quasars are the cores of newly formed galaxies. Quasars tend to be distant because most galaxies were formed when the universe was relatively young. If all quasars we see are very far away, the light from them has taken a long time to reach us, thus they must have formed a long time ago. The fact that we see very few nearby quasars means that most galaxy formation has already ceased.

Although we have photographed galaxies for over a hundred years, we know surprisingly little about them, especially what is at the core of galaxies. Recent evidence indicates that at the core of the galaxy M87 there is a black hole whose mass is of the order of millions of suns. The formation of such a black hole would produce brilliant radiation from a very small region of space, the kind of intense localized radiation seen in quasars. At this point we only have proof for one black hole at the center of one galaxy, but the pieces are beginning to fit together. Something quite spectacular may be at the center of most galaxies, and quasars are probably giving us a view of the formation of these centers.

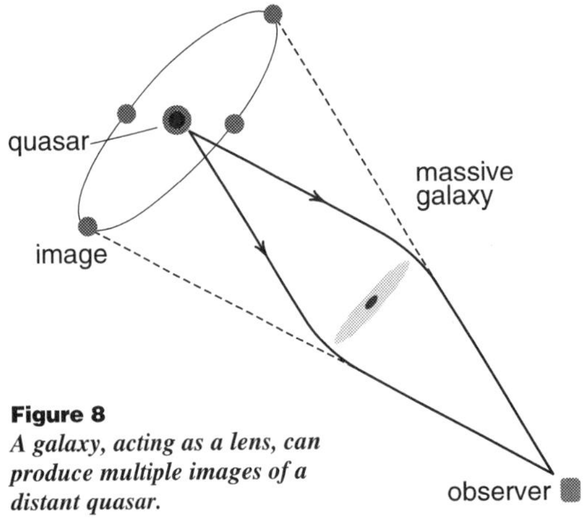

Figure 8
A galaxy, acting as a lens, can produce multiple images of a distant quasar.

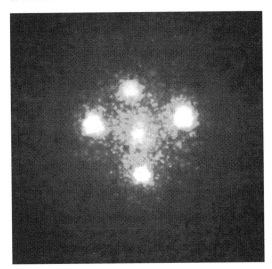

Figure 9
Hubble telescope photograph of a distant quasar surrounded by 4 images of the quasar. This is known as the **Einstein cross.**

EVOLUTION OF THE UNIVERSE

The two basic physical ideas involved in understanding the early universe are its expansion, and the idea that the universe was in thermal equilibrium. Before we see how these concepts are applied, we wish to develop a slightly different perspective of these two concepts. First we will see how the red shift of light can be interpreted in terms of the expansion of the universe. Then we will see that blackbody radiation can be viewed as a gas of photons in thermal equilibrium. With these two points of view, we can more easily follow the evolution of the universe.

Red Shift and the Expansion of the Universe

The original clue that we live in an expanding universe was from the red shift of light from distant galaxies. We have explained this red shift as being caused by the Doppler effect. The distant galaxies are moving away from us, and it is the recessional motion that stretches the wavelengths of the radiated waves, as seen in the ripple tank photograph back in Figure (33-29) reproduced here.

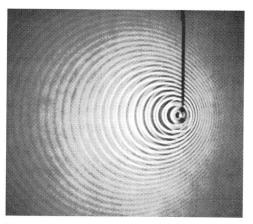

Figure 33-29
The doppler effect

There is another way to view the red shift that gives the same results but provides a more comprehensive picture of the evolution of the universe. Consider a galaxy that is, for example, receding from us at 10% the speed of light. According to the Doppler effect, the wavelength of the light from that galaxy will be lengthened by a factor of 10%.

Where is that galaxy now? If the galaxy were moving away from us at 10% the speed of light, it has traveled away from us 1/10th as far as the light has traveled in reaching us. In other words the galaxy is 10% farther away now than when it emitted the light. If the recessional motion of the galaxy is due to the expansion of the universe, then the universe is now 10% bigger than it was when the galaxy emitted the light.

In this example, the universe is now 10% bigger and the wavelength of the emitted light is 10% longer. We can take the point of view that *the wavelength of the light was stretched 10% by the expansion of the universe.*

In other words it makes no difference whether we say that the red shift was caused by the 10% recessional velocity of the galaxy, or the 10% expansion of the universe. Both arguments give the same answer. When we are studying the evolution of the universe, it is easier to use the idea that the universe's expansion stretches the photon wavelengths. This is especially true for discussions of the early universe where recessional velocities are close to the speed of light and relativistic Doppler calculations would be required.

Another View of Blackbody Radiation

The surface of the sun provides an example of a hot gas more or less in thermal equilibrium. Not only are the ordinary particles, the electrons, the protons, and other nuclei in thermal equilibrium, so are the photons, and this is why the sun emits a blackbody spectrum of radiation. Blackbody radiation at a temperature T can be viewed as a gas of photons in thermal equilibrium at that temperature.

In our derivation of the ideal gas law, we were surprisingly successful using the idea that the average gas molecule had a thermal kinetic energy 3/2 kT. In a similar and equally naive derivation, we can explain one of the main features of blackbody radiation from the assumption that the average or typical photon in blackbody radiation also has a kinetic energy 3/2 kT.

The main feature of blackbody radiation, that could not be explained using Maxwell's theory of light, was the fact that there was a peak in the blackbody spectrum. There is a predominant wavelength which we have called λ_{max} that is inversely proportional to the temperature T. The precise relationship given by Wein's displacement law is

$$\lambda_{max} = \frac{2.898}{T} \text{mm K} \qquad \text{(1) repeated}$$

a result we stated earlier. The blackbody radiation peaks around λ_{max} as seen in Figure (1) reproduced here.

If blackbody radiation consists of a gas of photons in thermal equilibrium at a temperature T, we can assume that the average photon should have a kinetic energy like 3/2 kT. (The factor 3/2 is not quite right for relativistic particles, but close enough for this discussion.) Some photons should have more energy, some less, but there should be a peak in the distribution of photons around this energy. Using Einstein's photoelectric effect formula we can relate the most likely photon energy to a most likely wavelength λ_{max}. We have

$$E_{photon} = \frac{3}{2}kT$$

$$E_{photon} = \frac{hc}{\lambda_{max}}$$

Combining these equations gives

$$\frac{hc}{\lambda_{max}} = \frac{3}{2}kT$$

$$\lambda_{max} = \frac{2hc}{3kT}$$

Putting in numbers gives

$$\lambda_{max} = \frac{2 \times 6.63 \times 10^{-34} \text{joule sec} \times 3 \times 10^8 \frac{m}{s}}{3 \times 1.38 \times 10^{-23} \frac{\text{joule}}{K} \times T}$$

$$= \frac{.0096 \text{ meter} \cdot K}{T}$$

Converting from meters to millimeters gives

$$\lambda_{max} = \frac{9.60 \text{ mm} \cdot K}{T} \qquad \begin{array}{l} \textit{our estimate} \\ \textit{for } \lambda_{max} \end{array} \qquad (16)$$

While this is not the exact result, it gives us the picture that there should be a peak in the blackbody spectrum around λ_{max}. The formula gives the correct temperature dependence, and the constant is only off by a factor of 3.3. Not too bad a result considering that we did not deal with relativistic effects and the distribution of energies in thermal equilibrium. None of these results can be understood without the photon picture of light.

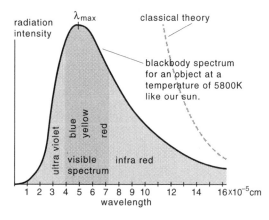

Figure 34-1 (reproduced)
Blackbody radiation spectrum showing the peak at λ_{max} (The classical curve goes up to infinity at $\lambda = 0$.)

MODELS OF THE UNIVERSE

As we saw in Chapter 33, Hubble was able to combine his new distance scale for stars and galaxies with Doppler shift measurements to discover that the universe is expanding, that the farther a galaxy is away from us, the faster it is moving away from us. Another property of the interaction of light with matter, the blackbody spectrum discussed at the beginning of this chapter, provided a critical clue to the role of this expansion in the history of the universe. To see why, it is instructive to look at the evolution of our picture of the universe, to see what led us to support or reject different models of its large scale structure.

Powering the Sun

In the 1860s, Lord Kelvin, for whom the absolute temperature scale is named, did a calculation of the age of the sun. Following a suggestion by Helmholtz, Kelvin assumed that the most powerful source of energy available to the sun was its gravitational potential energy. Noting the rate at which the sun was radiating energy, Kelvin estimated that the sun was no older than half a billion years. This was a serious problem for Darwin, whose theory of evolution required considerably longer times for the processes of evolution to have taken place. During their lifetimes neither Darwin or Kelvin could explain the apparent discrepancy of having fossils older than the sun.

This problem was overcome by the discovery that the main source of energy of the sun was not gravitational potential energy, but instead the nuclear energy released by the fusion of hydrogen nuclei to form helium nuclei. In 1938 Hans Beta worked out the details of how this process worked. The reaction begins when two protons collide with sufficient energy to overcome the Coulomb repulsion and get close enough to feel the very strong, but short range, attractive nuclear force. Such a strong collision is required to overcome the Coulomb barrier, that fusion is a rare event in the lifetime of any particular solar proton. On the average, a solar proton can bounce around about 30 million years before fusing. There are, of course, many protons in the sun, so that many such fusions are occurring at any one time.

Just after two protons fuse, electric potential energy is released when one of the protons decays via the weak interaction into a neutron, electron, and a neutrino. The electron and neutrino are ejected, leaving behind a deuterium nucleus consisting of a proton and a neutron. This reaction is the source of the neutrinos radiated by the sun.

Within a few seconds of its creation, the deuterium nucleus absorbs another proton to become a helium 3 nucleus. Since helium 3 nuclei in the sun are quite rare, it is on the average several million years before the helium 3 nucleus collides with another helium 3 nucleus. The result of this collision is the very stable helium 4 nucleus and the ejection of 2 protons. The net result of all these steps is the conversion of 4 protons into a helium 4 nucleus with the release of .6% of the proton's rest mass energy in the form of neutrinos and photons.

Not only did Beta's theory provide an explanation for the source of the sun's energy, it also demonstrated how elements can be created inside of a star. It raised the question of whether all the elements could be created inside stars. Could you start with stars initially containing only hydrogen gas and end up with all the elements we see around us?

Abundance of the Elements

From studies of minerals in the earth and in meteorites, and as a result of astronomical observations, we know considerable detail about the abundances of the elements around us. As seen in the chart of Figure (9), hydrogen and helium are the most abundant elements, followed by peaks at carbon, oxygen, iron and lead. There is a noticeable lack of lithium, beryllium and boron, and a general trailing off of the heavier elements. Is it possible to explain not only how elements could be created in stars, but also explain these observed abundances as being the natural result of the nuclear reactions inside stars?

The first problem is the fact that there are no stable nuclei with 5 or 8 nucleons. This means you cannot form a stable nucleus either by adding one proton to a helium 4 nucleus or fusing two helium 4 nuclei. How, then, would the next heavier element be formed in a star that consisted of only hydrogen and helium 4? The

answer was supplied by E. E. Salpeter in 1952 who showed that two helium 4 nuclei could produce an unstable beryllium 8 nucleus. In a dense helium rich stellar core, the beryllium 8 nucleus could, before decaying, collide with another helium 4 nucleus forming a stable carbon 12 nucleus. One result is that elements between helium and carbon are skipped over in the element formation process, explaining the exceptionally low cosmic abundances of lithium, beryllium and boron seen in Figure (10).

The biggest barrier to explaining element formation in stars is the fact that the iron 56 nucleus is the most stable of all nuclei. Energy is released if the small nuclei fuse together to create larger ones, but energy is also released if the very largest nuclei split up (as in the case of the fission of uranium in an atomic bomb or nuclear reactor). To put it another way, energy is released making nuclei up to iron, but it costs energy to build nuclei larger than iron. Iron is the ultimate ash of nuclear reactions. How then could elements heavier than iron be created in the nuclear furnaces of stars?

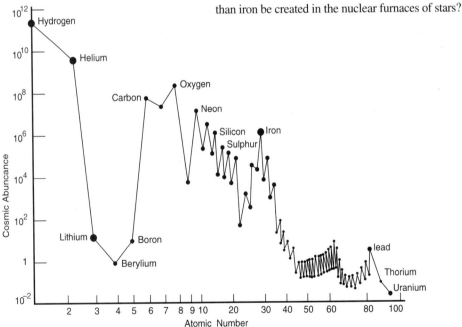

Figure 10
Abundance of the elements

In 1956 the element technetium 99 was identified in the spectra of a certain class of stars. Technetium 99, heavier than iron, is an unstable element with a half life of only two hundred thousand years. On a cosmic time scale, this element had to have been made quite recently. Thus elements heavier than iron are now being created in some kind of a process.

Soon after the observation of technetium, the British astronomer Geoffrey Burbidge, looking over recently declassified data from the Bikini Atoll hydrogen bomb tests, noticed that one of the elements created in the explosion, californium 254, had a half life of 55 days. Burbidge realized that this was also the half life of the intensity of a recently observed supernova explosion. This suggested that the light from the supernova was powered by decaying californium 254. That meant that it was the supernova explosion itself that created the very heavy californium 254, and probably all the other elements heavier than iron.

In 1957 Geoffrey and Margaret Burbidge, along with the nuclear physicist William Fowler at Caltech and the British astronomer Fred Hoyle, published a famous paper showing how the fusion process in stars could explain the abundances of elements up to iron, and how supernova explosions could explain the formation of elements heavier than iron. This was one of the important steps in the use of our knowledge of the behavior of matter on a small scale, namely nuclear physics, to explain what we see on a large scale—the cosmic abundance of the elements.

The Steady State
Model of the Universe

A model of the universe, proposed in 1948 by Fred Hoyle, Herman Bondi and Thomas Gold, fit very well with the idea that all the elements in the universe heavier than hydrogen, were created as a result of nuclear reactions inside stars. This was the so-called *steady state* model.

Knowing that the universe is expanding, it seems to be a contradiction to propose that the universe is steady state—i.e., that on the average, it is unchanging. If the universe is expanding and galaxies are flying apart, then in a few billion years the galaxies will be farther separated from each other than they are today. This is hardly a steady state picture.

The steady state theory got around this problem by proposing that matter was continually being created to replace that being lost due to the expansion. Consider, for example, a sphere a billion light years in diameter, centered on the earth. Over the next million years a certain number of stars will leave the sphere due to the expansion. To replace this matter flowing out of the sphere, the steady state theory assumed that hydrogen atoms were continually being created inside the sphere. All that was needed was about one hydrogen atom to be created in each cubic kilometer of space every year.

The advantage of constructing a model like the steady state theory is that the model makes certain definite predictions that can be tested. One prediction is that all the matter around us originated in the form of the hydrogen atoms that are assumed to be continually created. This implies that the heavier elements we see around us must be created by ongoing processes such as nuclear reactions inside stars. This provided a strong incentive for Hoyle and others to see if nuclear synthesis inside stars, starting from hydrogen, could explain the observed abundance of elements.

Another prediction of the steady state model is that galaxies far away must look much like nearby galaxies. When you look far away, you are also looking back in time. If you look at a galaxy one billion light years away, you are seeing light that started out a billion years ago. Light reaching us from a galaxy 10 billion light years away started out 10 billion light years ago. If the universe is really in a steady state, then galaxies 10 billion years ago should look much like galaxies do today.

THE BIG BANG MODEL

The discovery of the expansion of the universe suggests another model of the universe, namely that the universe started in one gigantic explosion, and that the expansion we now see is the result of the pieces from that explosion flying apart.

To see why you are led to the idea of an explosion, imagine that you take a motion picture of the expanding universe and then run the motion picture backwards. If the expansion is uniform, then in the reversed motion picture we see a uniform contraction. The particles in this picture are the galaxies which are getting closer and closer together. There is a time, call it $t = 0$, when all the galaxies come together at a point. Now run the motion picture forward and the galaxies all move out as if there were an explosion at that point.

The explosion of the universe was first proposed by the Belgian priest and mathematician Georges Lemaître in the late 1920s. It was, in fact, Lemaître who explained Hubble's red shift versus distance data as evidence for the expansion of the universe. In the late 1920s not much was known about nuclear physics, even the neutron had not yet been discovered. But in the 1940s after the development of the atomic fission bomb and during the design of the hydrogen fusion bomb, physicists gained considerable experience with nuclear reactions in hot, dense media, and some, George Gamov in particular, began to explore the consequences of the idea that the universe started in an initial gigantic explosion.

A rough picture of the early universe in the explosion model can be constructed using the concepts of the Doppler effect and thermal equilibrium. Let us see how this works.

We have seen that the red shift of the spectral lines of light from distant galaxies can be interpreted as being caused by the stretching of the wavelengths of the light due to the expansion of the universe. In a reverse motion picture of the universe, distant galaxies would be coming toward us and the wavelengths of the spectral lines would be blue shifted. We would say that the universe was contracting, shrinking the wavelength of the spectral lines. The amount of contraction would depend upon how far back toward the $t = 0$ origin we went. If we went back to when the universe was 1/10 as big as it is now, wavelengths of light would contract to 1/10 their original size.

In the Einstein photoelectric effect formula, $E_{photon} = hf = hc/\lambda$, the shorter the photon wavelength, the more energetic the photons become. This suggests that as we compress the universe in the time reversal moving picture, photon energies increase. If there is no limit to the compression, then there is no limit to how much the photon energies increase.

Now introduce the idea of thermal equilibrium. If we go back to a very small universe, we have very energetic photons. If these photons are in thermal equilibrium with other forms of matter, as they are inside of stars, then all of the matter has enormous thermal energy, and the temperature is very high. Going back to a zero sized universe means going back to a universe that started out at an infinite temperature. Fred Hoyle thought that this picture was so ridiculous that he gave the explosion model of the universe the derisive name the "Big Bang" model. The name has stuck.

The Helium Abundance

In the mid 1950s the cosmological theory taken seriously by most physicists was the steady state theory. In the late 1940s George Gamov had suggested that the elements had been created in the big bang when the universe was very small, dense and hot. But the work of Hoyle and Fowler was showing that the abundance of the elements could much more satisfactorily be explained in terms of nuclear synthesis inside of stars. This nuclear synthesis also explained the energy source in stars and the various stages of stellar evolution. What need was there to propose some gigantic, cataclysmic explosion?

Hoyle soon found a need. Most of the energy released in nuclear synthesis in stars results from the burning of hydrogen to form helium. By observing how much energy is released by stars, you can estimate how much helium should be produced. By the early 1960s Hoyle began to realize that nuclear synthesis could not produce enough helium to explain the observed cosmic abundance of 25%. In a 1964 paper with R. J. Taylor, Hoyle himself suggested that perhaps much of the helium was created in an initial explosion of the universe.

Cosmic Radiation

In a talk given at Johns Hopkins in early 1965, Princeton theoretician P. J. E. Peebles suggested that the early universe must have contained a considerable amount of radiation if the big bang model were correct. If there were little radiation, any hydrogen present in the early universe would have quickly fused to form heavier elements, and no hydrogen would be left today. This directly contradicts the observation that about 75% of the matter we see today consists of hydrogen. If, however, there were a large amount of radiation present in the early universe, the energetic photons would bust up the larger nuclei as they formed, leaving behind hydrogen.

Peebles proposed that this radiation, the cosmic photons which prevented the fusion of hydrogen in the early universe still exist today but in a very altered form. There should have been little change in the number of photons, but a great change in their energy. As the universe expanded, the wavelength of the cosmic photons should be stretched by the expansion, greatly reducing their energy. If the photons were in thermal equilibrium with very hot matter in the early universe, they should still have a thermal black body spectrum, but at a much lower temperature. He predicted that the temperature of the cosmic radiation should have dropped to around 10 kelvin. His colleagues at Princeton, P. G. Roll and D .T. Wilkinson were constructing a special antenna to detect such radiation. All of this work had been suggested by R. H. Dicke, inventor of the key microwave techniques needed to detect ten degree photons.

Peebles was not the first to suggest that there should be radiation left over from the big bang. That was first suggested in a 1948 paper by George Gamov and colleagues Ralph Alpher and Robert Herman in a model where all elements were to be created in the big bang. A more realistic model of the big bang proposed by Alpher and Herman in 1953 also led to the same prediction of cosmic radiation. In both cases, it was estimated that the thermal radiation should now have a temperature of 5 kelvin. In the early 1950s, Gamov, Alpher and Herman were told by radio astronomers that such radiation could not be detected by equipment then available, and the effort to detect it was not pursued. Peebles was unaware of these earlier predictions.

THE THREE DEGREE RADIATION

In 1964, two radio astronomers working for Bell Labs, Arno Penzias and Robert Wilson, began a study of the radio waves emitted from parts of our galaxy that are away from the galactic plane. They expected a faint diffuse radiation from this part of the galaxy and planned to use a sensitive low noise radio antenna shown in Figure (11), an antenna left over from the *Echo* satellite experiment. (In that early experiment on satellite communication, a reflecting balloon was placed in orbit. The low noise antenna was built to detect the faint radio signals that bounced off the balloon.)

Since the kind of signals Penzias and Wilson expected to detect would look a lot like radio noise, they had to be careful that the signals they recorded were coming from the galaxy rather than from noise generated by the antenna or by electronics. To test the system, they looked for signals at a wavelength of 7.35 cm, a wavelength where the galaxy was not expected to produce much radiation. They found, however, a stronger signal than expected. After removing a pair of pigeons that were living in the antenna throat, cleaning out the nest and other debris which Wilson referred to as "a white dielectric material", and taking other steps to eliminate noise, the extra signal persisted.

Figure 11
Penzias and Wilson, and the Holmdel radio telescope.

If the 7.35 cm wavelength signal were coming from the galaxy, there should be regions of the galaxy that produced a stronger signal than other regions. And the neighboring galaxy Andromeda should also be a localized source of this signal. However Penzias and Wilson found that the 7.35 cm signal was coming in uniformly from all directions. The radiation had to be coming in from a much larger region of space than our galaxy.

Studies of the signal at still shorter wavelengths showed that if the signal were produced by a blackbody spectrum of radiation, the effective temperature would be about 3.5 kelvin. Penzias talked with a colleague who had talked with another colleague who had attended Peebles' talk at Johns Hopkins on the possibility of radiation left over from the big bang. Penzias and Wilson immediately suspected that the signal they were detecting might be from this radiation.

Penzias and Wilson could detect only the long wavelength tail and of the three degree radiation. Three degree radiation should have a maximum intensity at a wavelength given by the Wein formula, Equation 1,

$$
\begin{aligned}
\lambda_{max} &= \frac{2.898 \text{ mm K}}{T} \\
&= \frac{2.898 \text{ mm K}}{3 \text{ K}} \approx 1 \text{ mm}
\end{aligned}
\tag{17}
$$

Radiation with wavelengths in the 1 mm region cannot get through the earth's atmosphere. As a result Penzias and Wilson, and others using ground based antennas, could not verify that the radiation had a complete blackbody spectrum. From 1965 to the late 1980s, various balloon and rocket based experiments, which lifted antennas above the earth's atmosphere, verified that the radiation detected by Penzias and Wilson was part of a complete blackbody spectrum of radiation at a temperature of 2.74 kelvin.

In 1989, NASA orbited the **COBE** (Cosmic Background Explorer) satellite to make a detailed study of the cosmic background radiation. The results from this satellite verified that this radiation has the most perfect blackbody spectrum ever seen by mankind. The temperature is 2.735 kelvin with variations of the order of one part in 100,000. The questions we have to deal with now are not whether there is light left over from the big bang, but why it is such a nearly perfect blackbody spectrum.

Thermal Equilibrium of the Universe

That the cosmic background radiation has nearly a perfect blackbody spectrum tells us that at some point in its history, the universe was in nearly perfect thermal equilibrium, with everything at one uniform temperature. That is certainly not the case today. The cosmic radiation is at a temperature of 2.735 kelvin, Hawaii has an average temperature of 295 kelvin, and the temperature inside of stars ranges up to billions of degrees. There must have been a dramatic change in the nature of the universe sometime in the past. That change occurred when the universe suddenly became transparent at an age of about 700,000 years.

To see why the universe suddenly became transparent, and why this was such an important event, it is instructive to reconstruct what the universe must have been like at still earlier times.

THE EARLY UNIVERSE

Imagine that we have a videotape recording of the evolution of the universe. We put the tape in our VCR and see that the tape has not been rewound. It is showing our current universe with stars, galaxies and the cosmic radiation at a temperature of 2.735 k. You can calculate the density of photons in the cosmic radiation, and compare that with the average density of protons and neutrons (nucleons) in the stars and galaxies. You find that the photons outnumber the nucleons by a factor of about 10 billion to 1. Although there are many more photons than nucleons, the rest energy of a proton or neutron is so much greater than the energy of a three degree photon that the total rest energy of the stars and galaxies is about 100 times greater than the total energy in the cosmic radiation.

Leaving the VCR on play, we press the rewind button. The picture is not too clear, but we can see general features of the contracting universe. The galaxies are moving together and the wavelength of the cosmic radiation is shrinking. Since the energy of the cosmic photons is given by Einstein's formula $E = hc/\lambda$, the shrinking of the photon wavelengths increases their energy. On the other hand the rest mass energy of the stars and galaxies is essentially unaffected by the contraction of the universe. As a result the energy of the cosmic photons is becoming a greater and greater share of the total energy of the universe. When the universe has contracted to about 1/100th of its present size, when the universe is about 1/2 million years old, the cosmic photons have caught up to the matter particles. At earlier times, the cosmic photons have more energy than other forms of matter.

The Early Universe

As the tape rewinds our attention is diverted. When we look again at the screen, we see that the tape is showing a very early universe. The time indicated is .01 seconds! The temperature has risen to 100 billion degrees, and the thermal photons have an average energy of 40 million electron volts! We obviously missed a lot in the rewind. Stopping the tape, we then run it forward to see what the universe looks like at this very early stage.

There is essentially the same number of nucleons in this early universe as there are today. Since the thermal energy of 40 MeV is much greater than the 1.3 MeV mass difference between neutrons and protons, there is enough thermal energy to freely convert protons into neutrons, and vice versa. As a result there are about equal numbers of protons and neutrons. There is also about the same number of thermal photons in this early universe as there are today, about 100 billion photons for each nucleon.

While there is not much change from today in the number of nucleons or photons in our .01 second universe, there is a vastly different number of electrons. The thermal photons, with an average energy of 40 MeV, can freely create positron and electron pairs. The rest energy of a positron or an electron is only .5 MeV, thus only 1 MeV is required to create a pair. The result is that the universe at this time is a thermal soup of photons, positrons and electrons—about equal numbers. There are also many neutrinos left over from an earlier time. All of those species outnumber the few nucleons by a factor of about 100 billion to one.

Excess of Matter over Antimatter

If you look closely and patiently count the number of positrons and electrons in some region of space, you will find that for every 100,000,000,000 positrons, there are 100,000,000,001 electrons. The electrons outnumber the positrons by 1 in 100 billion. In fact, the excess number of negative electrons is just equal to the number of positive protons, with the result that the universe is electrically neutral.

The tiny excess of electrons over positrons represents an excess of matter over antimatter. In most particle reactions we study today, if particles are created, they are created in particle, antiparticle pairs. The question is then, why does this early universe have a tiny excess of matter particles over antimatter particles? What in the still earlier universe created this tiny imbalance? There is a particle reaction, caused by the weak interaction, that does not treat matter and antimatter symmetrically. This reaction, discovered by Val Fitch in 1964, could possibly explain how this tiny imbalance came about. It is not clear whether there was enough time in the very early universe for Fitch's reaction to create the observed imbalance.

An excellent guidebook for our video tape is Steven Weinberg's *The First Three Minutes*. Weinberg was one of the physicists who discovered the connection between the weak interaction and electromagnetism. Weinberg breaks up the first three minutes of the life of the universe into five frames. We happened to have stopped the tape recording at Weinberg's frame #1. To see what we missed in our fast rewind, we will now run the tape forward, picking up the other four frames in the first three minutes as well as important later events.

Frame #2 (.11 seconds)

As we run the tape forward, the universe is now expanding, the wavelength of the thermal photons is getting longer, and their temperature is dropping. When the time counter gets up to t = .11 seconds, the temperature has dropped to 30 billion kelvin and the average energy of the thermal photons has dropped to 10 MeV. Back at frame #1, when the thermal energy was 40 MeV, there were roughly equal numbers of protons and neutrons. However, the lower thermal energy of 10 MeV is not sufficiently greater than the 1.3 MeV proton-neutron mass difference to maintain the equality.

In the many rapid collisions where protons are being converted into neutrons and vice versa (via the weak interaction), there is a slightly greater chance that the heavier neutron will decay into a lighter proton rather than the other way around. As a result the percentage of neutrons has dropped to 38% by the time t = .11 seconds.

Frame #3 (1.09 seconds)

Aside from the drop in temperature and slight decrease in the percentage of neutrons, not much else happened as we went from frame #1 at .01 seconds to frame #2 at .11 seconds. Starting up the tape player again, we go forward to t = 1.09 seconds, Weinberg's third frame. The temperature has dropped to 10 billion kelvin, which corresponds to a thermal energy of 4 MeV. This is not too far above the 1 MeV threshold for creating positron electron pairs. As a result the positron electron pairs are beginning to annihilate faster than they are being created. Also by this time the percentage of neutrons has dropped to 24%.

Frame #4 (13.82 seconds)

At a time of 13.82 seconds, Weinberg's fourth frame, the temperature has dropped to 3 billion kelvin, corresponding to an average thermal energy of 1 MeV per particle. With any further drop in temperature, the average thermal photon will not have enough energy to create positron electron pairs. The result is that vast numbers of positrons and electrons are beginning to annihilate each other. Soon there will be equal numbers of electrons and protons, and the only particles remaining in very large numbers will be neutrinos and thermal photons.

By this fourth frame, the percentage of neutrons has dropped to 17%. The temperature of 3 billion degrees is low enough for helium nuclei to survive, but helium nuclei do not form because of the deuterium bottleneck. When a proton and neutron collide, they can easily form a deuterium nucleus. Although deuterium is stable, it is weakly bound. At a temperature of 3 billion kelvin, the thermal protons quickly break up any deuterium that forms. Without deuterium, it is not possible to build up still larger nuclei.

Frame #5 (3 minutes and 2 seconds)

Going forward to a time of 3 minutes and 2 seconds, the universe has cooled to a billion kelvin, the positrons and most electrons have disappeared, and the only abundant particles are photons, neutrinos and antineutrinos. The neutron proton balance has dropped to 14% neutrons. While tritium (one proton and two neutrons) and helium 4 are stable at this temperature, deuterium is not, thus no heavier nuclei can form.

A short time later, the temperature drops to the point where deuterium is stable. When this happens, neutrons can combine with protons to form deuterium and tritium, and these then combine to form helium 4. Almost immediately the remaining nearly 13% neutrons combine with an equal number of protons to form most of the 25% abundance of cosmic helium we see today. This is where the helium came from that Hoyle could not explain in terms of nuclear synthesis inside of stars.

Because there are no stable nuclei with 5 or 8 nucleons, there is no simple route to the formation of still heavier elements. At a temperature of a billion degrees, the universe is only about 70 times hotter than the core of our sun, cooler than the core of hot stars around today that are fusing the heavier elements. As a result, nuclear synthesis in the early universe stops at helium 4 with a trace of lithium 7. One of the best tests of the big bang theory is a rather precise prediction of the relative abundances of hydrogen, deuterium, helium 4 and lithium 7, all left over from the early universe. When the formation of these elements is complete, the universe is 3 minutes and 46 seconds old.

Decoupling (700,000 years)

Continue running the tape forward, and nothing of much interest happens for a long time. The thermal photons still outnumber the nucleons and electrons by a factor of about 10 billion to one, and the constant collisions between these particles prevent the formation of atoms. What we see is a hot, ionized, nearly uniform plasma consisting of photons, charged nuclei and separate electrons. As time goes on, the plasma is expanding and cooling.

When you look at the sun, you see a round ball with an apparently sharp edge. But the sun is not a solid object with a well defined surface. Instead, it is a bag of mostly hydrogen gas held together by gravity. It is hottest at the center and cools off as you go out from the center. At what appears to us to be the surface, the temperature has dropped to about 3,000 kelvin.

At a temperature above 3,000 kelvin, hydrogen gas becomes ionized, a state where an appreciable fraction of the electrons are torn free from the proton nuclei. When the gas is ionized, it is opaque because photons can interact directly with the free charges present in the gas. Below a temperature of 3,000 kelvin, hydrogen consists essentially of neutral atoms which are unaffected by visible light. As a result the cooler hydrogen gas is transparent. The apparent surface of the sun marks the abrupt transition from an opaque plasma, at temperatures above 3,000K, to a transparent gas at temperatures below 3,000K.

A similar transition takes place in the early universe. By the age of about 700,000 years, the universe cools to a temperature of 3,000K. Before that the universe is an opaque plasma like the inside of the sun. The photons in thermal equilibrium with the matter particles have enough energy to bust up any complete atoms and any gravitational clumps that are trying to form.

When the universe drops to a temperature below 3,000 kelvin, the hydrogen gas forms neutral atoms and becomes transparent. (The 25% helium had already become neutral some time earlier). As a result the universe suddenly becomes transparent, and the thermal photons decouple from matter.

From this decoupling on, there is essentially no interaction between the thermal photons and any form of matter. All that happens to the photons is that their wavelength is stretched by the expansion of the universe. This stretching preserves the blackbody spectrum of the photons while lowering the effective blackbody temperature. This blackbody spectrum is now at the temperature of 2.735K, as observed by the COBE satellite.

When the matter particles are decoupled, freed from the constant bombardment of the cosmic photons, gravity can begin the work of clumping up matter to form stars, globular clusters, black holes and galaxies. All these structures start to form after the decoupling, after the universe is 700,000 years old. It is this formation of stars and galaxies that we see as we run the tape forward to our present day.

Looking out with ever more powerful telescopes is essentially equivalent to running our videotape backwards. The farther out we look, the farther back in time we see. Images from the Hubble telescope are giving us a view back toward the early universe when galaxies were very much younger and quite different than they are today. The most distant galaxy we have identified so far emitted light when the universe was 5% of its current size.

What happens when we build still more powerful telescopes and look still farther back? When we look out so far that the universe is only 700,000 years old, we are looking at the universe that has just become transparent. *We can see no farther!* To look farther is like trying to look down inside the surface of the sun.

In fact we do not need a more powerful telescope to see this far back. The three degree cosmic background radiation gives us a fantastically clear, detailed photograph of the universe at the instant it went transparent.

The horn antenna used by Penzias and Wilson was the first device to look at a small piece of this photograph. The COBE satellite looked at the whole photograph, but with rather limited resolution. COBE detected some very tiny lumpiness, temperature variations of about one part in 100,000. This lumpiness may have been what gravity needed to start forming galaxies. A higher resolution photograph will be needed to tell for sure.

Guidebooks

We ran the videotape quite rapidly without looking at many details. Our focus has been on the formation of the elements and the three degree radiation, two of the main pieces of evidence for the existence of a big bang. We have omitted a number of fascinating details such as how dense was the early universe, when did the neutrinos decouple from matter, and what happened before the first frame? There are excellent guidebooks that accompany this tape where you can find these details. There is Weinberg's *The First Three Minutes* which we have mentioned. The 1993 edition has an addendum that introduces some ideas about the very, very early universe, when the universe was millions of times younger and hotter than the first frame. Perhaps the best guidebook to how mankind came to our current picture of the universe is the book by Timothy Ferris *Coming of Age in the Milky Way*. Despite the title, this is one of the most fascinating and readable accounts available. In our discussion we have drawn much from Weinberg and Ferris.

Chapter 35
Bohr Theory of Hydrogen

The hydrogen atom played a special role in the history of physics by providing the key that unlocked the new mechanics that replaced Newtonian mechanics. It started with Johann Balmer's discovery in 1884 of a mathematical formula for the wavelengths of some of the spectral lines emitted by hydrogen. The simplicity of the formula suggested that some understandable mechanisms were producing these lines.

The next step was Rutherford's discovery of the atomic nucleus in 1912. After that, one knew the basic structure of atoms—a positive nucleus surrounded by negative electrons. Within a year Neils Bohr had a model of the hydrogen atom that "explained" the spectral lines. Bohr introduced a new concept, the energy level. The electron in hydrogen had certain allowed energy levels, and the sharp spectral lines were emitted when the electron jumped from one energy level to another. To explain the energy levels, Bohr developed a model in which the electron had certain allowed orbits and the jump between energy levels corresponded to the electron moving from one allowed orbit to another.

Bohr's allowed orbits followed from Newtonian mechanics and the Coulomb force law, with one small but crucial modification of Newtonian mechanics. The angular momentum of the electron could not vary continuously, it had to have special values, be quantized in units of Planck's constant divided by 2π, $h/2\pi$. In Bohr's theory, the different allowed orbits corresponded to orbits with different allowed values of angular momentum.

Again we see Planck's constant appearing at just the point where Newtonian mechanics is breaking down. There is no way one can explain from Newtonian mechanics why the electrons in the hydrogen atom could have only specific quantized values of angular momentum. While Bohr's model of hydrogen represented only a slight modification of Newtonian mechanics, it represented a major philosophical shift. Newtonian mechanics could no longer be considered the basic theory governing the behavior of particles and matter. Something had to replace Newtonian mechanics, but from the time of Bohr's theory in 1913 until 1924, no one knew what the new theory would be.

In 1924, a French graduate student, Louis de Broglie, made a crucial suggestion that was the key that led to the new mechanics. This suggestion was quickly followed up by Schrödinger and Heisenberg who developed the new mechanics called **quantum mechanics**. In this chapter our focus will be on the developments leading to de Broglie's idea.

THE CLASSICAL HYDROGEN ATOM

With Rutherford's discovery of the atomic nucleus, it became clear that atoms consisted of a positively charged nucleus surrounded by negatively charged electrons that were held to the nucleus by an electric force. The simplest atom would be hydrogen consisting of one proton and one electron held together by a Coulomb force of magnitude

$$\left|\vec{F}_e\right| = \frac{e^2}{r^2} \qquad (1)$$

(For simplicity we will use CGS units in describing the hydrogen atom. We do not need the engineering units, and we avoid the complicating factor of $1/4\pi\varepsilon_0$ in the electric force formula.) As shown in Equation 1, both the proton and the electron attract each other, but since the proton is 1836 times more massive than the electron, the proton should sit nearly at rest while the electron orbits around it.

Thus the hydrogen atom is such a simple system, with known masses and known forces, that it should be a straightforward matter to make detailed predictions about the nature of the atom. We could use the orbit program of Chapter 8, replacing the gravitational force GMm/r^2 by e^2/r^2. We would predict that the electron moved in an elliptical orbit about the proton, obeying all of Kepler's laws for orbital motion.

There is one important point we would have to take into account in our analysis of the hydrogen atom that we did not have to worry about in our study of satellite motion. The electron is a charged particle, and accelerated charged particles radiate electromagnetic waves. Suppose, for example, that the electron were in a circular orbit moving at an angular velocity ω as shown in Figure (1a). If we were looking at the orbit from the side, as shown in Figure (1b), we would see an electron oscillating up and down with a velocity given by $v = v_0 \sin \omega t$.

In our discussion of radio antennas in Chapter 32, we saw that radio waves could be produced by moving electrons up and down in an antenna wire. If electrons oscillated up and down at a frequency ω, they produced radio waves of the same frequency. Thus it is a prediction of Maxwell's equations that the electron in the hydrogen atom should emit electromagnetic radiation, and the frequency of the radiation should be the frequency at which the electron orbits the proton.

For an electron in a circular orbit, predicting the motion is quite easy. If an electron is in an orbit of radius r, moving at a speed v, then its acceleration $\vec{a}$ is directed toward the center of the circle and has a magnitude

$$\vec{a} = \frac{v^2}{r} \qquad (2)$$

Using Equation 1 for the electric force and Equation 2 for the acceleration, and noting that the force is in the same direction as the acceleration, as indicated in Figure (2), Newton's second law gives

$$\left|\vec{F}\right| = m\left|\vec{a}\right|$$

$$\frac{e^2}{r^2} = m\frac{v^2}{r} \qquad (3)$$

One factor of r cancels and we can immediately solve for the electron's speed v to get $v^2 = e^2/mr$, or

$$v_{electron} = \frac{e}{\sqrt{mr}} \qquad (4)$$

The period of the electron's orbit should be the distance $2\pi r$ travelled, divided by the speed v, or $2\pi r/v$ seconds per cycle, and the frequency should be the inverse of that, or $v/2\pi r$ cycles per second. Using Equation 4 for v, we get

$$\frac{\text{frequency of}}{\text{electron in orbit}} = \frac{v}{2\pi r} = \frac{e}{2\pi r\sqrt{mr}} \qquad (5)$$

According to Maxwell's theory, this should also be the frequency of the radiation emitted by the electron.

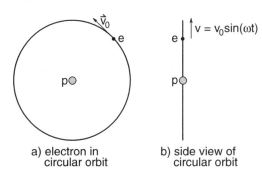

a) electron in circular orbit

b) side view of circular orbit

Figure 1
The side view of circular motion is an up and down oscillation.

Electromagnetic radiation carries energy. Thus, to see what effect this has on the electron's orbit, let us look at the formula for the energy of an orbiting electron.

From Equation 3 we can immediately solve for the electron's kinetic energy. The result is

$$\frac{1}{2}mv^2 = \frac{e^2}{2r} \quad \begin{array}{l} electron \\ kinetic \\ energy \end{array} \qquad (6)$$

The electron also has electric potential energy just as an earth satellite had gravitational potential energy. The formula for the gravitational potential energy of a satellite was

$$\begin{array}{l} potential\ energy \\ of\ an\ earth\ satellite \end{array} = -\frac{GMm}{r} \qquad (10\text{-}50a)$$

where M and m are the masses of the earth and the satellite respectively. This is the result we used in Chapter 8 to test for conservation of energy (Equations 8-29 and 8-31) and in Chapter 10 where we calculated the potential energy (Equations 10-50a and 10-51). The minus sign indicated that the gravitational force is attractive, that the satellite starts with zero potential energy when r = ∞ and loses potential energy as it falls in toward the earth.

We can convert the formula for gravitational potential energy to a formula for electrical potential energy by comparing formulas for the gravitational and electric forces on the two orbiting objects. The forces are

$$\left|\vec{F}_{gravity}\right| = \frac{GMm}{r^2} \quad ; \quad \left|\vec{F}_{electric}\right| = \frac{e^2}{r^2}$$

Since both are $1/r^2$ forces, we can go from the gravitational to the electric force formula by replacing the

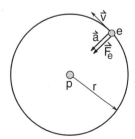

Figure 2
For a circular orbit, both the acceleration $\vec{a}$ and the force $\vec{F}$ point toward the center of the circle. Thus we can equate the magnitudes of F and ma.

constant GMm by e^2. Making this same substitution in the potential energy formula gives

$$PE = \frac{-e^2}{r} \quad \begin{array}{l} electrical\ potential\ energy \\ of\ the\ electron\ in\ the \\ hydrogen\ atom \end{array} \qquad (7)$$

Again the potential energy is zero when the particles are infinitely far apart, and the electron loses potential energy as it falls toward the proton. (We used this result in the analysis of the binding energy of the hydrogen molecule ion, explicitly in Equation 18-15.)

The formula for the total energy E_{total} of the electron in hydrogen should be the sum of the kinetic energy, Equation 6, and the potential energy, Equation 7.

$$E_{total} = \begin{array}{c} kinetic \\ energy \end{array} + \begin{array}{c} potential \\ energy \end{array}$$

$$= \frac{e^2}{2r} \qquad - \frac{e^2}{r}$$

$$\boxed{E_{total} = \frac{-e^2}{2r}} \quad \begin{array}{l} total\ energy \\ of\ electron \end{array} \qquad (8)$$

The significance of the minus (–) sign is that the electron is bound. Energy is required to pull the electron out, to ionize the atom. For an electron to escape, its total energy must be brought up to zero.

We are now ready to look at the predictions that follow from Equations 5 and 8. As the electron radiates light it must lose energy and its total energy must become more negative. From Equation 8 we see that for the electron's energy to become more negative, the radius r must become smaller. Then Equation 5 tells us that as the radius becomes smaller, the frequency of the radiation increases. We are lead to the picture of the electron spiraling in toward the proton, radiating even higher frequency light. There is nothing to stop the process until the electron crashes into the proton. It is an unambiguous prediction of Newtonian mechanics and Maxwell's equations that the hydrogen atom is unstable. It should emit a continuously increasing frequency of light until it collapses.

Energy Levels

By 1913, when Neils Bohr was trying to understand the behavior of the electron in hydrogen, it was no surprise that Maxwell's equations did not work at an atomic scale. To explain blackbody radiation and the photoelectric effect, Planck and Einstein were led to the picture that light consists of photons rather than Maxwell's waves of electric and magnetic force.

To construct a theory of hydrogen, Bohr knew the following fact. Hydrogen gas at room temperature emits no light. To get radiation, it has to be heated to rather high temperatures. Then you get distinct spectral lines rather than the continuous radiation spectrum expected classically. The visible spectral lines are the H_α, H_β and H_γ lines we saw in the hydrogen spectrum experiment. These and many infra red lines we saw in the spectrum of the hydrogen star, Figure (33-28) reproduced below, make up the Balmer series of lines. Something must be going on inside the hydrogen atom to produce these sharp spectral lines.

Viewing the light radiated by hydrogen in terms of Einstein's photon picture, we see that the hydrogen atom emits photons with certain precise energies. As an exercise in the last chapter you were asked to calculate, in eV, the energies of the photons in the H_α, H_β and H_γ spectral lines. The answers are

$$E_{H_\alpha} = 1.89 \text{ eV}$$

$$E_{H_\beta} = 2.55 \text{ eV}$$

$$E_{H_\gamma} = 2.86 \text{ eV} \qquad (9)$$

The question is, why does the electron in hydrogen emit only certain energy photons? The answer is Bohr's main contribution to physics. Bohr assumed that the electron had, for some reason, only certain allowed energies in the hydrogen atom. He called these *allowed energy levels*. When an electron jumped from one energy level to another, it emitted a photon whose energy was equal to the difference in the energy of the two levels. The red 1.89 eV photon, for example, was radiated when the electron fell from one energy level to another level 1.89 eV lower. There was a bottom, lowest energy level below which the electron could not fall. In cold hydrogen, all the electrons were in the bottom energy level and therefore emitted no light.

When the hydrogen atom is viewed in terms of Bohr's energy levels, the whole picture becomes extremely simple. The lowest energy level is at -13.6 eV. This is the total energy of the electron in any cold hydrogen atom. It requires 13.6 eV to ionize hydrogen to rip an electron out.

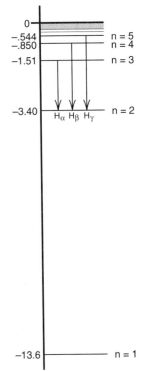

Figure 3
Energy level diagram for the hydrogen atom. All the energy levels are given by the simple formula $E_n = -13.6/n^2 \, eV$. All Balmer series lines result from jumps down to the n = 2 level. The 3 jumps shown give rise to the three visible hydrogen lines.

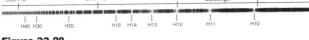

Figure 33-28
Spectrum of a hydrogen star

The first energy level above the bottom is at –3.40 eV which turns out to be (–13.6/4) eV. The next level is at –1.51 eV which is (–13.6/9) eV. All of the energy levels needed to explain every spectral line emitted by hydrogen are given by the formula

$$E_n = \frac{-13.6 \text{ eV}}{n^2} \tag{10}$$

where n takes on the integer values 1, 2, 3, ... These energy levels are shown in Figure (3).

Exercise 1

Use Equation 10 to calculate the lowest 5 energy levels and compare your answer with Figure 3.

Let us see explicitly how Bohr's energy level diagram explains the spectrum of light emitted by hydrogen. If, for example, an electron fell from the n=3 to the n=2 level, the amount of energy E_{3-2} it would lose and therefore the energy it would radiate would be

$$\begin{aligned} E_{3-2} &= E_3 - E_2 \\ &= -1.51 \text{ eV} - (-3.40 \text{ eV}) \\ &= 1.89 \text{ eV} \end{aligned} \tag{11}$$

$$= \begin{array}{l} \text{energy lost in falling} \\ \text{from n = 3 to n = 2 level} \end{array}$$

which is the energy of the red photons in the H_α line.

Exercise 2

Show that the H_β and H_γ lines correspond to jumps to the n = 2 level from the n = 4 and the n = 5 levels respectively.

From Exercise 2 we see that the first three lines in the Balmer series result from the electron falling from the third, fourth and fifth levels down to the second level, as indicated by the arrows in Figure (3).

All of the lines in the Balmer result from jumps down to the second energy level. For historical interest, let us see how Balmer's formula for the wavelengths in this series follows from Bohr's formula for the energy levels. For Balmer's formula, the lines we have been calling H_α, H_β and H_γ are H_3, H_4, H_5. An arbitrary line in the series is denoted by H_n, where n takes on the values starting from 3 on up. The Balmer formula for the wavelength of the H_n line is from Equation 33-6

$$\lambda_n = 3.65 \times 10^{-5} \text{cm} \times \left(\frac{n^2}{n^2 - 4} \right) \tag{33-6}$$

Referring to Bohr's energy level diagram in Figure (3), consider a drop from the nth energy level to the second. The energy lost by the electron is $(E_n - E_2)$ which has the value

$$E_n - E_2 = \frac{13.6 \text{ eV}}{n^2} - \frac{13.6 \text{ eV}}{2^2} \quad \begin{array}{l} \textit{energy lost by} \\ \textit{electron going} \\ \textit{from nth to} \\ \textit{second level} \end{array}$$

This must be the energy $E(H_n)$ carried out by the photon in the H_n spectral line. Thus

$$E(H_n) = 13.6 \text{ eV} \left(\frac{1}{4} - \frac{1}{n^2} \right)$$

$$= 13.6 \text{ eV} \left(\frac{n^2 - 4}{4n^2} \right) \tag{12}$$

We now use the formula

$$\lambda = \frac{12.4 \times 10^{-5} \text{cm} \cdot \text{eV}}{E_{photon} (\text{in eV})} \tag{34-8}$$

relating the photon's energy to its wavelength. Using Equation 12 for the photon energy gives

$$\lambda_n = \frac{12.4 \times 10^{-5} \text{cm} \cdot \text{eV}}{13.6 \text{ eV}} \left(\frac{4n^2}{n^2 - 4} \right)$$

$$\lambda_n = 3.65 \times 10^{-5} \text{cm} \left(\frac{n^2}{n^2 - 4} \right)$$

which is Balmer's formula.

It does not take great intuition to suspect that there are other series of spectral lines beyond the Balmer series. The photons emitted when the electron falls down to the lowest level, down to -13.6 eV as indicated in Figure (4), form what is called the *Lyman series*. In this series the least energy photon, resulting from a fall from -3.40 eV down to -13.6 eV, has an energy of 10.2 eV, well out in the ultraviolet part of the spectrum. All the other photons in the Lyman series have more energy, and therefore are farther out in the ultraviolet.

It is interesting to note that when you heat hydrogen and see a Balmer series photon like H_α, H_β or H_γ, eventually a 10.2 eV Lyman series photon must be emitted before the hydrogen can get back down to its ground state. With telescopes on earth we see many hydrogen stars radiating Balmer series lines. We do not see the Lyman series lines because these ultraviolet photons do not make it down through the earth's atmosphere. But the Lyman series lines are all visible using orbiting telescopes like the Ultraviolet Explorer and the Hubble telescope.

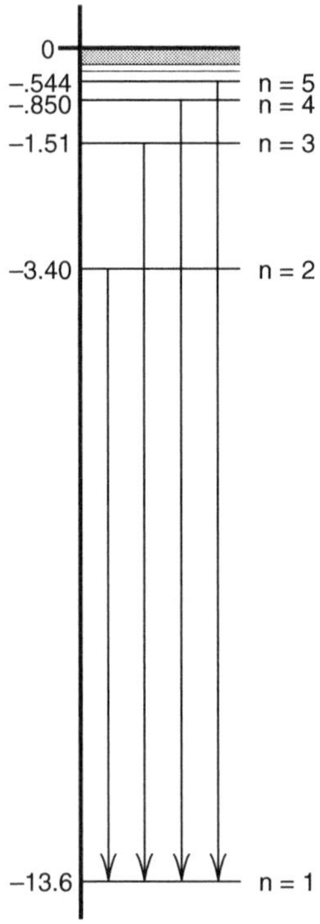

Another series, all of whose lines lie in the infra red, is the Paschen series, representing jumps down to the n = 3 energy level at -1.55 eV, as indicated in Figure (5). There are other infra red series, representing jumps down to the n = 4 level, n = 5 level, etc. There are many series, each containing many spectral lines. And all these lines are explained by Bohr's conjecture that the hydrogen atom has certain allowed energy levels, all given by the simple formula $E_n = (-13.6/n^2)\,eV$. This one simple formula explains a huge amount of experimental data on the spectrum of hydrogen.

Exercise 3

Calculate the energies (in eV) and wavelengths of the 5 longest wavelength lines in

(a) the Lyman series

(b) the Paschen series

On a Bohr energy level diagram show the electron jumps corresponding to each line.

Exercise 4

In Figure (33-28), repeated 2 pages back, we showed the spectrum of light emitted by a hydrogen star. The lines get closer and closer together as we get to H_{40} and just beyond. Explain why the lines get closer together and calculate the limiting wavelength.

Figure 4
The Lyman series consists of all jumps down to the –13.6 eV level. (Since this is as far down as the electron can go, this level is called the "ground state".)

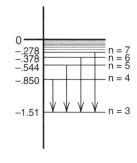

Figure 5
The Paschen series consists of all jumps down to the n = 3 level. These are all in the infra red.

THE BOHR MODEL

Where do Bohr's energy levels come from? Certainly not from Newtonian mechanics. There is no excuse in Newtonian mechanics for a set of allowed energy levels. But did Newtonian mechanics have to be rejected altogether? Planck was able to explain the blackbody radiation formula by patching up classical physics, by assuming that, for some reason, light was emitted and absorbed in quanta whose energy was proportional to the light's frequency. The reason why Planck's trick worked was understood later, with Einstein's proposal that light actually consisted of particles whose energy was proportional to frequency. Blackbody radiation had to be emitted and absorbed in quanta because light itself was made up of these quanta.

By 1913 it had become respectable, frustrating perhaps, but respectable to modify classical physics in order to explain atomic phenomena. The hope was that a deeper theory would come along and naturally explain the modifications.

What kind of a theory do we construct to explain the allowed energy levels in hydrogen? In the classical picture we have a miniature solar system with the proton at the center and the electron in orbit. This can be simplified by restricting the discussion to circular orbits. From our earlier work with the classical model of hydrogen, we saw that an electron in an orbit of radius r had a total energy $E(r)$ given by

$$E(r) = \frac{-e^2}{2r} \quad \begin{array}{l} \textit{total energy of} \\ \textit{an electron in} \\ \textit{a circular orbit} \\ \textit{of radius r} \end{array} \quad \text{(8 repeated)}$$

If the electron can have only certain allowed energies $E_n = -13.6/n^2$ eV, then if Equation (8) holds, the electron orbits can have only certain allowed orbits of radius r_n given by

$$E_n = \frac{-e^2}{2r_n} \quad \text{(13)}$$

The r_n are the radii of the famous Bohr orbits. This leads to the rather peculiar picture that the electron can exist in only certain allowed orbits, and when the electron jumps from one allowed orbit to another, it emits a photon whose energy is equal to the difference in energy between the two orbits. This model is indicated schematically in Figure (6).

Exercise 5

From Equation 13 and the fact that $E_1 = -13.6$ eV, calculate the radius of the first Bohr orbit r_1. [Hint: first convert eV to ergs.] This is known as the *Bohr radius* and is in fact a good measure of the actual radius of a cold hydrogen atom. [The answer is $r_1 = .529 \times 10^{-8}$ cm $= .529$Å .] Then show that $r_n = n^2 r_1$.

Figure 6

The Bohr orbits are determined by equating the allowed energy $E_n = -13.6/n^2$ to the energy $E_n = -e^2/2r_n$ for an electron in an orbit of radius r_n. The Lyman series represents all jumps down to the smallest orbit, the Balmer series to the second orbit, the Paschen series to the third orbit, etc. (The radii in this diagram are not to scale, the radii r_n increase in size as n^2, as you can easily show by equating the two values for E_n.)

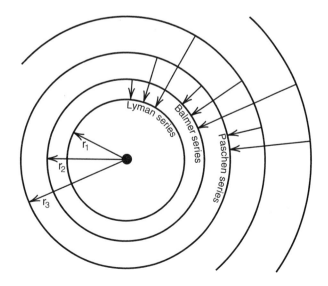

Angular Momentum in the Bohr Model

Nothing in Newtonian mechanics gives the slightest hint as to why the electron in hydrogen should have only certain allowed orbits. In the classical picture there is nothing special about these particular radii.

But ever since the time of Max Planck, there was a special unit of angular momentum, the amount given by Planck's constant h. Since Planck's constant keeps appearing whenever Newtonian mechanics fails, and since Planck's constant has the dimensions of angular momentum, perhaps there was something special about the electron's angular momentum when it was in one of the allowed orbits.

We can check this idea by re expressing the electron's total energy not in terms of the orbital radius r, but in terms of its angular momentum L.

We first need the formula for the electron's angular momentum when in a circular orbit of radius r. Back in Equation 4, we found that the speed v of the electron was given by

$$v = \frac{e}{\sqrt{mr}} \qquad \text{(4 repeated)}$$

Multiplying this through by m gives us the electron's linear momentum mv

$$mv = \frac{me}{\sqrt{mr}} = e\sqrt{\frac{m}{r}} \qquad (14)$$

The electron's angular momentum about the center of the circle is its linear momentum mv times the lever arm r, as indicated in the sketch of Figure (7). The result is

$$L = (mv)r = \left(e\sqrt{\frac{m}{r}}\right)r$$
$$= e\sqrt{mr} \qquad (15)$$

where we used Equation 14 for mv.

The next step is to express r in terms of the angular momentum L. Squaring Equation 13 gives

$$L^2 = e^2 mr$$

or

$$r = \frac{L^2}{e^2 m} \qquad (16)$$

Finally we can eliminate the variable r in favor of the angular momentum L in our formula for the electron's total energy. We get

$$\begin{aligned} \text{\textit{total energy}} \atop \text{\textit{of the electron}} \quad E &= \frac{-e^2}{2r} \\ &= \frac{-e^2}{2\left(L^2/e^2 m\right)} \\ &= \frac{-e^2}{2}\left(\frac{e^2 m}{L^2}\right) \\ &= \frac{-e^4 m}{2L^2} \qquad (17) \end{aligned}$$

In the formula $-e^4 m/2L^2$ for the electron's energy, only the angular momentum L changes from one orbit to another. If the energy of the nth orbit is E_n, then there must be a corresponding value L_n for the angular momentum of the orbit. Thus we should write

$$E_n = \frac{-e^4 m}{2L_n^2} \qquad (18)$$

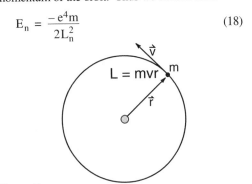

Figure 7
Angular momentum of a particle moving in a circle of radius r.

At this point, Bohr had the clue as to how to modify Newtonian mechanics in order to get his allowed energy levels. Suppose that angular momentum is quantized in units of some quantity we will call L_0. In the smallest orbit, suppose it has one unit, i.e., $L_1 = 1 \times L_0$. In the second orbit assume it has twice as much angular momentum, $L_2 = 2 L_0$. In the nth orbit it would have n units

$$L_n = nL_0 \quad \begin{array}{l} \textit{quantization} \\ \textit{of angular} \\ \textit{momentum} \end{array} \quad (19)$$

Substituting Equation 19 into Equation 18 gives

$$E_n = -\left(\frac{e^4 m}{2L_0^2}\right)\frac{1}{n^2} \quad (20)$$

as the total energy of an electron with n units of angular momentum. Comparing Equation 20 with Bohr's energy level formula

$$E_n = -13.6 \text{ eV}\left(\frac{1}{n^2}\right) \quad (10 \text{ repeated})$$

we see that we can explain the energy levels by assuming that the electron in the nth energy level has n units of quantized angular momentum L_0. We can also evaluate the size of L_0 by equating the constant factors in Equations 10 and 20. We get

$$\frac{e^4 m}{2L_0^2} = 13.6 \text{ eV} \quad (21)$$

Converting 13.6 eV to ergs, and solving for L_0 gives

$$\frac{e^4 m}{2L_0^2} = 13.6 \text{ eV} \times 1.6 \times 10^{-12}\frac{\text{ergs}}{\text{eV}}$$

With $e = 4.8 \times 10^{-10}$ esu and $m = .911 \times 10^{-27}$ gm in CGS units, we get

$$L_0 = 1.05 \times 10^{-27}\frac{\text{gm cm}^2}{\text{sec}} \quad (22)$$

which turns out to be Planck's constant divided by 2π.

$$L_0 = \frac{h}{2\pi} = \frac{6.63 \times 10^{-27}\text{gm cm/sec}}{2\pi}$$

$$= 1.05 \times 10^{-27}\frac{\text{gm cm}}{\text{sec}}$$

This quantity, Planck's constant divided by 2π, appears so often in physics and chemistry that it is given the special name *"h bar"* and is written $\hbar$

$$\hbar \equiv \frac{h}{2\pi} \quad \text{"h bar"} \quad (23)$$

Using $\hbar$ for L_0 in the formula for E_n, we get Bohr's formula

$$E_n = -\left(\frac{e^4 m}{2\hbar^2}\right)\frac{1}{n^2} \quad (24)$$

where $e^4 m/2\hbar^2$, expressed in electron volts, is 13.6 eV. This quantity is known as the **Rydberg constant**. *[Remember that we are using CGS units, where e is in esu, m in grams, and h is erg-sec.]*

Exercise 6

Use Equation 21 to evaluate L_0.

Exercise 7

What is the formula for the first Bohr radius in terms of the electron mass m, charge e, and Planck's constant $\hbar$. Evaluate your result and show that $r_1 = .51 \times 10^{-8}$cm $= .51$Å . (Answer: $r_1 = \hbar^2/e^2 m$.)

Exercise 8

Starting from Newtonian mechanics and the Coulomb force law $F = e^2/r^2$, write out a clear and concise derivation of the formula

$$E_n = \frac{-e^4 m}{2\hbar^2}\frac{1}{n^2}$$

Explain the crucial steps of the derivation.

A day or so later, on an empty piece of paper and a clean desk, see if you can repeat the derivation without looking at notes. When you can, you have a secure knowledge of the Bohr theory.

Exercise 9

An ionized helium atom consists of a single electron orbiting a nucleus containing two protons as shown in Figure (8). Thus the Coulomb force on the electron has a magnitude

$$|F_e| = \frac{(e)(2e)}{r^2} = \frac{2e^2}{r^2}$$

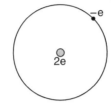

Figure 8
*Ionized helium has a
nucleus with two protons,
surrounded by one electron.*

a) Using Newtonian mechanics, calculate the total energy of the electron. (Your answer should be $-e^2/r$. Note that the r is not squared.)

b) Express this energy in terms of the electron's angular momentum L. (First calculate L in terms of r, solve for r, and substitute as we did in going from Equations 16 to 17.)

c) Find the formula for the energy levels of the electron in ionized helium, assuming that the electron's angular momentum is quantized in units of $\hbar$.

d) Figure out whether ionized helium emits any visible spectral lines (lines with photon energies between 1.8 eV and 3.1 eV.) How many visible lines are there and what are their wavelengths?) ·

Exercise 10

You can handle all single electron atoms in one calculation by assuming that there are z protons in the nucleus. (z = 1 for hydrogen, z = 2 for ionized helium, z = 3 for doubly ionized lithium, etc.) Repeat parts a), b), and c) of Exercise 9 for a single electron atom with z protons in the nucleus. (There is no simple formula for multi electron atoms because of the repulsive force between the electrons.)

DE BROGLIE'S HYPOTHESIS

Despite its spectacular success describing the spectra of hydrogen and other one-electron atoms, Bohr's theory represented more of a problem than a solution. It worked only for one electron atoms, and it pointed to an explicit failure of Newtonian mechanics. The idea of correcting Newtonian mechanics by requiring the angular momentum of the electron be quantized in units of $\hbar$, while successful, represented a bandaid treatment. It simply covered a deeper wound in the theory. For two centuries Newtonian mechanics had represented a complete, consistent scheme, applicable without exception. Special relativity did not harm the integrity of Newtonian mechanics—relativistic Newtonian mechanics is a consistent theory compatible with the principle of relativity. Even general relativity, with its concepts of curved space, left Newtonian mechanics intact, and consistent, in a slightly altered form.

The framework of Newtonian mechanics could not be altered to include the concept of quantized angular momentum. Bohr, Sommerfield, and others tried during the decade following the introduction of Bohr's model, but there was little success.

In Paris, in 1923, a graduate student Louis de Broglie, had an idea. He noted that light had a wave nature, seen in the 2-slit experiment and Maxwell's theory, and a particle nature seen in Einstein's explanation of the photoelectric effect. Physicists could not explain how light could behave as a particle in some experiments, and a wave in others. This problem seemed so incongruous that it was put on the back burner, more or less ignored for nearly 20 years.

De Broglie's idea was that, if light can have both a particle and a wave nature, perhaps electrons can too! Perhaps the quantization of the angular momentum of an electron in the hydrogen atom was due to the wave nature of the electron.

The main question de Broglie had to answer was how do you determine the wavelength of an electron wave?

An analogy with photons might help. There is, however, a significant difference between electrons and photons. Electrons have a rest mass energy and photons do not, thus there can be no direct analogy between the total energies of the two particles. But both particles have mass and carry linear momentum, and the amount of momentum can vary from zero on up for both particles. Thus photons and electrons could have similar formulas for linear momentum.

Back in Equation 34-13 we saw that the linear momentum p of a photon was related to its wavelength λ by the simple equation

$$\lambda = \frac{h}{p} \qquad \begin{pmatrix} de\ Broglie \\ wavelength \end{pmatrix} \qquad (34\text{-}13)$$

De Broglie assumed that this same relationship also applied to electrons. An electron with a linear momentum p would have a wavelength $\lambda = h/p$. This is now called the **de Broglie wavelength**. This relationship applies not only to photons and electrons, but as far as we know, to all particles!

With a formula for the electron wavelength, de Broglie was able to construct a simple model explaining the quantization of angular momentum in the hydrogen atom. In de Broglie's model, one pictures an electron wave chasing itself around a circle in the hydrogen atom. If the circumference of the circle, $2\pi r$ did not have an exact integral number of wavelengths, then the wave, after going around many times, would eventually cancel itself out as illustrated in Figure (9).

But if the circumference of the circle were an exact integral number of wavelengths as illustrated in Figure (10), there would be no cancellation. This would therefore be one of Bohr's allowed orbits shown in Figure (6).

Suppose (n) wavelengths fit around a particular circle of radius r_n. Then we have

$$n\lambda = 2\pi r_n \qquad (25)$$

Using the de Broglie formula $\lambda = h/p$ for the electron wavelength, we get

$$n\left(\frac{h}{p}\right) = 2\pi r_n \qquad (26)$$

Multiplying both sides by p and dividing through by 2π gives

$$n\frac{h}{2\pi} = pr_n \qquad (27)$$

Now $h/2\pi$ is just $\hbar$, and pr_n is the angular momentum L_n (momentum times lever arm) of the electron. Thus Equation 27 gives

$$n\hbar = pr_n = L_n \qquad (28)$$

Equation 28 tells us that for the allowed orbits, the orbits in which the electron wave does not cancel, the angular momentum comes in integer amounts of the angular momentum $\hbar$. The quantization of angular momentum is thus due to the wave nature of the electron, a concept completely foreign to Newtonian mechanics.

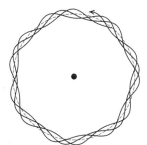

Figure 9
De Broglie picture of an electron wave cancelling itself out.

Figure 10
If the circumference of the orbit is an integer number of wavelengths, the electron wave will go around without any cancellation.

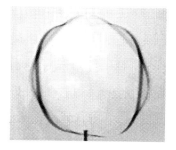

Figure 10a--Movie
The standing waves on a circular metal band nicely illustrate de Broglie's waves

When a graduate student does a thesis project, typically the student does a lot of work under the supervision of a thesis advisor, and comes up with some new, hopefully verifiable, results. What do you do with a student that comes up with a strange idea, completely unverified, that can be explained in a few pages of algebra? Einstein happened to be passing through Paris in the summer of 1924 and was asked if de Broglie's thesis should be accepted. Although doubtful himself about a wave nature of the electron, Einstein recommended that the thesis be accepted, for de Broglie *just might be right*.

In 1925, two physicists at Bell Telephone Laboratories, C. J. Davisson and L. H. Germer were studying the surface of nickel by scattering electrons from the surface. The point of the research was to learn more about metal surfaces in order to improve the quality of switches used in telephone communication. After heating the metal target to remove an oxide layer that accumulated following a break in the vacuum line, they discovered that the electrons scattered differently. The

metal had crystallized during the heating, and the peculiar scattering had occurred as a result of the crystallization. Davisson and Germer then prepared a target consisting of a single crystal, and studied the peculiar scattering phenomena extensively. Their apparatus is illustrated schematically in Figure (11), and their experimental results are shown in Figure (12). For their experiment, there was a marked peak in the scattering when the detector was located at an angle of 50° from the incident beam.

Davisson presented these results at a meeting in London in the summer of 1927. At that time there was a considerable discussion about de Broglie's hypothesis that electrons have a wave nature. Hearing of this idea, Davisson recognized the reason for the scattering peak. The atoms of the crystal were diffracting electron waves. The enhanced scattering at 50° was a diffraction peak, a maximum similar to the reflected maxima we saw back in Figure (33-19) when light goes through a diffraction grating. Davisson had the experimental evidence that de Broglie's idea about electron waves was correct after all.

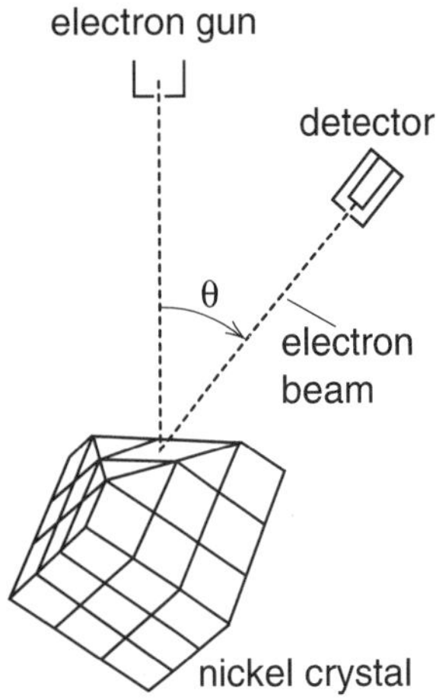

electron gun

detector

Reflected maximum

θ

electron beam

transmitted maximum

nickel crystal

Figure 11
Scattering electrons from the surface of a nickel crystal.

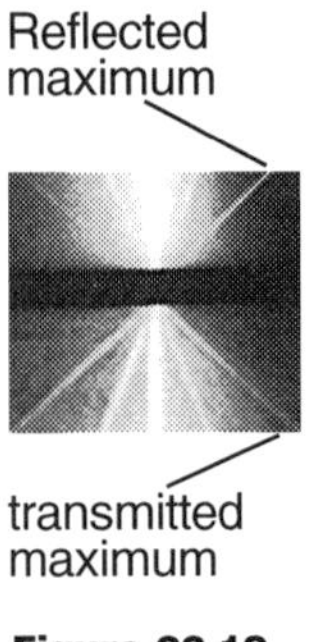

Figure 33-19
Laser beam impinging on a diffraction grating.

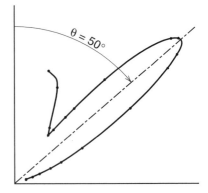

Figure 12
Plot of intensity vs. angle for electrons scattered by a nickel crystal, as measured by Davisson and Germer. The peak in intensity at 50° was a diffraction peak like the ones produced by diffraction gratings. (The intensity is proportional to the distance out from the origin.)

Chapter 36

Scattering of Waves

We will briefly interrupt our discussion of the hydrogen atom and study the scattering of waves by atoms. It was the scattering of electron waves from the surface of a nickel crystal that provided the first experimental evidence of the wave nature of electrons. Earlier experiments involving the scattering of x rays had begun to yield detailed information about the atomic structure of crystals.

Our main focus in this chapter will be an experiment developed in the early 1960s by Harry Meiners at R.P. I., that makes it easy for students to study electron waves and work with de Broglie's formula $\lambda = h/p$. The apparatus involves the scattering of electrons from a graphite crystal. The analysis of the resulting diffraction pattern requires nothing more than a combination of the de Broglie formula with the diffraction grating formula discussed in Chapter 33. We will use Meiner's experiment as our main demonstration of the wave nature of the electron.

SCATTERING OF A WAVE BY A SMALL OBJECT

The first step in studying the scattering of waves by atoms is to see what happens when a wave strikes a small object, an object smaller in size than the wavelength of the wave. The result can be seen in the ripple tank photographs shown in Figure (1). In (1a), an incident wave is passing over a small object. You can see scattered waves emerging from the object. In (1b), the incident wave has passed, and you can see that the scattered waves are a series of circular waves, the same pattern you get when you drop a stone into a quiet pool of water.

If the scattering object is smaller in size than the wavelength of the wave, as in Figure (1), the scattered waves contain essentially no information about the shape of the object. For this reason, you cannot study the structure of something that is much smaller than the wavelength of the wave you are using for the study. Optical microscopes, for example, cannot be used to study viruses, because most viruses are smaller than the wavelength of visible light. (Very clever work with optical microscopes allows one to see down to about 1/10th of the wavelength of visible light, to see objects like microtubules.)

incident wave →

incident wave →

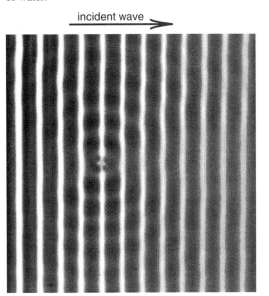

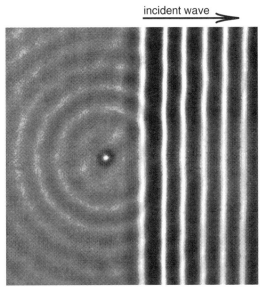

a) Incident and scattered wave together.

b) After the incident wave has passed.

Figure 1
If the scattering object is smaller than a wavelength, we get circular scattered waves that contain little or no information about the shape of the object.

REFLECTION OF LIGHT

Using the picture of scattering provided by Figure (1), we can begin to understand the reflection of visible light from a smooth metal surface. Suppose we have a long wavelength wave impinging on a metal surface represented by a regular array of atoms, as illustrated in Figure (2). As the wave passes over the array of atoms, circular scattered waves emerge. As seen in Figure (2a), the scattered waves add up to produce a reflected wave coming back out of the surface. The angles labeled θ_i and θ_r in Figure (2b) are what are called the *angle of incidence* and *angle of reflection* , respectively. Since the scattered waves emerge at the same speed as the incident wave enters, it is clear from the geometry that the angle of incidence is equal to the angle of reflection. That is the main rule governing the reflection of light.

What happens inside the material depends upon details of the scattering process. Note that the reflected wavefront inside the material coincides with the incident wave. For a metal surface, the phases of the scattered waves are such that the reflected wave inside just cancels the incident wave and there is no wave inside. All the radiation is reflected. For other types of material that are not opaque, the incident and scattered waves do not cancel. Instead they add up to produce a new, transmitted wave whose crests move slower than the speed of light. This apparent slowing of the speed of light, due to the interference of transmitted and scattered waves, leads to the bending of a beam of light as it enters or leaves a transparent medium. *It is this bending that allows one to construct lenses and optical instruments*.

Exercise 1

Using Figure (2), prove that the angle of incidence equals the angle of reflection.

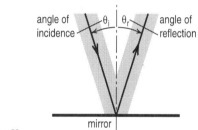

Figure 2b
When light reflects from a mirror, the angle of incidence equals the angle of reflection.

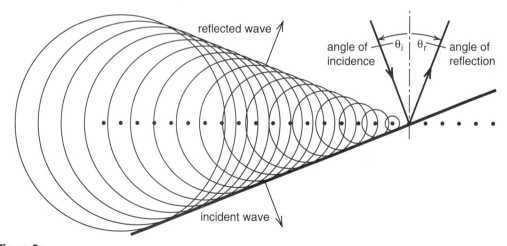

Figure 2a
A reflected wave is produced when the incident wave is scattered by many atoms. From this diagram, you can see why the angle of incidence equals the angle of reflection.

X RAY DIFFRACTION

If the wavelength of the light striking a crystal becomes comparable to the spacing between atoms, we get a new effect. The scattered waves from adjacent atoms begin to interfere with each other and we get diffraction patterns.

The spacing between atoms in a crystal is of the order of a few angstroms. (An angstrom, abbreviated Å, is 10^{-8}cm. An angstrom is essentially the diameter of a hydrogen atom.) Light with this wavelength is in the x ray region. Using Einstein's formula $E = hf = hc/\lambda$, but in the form

$$E\ (in\ eV)\ =\ \frac{12.4 \times 10^{-5}\ eV \cdot cm}{\lambda\ (in\ cm)}$$

we see that photons with a wavelength of Å 2 have an energy

$$E\left(\begin{array}{c} photon\ with\ 2Å \\ wavelength \end{array}\right)\ =\ \frac{12.4 \times 10^{-5} eV \cdot cm}{2 \times 10^{-8} cm}$$

$$=\ 6{,}200\ eV \qquad (1)$$

This is a considerably greater energy than the 2 to 3 eV of visible photon.

When a beam of x rays is sent through a crystal structure, the x rays will reflect from the planes of atoms within the crystal. The process, called ***Bragg reflection***, is illustrated for the example of a cubic lattice in Figure (3). The dotted lines connect lines of atoms, which are actually planes of atoms if you consider the depth of the crystal. An incident wave coming into the crystal can be reflected at various angles by various planes, with the angle of incidence equal to the angle of reflection in each case.

When the wavelength of the incident radiation is comparable to the spacing between atoms, we get a strong reflected beam when the reflected waves from one plane of atoms are an integral number of wavelengths behind the reflected waves from the plane above as illustrated in Figure (4). If it is an exact integral wavelength, then the reflected light from all the parallel planes will interfere constructively giving us an intense reflected wave. If, instead, there is a slight mismatch, then light from relatively distant planes will cancel in pairs and we will not get constructive interference. The argument is similar to the one used to find the maxima in a diffraction grating.

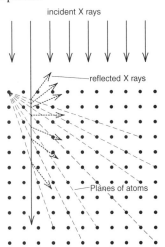

Figure 3
Planes of atoms act like mirrors reflecting X rays.

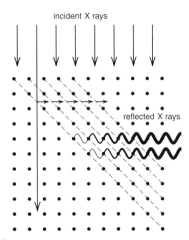

Figure 4
When the incident X ray wavelength equals the spacing between one of the sets of planes, the reflected waves add up to produce a maxima.

Thus with Bragg reflection you get an intense reflection only from planes of atoms, and only if the wavelength of the x ray is just right to produce the constructive interference described above. As a result, if you send an x ray beam through a crystal, you get diffraction pattern consisting of a series of dots surrounding the central beam, like those seen in Figure (5). Figure (5a) is a sketch of the setup and (5b) the resulting diffraction pattern for x rays passing through a silver bromide crystal whose structure is shown in (5c).

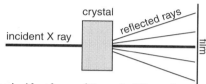

a) *An incident beam of X rays is diffracted by the atoms of the crystal.*

b) *X ray diffraction pattern produced by a silver bromide crystal. (Photograph courtesy of R. W. Christy.)*

c) *The silver bromide crystal is a cubic array with alternating silver and bromine atoms.*

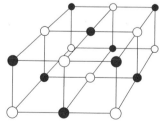

Figure 5
X ray diffraction study of a silver bromide crystal.

The main use of x ray diffraction has been to determine the structure of crystals. From the location of the dots in the x rays' diffraction photograph, and a knowledge of the wavelength of the x rays, you can figure out the orientation of and spacing between the planes of atoms. By using various wavelength x rays, striking the crystal at different angles, it is possible to decipher complex crystal structures. Figure (6) is one of many x ray diffraction photographs taken by J. C. Kendrew of a crystalline form of myoglobin. Kendrew used these x ray diffraction pictures to determine the structure of the myoglobin molecule shown in Figure (17-3). Kendrew was awarded the 1962 Nobel prize in chemistry for this work.

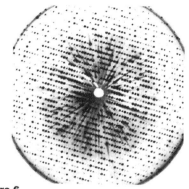

Figure 6
One of the X ray diffraction photographs used by Kendrew to determine the structure of the Myoglobin molecule.

Figure 17-3
The Myoglobin molecule, whose structure was determined by X ray diffraction studies.

Diffraction by Thin Crystals

The diffraction of waves passing through relatively thin crystals can also be analyzed using the diffraction grating concepts discussed in Chapter 33. Suppose for example, we had a thin crystal consisting of a rectangular array of atoms as shown in Figure (7a). The edge view of the array is shown in (7b). Here each dot represents the end view of a line of atoms.

Figure 7a
Front view of a rectangular array of atoms in a thin crystal.

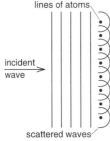

Figure 7b
Edge view with an incident wave. Each dot now represents one of the line of atoms in Figure (7a).

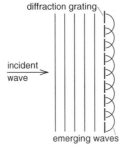

Figure 8
The waves emerging from a diffraction grating have a similar structure as waves scattered by a line of atoms.

Now suppose a beam of waves is impinging upon the crystal as indicated in Figure (7b). The impinging waves will scatter from the lines of atoms, producing an array of circular waves as shown.

Compare this with Figure (8), a sketch of waves emerging from a diffraction grating. The scattered waves from the lines of atoms, and the waves emerging from the narrow slits have a similar structure and therefore should produce similar diffraction patterns.

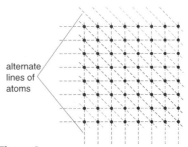

Figure 9
Various lines of atoms can imitate slits in a diffraction grating.

Figure 10a
A laser beam sent through a single grating. The lines of the grating were 25 microns wide, spaced 150 microns apart.

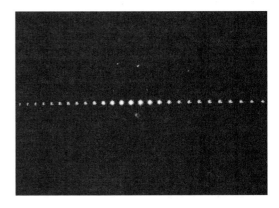

There is one major difference between the array of atoms in Figure (7) and the diffraction grating of Figure (8). In the crystal structure there are numerous sets of lines of atoms, some of which are indicated in Figure (9). Each of these sets of lines of atoms should act as an independent diffraction grating, producing its own diffraction pattern. The main sets of lines are horizontal and vertical, thus the main diffraction pattern we should see should look like that produced by two diffraction gratings crossed at right angles. Sending a laser beam through two crossed diffraction gratings produces the image shown in Figure (10). In Figure (10a), the laser beam is sent through a single grating. In (10b) we see the effect of adding another grating crossed at right angles.

Exercise 2

In Figure (10a) the maxima seen in the photograph are 1.68 cm apart and the distance from the grating to the screen is 4.00 meters. The wavelength of the laser beam is 6.3×10^{-5} cm. What is the spacing between the slits of the diffraction grating?

Exercise 3

In Figure (11), a laser beam is sent through two crossed diffraction gratings of different spacing. Which image, (a) or (b) is oriented correctly? (What happens to the spacing of the maxima when you make the grating lines closer together?)

Figure 11
Two diffraction gratings with different spacing are crossed. As shown, the vertical lines are farther apart than the horizontal ones. Which of the two images of the resulting diffraction pattern has the correct orientation?

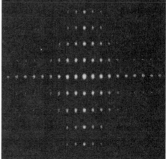

a)

Figure 10b
A laser beam sent through crossed diffraction gratings. Again the lines of the grating were 25 microns wide, spaced 150 microns apart.

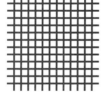

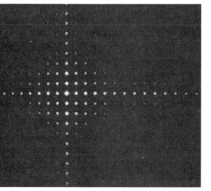

b)

THE ELECTRON DIFFRACTION EXPERIMENT

One of the main differences between the scattering of x rays and of electrons is that x ray photons interact less strongly with atoms, with the result that x rays can penetrate deeply into matter. This enables doctors to photograph through flesh to observe broken bones, or engineers to photograph through metal looking for hidden flaws. Electrons interact strongly with atoms, do not penetrate nearly as deeply, and therefore are well suited for the study of the structure of surfaces or thin crystals where you get considerable scattering from a few layers of atoms.

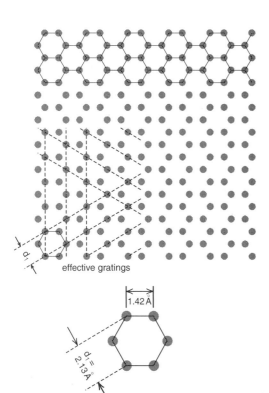

effective gratings

1.42 Å

$d_1 = 2.13$ Å

Figure 12
The hexagonal array of atoms in one layer of a graphite crystal. Lines of atoms in this crystal act as crossed diffraction gratings.

The Graphite Crystal

Graphite makes an ideal substance to study by electron scattering because graphite crystals come in thin sheets. A graphite crystal consists of a series of planes of carbon atoms. Within one plane the atoms have the hexagonal structure shown in Figure (12), reminiscent of the tiles often seen on bathroom floors. The spacing between neighboring atoms in each hexagon is 1.42 Å as indicated at the bottom of Figure (12).

The atoms within a plane are very tightly bound together. The hexagonal array forms a very strong framework. The planes themselves are stacked on top of each other at the considerable distance of 3.63 Å as indicated in Figure (13). The forces between these planes are weak, allowing the planes to easily slide over each other. The result is that graphite is a slippery substance, making an excellent dry lubricant. In contrast, the strength within a plane makes graphite an excellent strengthening agent for epoxy. The resulting carbon filament epoxies, used for constructing racing boat hulls, light airplanes and stayless sailboat masts, is one of the strongest plastics available.

plane separation = 3.63 Å

Figure 13
Edge view of the graphite crystal, showing the planes of atoms. The planes can easily slide over each other, making the substance slippery.

The Electron Diffraction Tube

The electron diffraction experiment where we sent a beam of electrons through a graphite crystal, can be viewed either as an experiment to demonstrate the wave nature of electrons or as an experiment to study the structure of a graphite crystal. Perhaps both.

The apparatus, shown in Figure (14), consists of an evacuated tube with an electron gun at one end, a graphite target in the middle, and a phosphor screen at the other end. A finely collimated electron beam can be aimed to strike an individual flake of graphite, producing a single crystal diffraction pattern on the phosphor screen. Usually you hit more than one crystal and get a multiple image on the screen, but with some adjustment you can usually obtain a single crystal image.

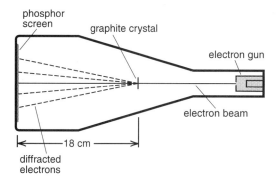

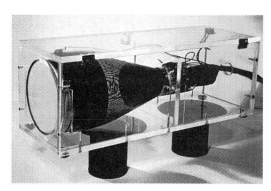

Figure 14
Electron diffraction apparatus. An electron beam, produced by an electron gun, strikes a graphite crystal located near the center of the evacuated tube. The original beam and the scattered electrons strike a phosphor screen located at the end of the tube.

Electron Wavelength

The accelerating voltage required to produce a good diffraction pattern is in the range of 6,000 volts. As our first step in the analysis, let us use the de Broglie wavelength formula to calculate the wavelength of 6,000 eV electrons.

The rest energy of an electron is .51 MeV, or 510,000 eV, far greater than the 6,000 eV we are using in this experiment. Since the 6,000 eV kinetic energy is much less than the rest energy, we can use the nonrelativistic formula $1/2\, mv^2$ for kinetic energy. First converting 6,000 eV to ergs, we can equate that to $1/2\, mv^2$ to calculate the speed v of the electron. We get

$$6000\ eV \times 1.6 \times 10^{-12} \frac{ergs}{eV} = 1/2\, m_e v^2 \quad (2)$$

With the electron mass $m_e = .911 \times 10^{-27} gm$, we get

$$v^2 = \frac{2 \times 6000 \times 1.6 \times 10^{-12} ergs}{.911 \times 10^{-27} gm}$$
$$= 21.1 \times 10^{18} \frac{cm^2}{sec^2}$$
$$v = 4.59 \times 10^9 cm/sec \quad (3)$$

which is slightly greater than 10% the speed of light.

The next step is to calculate the momentum of the electron for use in de Broglie's formula. We have

$$p = mv$$
$$= .911 \times 10^{-27} gm \times 4.59 \times 10^9 \frac{cm}{sec} \quad (4)$$
$$= 4.18 \times 10^{-18} \frac{gm\ cm}{sec}$$

Finally using de Broglie's formula we have

$$\lambda = \frac{h}{p} = \frac{6.63 \times 10^{-27} gm\ cm^2/sec}{4.18 \times 10^{-18} gm\ cm/sec}$$

$$\boxed{\lambda_{electron} = 1.59 \times 10^{-9} cm = .159\ \overset{\circ}{A}} \quad (5)$$

Thus the wavelength of the electrons we are using in this experiment is about one tenth the spacing between atoms in the hexagonal array.

Exercise 4

Calculate the wavelength of a 6000 eV photon. What would cause such a difference in the wavelengths of a photon and an electron of the same energy?

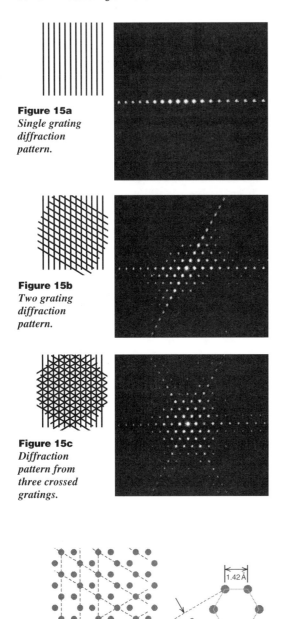

Figure 15a
Single grating diffraction pattern.

Figure 15b
Two grating diffraction pattern.

Figure 15c
Diffraction pattern from three crossed gratings.

Figure 12 (section)
Three sets of lines of atoms act as three crossed diffraction gratings with 2.13 angstrom spacing.

The Diffraction Pattern

What should we see when a beam of waves is diffracted by the hexagonal array of atoms in a graphite crystal? Looking back at the drawing of the graphite crystal, Figure (12), we see that there are prominent sets of lines of atoms in the hexagonal array. To make an effective diffraction grating, the lines of atoms have to be equally spaced. We have marked three sets of equally-spaced lines of atoms, each set being at an angle of 60° from each other. We expect that these lines of atoms should produce a diffraction pattern similar to three crossed diffraction gratings.

In Figure (15), we are looking at the diffraction we get when a laser beam is sent through three crossed diffraction gratings. In (15a), we have 1 diffraction grating. In (15b) a second grating at an angle of 60° has been added. In (15c) we have all three gratings, and see a hexagonal array of dots surrounding the central beam, the central maximum.

Figure (16) is the electron diffraction pattern photographed from the face of the electron diffraction tube shown in Figure (14). We clearly see an hexagonal array of dots expected from our diffraction grating analysis. On the photograph we have superimposed a centimeter scale so that measurements may be made from this photograph.

Figure 16
Diffraction pattern produced by a beam of electrons passing through a single graphite crystal. The energy of the electrons was 6000 eV.

The electron diffraction apparatus allows us to move the beam around, so that we can hit different parts of the target. In Figure (16), we have essentially hit a single crystal. When the electron beam strikes several graphite crystals at the same time, we get the more complex pattern seen in Figure (17).

Analysis of the Diffraction Pattern

Let us begin our analysis of the diffraction pattern by selecting one set of dots in the pattern that would be produced by one set of lines of atoms in the crystal. The dots and the corresponding lines of atoms are shown in Figure (18). In (18a) we see that the spacing Y_{max} between the dots on the screen is 1.33 cm. These horizontal dots correspond to the maxima for a vertical set of lines of atoms indicated in (18c). In (18b) we are reminded that the distance from the target to the screen is 18 cm. Using the diffraction grating formula, we can calculate the wavelength of the electron waves that produce this set of maxima.

Using the diffraction formula, Equation 33-3, and noting that $Y_{max} \ll D$, we have

$$\lambda = Y_{max}\frac{d}{\sqrt{D^2 + Y_{max}^2}} \approx Y_{max}\frac{d}{D}$$

$$= \frac{1.33 \text{ cm} \times 2.13 \times 10^{-8}\text{cm}}{18 \text{ cm}}$$

$$\boxed{\lambda = 1.57 \times 10^{-9}\text{cm}} \tag{6}$$

which agrees well with Equation 5, the calculation of the electron wavelength using the de Broglie wavelength formula.

Figure 18a
The diffraction grating maxima from one set of lines in the graphite crystal. You can see that $3y_{max} = 4cm$, so that $y_{max} = 1.33$ cm.

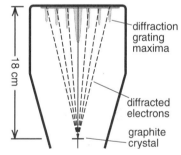

Figure 18b
Top view of the electron diffraction apparatus..

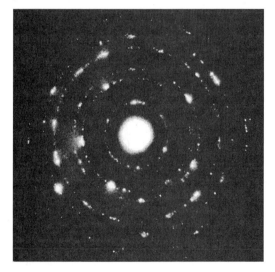

Figure 17
Diffraction pattern produced by a beam of electrons passing through multiple graphite crystals.

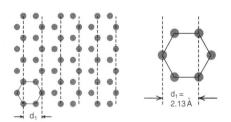

Figure 18c
The vertical lines of atoms in the graphite crystal that produce the horizontal row of dots seen in (a).

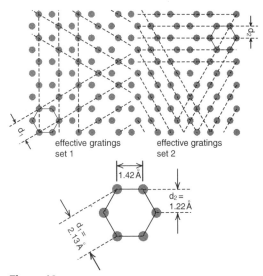

Figure 19
It is easy to find a second set of effective gratings, rotated 30° from the first set, and with a narrower spacing.

Figure 20
We have highlighted the maxima produced by this set of lines. Note that the more narrowly spaced lines produce more widely spaced maxima.

Other Sets of Lines

With a careful analysis of the lines of atoms in the hexagonal ray of atoms, one can explain all the dots of the diffraction pattern of Figure (16). For example, in Figure (19) we see that there is another set of lines that are rotated at an angle of 30° and more closely spaced than our original set. In Figure (20), we have highlighted a set of dots in the diffraction pattern that are rotated by an angle of 30° and more widely spaced than the dots we have been analyzing. Since more closely spaced lines in a grating produce more widely spaced maxima, we should suspect that the highlighted maxima result from this new set of lines. The point of Exercise 5 is to see if this is true.

Exercise 5

(a) Explain why more closely spaced atoms should produce more widely spaced dots in the diffraction pattern.

(b) Assuming that the dots highlighted in Figure (20) are produced by the lines of atoms shown by dotted lines in set 2 of the effective gratings, calculate the wavelength of the waves producing the dots. Compare your results with our previous analysis.

Exercise 6

Suppose that a beam of neutrons rather than electrons were fired at the graphite crystal. Assuming that neutrons also obey the de Broglie relationship $\lambda = h/p$, what should be the kinetic energy, in eV, of the neutrons in order to produce the same diffraction pattern with the same spacing between dots?

Student Projects

The crossed diffraction gratings used to obtain the various laser diffraction patterns in this chapter, were created using the Adobe Illustrator program, and then printed on film using a Linatronic imagesetter at a local desktop publishing company. The one micron resolution of the imagesetter allowed us to construct various grating and dot patterns that produced reasonable diffraction patterns with a laser.

Several students doing project work with these gratings and dot patterns suspected that some patterns were not as good as they should be and took microscope photographs of them. They found that lines or dots as small as 10 microns wide tended to be filled in and blotchy, but lines or dots 25 microns wide came out fairly well as can be seen in Figures (21) and (22). Figure (23) is the laser diffraction pattern produced by a laser beam passing through the hexagonal dot pattern of Figure (22).

Figure 21
Microphotograph of the three crossed diffraction gratings. The lines are 25 microns wide and 100 microns apart, on centers. (Student project by Brady Beale and Amy Coughlin.)

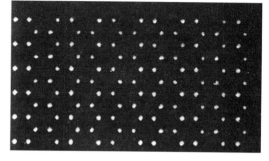

Figure 22
Microphotograph of a hexagonal dot pattern. The dots are 25 microns in diameter and 100 microns apart. (Student project by Brady Beale and Amy Coughlin.)

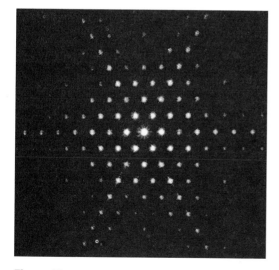

Figure 15c
Diffraction pattern produced by a laser beam going through the three crossed gratings of Figure 21.

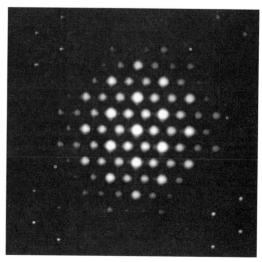

Figure 23
Diffraction pattern produced by a laser beam going through the hexagonal dot pattern of Figure 22.

Student project by Gwendylin Chen

In our discussion of the diffraction of waves by the atoms of a crystal, we pointed out that waves should emerge from a line of atoms in much the same wah that they do from the slits of a diffraction grating. The two situations were illustrated in Figures (7b and 8) reproduced below.

That a slit and a line produce similar diffraction patterns was clearly illustrated in a project by Gwendylin Chen. While working with a laser, she observed that when the beam passed over a strand of hair it produced a single slit diffraction pattern superimposed on the image of the beam itself. Here we have reproduced Gwendylin's experiment. Figure (24) is a photograph of a slit made from two scapel blades, and a strand of Gwendylin's hair. We tried to make the width of the slit the same as the width of the hair. The two circles indicate where we aimed the laser for the two diffraction patterns.

The results are seen in Figure (25).The diffraction patterns are almost identical. The only difference is that when the beam passes over the hair, it continues on landing in the center of the diffraction pattern.

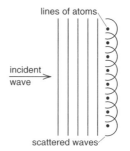

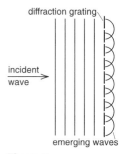

Figure 7b
Edge view of a thin crystal with an incident wave. Each dot now represents one of the line of atoms in the crystal.

Figure 8
The waves emerging from a diffraction grating have a structure similar to the waves scattered by a line of atoms.

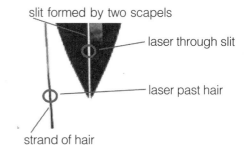

Figure 24
Slit and hair used to produce diffractiopn patterns. The circles indicate where we aimed the laser.

a) single slit diffraction pattern

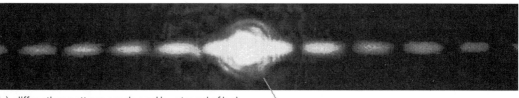

b) diffraction pattern produced by strand of hair

Figure 25
Comparason of diffeaction patterns.

Chapter 37

Lasers, a Model Atom and Zero Point Energy

"Once at the end of a colloquium I heard Debye saying something like: 'Schrödinger, you are not working right now on very important problems...why don't you tell us some time about that thesis of de Broglie, which seems to have attracted some attention?' So in one of the next colloquia, Schrödinger gave a beautifully clear account of how de Broglie associated a wave with a particle, and how he could obtain the quantization rules ... by demanding that an integer number of waves should be fitted along a stationary orbit. When he had finished, Debye casually remarked that he thought this way of talking was rather childish ... To deal properly with waves, one had to have a wave equation."

FELIX BLOCK, in an address to the American Physical Society in 1976.

Schrödinger took Debye's advice, and in the following months devised a wave equation for the electron wave, an equation from which one could calculate the electron energy levels. The structure of the hydrogen atom was a prediction of the equation without arbitrary assumptions like those needed for the Bohr theory. The wave nature of the electron turned out to be the key to the new mechanics that was to replace Newtonian mechanics as the fundamental theory.

In the next chapter we will take a look at some of the electron wave patterns determined by Schrödinger's equation, and see how these patterns, when combined with the Pauli exclusion principle and the concept of electron spin, begin to explain the chemical properties of atoms and the structure of the periodic table.

The problem one encounters when discussing the application of Schrödinger's equation to the hydrogen atom, is that relatively complex mathematical steps are required in order to obtain the solutions. These steps are usually beyond the mathematical level of most

introductory physics and chemistry texts, with the result that students must simply be shown the solutions without being told how to get them. We will have to do the same in the next chapter.

In this chapter we will study a model atom, one in which we can see how the particle-wave nature of the electron leads directly to quantized energy levels and atomic spectra. The basic idea, which we illustrate with the model atom, is that whenever you have a wave confined to some region of space, there will be a set of allowed standing wave patterns for that wave. Whether the patterns are complex or simple depends upon the way the wave is confined. If the wave is also a particle, like an electron or photon, you can then use the particle wave nature to calculate the energy of the particle in each of the allowed standing wave patterns. These energy values are the quantized energy levels of the particle.

An example of a set of simple standing waves that are easily analyzed is found in the laser. It is essentially the laser standing wave patterns that we use for our model atom. For this reason we begin the chapter with a discussion of the laser and how the photon standing waves are established. In the model atom the photon standing waves of the laser are replaced by electron standing waves.

An analysis of the model atom shows why any particle, when confined to some region of space, must have a non zero kinetic energy. The smaller the region of space, the greater this so called **zero point** *kinetic energy. When these ideas are applied to the atoms in liquid helium, we see why helium does not freeze even at absolute zero. We also see why the entropy definition of temperature must be used at these low temperatures.*

THE LASER AND STANDING LIGHT WAVES

The laser, the device that is at the heart of your CD player and fiber optics communications, provides a common example of a standing light wave. In most cases a laser consists of two parallel mirrors with standing light waves trapped between the mirrors as illustrated in Figure (1). The light comes from radiation emitted by excited atoms that are located within the standing wave.

How the light radiated by the excited atoms ends up in a standing wave is a story in itself. An atom excited to a high energy level can drop down to a lower level by emitting a photon whose energy is the difference in energy of the two levels. This photon will have the wavelength of the spectral line associated with those two levels.

Spectral lines are not absolutely sharp. For example, due to the Doppler effect, thermal motion slightly shifts the wavelength of the emitted radiation. If the atom is moving toward you when it radiates, the wavelength is shifted slightly towards the blue. If moving away, the shift is toward the red. In addition the photons are radiated in all directions, and waves from different photons have different phases. Even in a sharp spectral line the light is a jumble of directions and phases, giving what is called ***incoherent*** light.

In contrast the light in a laser beam travels in one direction, the phases of the waves are lined up and there is almost no spread in photon energies. This is the beam of ***coherent*** light which made it so easy for us to study interference effects like those we saw in the two slit and multiple slit diffraction patterns. These patterns would be much more difficult to observe if we had to use incoherent light.

The purity of the light in a laser beam depends upon the standing light wave pattern created by the two mirrors, and upon a quantum mechanical effect discovered by Einstein in 1915.

Einstein found that there were two distinct ways an excited atom could radiate light, either by ***spontaneous emission*** or ***stimulated emission***. An example of spontaneous emission is when an excited atom is all by itself and eventually drops down to a lower energy level. The emitted photon can come out in any direction and can be Doppler shifted.

If, however, a photon with the right energy passes by the excited atom, there is some chance that the atom will emit a photon ***exactly*** like the one passing by. This is called stimulated emission. (The energy of the passing photon has to be close to the energy the atom would naturally radiate.)

It is the process of stimulated emission that can lead to a laser beam. Suppose we have a gas of excited atoms located between parallel mirrors. At first the atoms radiate spontaneously in all directions. (We assume that there is some mechanism to excite the atoms). After a while one of the photons hits a mirror straight on and starts reflecting back and forth between the parallel mirrors. As the photon moves back and forth, it passes by an excited atom, stimulating that atom to emit an identical photon.

Now there are two identical photons bouncing back and forth. Each is likely to stimulate another atom to emit an identical photon, and we have four identical photons, etc. Soon there are so many identical photons moving through the excited atoms that there is little chance that an atom can radiate spontaneously. All the radiation is stimulated and all the photons are identical to the one that started bouncing back and forth between the mirrors.

The mirrors on the ends of the laser are not perfect reflectors, a few percent of the photons striking the mirror pass through, forming the beam produced by the laser. The photons lost to the laser beam are continually replaced by new identical photons being emitted by stimulated emission. One of the tricky technical parts of constructing a laser is to maintain a continuous supply of excited atoms. There are various ways of doing this that we need not discuss here.

optically   excited   standing   partially reflecting
flat mirror   gas atoms   light wave   optically flat mirror

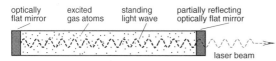

laser beam

Figure 1
Laser consisting of two parallel mirrors with standing light waves trapped between the mirrors

Photon Standing Waves

The photons bouncing back and forth between the mirrors in a laser are in an allowed standing wave pattern. Back in Chapter 15 in our discussion of standing waves on a guitar string, we saw that only certain standing wave patterns were allowed, those shown in Figure (15-15) reproduced here which had an integral or half integral number of wavelengths between the ends of the string. For photons trapped between two mirrors, the allowed standing wave patterns are also those with an integral or half integral number of wavelengths between the mirrors, as indicated schematically in Figure (2).

Because of the simple geometry, we do not need to solve a wave equation to determine the shape of these standing light waves. The waves are sinusoidal, and the allowed wavelength are given by the same formula as for the allowed waves on a guitar string, namely

$$\lambda_n = \frac{2D}{n} \qquad \begin{array}{l}\textit{wavelength of the}\\ \textit{nth standing wave}\end{array} \qquad (1)$$

where D is the separation between the mirrors.

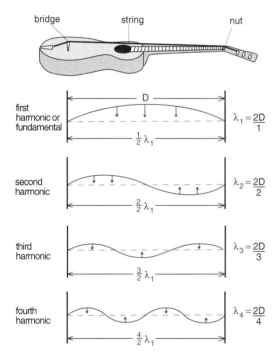

Figure 15-15 (reproduced)
On a guitar string only certain standing wave patterns which have an integral or half integral number of wavelengths between the ends of the string are allowed.

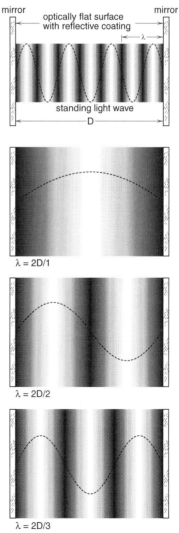

Figure 2
Three longest wavelength standing wave patterns for a light beam trapped between two mirrors.

Photon Energy Levels

The special feature of the standing light wave is that the light has both a wave and a particle nature. Equation 1, which tells us the allowed wavelengths, is all we need to know about the wave nature of the light. The particle nature is described by Einstein's photoelectric effect formula $E = hf = hc/\lambda$. Applying this formula to the photons in the standing wave, we find that a photon with an allowed wavelength λ_n has a corresponding energy E_n given by

$$E_n = \frac{hc}{\lambda_n} \tag{2}$$

Because only certain wavelengths λ_n are allowed, only certain energy photons, those with an energy E_n are allowed between the mirrors. We can say that the photon energies are quantized. If the separation of the mirrors is D, then from Equation 1 ($\lambda_n = 2D/n$), and Equation 2 ($E_n = hc/\lambda_n$), we find that the quantized values of E_n are

$$E_n = \frac{hc}{\lambda_n} = \frac{hc}{\frac{2D}{n}}$$

$$\boxed{E_n = n\frac{hc}{2D}} \tag{3}$$

From Equation 3 we can construct an energy level diagram for the photons trapped between the mirrors. In contrast to the energy level diagram for the hydrogen atom, the photon energies start at zero because there is no potential energy. We see that the levels are equally spaced, a distance hc/2D apart.

Exercise 1

If you could have two mirrors 1Å apart (the size of a hydrogen atom) what would be the energy, in eV, of the lowest 5 energy levels for a photon trapped between the mirrors?

A MODEL ATOM

Now imagine that we replace the photons trapped between two mirrors with an electron between parallel walls located a distance D apart, as shown in Figure (4). For this model, the allowed standing wave patterns are again similar to the guitar string standing waves. The allowed electron wavelengths are

$$\lambda_n = \frac{2D}{n} \quad \begin{array}{l}\textit{allowed wavelength}\\\textit{of an electron trapped}\\\textit{between two walls}\end{array} \tag{1a}$$

The difference between having a photon trapped between mirrors and an electron between walls, is the formula for the energy of the particle. If the energy of the electron is non relativistic, then the formula for its

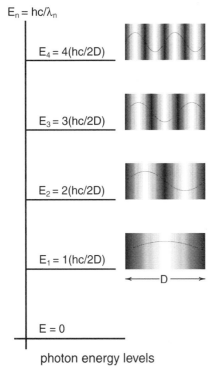

$E_n = hc/\lambda_n$

$E_4 = 4(hc/2D)$

$E_3 = 3(hc/2D)$

$E_2 = 2(hc/2D)$

$E_1 = 1(hc/2D)$

$\longleftarrow$D$\longrightarrow$

$E = 0$

photon energy levels

Figure 3
Energy level diagram for a photon trapped between two mirrors.

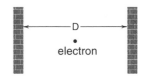

D

• electron

Figure 4
Electron trapped between two walls

kinetic energy is $1/2\,mv^2$, not the Einstein formula $E = hc/\lambda$ that applies to photons. The difference arises because the electron has a rest mass while the photon does not.

For the electron trapped between walls, there is no electric potential energy like there was in the hydrogen atom. Thus we can take $1/2\,mv^2$ as the formula for the electron's total energy, ignoring the electron's rest mass energy as we usually do in non relativistic calculations.

To relate the kinetic energy to the electron's allowed wavelength λ_n, we use de Broglie's formula $p = h/\lambda$. The easy way to do this is to express the energy $1/2\,mv^2$ in terms of the electron's momentum $p = mv$. We get

$$E = 1/2\,mv^2 = \frac{1}{2m}(mv)^2$$

$$E = \frac{p^2}{2m} \qquad (4)$$

Next use the de Broglie formula $p = h/\lambda$ to give us

$$E_n = \frac{(h/\lambda_n)^2}{2m} = \frac{h^2}{2m\lambda_n^2} \qquad (5)$$

as the formula for the energy of an electron of wavelength λ_n.

Finally use Equation 1, $\lambda_n = 2D/n$, for allowed electron wavelengths to get

$$E_n = \frac{h^2}{2m\left(\frac{2D}{n}\right)^2}$$

$$\boxed{E_n = n^2\left(\frac{h^2}{8mD^2}\right)} \qquad (6)$$

This is our equation for the energy levels of an electron trapped between two plates separated by a distance D. The corresponding energy level diagram is shown in Figure (5). The energy levels go up as n^2 instead of being equally spaced as they were in the case of a photon trapped between two mirrors.

If the electron is in one of the higher levels and falls to a lower one, it will get rid of its energy by emitting a photon whose energy is equal to the difference in the energy of the two levels. Thus the trapped electron should emit a spectrum of radiation with sharp spectral lines, where the lines correspond to energy jumps between levels just as in the hydrogen atom. Thus the electron trapped between plates is effectively a model atom, complete with an energy level diagram and spectral lines.

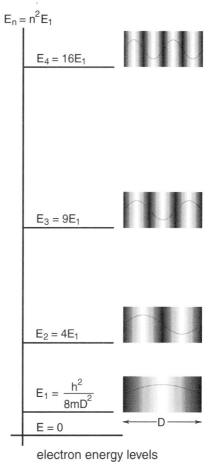

$E_n = n^2 E_1$

$E_4 = 16E_1$

$E_3 = 9E_1$

$E_2 = 4E_1$

$E_1 = \dfrac{h^2}{8mD^2}$

$E = 0$

D

electron energy levels

Figure 5
Energy level diagram for an electron trapped between two walls.

Our model atom is not just a fantasy. With the techniques used to fabricate microchips, it has been possible to construct tiny boxes, the order of a few angstroms across, and trap electrons inside. An electron microscope photograph of these *quantum dots* as they are called, is shown in Figure (6). The allowed standing wave patterns are reasonably well represented by the sine wave patterns of Figure (5), where D is the smallest dimension of the box. Thus we predict that electrons trapped in these boxes should have allowed energies E_n close to those given by Equation (6), and emit discrete line spectra like an atom. This is precisely what they do. (Some of the low energy jumps are shown in Figure 7.)

In calculating with the model atom we have not fudged in any way by modifying Newtonian mechanics or even picturing a wave chasing itself around in a circle. We see a spectrum resulting purely from a combination of the particle nature and the wave nature of electrons and photons, where the connection between the two points of view is de Broglie's formula $p = h/\lambda$.

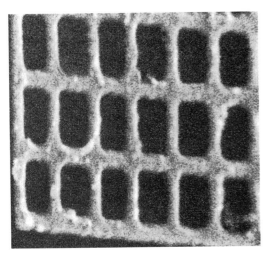

Figure 6
Grid of quantum dots. These cells are made on a silicon wafer with the same technology used in making electronic chips. An electron trapped in one of these cells has energy levels similar to those of our model atom. (See Scientific American, Jan. 1993, p118.)

Exercise 2

Assume that an electron is trapped between two walls a distance D apart. The distance D has been adjusted so that the lowest energy level is $E_1 = 0.375\,eV$.

(a) What is D?

(b) What are the energies, in eV, of the photons in the six longest wavelength spectral lines radiated by this system? Draw the energy level diagram for this system and show the electron jumps corresponding to each spectral line.

(c) What are the corresponding wavelengths, in cm, of these six spectral lines?

(d) Where in the electromagnetic spectrum (infra red, visible, or ultra violet) do each of these spectral lines lie? If any of these lines are visible, what color are they? (Partial answer: the photon energies are 1.125, 1.875, 2.625, 3.00, 3.375, and 4.125 eV)

Exercise 3

Explain why an electron, confined in a box, cannot sit at rest. This is an important result whose consequences will be discussed next. Try to answer it now.

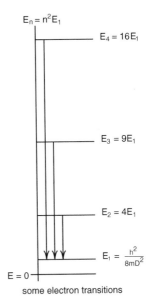

$$E_n = n^2 E_1$$

$$E_4 = 16E_1$$

$$E_3 = 9E_1$$

$$E_2 = 4E_1$$

$$E_1 = \frac{h^2}{8mD^2}$$

$$E = 0$$

some electron transitions

Figure 7
When an electron falls from one energy level to another, the energy of the photon it emits equals the energy lost by the photon.

ZERO POINT ENERGY

One of the immediate consequences of the particle-wave nature of the electron is that a confined electron can never be at rest. The smaller the confinement, the greater the kinetic energy the electron must have. This follows from the fact that at least half a wavelength of the electron's wave must fit within the confining region. If D is the length of the smallest dimension of the confining region, then the electron's wavelength cannot be greater than 2D. But the smaller D is, the shorter the electron's wavelength is, the greater its kinetic energy.

The de Broglie wavelength formula $\lambda = h/p$ applies not only to photons and electrons, but to any particle, even an entire atom. As a result, an atom confined to a region of size D should have a wavelength no greater than $\lambda_1 = 2D$, and thus a minimum kinetic energy

$$E_{min} = \frac{h^2}{8m_{atom}D^2} \qquad (7)$$

where we simply replaced the electron's mass by the atom' mass in Equation 7. Equation 7 is somewhat approximate if the atom is confined on all sides in a three dimensional box, but it is reasonably accurate if D is the smallest dimension of the box.

An atom in a solid or a liquid is an example of a particle confined in a box. The atom is confined by its neighboring atoms as illustrated in Figure (8). We may think of its neighbors as forming a box of size D where D is the average spacing between atoms. Thus atoms in solids or liquids have a minimum kinetic energy given by Equation 7, and the atoms must be in continual motion *no matter how low the temperature*! Cooling the solid cannot get rid of this so-called *zero point energy*.

Figure 8
A helium atom in liquid helium is confined by its neighbors. As a result it has a zero point energy like an electron confined between walls.

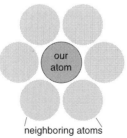

neighboring atoms

Exercise 4

In liquid helium, the helium atoms are about 3Å apart and the atoms have a mass essentially equal to 4 times the mass of a proton.

(a) what is the zero point energy, in ergs, of helium atoms in liquid helium?

(b) at what temperature T is the helium atom's thermal kinetic energy 3/2 kT equal to the zero point energy calculated in part (a)? [Answer: (a) 9.1×10^{-16} ergs, (b) 4.42 kelvin.]

Helium is an especially interesting substance to study at low temperatures because it is the only substance that remains a liquid all the way down to absolute zero. The only way you can freeze helium is to take it down to very low temperatures, and then squeeze it at relatively high pressure.

In all other substances, at low enough temperatures the atoms settle down to a solid array. To melt the solid, you have to add enough thermal energy to disrupt the molecular bonds that hold the atoms in a more or less fixed array.

Why can't helium atoms be cooled to the point where molecular forces dominate and the atoms form a solid array? Part of the answer is that the molecular forces between helium atoms are very weak, the weakest there is between any atoms. Consequently you have to go to very low temperatures before helium gas even becomes a liquid. At atmospheric pressure, helium becomes a liquid at 4.5 kelvins. To turn liquid helium into a solid you should have to go to still lower temperatures.

From Exercise 4, you saw that, in one sense, you cannot get helium to a lower temperature, at least as far as the kinetic energy of the atoms is concerned. The zero point energy of the atoms is as big as the thermal energy that the atoms would have at a few kelvin-- 4.4 kelvin by our rough estimate in Exercise 4. As a result, cooling the helium further cannot remove enough kinetic energy to allow the helium liquid to freeze. Helium thus remains a liquid all the way down to absolute zero.

Definition of Temperature

This discussion raises interesting questions about the very concept of temperature. Our initial experimental definition of temperature was the ideal gas thermometer, which, as we saw from the derivation of the ideal gas law, is based on the thermal kinetic energy of the particles. The simple idea of absolute zero was the point where all the thermal kinetic energy was gone and the atoms were at rest. Now we see that no matter how much thermal kinetic energy we try to remove, zero point or "quantum kinetic energy" remains. This is not a problem at ordinary temperatures, but it can significantly affect the behavior of matter at temperatures close to absolute zero.

At low temperatures, the ideal gas thermometer is not adequate, and a new definition of temperature is needed. That new definition is provided by the efficiency of Carnot's heat engine. As we suggested in Chapter 17, this gives us a definition of temperature based, not on the kinetic energy of the molecules, but upon the degree of randomness or disorder. A system at absolute zero is as perfectly ordered as it can be. If zero point energy is required by the particle wave nature of the atoms, if it cannot be removed, then the most organized, least disordered state of the system must include this zero point energy. Helium can go to its most ordered state at absolute zero, retain its zero point energy, and remain a liquid.

TWO DIMENSIONAL STANDING WAVES

In our discussion of percussion instruments in Chapter 16, we saw that a drumhead has a set of allowed standing wave patterns or normal modes, in some ways like the standing waves or normal modes on a guitar string. On a guitar string we have one dimensional waves, while the drumhead has the two dimensional wave patterns. The six lowest frequency patterns are shown in Figure (16-41) repeated here. We could excite and observe individual standing waves using the apparatus shown in Figure (16-40) also shown again here.

That we get the same kind of standing wave patterns on an atomic scale is seen in Figure (9), which is a recent tunneling microscope image of an electron standing wave on the surface of a copper crystal. The standing wave, which is formed inside a corral of 48 iron atoms, has the same shape as one of the allowed standing waves on a drumhead. (This particular standing wave pattern is excited because the average wavelength in the standing wave is closest to the wavelength of the conduction electrons at the surface of the copper.)

A colleague Geoff Nunes, who works with scanning microscopes, describes the image: "The incredible power of today's personal computers has been made possible by our ability to make smaller and smaller transistors. The smallest transistor one could imagine building would be made up of single atoms. In a dramatic series of experiments at IBM, Don Eigler and his co-workers have shown how to use a tunneling microscope to move and arrange single atoms."

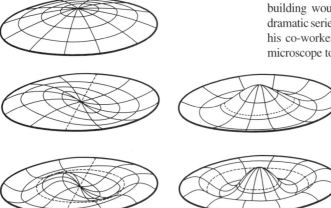

Figure 16-41 (reproduced)
Standing waves on a drumhead.

"This picture (Figure 9) shows a ridge of 48 iron atoms arranged in a circle on the surface of a copper crystal. Electrons in the copper are reflected from these iron atoms much as the waves on the surface of a pond are reflected from anything at the surface: rocks, weeds, the shoreline. Inside the ring, the electron waves form a beautifully symmetric pattern. This pattern occurs often in the physical world. For example it is the shape that the head of a drum forms when struck. You can easily observe a similar pattern by gently skidding the base of a Styrofoam cup full of coffee across the surface of a table."

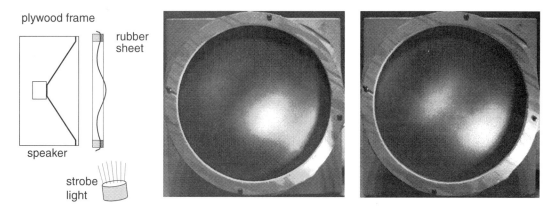

Figure 16-40 (reproduced)
Exciting and observing the standing waves on a drumhead.

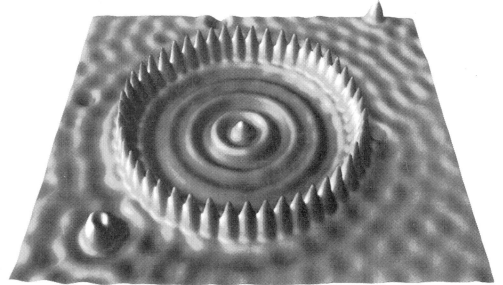

Figure 9
Conduction electrons on the surface of a copper crystal, forming a standing wave inside a corral of 48 iron atoms. The shape is the same as one of the symmetric standing waves on a drumhead. (Photo credits: Crommie and Eigler/IBM.)

Chapter 38
Atoms

The focus of this chapter will be on the allowed electron standing wave patterns in hydrogen, as calculated by Schrödinger's wave equation. We will not work with Schrödinger's equation itself, which involves derivatives in both space and time, and requires fairly advanced mathematical techniques to handle. But this is not a terrible loss, because the resulting wave patterns are well known, and are all we need in order to understand much of the structure and behavior of atoms.

As we saw at the end of the last chapter, when we go to two dimensions, the standing wave patterns become more complex. For example, to find the drumhead standing wave patterns, we either had to do an experiment to observe the patterns, or solve a wave equation to calculate them.

To determine electron waves in hydrogen, our only option is to rely on Schrödinger's equation. The resulting standing waves are three dimensional in shape, and do not have sharp edges like the drumhead waves. The electron in hydrogen is confined by the electric force of the nucleus, in what physicist Jay Orear called a "fuzzy walled box". Even though the walls are not rigid, the standing waves have precise shapes.

Although the standing wave patterns we will discuss were calculated for the hydrogen atom, the general features of these patterns apply to the electrons in larger atoms. We will find that when we include Pauli's exclusion principle and the concept of electron spin, we can begin to see how the electron wave patterns determine the chemical properties of atoms and the structure of the periodic table.

SOLUTIONS OF SCHRÖDINGER'S EQUATION FOR HYDROGEN

In the Bohr theory, the energy levels for hydrogen were determined by the assumption that the electron's angular momentum L was quantized in units of $\hbar$. The electron's angular momentum was $L_1 = \hbar$ in the lowest energy level, $L_2 = 2\hbar$ in the second level, etc. Assuming circular orbits, and applying classical mechanics, this led to the set of energy levels E_n given by

$$E_1 = \frac{-e^4 m}{2\hbar^2} = -13.6 \text{ eV}$$

$$E_n = \frac{E_1}{n^2} \tag{1}$$

which gave us the values $E_1 = -13.6 \text{ eV}$, $E_2 = -3.40 \text{ eV}$, etc., that explained the hydrogen spectra.

De Broglie's contribution was to show that one could understand the reason for quantization of angular momentum by assuming that the electron had a wave nature, with the electron's wavelength λ related to its momentum p by the formula

$$\lambda = \frac{h}{p} \quad \textit{de Broglie formula} \tag{2}$$

The quantization of angular momentum came from the picture that an integral number of wavelengths fit around one of the allowed circular orbits.

Following Debye's suggestion (see the introduction to Chapter 37), Schrödinger developed a wave equation with which he was able to solve for the allowed standing wave patterns of the electron in a hydrogen atom. Doing this required no arbitrary assumptions like circular orbits or the quantization of angular momentum. The wave patterns are simply solutions of the wave equation.

Schrödinger's equation has a surprisingly large number of solutions for the allowed standing waves of an electron in hydrogen. These waves are characterized by three numbers commonly given the names "n", "ℓ", and "m". It turns out that for the wave to be an acceptable solution, a solution that does not have an infinite value at some point, the numbers n, ℓ, and m have to have integer values. These integer numbers have become known as *quantum numbers*.

Each of the allowed standing wave patterns in hydrogen has a distinct set of values of the quantum numbers n, ℓ, and m. Figure (1) shows six of the allowed patterns. What we have drawn is the intensity of the wave pattern as it would be seen if we looked through the wave. When the side and top view are different we show both to help visualize the three dimensional structure of the wave.

The pattern on the bottom row labeled by the quantum numbers (n = 1, ℓ = 0, m = 0) is a spherical ball with a fuzzy edge. The radius of the ball is about equal to the Bohr radius of .529 angstroms. Schrödinger's equation allows us to calculate the energy of the electron in this pattern and the result is -13.6 eV, the same as the lowest energy state of the electron in the Bohr theory. This is the standing wave pattern for an electron in the ground state – cool, transparent hydrogen.

On the second row in Figure (1) there are four distinct patterns, all with n = 2 but with different values of ℓ and m. Schrödinger's equation predicts that the energy of an electron in hydrogen is given, in general, by the formula

$$E_n = \frac{E_1}{n^2} ; \quad E_1 = \frac{-e^4 m}{2\hbar^2} \tag{3}$$

where n is the "n" quantum number we have been discussing. Since these are the same values we got from the Bohr theory, Schrödinger's equation predicts all the energy levels needed to explain the entire spectrum of light radiated by hydrogen. Because of Equation 3, it is reasonable to call n the *energy quantum number* for hydrogen.

The first big surprise from Schrödinger's equation is that we can have several standing wave patterns all representing an electron with the same energy. In the n = 2 energy level, there are four distinct patterns representing an electron with the energy $E_2 = -3.40 \text{eV}$. One of these patterns has quantum numbers $\ell = 0$, m = 0. The other three have quantum numbers $\ell = 1$, m = 1, m = 0, and m = –1.

When we get up to the third energy level, n = 3, there are nine patterns all with an energy of -1.51 eV. There is one pattern with $\ell = 0$, m = 0; three patterns with $\ell = 1$, m = (1, 0, –1); and five patterns with $\ell = 2$, m = (2, 1, 0, –1, –2). As we go up in energy, we get an ever increasing number of patterns. The general rule is that ℓ can range from zero up to n –1, and the m values can range from + ℓ to – ℓ.

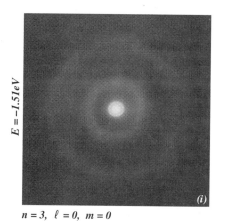

$E = -1.51eV$

n = 3, ℓ = 0, m = 0 *(i)*

There are 8 more n = 3 patterns
in addition to the one shown.
The ℓ and m quantum numbers
are
ℓ = 1; m = 1, 0, –1
ℓ = 2; m = 2, 1, 0, –1, –2.

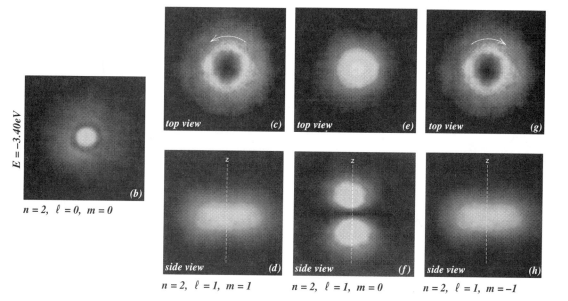

$E = -3.40eV$

n = 2, ℓ = 0, m = 0 *(b)*

top view *(c)*   *top view* *(e)*   *top view* *(g)*

side view *(d)*   *side view* *(f)*   *side view* *(h)*

n = 2, ℓ = 1, m = 1   *n = 2, ℓ = 1, m = 0*   *n = 2, ℓ = 1, m = –1*

$E = -13.6eV$

n = 1, ℓ = 0, m = 0 *(a)*

Figure 1
*The lowest energy standing wave patterns in hydrogen. The intensity
is what you would see looking through the wave.*

Exercise 1

a) For the n = 4 energy level, where $E_4 = -.85eV$, there are 16 allowed standing wave patterns. What are the values of the quantum numbers ℓ and m for these patterns?

b) How many allowed patterns are there, what are the values of ℓ and m, and what is the energy, for the n = 5 standing wave patterns?

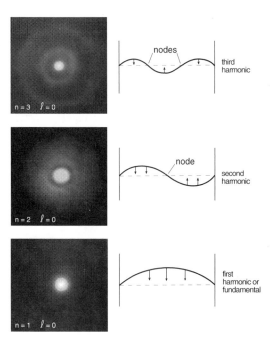

Figure 2
Comparison of the L = 0 electron standing wave patterns with the guitar standing waves. Each step up in energy level or harmonic introduces one more node.

The ℓ = 0 Patterns

For each energy level n, there is one wave pattern with $\ell = 0$. We have shown the first three $\ell = 0$ patterns in Figure (1). All $\ell = 0$ wave patterns are spherically symmetric. The n = 1, $\ell = 0$ pattern, for the ground state electron, is a fuzzy spherical ball with a diameter of about one angstrom. The n = 2, $\ell = 0$ pattern is a spherical ball surrounded by a spherical shell. Between the ball and the shell, at a radius r = 1.06 angstroms, the wave has value of zero. We can call this a spherical node.

In the n = 3, $\ell = 0$ pattern we have a spherical ball surrounded by two spherical shells. There are now two spherical nodes, the inner one located at r = 1.00 angstroms and the outer one at r = 3.75 angstroms. As we go up higher in energy, we get one more spherical node for each step up in energy level. In Figure (2), we compare the three lowest energy $\ell = 0$ wave patterns with the three lowest frequency standing wave patterns on a guitar string. While the patterns look quite different, both have the feature that as we go up one step in energy or frequency, we get one more node in the wave pattern. The first harmonic (n = 1) has no nodes between the ends of the string. The second harmonic (n = 2) has one node, while the third harmonic (n = 3) has two nodes, etc.

One of the key features of angular momentum is that it represents a rotation about an axis. In Newtonian mechanics, we defined the direction of the angular momentum vector $\vec{L}$ as being the direction of the axis of rotation. Because all the hydrogen wave patterns with $\ell = 0$ are spherically symmetric, they have no preferred axis about which the electron could have angular momentum.

The $\ell \neq 0$ Patterns

In all the $\ell \neq 0$ patterns, like the three $n = 1, \ell = 1$ patterns shown in the middle row of Figure (1), there is a special axis about which the electron can have angular momentum. This suggests that the ℓ quantum number is related to the electron's angular momentum, and that when $\ell = 0$, the electron has no angular momentum. This is different from the Bohr picture where the electron's angular momentum started with one unit $L = \hbar$ in the lowest energy level, two units $L = 2\hbar$ in the second level, etc. The Bohr theory did not allow for zero orbital angular momentum orbits while Schrödinger's equation tells us that there is a zero angular momentum wave pattern in each energy level.

Intensity at the Origin

Another general feature of the hydrogen wave patterns is that all $\ell = 0$ patterns have a maximum intensity at the origin, at the nucleus, while all the $\ell \neq 0$ patterns have a node there. The node at the origin for $\ell \neq 0$ patterns has a simple classical explanation. The classical formula for angular momentum is *the linear momentum p times the lever arm $r_\perp$*. In order for the electron to have non zero angular momentum about the nucleus, it must have a non zero lever arm $r_\perp$ and therefore cannot be at the nucleus. (One has to be careful applying Newtonian arguments to atomic phenomena. In the next chapter we will see a similar argument fail when we discuss electron spin).

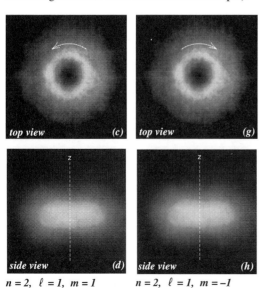

top view   (c)   *top view*   (g)

side view   (d)   *side view*   (h)

$n = 2, \ \ell = 1, \ m = 1$ $n = 2, \ \ell = 1, \ m = -1$

Figures 1c,d,g, and h repeated

Later in this chapter we will see that the fact that $\ell = 0$ patterns have a maximum at the nucleus while the $\ell \neq 0$ patterns have a node there, plays an important role in the electron structure and chemical properties of atoms.

Quantized Projections of Angular Momentum

A clue to understanding the $\ell \neq 0$ wave patterns can be obtained from a more detailed look at the two doughnut shaped patterns in Figure (1), the patterns labeled by the quantum numbers $n = 2$, $\ell = 1$, $m = +1$ and $m = -1$. While the $m = +1$ and $m = -1$ patterns look the same, a more detailed calculation with the Schrödinger equation shows in the $m = +1$ pattern the electron is traveling around the doughnut in a *counterclockwise direction*, while in the $m = -1$ pattern the electron is traveling *clockwise*.

These two patterns have an axis of symmetry which we have labeled the z axis, that passes up through the center of the doughnut. (These axes are shown as white dotted lines in the side views of these patterns, Figures 1d and 1h.) Further calculation with Schrödinger's equation shows that the electron in the $\ell = 1$, $m = 1$ pattern (counterclockwise motion), the electron has a *z component of angular momentum* precisely equal to one unit $\hbar$.

$$L_z \left(\begin{array}{c} \text{for the } \ell = 1 \\ m = 1 \text{ pattern} \end{array} \right) = \hbar \qquad (4a)$$

For the clockwise motion, the $\ell = 1$, $m = -1$ pattern, the z component of angular momentum is minus one unit $\hbar$

$$L_z \left(\begin{array}{c} \text{for the } \ell = 1 \\ m = -1 \text{ pattern} \end{array} \right) = -\hbar \qquad (4b)$$

The pattern in between, the one that looks like two tennis balls, one on top of the other, described by the quantum numbers $\ell = 1$, $m = 0$ turns out to have no angular momentum in the z direction.

$$L_z \left(\begin{array}{c} \text{for the } \ell = 1 \\ m = 0 \text{ pattern} \end{array} \right) = 0 \qquad (4c)$$

We see that the "m" quantum number tells us how many units of angular momentum the electron has in the z direction.

There is somewhat of an analogy between the three $n = 2$, $\ell = 1$ patterns in Figure (1) and the bicycle wheel demonstration we discussed in Chapter 7 (Figures 7-15 and 7-16). In Figure (3) we compare the $m = 1$ pattern with a bicycle wheel whose angular momentum $\vec{L}$ points in the $+ z$ direction, the $m = -1$ pattern with a bicycle wheel whose angular momentum points in the $- z$ direction, and the $m = 0$ pattern with a bicycle wheel that has no component of angular momentum in the z direction.

The analogy shown in Figure (3) actually demonstrates how different angular momentum is on an atomic scale from what we are familiar with on a human scale. The most striking difference is that you can point a bicycle wheel in any direction you want. By turning the wheel

over, you can change the z component L_z from $+L$ when it is pointing up to any value down to $-L$ when the wheel is pointing down. Any value between $+L$ and $-L$ is allowed.

For the hydrogen atom, an electron in the second energy level has only three $\ell = 1$ wave patterns, only three distinct z projections of angular momentum, each differing by one unit of angular momentum $\hbar$. There is no wave pattern for L_z equal to some fractional value of $\hbar$—*the projections of angular momentum are quantized!* There is absolutely nothing in Newtonian mechanics that prepares us for understanding how projections of angular momentum can be quantized. It is strictly a consequence of the wave nature of the electron, and the fact that a confined wave has only certain allowed standing wave patterns.

Figure 3
There are three standing wave patterns for a second energy level, unit angular momentum electron. Schrödinger's equation tells us that the z axis projection of angular momentum in the three patterns are 1 unit, 0 units, and –1 units . There are no intermediate values, because there are no other wave patterns. In comparison, a bicycle wheel has not only the three projections of angular momentum shown, but also many intermediate values.

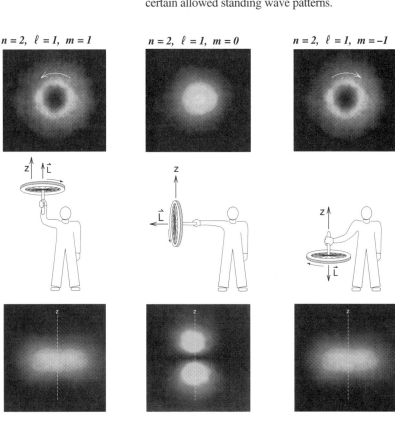
$n = 2$, $\ell = 1$, $m = 1$ $n = 2$, $\ell = 1$, $m = 0$ $n = 2$, $\ell = 1$, $m = -1$

The Angular Momentum Quantum Number

In Figure (3) we showed the different orientations of a bicycle wheel with a total angular momentum of magnitude L. The z component of the bicycle wheel's angular momentum ranges from +L when the axis is pointing up to –L when the axis is pointing down. For the hydrogen electron in Figure (3), $\ell = 1$ for all the patterns, and the z projection of the electron's angular momentum ranges from +1 unit for the m = 1 pattern down to –1 unit for the m = –1 pattern. This suggests that the quantum number ℓ represents the total angular momentum of the electron while m represents the allowed z projections.

This interpretation is almost right. The ℓ quantum number is related to the electron's total angular momentum, but the value of the total angular momentum is not quite equal to ℓ units of angular momentum as one might expect. Solving Schrödinger's equation for the magnitude L of the electron's orbital angular momentum about the proton gives the result

$$L = \sqrt{\ell(\ell+1)}\ \hbar \qquad \begin{array}{l}\textit{total angular}\\\textit{momentum of}\\\textit{the electron}\end{array} \qquad (5)$$

For large values of ℓ, the difference between ℓ and $\sqrt{\ell(\ell+1)}$ is slight and the z projections of angular momentum can range essentially from $+\ell\hbar$ to $-\ell\hbar$ as one would expect from the bicycle wheel analogy. But for small (non zero) values of ℓ, the total angular momentum is significantly larger than the maximum z projection. For $\ell = 1$, the maximum z projection is $\hbar$, while the total angular momentum is $L = \sqrt{1(1+1)}\ \hbar = \sqrt{2}\ \hbar$.

There was no guarantee that angular momentum had to behave on an atomic scale, just the way we expected it to from our experience with large scale phenomena. All we need to do is understand the transition from large to small scale phenomena. In the case of angular momentum, we can picture the bicycle wheel as having a huge angular momentum quantum number ℓ. As a result there are a huge number of allowed projections, with m ranging from $+\ell$ to $-\ell$, which allows us to rotate the bicycle wheel axis in an apparently continuous fashion. And there is essentially no difference between ℓ and $\sqrt{\ell(\ell+1)}$, thus the maximum z projection of angular momentum essentially equals the total angular momentum L.

Other notation

Further notation that some readers may have encountered, are the names (*s waves*) for the $\ell = 0$ patterns, (*p waves*) for $\ell = 1$ patterns and (*d waves*) for $\ell = 2$ patterns. These names, which are fairly common, have a rather obscure historical origin. Using this notation, one can, for example, refer to an electron in an n = 3, $\ell = 2$ pattern as a 3d wave. The ground state of hydrogen is a 1s wave.

Exercise 2

Using the s,p,d notation, what would we call the waves shown in Figure (3) ?

An Expanded Energy Level Diagram

In our discussion of the Bohr theory, we drew an energy level diagram so that we could study transitions from one level to another in order to predict the energy of the photons the atom could emit. The diagram, like the one in Figure (35-3), is quite simple with one line for each energy level.

With the Schrödinger equation we discover that there are numerous standing wave patterns for each energy level. The simple energy level diagram of Figure (35-3) does not give a hint of the multiple wave patterns. Only the energy quantum number n is shown, there being no indication of the ℓ and m quantum numbers (which were unknown when Bohr developed his theory).

It is traditional (and convenient) to expand the energy level diagram as we have done in Figure (4), to distinguish not only the energy quantum numbers n, but also the angular momentum quantum numbers ℓ. We might be tempted to expand the diagram further and display the separate projections m, but this would make the diagram too complex. (In Figure (4) we indicated the z projections by including some sketches of the lower energy wave patterns. Such sketches are not usually included in energy level diagrams.)

One advantage of the expanded energy level diagram is that it illustrates graphically that the maximum value of ℓ goes up only to n–1. It shows that there is one $\ell = 0$ pattern for n = 1, an $\ell = 0$ and an $\ell = 1$ pattern for n = 2, etc. When you look at this diagram, you have to remember that for each line, the z projections m can range from m = + ℓ down to m = – ℓ in unit steps.

Another advantage of the energy level diagram of Figure (4) is related to the fact that when an electron in an atom radiates a photon, the electron's ℓ value almost always changes by one unit. This is because a photon carries out angular momentum, and to conserve angular momentum, the electron's angular momentum has to change. The common transitions represent not only steps up and down, but one step sideways.

(It is not impossible for an electron to emit a photon and not change its angular momentum ℓ, it just a much less likely event. We only see such $\Delta\ell = 0$ transitions, called "forbidden transitions", when the electron has no where to jump and change its ℓ value by one unit. For example, if the electron, for some reason, ends up in the n = 2, $\ell = 0$ state, the only lower energy state is the n = 1, $\ell = 0$ state. The electron cannot fall there and change ℓ by one unit. As a result the electron hangs up in the n = 2, $\ell = 0$ state for a much longer time than it would if a $\Delta\ell = 1$ transition were available.)

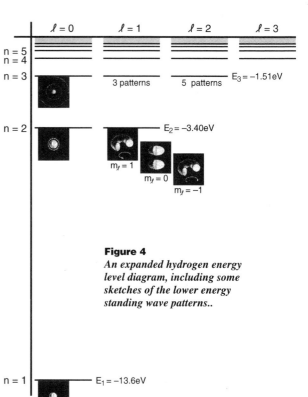

Figure 4
An expanded hydrogen energy level diagram, including some sketches of the lower energy standing wave patterns..

MULTI ELECTRON ATOMS

Straightforward techniques can be used to solve Schrödinger's equation for one electron atoms like hydrogen. To deal with a two electron atom like helium, we have to take into account not only the attraction between the electrons and the nucleus, but also the repulsion between electrons. This makes Schrödinger's equation more difficult to solve. One has to either use approximation techniques or a computer.

However, for all atoms, there are certain properties that can be understood in terms of the general structure of the hydrogen standing wave patterns, rather than from detailed calculations. We can learn enough from these general properties to begin to see why atoms behave as they do in chemical reactions.

To study multi electron atoms, imagine that we start with hydrogen and add electrons one at a time (also increasing the number of protons and neutrons in the nucleus to keep the atom electrically neutral and the nucleus stable). We will assume that as we add each electron, it falls down to the lowest energy wave pattern available.

If we start with a nucleus with one proton, and drop in one electron, the electron eventually falls down to the E_1, $\ell = 0$ standing wave pattern shown in Figure (1a).

Add a proton to form a helium nucleus, drop in another electron, and we can expect the electron to also fall down to the lowest energy E_1 standing wave pattern. The extra Coulomb attractive force of the two protons in the nucleus strengthens the binding of the electrons, but the repulsive force between the two electrons weakens it. Experimentally, it takes 24.6 eV to remove an electron from helium, while only 13.6 eV are needed for hydrogen. Thus the electrons are more tightly bound in helium, and we see that the extra Coulomb attraction to the nucleus is more important than the repulsion between electrons.

Using helium as a guide, we should expect that when we go to lithium with 3 protons in the nucleus, the increased Coulomb attraction to the nucleus should cause lithium's three electrons to be even more tightly bound than helium's two. This would lead us to predict that it takes even more than 24.6 eV to pull one of the electrons out of lithium.

This is not a good prediction. Experimentally, the amounts of energy needed to remove electrons from lithium one at a time are *5.39 eV, 75.26 eV and 121.8 eV*. While two of lithium's electrons are tightly bound, one is very loosely bound, requiring less than half the energy to remove than the hydrogen electron. A possible explanation for the loose binding of lithium's third electron is that, for some reason, that electron did not fall down to the lowest energy E_1 type of standing wave pattern. It appears to be hung up in the much higher energy, less tightly bound E_2 type of standing wave, one of the four E_2 patterns seen in Figure (1).

Pauli Exclusion Principle

But why couldn't the third lithium electron fall down to the low energy E_1 pattern? In 1925, two separate ideas provided the explanation. Wolfgang Pauli proposed that no two electrons were allowed to be in exactly the same state. This is known as the *Pauli exclusion principle*. But the exclusion principle seems to go too far, because in helium, both electrons are in the same E_1, $\ell = 0$ standing wave pattern. If you cannot have two electrons in exactly the same state in an atom, then something must be different about the two electrons in helium.

Electron Spin

To explain what the difference between the two electrons might be, two graduate students, Samuel Goudsmit and George Uhlenbeck, proposed that the electron was like a spinning top with its own internal angular momentum. This became known as *spin* angular momentum. The special feature of the electron's spin is that it has *two allowed projections*, which we call *spin up* and *spin down*. In helium you could have two electrons in the same E_1 wave pattern if they had different spin projections, for then they would not be in identical states.

Because the electron spin has only two allowed projections, we cannot add a third electron to the E_1 wave pattern. Lithium's third electron must stop at one of the higher energy E_2 standing wave patterns. Its energy is much less negative and therefore this electron is much less tightly bound than the first two electrons that went down to the E_1 wave pattern.

THE PERIODIC TABLE

As we go to larger atoms, adding electrons one at a time, the E_2 standing wave patterns begin to fill up. Since there are four E_2 patterns, each with two allowed spin states, up to 8 electrons can fit there. When the E_2 patterns are full, when we get to the element neon with two E_1 electrons and eight E_2 electrons, we have an inert noble gas that is chemically similar to helium. Adding one more electron by going to sodium, the eleventh electron has to go up into the E_3 energy level since both the E_1 and E_2 patterns are full. This eleventh electron in sodium is loosely bound like the third electron in lithium, with the result that both lithium and sodium have similar chemical properties. They are both strongly reactive metals.

Table 1 shows the electron structure and the binding energy of the last electron for the first 36 elements in the periodic table. The general features of this table are that the lowest energy levels fill up first, and there is a large drop in binding energy when we start filling a new energy level. We see these drops when we go from the inert gases helium, neon, and argon to the reactive metals lithium, sodium and potassium. We can see that this sudden change in binding energy leads to a significant change in the chemical properties of the atom.

A closer look at Table 1 shows that there is a relatively steady uniform increase in the electron binding as the energy level fills up. The binding energy goes from 5.39 eV for lithium in fairly equal steps up to 21.56 eV for neon as the E_2 energy level fills. The pattern is more or less repeated as we go from 5.14 eV for sodium up to 15.76 eV for argon while filling the E_3 energy level. It repeats again in going from 4.34 eV for potassium up to the 14.00 eV for krypton.

A closer look also uncovers some exceptions to the rule that the lower energy levels fill first. The most notable exception is at potassium, where the E_4 patterns with $\ell = 0$ begin to fill before the E_3 patterns with $\ell = 2$.

To understand why the binding energy gradually increases as an energy level fill up, and why the E_3, $\ell = 2$ patterns fill up late, we have to take a closer look at the structure of the electron wave patterns and see how this structure affects the binding energy. To do this it is useful to introduce the concepts of electron screening and effective nuclear charge.

Electron Screening

In our discussion of the binding energy of the two electrons in helium, we pointed out that there was a competition between the increased Coulomb attractive force to the nucleus and the repulsion between the electrons. We could see that the increased attraction was more important because helium's two electrons are each more tightly bound to the nucleus than hydrogen's one. It requires 24.5 eV to remove an electron from helium and only 13.6 eV from hydrogen.

The following argument provides an explanation of this increased binding of helium's electron. Since the two electrons are in the same E_1 wave pattern, half the time a given electron is closer to the nucleus than its partner and feels the full force of the nuclear charge +2e. But half the time it is farther away, and the net charge attracting it toward the nucleus is +2e reduced by the other electron's charge –1e for a total +1e. Thus, on the average the electron sees an effective charge of approximately 1.5 e. This is greater than the charge +1e seen by the single electron in hydrogen, and thus results in a stronger binding energy. What we have done is to account for the repulsion of the other electron by saying that the other electron screens the nucleus, reducing the nuclear charge from +2e to an effective value of approximately 1.5e.

Z		Element	Binding energy of last electron in eV	E_1	E_2		E_3			E_4		Energy level E_n
				0	0	1	0	1	2	0	1	Angular momentum quantum number ℓ
1	H	Hydrogen	13.60	1								
2	He	Helium	24.58	2								
3	Li	Lithium	5.39		1							
4	Be	Beryllium	9.32		2							
5	B	Boron	8.30		2	1						
6	C	Carbon	11.26		2	2						
7	N	Nitrogen	14.54		2	3						
8	O	Oxygen	13.61		2	4						
9	F	Fluorine	17.42		2	5						
10	Ne	Neon	21.56		2	6						
11	Na	Sodium	5.14				1					
12	Mg	Magnesium	7.64				2					
13	Al	Aluminum	5.98				2	1				
14	Si	Silicon	8.15				2	2				
15	P	Phosphorus	10.55				2	3				
16	S	Sulfur	10.36				2	4				
17	Cl	Chlorine	13.01				2	5				
18	A	Argon	15.76				2	6				
19	K	Potassium	4.34							1		
20	Ca	Calcium	6.11							2		
21	Sc	Scandium	6.56						1	2		
22	Ti	Titanium	6.83						2	2		
23	V	Vanadium	6.74						3	2		
24	Cr	Chromium	6.76						5	1		
25	Mn	Manganese	7.43						5	2		
26	Fe	Iron	7.90						6	2		
27	Co	Cobalt	7.86						7	2		
28	Ni	Nickel	7.63						8	2		
29	Cu	Copper	7.72						10	1		
30	Zn	Zinc	9.39						10	2		
31	Ga	Gallium	6.00						10	2	1	
32	Ge	Germanium	7.88						10	2	2	
33	As	Arsenic	9.81						10	2	3	
34	Se	Selenium	9.75						10	2	4	
35	Br	Bromide	11.84						10	2	5	
36	Kr	Krypton	14.00						10	2	6	

(Labels within the table: "Helium core" spans the E_1 region for Li–Ne; "Neon core" spans the E_1–E_2 region for Na–A; "Argon core" spans the E_1–E_3(ℓ=0,1) region for K–Kr.)

Table 1
Electron binding energies. Adapted from Charlotte E Moore, **Atomic Energy Levels,** *Vol II, National Bureau of Standards Circular 467, Washington, D.C., 1952.*

Effective Nuclear Charge

To see what effect changing the nuclear charge has on the binding energy, we can go back to our Bohr theory calculations for different one electron atoms. In Exercises 9 and 10 of Chapter 35 we found that the ground state energy of a single electron in an atom where the nucleus had z *protons* was

$$E_1 = z^2 \times (-13.6 \text{ eV}) \quad \begin{matrix} \textit{ground state} \\ \textit{energy in a single} \\ \textit{electron atom} \end{matrix} \quad (6)$$

While Equation 6 was derived from the Bohr theory, it gave results in excellent agreement with experiment as one can easily see from working Exercise 3.

Exercise 3

Table 2 lists the binding energy for the last electron for the elements hydrogen through boron. This is the binding energy when all the other electrons have already been removed. For each element, check the prediction that the binding energy is given by Equation 6.

z	Element	Binding energy of last electron
1	Hydrogen	13.6 eV
2	Helium	54.14 eV
3	Lithium	121.8 eV
4	Beryllium	216.6 eV
5	Boron	338.5 eV

Table 2

Equation 6 suggests that if an electron in a multi electron atom sees an effective nuclear charge $(z_{eff}e)$, the electron binding energy should be approximately z_{eff}^2 times the energy the electron would have in the same energy level in hydrogen. Trying out this idea on helium, where we estimated z_{eff} to be about 1.5e, we get

$$E_1 \begin{pmatrix} neutral \\ helium \end{pmatrix} = z_{eff}^2 \times (-13.6 \text{ eV})$$

$$= (1.5)^2 \times (-13.6 \text{ eV})$$

$$= -30.6 \text{ eV}$$

$$\begin{matrix} \textit{estimated} \\ = \textit{binding energy} \\ \textit{of helium} \end{matrix} \quad (7)$$

This estimate of 30.6 eV is about 25% too high since the experimental value is only 24.6 eV. We can take this to imply that our estimate of $z_{eff} = 1.5$ e for helium was a bit too crude. Our simple arguments about screening are not a substitute for an accurate calculation using Schrödinger's equation.

What we can do, however, is to turn our approach around and use the experimental values of the binding energy to calculate an effective nuclear charge (z_{eff}). Doing this for helium gives

$$E_1 \begin{pmatrix} neutral \\ helium \end{pmatrix} = z_{eff}^2 \times (-13.6 \text{ eV})$$

$$-24.6 \text{ eV} = z_{eff}^2 \times (-13.6 \text{ eV})$$

$$z_{eff} = \sqrt{\frac{24.6}{13.6}} = 1.34 \quad (8)$$

The value of 1.34 is not too far off our original guess of 1.5. The result tells us that the electron screening is a bit more effective than we had predicted.

Lithium

We will now see that various features of the periodic table begin to make sense when viewed in terms of electron screening and the structure of the electron wave patterns. Let us start off with lithium where the last electron is in the E_2, $\ell = 0$ pattern and has a binding energy of 5.39 eV. Since this electron is in an E_2 energy level, our formula for z_{eff} should be

$$E_1 (lithium) = z_{eff}^2 \times (-3.40 \text{ eV})$$

$$-5.39 \text{ eV} = z_{eff}^2 \times (-3.40 \text{ eV})$$

$$z_{eff} = \sqrt{\frac{5.39}{3.40}} = 1.26 \quad (9)$$

where we used - 3.40 eV rather than - 13.6 eV because we are discussing an E_2 electron.

In a nucleus with 3 protons, why does the E_2 electron only see an effective charge of 1.26e? The answer lies in the shape of the E_1 and E_2 wave patterns. The first two electrons in lithium are in the E_1 pattern of Figure (1a) reproduced below. It consists of a small spherical ball centered on the nucleus. The third electron, the one whose binding energy we are discussing, is in the E_2, $\ell = 0$ pattern of Figure (1b). This pattern consists of a larger spherical ball surrounded by a spherical shell. The electron in this pattern spends a considerable amount of the time outside the smaller spherical ball of the two E_1 electrons. Thus much of the time the third electron sees only an effective nuclear charge of about (1.0e). Some of the time, however, the third electron is also down at the nucleus feeling the full nuclear charge of (3e). That the average nuclear charge seen by the third electron is (1.26e) is not too difficult to believe.

Figure 1a

E_1, $\ell = 0$

Figure 1b

E_2, $\ell = 0$

Beryllium

When we went from one E_1 electron in hydrogen to two E_1 electrons in helium, the binding energy about doubled, from 13.6 eV to 24.6 eV. In going from one E_2 electron in lithium to two E_2 electrons in beryllium, the binding energy increases from 5.39 eV to 9.32 eV. Again the electron binding energy almost doubled as we went from one to two electrons in the same energy level.

Boron

When we go from beryllium to boron, we add a third electron to the E_2 energy level. From our experience with beryllium, we expect another significant increase in binding energy, up to perhaps 13 eV or 14 eV. But instead the binding energy drops from 9.32 eV down to 8.30 eV. Something broke the pattern and caused this drop.

At boron, both the E_2, $\ell = 0$ wave patterns are already full and the electron has to go into one of the E_2, $\ell = 1$ patterns. All the electron standing wave patterns with a non zero amount of angular momentum have a node at the origin. The more the angular momentum, the more spread out the node and the farther the electron is kept away from the nucleus. An electron in an $\ell \neq 0$ wave pattern will thus be effectively screened by electrons in $\ell = 0$ wave patterns where the electron spends a lot of time right down at the nucleus. Thus we expect that electrons in $\ell \neq 0$ patterns to be less tightly bound than those in the $\ell = 0$ pattern of the same energy level. This shows up with the drop in binding energy in going from beryllium to boron.

Up to Neon

For all atoms beyond helium, there is a core consisting of the nucleus and the two tightly bound E_1 electrons. As the charge on the nucleus increases, the size of the E_1 patterns shrink, and are penetrated less and less by the outer electrons. We can think of this helium core as acting as the effective nucleus for the larger atoms.

As we go from boron to neon, the charge on the helium core increases from 3e to 8e as the E_2, $\ell = 1$ patterns fill up. This increase in the charge of the core causes a more or less gradual increase in the binding energy, from 8.30 eV up to 21.56 eV. The one exception is the slight drop in binding energy as we go from nitrogen to oxygen. The arguments we have made so far are not detailed enough to explain this drop.

Sodium to Argon

We get the expected large drop in binding energy as we go from neon to sodium and start filling the E_3 patterns. The E_3, $\ell = 0$ patterns are full at magnesium and we get a small drop in binding energy as the non zero angular momentum patterns E_3, $\ell = 1$ start to fill at aluminum. Again the angular momentum keeps the electrons away from the nucleus and increases the screening. As the E_3, $\ell = 1$ patterns fill up, they are building a structure on the ever shrinking neon core. The charge on the neon core increases from (3e) at aluminum to (8e) at argon, again causing a gradual buildup of the electron binding energy from 5.98 eV to 15.76 eV. There is even the slight glitch going from phosphorous to sulfur that mirrors the glitch from nitrogen to oxygen.

Potassium to Krypton

The first major break in the pattern of filling the lower energy levels first occurs at potassium. At potassium the E_3, $\ell = 2$ levels remain unfilled while the last electron goes into the higher energy level E_4, $\ell = 0$ pattern. At this point the screening due to angular momentum has become more important than the energy level. The $\ell = 2$ patterns have such a big fat node at the nucleus that an $\ell = 2$ electron cannot get near the nucleus to feel the now large nuclear charge. Even though an E_4, $\ell = 0$ electron is in a higher energy level, its wave pattern has a non zero value right down at the nucleus. Some of the time this electron feels the full charge of (19e) for potassium, and this increases the binding beyond that of the E_3, $\ell = 2$ patterns.

At calcium, the E_4, $\ell = 0$ pattern is full, and now the five E_3, $\ell = 2$, $m_\ell = +2, +1, 0, -1, -2$ patterns begin to fill up. There is room for 10 electrons in these 5 patterns, and that takes us down to zinc. As the E_3, $\ell = 2$ patterns fill up underneath the E_4, $\ell = 0$ pattern, there is little change in the outer electron structure and the binding energy increases slowly. The result is that the 10 elements from scandium to zinc have similar chemical properties—all are metals. In some periodic tables, these elements are shown as the first set of *transition elements*.

As we go from gallium to krypton we have the familiar pattern of the E_4, $\ell = 1$ states filling up. There is a gradual increase in binding energy from 6.00 eV at gallium to 14.00 eV at the noble gas krypton. There is even the slight glitch in binding energy going from arsenic to selenium that mirrors the glitches from phosphorus to sulfur, and from nitrogen to oxygen.

Summary

At this point it should be clear that the structure of the periodic table of the elements arises from the allowed electron standing wave patterns. Because of the exclusion principle, no two electrons can be in the same state. But because electron spin has two allowed states, up to two electrons can fit into each standing wave pattern.

In general, as we go to atoms with more electrons, the lowest energy patterns fill up first, and there is a significant change in chemical properties when a new energy level begins to fill. But the angular momentum of the wave pattern also plays a significant role. The $\ell = 0$ patterns can penetrate down to the nucleus, where the electron feels the full strength of the nuclear charge. The $\ell \neq 0$ patterns have a node at the nucleus, and the full nuclear charge is screened by $\ell = 0$ electrons.

The effect of angular momentum shows up most noticeably at potassium and calcium, where the two E_4, $\ell = 0$ patterns fill before the E_3, $\ell = 2$ patterns. Because of the extra angular momentum of the $\ell = 2$ electrons, the $\ell = 2$ patterns have an extra large node at the nucleus, keeping these electrons farther away and more effectively screened.

As we get to the heavier elements in the periodic table, those beyond krypton, the energy levels get closer together and the binding energy depends more on the detailed structure of the wave patterns. As a result it becomes more difficult to predict how the wave patterns will be filled and to estimate what the binding energies should be. But despite this, we have been able to go a long way in explaining the structure of the periodic table from a few simple arguments about the shape of the electron standing waves in hydrogen, and the idea of electron screening.

IONIC BONDING

In 1871 the Russian chemist Dimitri Mendeleyev worked out the periodic table of the elements from an analysis of the atomic weights and chemical reactions of the elements. Here we will reverse Mendeleyev's approach and use Table 1, our shortened version of the periodic table, to explain some of the typical chemical reactions and chemical compounds.

As an example, suppose we placed a sodium atom next to a chlorine atom, what would happen? The sodium atom has one loosely bound electron in the E_2, $\ell = 0$ wave pattern. The binding energy of this electron is 5.14 eV. The chlorine atom has seven E_2 electrons all tightly bound because of the increase in the effective nuclear charge seen by these electrons. It requires 13.01 eV to remove an electron from chlorine.

If the sodium and the chlorine atom are brought close enough together, the loosely bound outer sodium electron can lose energy by moving into the remaining E_2 wave pattern in the chlorine atom. We end up with a negative chlorine *ion* Cl^-, where all the E_2 patterns are full, and a positively charged sodium ion Na^+ which has lost its outer electron. These charged ions then attract each other electrically to form a sodium chloride molecule $NaCl$ which is common table salt.

Sodium chloride is a typical example of *ionic bonding*. The class of elements like lithium, sodium, magnesium, aluminum, etc. that have one, two, or even three loosely bound electrons, tend to give up these electrons in a chemical reaction. These are called *metals*. Those elements like oxygen, fluorine and chlorine, which have nearly full wave patterns and tightly bound electrons, tend to take up electrons in a chemical reaction and are called *non metals*. When metals and non metals combine, held together by ionic bonding, you get a compound called a *salt*.

By looking at the number of loosely bound electrons in a metal, or the number of empty slots in a non metal (the number of electrons required to get to the next noble gas), you can predict the kind of compounds an element can form. For example, sodium, magnesium, and aluminum have one, two and three loosely bound electrons respectively, while oxygen has two empty slots. (Oxygen has six E_2 electrons, and needs two more to fill up the E_2 standing wave patterns). When you completely burn the three metals, the oxides you end up with are Na_2O, MgO and Al_2O_3. In Na_2O, two sodium atoms each contribute one electron to fill oxygen's two slots. In MgO magnesium's two loosely bound electrons are taken up by one oxygen atom. In Al_2O_3, two aluminum atoms each supply three electrons, these six electrons are then taken up by three oxygen atoms. There is no simpler way for all the aluminum's loosely bound electrons to completely fill all of oxygen's empty slots.

Hydrogen has one moderately bound electron which it can give up in some chemical reaction and act like a metal. An example is hydrochloric acid, HCl, where the chlorine ion has grabbed the hydrogen electron.

Hydrogen can also behave as a non metal. When it combines with active metals like lithium and sodium, hydrogen grabs the metal's loosely bound electron to complete its E_1 standing wave pattern. The results are the compounds lithium and sodium hydride, LiH and NaH.

More important to life are the bonds like those between hydrogen and carbon atoms which are not ionic in nature. Neither atom has a strong preference to give up or grab electrons. Instead the bonding results from the sharing of electrons. This is the *covalent bonding* that we described in our discussion of the hydrogen molecule in Chapter 18.

Chapter 39
Spin

In the last chapter we saw that the basic structure of the periodic table follows from the idea that up to two electrons can fit into any given standing electron wave pattern. If you consider the Pauli exclusion principle which says that two electrons cannot be in exactly the same state, then you have to find some difference between the two electrons that can occupy the same wave pattern. Gaudsmit and Uhlenbeck introduced the concept of electron spin to explain this difference. They proposed that the electron had an inherent angular momentum or spin that had two allowed projections, and that the difference between the two electrons in one wave pattern was their spin projections. We commonly call these two allowed projections **spin up** *and* **spin down**.

In this chapter we take a more detailed look at electron and nuclear spin and the interaction of spin with a magnetic field. An electron with its spin projection parallel to the magnetic field gains magnetic energy, while the opposite projection loses it. The amount of energy gained or lost is proportional to the strength of the magnetic field.

The most accurate way to measure the spin magnetic energy, is to start with an electron in the low energy state and strike it with a photon. If the energy of the photon is precisely equal to the energy required to raise the electron from the low energy spin projection to the high energy spin projection, the photon can be absorbed. We say that the photon flips the spin of the electron. Since the energy of a photon is proportional to its frequency according to Einstein's photoelectric effect formula E=hf, measuring the frequency of the electromagnetic radiation that causes a spin flip tells you how big the magnetic energy is.

The energy required to flip the spin of an electron is usually not very large. If the electron is in a magnetic field of around 10 gauss, typical fields produced by the Helmholtz coils used in several of our laboratory experiments, then photons in radio waves whose frequency is of the order of 30 megacycles have enough energy to flip the electron spin. Since it is not hard to generate electromagnetic waves of this frequency, we can observe electron spin flip using much of our standard laboratory equipment.

When we talk of electromagnetic waves with frequencies of the order of 30 megacycles, we are talking about radio waves between the AM and FM broadcast bands. It is such a low frequency that individual photons should be very hard to detect. Yet the spin flip experiment does just that.

At radio wave frequencies, Maxwell's theory and the ideas of classical electric and magnetic fields should work as well as the photon picture. If we treat the spinning electron as a classical gyroscope with a magnetic moment, we find that a magnetic field can exert a torque on the gyroscope, causing the gyroscope to precess. If we add an oscillating magnetic field,

oscillating at the frequency of precession, essentially pushing on the gyroscope once each time it comes around, the gyroscope can gain energy from the oscillating field. This is a resonance phenomena; it is like pushing a kid on a swing. You have to push the kid in time with the swing in order to increase the amplitude of the motion.

It turns out that the frequency with which we have to oscillate the magnetic field is the same as the frequency of the photon that can cause the electron spin to flip. The quantum picture of a photon flipping a spin, and the classical picture of a precessing gyroscope in resonance with an oscillating magnetic field, gives the same results. Thus it is a matter of convenience whether you use the classical or quantum picture.

Because the classical picture involves a resonance, this spin flip process is called **electron spin resonance**. In this chapter we discuss an electron spin resonance experiment.

For a given magnetic field, much less energy is required to flip the spin of a nucleus than of an electron. Measurements of nuclear spin flip energies, in the so-called **nuclear magnetic resonance** experiments, can be done so accurately that one can study not only the spin of the nucleus but also the magnetic environment in which the nucleus sits. Nuclear magnetic resonance forms the basis of magnetic resonance imaging which has become such an important diagnostic tool in medicine.

THE CONCEPT OF SPIN

A spinning top has an inherent angular momentum, but if you try to picture an electron as a spinning top, you run into conceptual problems. First of all, if you have a spinning top, you can orient the top in any direction you please. The top's angular momentum vector can point up, down, sideways to the left, sideways to the right. But when we describe the electron's spin, we have only two orientations, up and down. We ran into the puzzling idea of quantized projections of angular momentum in our interpretation of the allowed standing wave patterns of the hydrogen atom. But the idea of the electron's spin or rotational axis only pointing up or down seems even more counter intuitive.

Another blow at our classical intuition for angular momentum is our current theoretical picture of the electron as a point particle. No experiment has demonstrated any finite size to the electron, and the theory that treats the electron as a point particle, quantum electrodynamics, is the most accurately tested theory in all of physics. (String theory allows for some size for an electron, radii of the order of 10^{-72} cm, but there are no experimental tests of string theory.)

If an electron has no radius, how can it have an inherent angular momentum? Angular momentum is linear momentum times a *lever arm*. How can there be angular momentum if the particle has no radius, no lever arm? The classical picture of electron spin resembling that of a spinning tops leaves a lot to be desired. However, despite the problems one encounters, this picture does lead to some useful insights which we will mention shortly.

Perhaps the best way to view electron spin is to realize that we are dealing with a wave equation, and wave equations have specific allowed standing waves as solutions. While electron spin does not come from Schrödinger's wave equation, it does from Dirac's more accurate relativistic wave equation. From Dirac's equation, we find that the electron has an inherent angular momentum of $\hbar/2$, with two possible projections along the z axis, $+\hbar/2$ and $-\hbar/2$. These are the two allowed states of the electron. They are not different standing wave shapes like the hydrogen standing waves of Figure (1) of the last chapter, but they are different solutions to Dirac's wave equation.

One of the surprises is that the spin angular momentum of the electron is half a unit $\hbar/2$. The quantity $\hbar$ is not the smallest amount of quantized angular momentum, $\hbar/2$ is. The standard terminology is to say that the electron has half a unit of angular momentum, that it is a "spin 1/2" particle. The orbital angular momentum, representing the motion of the electron around a nucleus, is quantized in units of $\hbar$. Only spin angular momentum can come in half integer units.

While the electron's spin has a half integer value, its projection along the z axis changes by an integer value. In our discussion of the angular momentum of the hydrogen standing waves, we saw that an electron in a certain energy level E_n, with a total angular momentum ℓ, could have z projections ranging from ℓ, $\ell - 1$, $\ell - 2$ down to $-\ell$. The allowed projections changed in units. The same is true for the electron spin, the allowed projections are $+1/2$ and $-1/2$, a change of one unit.

INTERACTION OF THE MAGNETIC FIELD WITH SPIN

One of the predictions of the Dirac equation for electrons is that the electron spin interacts with a magnetic field. The state with the spin parallel to the magnetic field gains magnetic energy while the other state loses it. The amount of energy gained or lost is proportional to the strength B of the magnetic field, and the proportionality constant turns out to be a quantity called the **Bohr magneton,** designated by the symbol μ_B.

$$\begin{matrix} \text{magnetic} \\ \text{energy of} \\ \text{electron} \\ \text{spin} \end{matrix} \quad E_{mag} = \begin{cases} +\mu_B B & \text{spin parallel to } \vec{B} \\ -\mu_B B & \text{spin opposite to } \vec{B} \end{cases} \quad (1)$$

$$\mu_B = e\hbar/2m \quad \textit{Bohr magneton}$$

$$= 5.79 \times 10^{-5} \text{eV/tesla} \quad (2)$$

The amount of energy required to flip an electron from its low energy state to the high energy state is thus

$$\boxed{\Delta E_{mag} = 2\mu_B B} \quad \begin{matrix} \textit{energy required to} \\ \textit{flip the electron spin} \\ \textit{in a magnetic field} \end{matrix} \quad (3)$$

Since a Bohr magneton is 5.79×10^{-5}eV/tesla we can express Equation 3 numerically as

$$\Delta E_{mag} = \left(11.6 \times 10^{-5} B\right) eV \quad (4)$$

where B has to be expressed in tesla.

If you wish to measure magnetic fields in gauss, then convert μ_B to eV/gauss:

$$\mu_B = 5.79 \times 10^{-5} \frac{eV}{tesla} \times \frac{1}{10^4 \frac{gauss}{tesla}}$$

$$\boxed{\mu_B = 5.79 \times 10^{-9} \frac{eV}{gauss}} \quad (5)$$

Magnetic Moments and the Bohr Magneton

The formula for the Bohr magneton has its origin in a combination of classical physics with the Bohr theory. Back in Chapter 31 we observed that if you place a loop of wire in a magnetic field $\vec{B}$, and then run an electric current i through the wire, the magnetic field can exert a torque on the loop. We found that if you curled the fingers of your right hand in the direction the current is going around the loop, then the magnetic torque tended to orient the loop so that your thumb pointed parallel to the magnetic field. We called this the low energy orientation of the loop. To turn the loop over to the high energy orientation required an amount of work that was proportional to the current i, the strength of the magnetic field B, and to the area A of the loop. The explicit formula for the amount of work required was

$$\begin{matrix} \text{energy required} \\ \text{to turn loop over} \end{matrix} \Big\} = 2(iA)B$$

We defined the product of the current i times the area A as the **magnetic moment** μ of the current loop.

$$\mu \equiv iA \quad \text{(see 31-34)}$$

which gave us the formula

$$\begin{matrix} \text{energy required} \\ \text{to turn loop over} \end{matrix} \Big\} = 2\mu B \quad (31\text{-}36a)$$

Later in the chapter we considered a special kind of current loop consisting of a charge q moving at a speed v in a circular orbit of radius r. We found that the magnetic moment $\mu = iA$ of this special current loop could be written in the form

$$\mu = \frac{q}{2m}(mvr) \quad (31\text{-}38a)$$

However mvr is the angular momentum L (we called it J back there because we were using L for inductance). Thus the formula for the magnetic moment μ of an orbiting charge can be written

$$\mu = \frac{q}{2m}L \quad (31\text{-}39)$$

The above result is strictly classical. If we jump ahead to the Bohr picture where we are dealing with an electron whose charge is q = –e, and angular momentum L is quantized in units of $\hbar$, then we find that the magnetic moment is quantized in units of

$$\mu_B = \frac{e}{2m}\hbar \qquad (6)$$

where this unit of magnetic moment μ_B is called a *Bohr magneton*. This is where the name and formula for the Bohr magneton originated. The same constant appeared in Dirac's equation for the energy required to flip the spin of the electron.

The minus sign of the electron charge means that the high energy orientation of the electron is when the spin is parallel to the magnetic field.

Exercise 1

The formulas

$$E_{magnetic} = 2\mu_B B$$

$$\mu_B = \frac{e}{2m}\hbar$$

came mostly from Chapter 31 where we were working in MKS units. As a result, we need to use MKS units to evaluate μ_B. (The constant μ_B has a different formula in CGS units.)

We can get the dimensions of μ_B from the equation $E_{magnetic} = 2\mu_B B$, or

$$\mu_B = \frac{E_{magnetic}}{2B} \frac{joules}{tesla}$$

Thus when you use MKS units to evaluate $\mu_B = e\hbar/2m$, your answer comes out in joules/tesla rather than eV/tesla. To get the final answer in eV/tesla, you then use the conversion factor 1.6×10^{-19} joules/eV.

With this background, show that

$$\mu_B = 5.79 \times 10^{-5} \frac{eV}{tesla}$$

(You will get a value of $\mu_B = 5.82 \times 10^{-5}$ *eV/tesla*, which differs slightly due to the way we have rounded off the constants.)

Electron Spin Resonance Experiment

The basic idea of the electron spin resonance experiment is to flip the spin of an electron by striking the electron with a photon. The electron's spin will flip only if the photon's energy hf is equal to the magnetic spin flip energy $2\mu_B B$. Thus we wish to test the relationship

$$hf = 2\mu_B B \quad \textit{spin flip requirement} \qquad (7)$$

where f is the frequency of the photon.

Exercise 2

a) An electron is placed in a 10 gauss magnetic field. How much energy, in eV, is required to flip the electron from its low energy to its high energy state? (Answer: 1.15×10^{-7} eV).

b) You wish to flip the spin of the electron in part (a) by striking it with a photon. Assume that the photon is absorbed by the electron, and that all the photon's energy goes into flipping the electron's spin. What wavelength photon should you use? (Answer: $\lambda = 1071$ cm)

c) What is the frequency of the photon in part (b)? (Answer: 28 megacycles).

Exercise 3

The student FM radio station at Dartmouth College broadcasts on a frequency of 99.4 megacycles. If you wished to use this frequency radiation to flip the spin of an electron in a magnetic field B, what should be the strength of B? Give the answer in gauss.

In our discussion of the particle nature of light, we pointed out that because a radio wave consists of so many photons of such low energy, it would be difficult to detect individual photons, and thus the wave nature of radio waves should predominate. However the spin of the electron is just the right detector for these low energy photons. An electron spin flip experiment can be viewed as an experimental detection of the individual photons in a radio wave.

Nuclear Magnetic Moments

Both the proton and neutron are spin 1/2 particles, which means that they each have a spin angular momentum with two allowed spin states, spin up and spin down. If you place either of these particles in a magnetic field, one of the projections will gain magnetic energy while the other loses it.

For an electron, the high magnetic energy state was when the spin pointed parallel to the magnetic field. Because the proton has the opposite charge from the electron, the opposite orientation is the high magnetic energy state.

If the Dirac equation is applied accurately to a proton, then the formula for the proton's magnetic moment would be one *nuclear magneton* μ_N defined by the equation

$$\mu_N = \frac{e\hbar}{2m_{proton}} \qquad \begin{array}{l}\textit{definition of the}\\ \textit{nuclear magneton}\end{array} \qquad (8)$$

which is the Bohr magneton formula with the electron mass replaced by the proton mass. Since the proton is 1836 times heavier than an electron, a nuclear magneton is 1/1836 times smaller than a Bohr magneton.

The Dirac equation, however, does not give the correct value for the proton's magnetic moment μ_p. The experimental value is

$$\boxed{\mu_p = 2.79\,\mu_N} \qquad (9)$$

The fact that the Dirac equation is off by a factor of 2.79 is one indication that the proton is a more complex object than the electron.

(The Dirac equation is not exact even for the electron. The experimental value for the electron's magnetic moment is 1.00114 Bohr magnetons. The correction of .00114 Bohr magnetons is accurately explained by the theory of quantum electrodynamics.)

Sign Conventions

To handle the fact that the electrons and protons have opposite charges and therefore opposite magnetic moments, the following sign conventions are generally used

$$\begin{array}{l}\text{magnetic}\\ \text{energy}\\ \text{of spin}\end{array} \quad E_{magnetic} = -\vec{\mu} \cdot \vec{B}$$

$$= \begin{cases} -\mu B & \text{spin parallel to B} \\ +\mu B & \text{spin opposite to B} \end{cases} \qquad (10)$$

where the electron, proton and neutron have the following magnetic moments

$$\mu_e = -1.00114\,\mu_B \qquad \begin{array}{l}\textit{electron}\\ \textit{magnetic moment}\end{array} \qquad (11)$$

$$\mu_p = 2.79\,\mu_N \qquad \begin{array}{l}\textit{proton}\\ \textit{magnetic moment}\end{array} \qquad (9)$$

$$\mu_n = -1.19\,\mu_N \qquad \begin{array}{l}\textit{neutron}\\ \textit{magnetic moment}\end{array} \qquad (12)$$

where the Bohr magneton μ_B is

$$\mu_B = \frac{e\hbar}{2m_{electron}} = 5.79 \times 10^{-5}\,\frac{eV}{tesla} \qquad (13a)$$

and the nuclear magneton μ_N is

$$\mu_N = \frac{e\hbar}{2m_{proton}} = 3.15 \times 10^{-8}\,\frac{eV}{tesla} \qquad (13b)$$

Note that by putting a – (minus) sign in the formula for $E_{magnetic}$, and making the electron's magnetic moment negative, we still have the result that the electron's spin magnetic energy is positive when the electron's spin is parallel to $\vec{B}$.

Where the Dirac equation completely fails is in the case of the neutron. If the neutron were a simple uncharged particle, it would have no magnetic moment. The fact that it does have a magnetic moment suggests that, while it has no net charge, it must be some kind of composite object with charged particles inside. We now know that this suggestion is correct. The neutron is made up of three quarks, one *up quark* with a charge +2/3 e and two *down quarks* with a charge -1/3 e. While there is no net charge, the quarks contribute to magnetic energy. (The proton, which consists of two up quarks and one down quark, has a total charge of - 2/3e + 2/3e - 1/3e = +e.)

As we mentioned, the difference between the Bohr magneton μ_B and the nuclear magneton μ_N is due to the mass difference between the electron and the proton. Since a proton is 1836 times as massive as an electron, the nuclear magneton is 1836 times smaller than the Bohr magneton. The result is that the magnetic moments of protons, neutrons, and nuclei in general are typically an order of a thousand times smaller than the electron magnetic moment. To get the same magnetic spin energies as you do for electrons, you thus need magnetic fields of the order of a thousand times stronger when working with nuclei.

Exercise 4

(a) Express the magnetic moment of the proton in eV/tesla.

(b) A proton is in a 1 tesla magnetic field. How much energy, in eV, is required to flip the spin of the proton?

(c) What is the wavelength and frequency of a photon that can flip the spin of the proton in part (b)? (Answer: (a) 8.79×10^{-8} eV/tesla, (b) 17.6×10^{-8} eV, (c) 706 cm and 42.5 megacycles.)

Exercise 5

What strength magnetic field should you use so that photons from the student FM radio station (99.4 megacycles) can flip the spin of the proton? (Answer: 2.34 tesla.)

Classical Picture of Magnetic Resonance

In the appendix to this chapter we work out the classical picture of the interaction of the electron's spin with a magnetic field. One pictures the spinning electron as acting as a tiny current loop with a magnetic moment μ as described at the end of Chapter 31. The current loop also has an angular momentum $\vec{L}$ which makes it act like a gyroscope. If you place the current loop in a magnetic field, as shown in Figure (1), the magnetic field exerts a torque and one predicts that the current loop should precess about the magnetic field lines. The precession is analogous to the precession of the bicycle wheel gyroscope studied in Chapter 12.

In this classical picture, if you subject the precessing current loop to the electromagnetic field of a radio wave, whose frequency f is equal to the precessional frequency f_p of the loop, the loop can gain energy from the radio wave. This is a resonance phenomena, where the push of the fields of the radio wave have to match the timing of the precession of the loop. It is analogous to pushing a child on a swing, where you have to time your pushes with the motion of the child in order to add energy to the motion.

The classical picture of a precessing current loop gradually gaining energy from a radio wave, and the quantum picture of an electron spin being flipped by a photon, happen to lead to nearly the same predictions. If we start with the condition $hf = 2\mu_B B$ for the photon energy to match the spin flip energy, then replace μ_B by $e\hbar/2m = e(h/2\pi)/2m = eh/4\pi m$, we get

$$hf = 2\mu_B B = 2\frac{eh}{4\pi m}B \qquad (14)$$

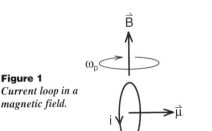

Figure 1
Current loop in a magnetic field.

In Equation 14, Planck's constant h cancels and we get

$$f = \frac{e}{2\pi m}B \qquad \begin{array}{l}\textit{frequency of radio wave}\\ \textit{photon that can flip}\\ \textit{an electron spin}\end{array} \qquad (15)$$

as the relationship between the frequency f of the radio wave and the strength of the magnetic field B.

The fact that Planck's constant cancelled in Equation 15 suggests that a classical analysis might give similar results. In the appendix, we analyze the behavior of a current loop consisting of a particle of charge q and mass m, travelling in a circular orbit. The magnetic moment $\vec{\mu}$ of the loop points along the axis of the orbit. If the loop is placed in a magnetic field $\vec{B}$ oriented perpendicular to $\vec{\mu}$, then the loop will precess around the magnetic field line at a precessional frequency given by the formula

$$f = \frac{e}{2\pi m}B \qquad \begin{array}{l}\textit{precessional frequency}\\ \textit{of a current loop in}\\ \textit{a magnetic field}\end{array} \qquad (16)$$

which is the same formula as Equation 15 for the frequency of a radio wave photon that can flip the spin of an electron. In the classical picture, if we superimpose a radio wave at this frequency, we get a resonance between the frequency of the radio wave and the precessional frequency of the current loop, enabling the radio wave to add energy to the current loop.

The classical calculations have certain errors that have to be corrected on an ad hoc basis. The current loop model leads to a relationship between the loop's angular momentum L and its magnetic moment μ. If we evaluate μ experimentally from the relationship $E_{mag} = -\vec{\mu} \cdot \vec{B}$, and set $L = \hbar/2$ for a spin one half particle, the classical relationship is off by a factor of 2 for electrons and 2 x 2.79 for protons. These errors are accounted for by introducing a fudge factor called the ***Laudé g factor*** to correct the value of μ. Since the classical picture has fundamental problems, such as no hint of quantization of angular momentum, and no explanation of how a particle of zero radius can have angular momentum, it is surprising that the semi classical picture works as well as it does.

ELECTRON SPIN RESONANCE EXPERIMENT

The point of the electron spin resonance experiment is to detect the electron spin flip energy $\Delta E = 2\mu_B B$ predicted by the Dirac equation using photons of energy $E = hf$. You might try to do this by placing a container of hydrogen in a magnetic field B and radiating the hydrogen with a radio wave. If the frequency f of the radio wave were such that the photon energies hf equalled the spin flip energy $2\mu_B B$, then perhaps we could detect radio wave energy being absorbed as hydrogen atom electrons in the low energy spin state were flipped over to the high energy spin state.

Such an experiment will not work because of an interesting quantum mechanical effect. Hydrogen atoms form hydrogen molecules consisting of 2 protons surrounded by 2 electrons. In the ground state of the hydrogen molecule, both electrons are in the lowest energy standing wave pattern allowed for the molecule. This is the electron cloud we sketched in Figure (19-8) in our discussion of the molecular forces between hydrogen atoms.

The Pauli exclusion principle requires that no two electrons be in exactly the same state. If the two electrons in the hydrogen molecule are in the same standing wave pattern, then they must have opposite spins in order to satisfy the exclusion principle.

If we try to flip one of the electron spins with a radio wave, the spin flipped electron cannot stay in the low energy standing wave pattern, for then we would have two electrons with the same spin in the same wave pattern. In order to flip the spin of one of the electrons, we must supply not only the spin flip energy $2\mu_B B$, but also enough energy to raise the electron into a higher energy standing wave pattern. Going to a higher energy standing wave requires much more energy than flipping a spin, thus photons with an energy hf equal to $2\mu_B B$ will have no effect on hydrogen molecules.

When two electrons are in the same standing wave, we say that the electrons are *paired*. In order to see the spin flip energy, we need a substance with an *unpaired* electron, a substance where the electron spin can be flipped without otherwise disturbing the structure of the substance. Such unpaired electrons can be quite chemically active and are known as *free radicals*. An example of a substance with such an unpaired electron is the crystalline organic chemical diphenyl-picryl-hydrazyl or DPPH for short. Since free radicals cause cancer, when we use this substance in our electron spin resonance experiment, we seal it in a small glass vial to keep from coming in contact with it.

To perform the magnetic resonance experiment, we need to both create the radio waves and detect the energy lost to the electrons being flipped. Both of these steps can be accomplished by placing the glass vial containing the DPPH inside the coil of a resonant LC circuit. The circuit, oscillating at its resonant frequency (f) is the source of the photons of energy hf. Detecting the drain of energy from the coil when hf equals $2\mu_B B$ is the way we detect the spin flips.

The easiest way to perform the experiment is to get the LC circuit oscillating at some frequency (f), and then change the magnetic field strength B until $2\mu_B B = hf$. To detect the loss of energy at this point, we use a specially designed LC circuit that is barely oscillating. The circuit is designed to stop oscillating if any energy is being drained from the circuit. To detect whether or not the circuit is oscillating, another circuit detects the amplitude of the oscillation and puts out a DC voltage proportional to that amplitude. The DC voltage can then be displayed on an oscilloscope.

With this arrangement we can sweep the magnitude of B through the value $2\mu_B B = hf$ and watch on an oscilloscope the amplitude of the oscillation of the circuit. The result is shown in the solid curve of Figure (2). We see six peaks because the conditions $2\mu_B B = hf$ was met six times during the 40 milliseconds shown in the diagram.

To produce the magnetic field B, the probe containing the vial of DPPH was placed at the center of our familiar Helmholtz coils as shown in Figure (4). As in our magnetic field mapping experiment *[Figure (24) on page 30-24]*, we power the helmholtz coils with a 60Hz current i(t) to produce a sinusoidally varying magnetic field. This current passes through a 0.1Ω resistor so that we can measure the strength of the magnetic field by plotting the voltage V(t) across the resistor. The result is the dashed curve seen in Figure (2). We can estimate the strength of B by using $i(t) = V(t)/0.1\Omega$ and then remembering that for these coils the magnetic field B in gauss is about equal to 8i(t).

As the magnetic field went through somewhat more than one cycle in Figure (2), the condition $2\mu_B B = hf$

was met six times producing the six resonant peaks. To see how the condition was met, consider the detailed diagram of the center peaks shown in Figure (3). The first peak, at the time of 18.7 milliseconds, occurred when the magnetic field had a magnitude of 8.4 gauss. *[We calculated this from 105×10^{-3} volts $/0.1\Omega$ $= 1.05$ amps, and then $B(gauss) = 8i(t) = 8 \times 1.05$ amps $= 8.4$ gauss.]* At time t = 20.1 milliseconds, the magnitude of B goes down through zero, and then reaches a magnitude of –8.4 gauss at a time t = 21.5 milliseconds. We get a peak at both +8.4 gauss and –8.4 gauss because the resonance does not depend upon which of the two ways the magnetic field was pointing. Looking back at Figure (2), we see that B had a magnitude of + or – 8.4 gauss six times, which is why we got the six peaks.

For the experiment shown in Figures (2) and (3), the photons in the resonant LC circuit, the photons flipping the spin of the DPPH electrons, had a frequency f = 28 megacycles. We can use the fact that these photons flipped the spins when the magnetic field was about 8.4 gauss to calculate our value of the electron magnetic moment μ_B. We have

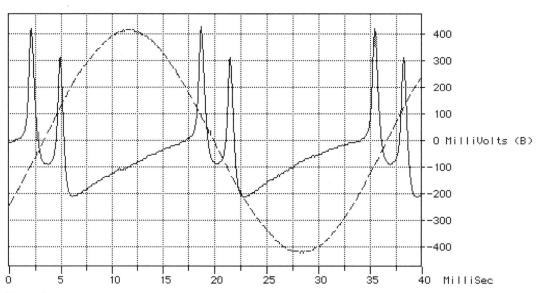

Figure 2
The solid curve shows the resonant peaks while the dashed curve is proportional to the strength of the magnetic field.

$$2\mu_B B = hf$$

$$\mu_B = \frac{hf}{2B}$$

Using the values

$$h = \frac{6.63\times10^{-34}\text{joule sec}}{1.6\times10^{-19}\dfrac{\text{joules}}{\text{eV}}} = 4.14\times10^{-15}\,\text{eV sec}$$

$$f = 28\times10^6\left(\frac{1}{\text{sec}}\right)$$

$$B = 8.4\text{ gauss}\times10^{-4}\left(\frac{\text{tesla}}{\text{gauss}}\right) = 8.4\times10^{-4}\text{tesla}$$

we get

$$\mu_B = \frac{4.14\times10^{-15}\,\text{eV sec}\times28\times10^6\left(\frac{1}{\text{sec}}\right)}{2\times8.4\times10^{-4}\text{tesla}}$$

$$\mu_{B\cdot} = 6.9\times10^{-5}\frac{\text{eV}}{\text{tesla}} \qquad \begin{matrix}\textit{our}\\\textit{result}\end{matrix} \qquad (17)$$

This result is nearly 20% above the known value

$$\mu_{B\cdot} = 5.79\times10^{-5}\frac{\text{eV}}{\text{tesla}} \qquad \begin{matrix}\textit{accepted}\\\textit{value}\end{matrix} \qquad (5)$$

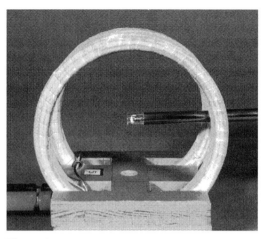

Figure 4
Electron spin resonance apparatus. The coil containing the vial of DPPH is at the tip of the probe, which is at the center of the Helmholtz coils. The capacitor of the LC circuit, and the controls, are at the other end of the probe. The Helmholtz coils are being driven by a 60Hz alternating current. As a result, the magnitude of the magnetic field sweeps back and forth through the resonant value.

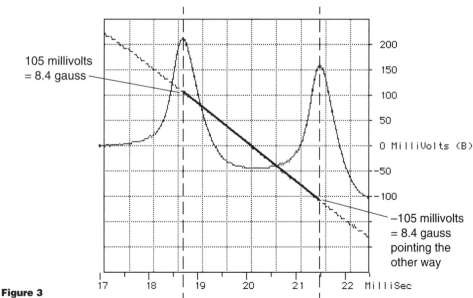

105 millivolts
= 8.4 gauss

−105 millivolts
= 8.4 gauss
pointing the
other way

Figure 3
We get a resonant peak for both orientations of the magnetic field.

The source of this error is in our measurement of the strength of the magnetic field. We have relied on the accuracy of the value of the 0.1Ω resistor through which the helmholtz coil current i(t) passes, and then used the approximate formula B(gauss) = 8i(t). One can obtain much more accurate results using the precision search coil shown in Figure (5). This is a 100 turn coil wound on a 1 inch (2.54cm) plastic rod. Using the techniques discussed in the magnetic mapping experiment, one can accurately relate the voltage V_R measured across the 0.1Ω resistor to the actual value of B. We have encouraged students who wish to do a project involving electron spin resonance to see how accurate a value of μ_B they can obtain using this precision search coil.

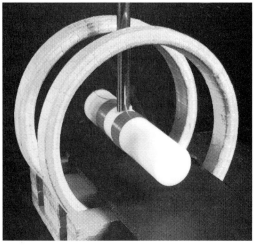

Figure 5
100 turn search coil for accurately determining the magnetic field.

Exercise 6

In our electron spin resonance experiment, we saw that an electron in a magnetic field had two energy states, and that the difference in the energy between the states was proportional to the strength B of the magnetic field. We measured this energy by placing the electrons in an oscillating electromagnetic field of a given frequency (around 30 megacycles) and observing at what values of the magnetic field we got a transition between the two states.

In Figure (6) we have a somewhat similar situation except the object being studied is a HD (Hydrogen-Deuterium) molecule. In this molecule the Hydrogen nucleus (a proton) weakly interacts with the Deuterium nucleus (a proton and a neutron). These two nuclei form a system with several energy states or levels. If you apply a magnetic field B to the HD molecule, the difference in the energy between the states is related to the strength of B. The energy difference between the states can be measured by applying an oscillating electromagnetic field of a given frequency and observing at what values of the magnetic field we get a transition between states.

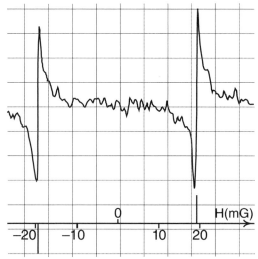

H. Benoit and P. Piejus, *Compt. Rend.* 265B, 101 (1967).
"Spectre de R.M.N. de HD á une Fréquence de 54 Hz."
Transition from the 3/2,1/2 to the 3/2,3/2 states of the nuclei in an HD molecules at 20 kelvins. The nuclei were pre alligned (polarized) in an 8 killogauss field.

Figure 6
NMR data on liquid Hydrogen-Deuterium.

Figure (6) is a nuclear magnetic resonance scan of the HD molecules. As in our electron spin resonance experiment, the molecules are placed in an oscillating electromagnetic field of a given frequency, and the strength of a uniform magnetic field B is varied. Two resonance peaks are observed, but they represent the same transition between the energy levels, since the transition does not depend upon the sign of the magnetic field.

One of the main differences between the electron system we studied in the lab and the HD molecule, is that the energy level splitting is really really small in the HD molecule experiment compared to the splitting we observed in the electron experiment. Instead of fields of tens of gauss and frequencies of around 30 Megacycles, in the HD experiment of Figure (6) , the frequency was **54 cycles per second** and the field B was just under **20 milligauss** (.020 gauss or .000002 tesla)! This example demonstrates the enormous range of applicability of the magnetic resonance experiments.

For this exercise, we want you to calculate the splitting between the two energy levels involved in the resonance transitions seen in Figure (6). Give the answer in ergs or joules, and in electron volts. Comment on the reasonableness of your answer—do you think your result is too big, too small, or perhaps OK.

APPENDIX

CLASSICAL PICTURE OF MAGNETIC INTERACTIONS

At the end of Chapter 31, we discussed the magnetic moment of a current loop, deriving the formulas

$\vec{A}$ = area of current loop

$$\vec{\mu} \equiv i\vec{A} \qquad \textit{magnetic moment} \qquad (31\text{-}34)$$

$$\vec{\tau} = \vec{\mu} \times \vec{B} \qquad \textit{magnetic torque} \qquad (31\text{-}35)$$

$$E_{mag} = -\vec{\mu} \cdot \vec{B} \quad \textit{magnetic energy} \qquad (31\text{-}37)$$

These equations applied to a current loop whose current i and area $|\vec{A}|$ were unaffected by the magnetic field. We only allowed the magnetic field to change the orientation of the loop via the torque $\vec{\tau}$.

We then treated a charged particle in a circular orbit as a current loop. Using the definition $\vec{\mu} = i\vec{A}$, we found that the orbiting particle had a magnetic moment $\vec{\mu}$ related to its angular momentum $\vec{L}$ by

$$\vec{\mu} = \frac{q}{2m}\vec{L} \qquad (31\text{-}39)$$

where q is the charge and m the mass of the particle.

To use the magnetic energy formula $E_{mag} = -\vec{\mu} \cdot \vec{B}$, we have to make the same assumptions about the orbiting particle as we did about the current loop. Namely we have to assume that the magnetic field alters only the orientation of the orbit and not the particle's speed v or orbital radius r. Since $|\vec{L}| = mvr$, we are thus assuming that the magnetic field does not affect the magnitude of the particle's angular momentum.

For a classical particle, such an argument is not reasonable. If we turn on a magnetic field, we change the magnetic flux through the orbit and thus by Faraday's law induce a voltage around the orbit. This induced voltage should affect both the particle's speed and orbital radius.

But if the angular momentum of the particle is quantized, if the magnitude of $E_{mag} = -\vec{\mu} \cdot \vec{B}$ cannot change, then our current loop analysis and magnetic energy formula $E_{mag} = -\vec{\mu} \cdot \vec{B}$ has a better chance of working. There is no reason to expect any classical formulas to apply to atomic or subatomic systems. What we are looking for are those that do.

To apply classical formulas to particle spins, we have to fudge the relationship between the particle's magnetic moment $\vec{\mu}$ and its spin angular momentum. As a general relationship between magnetic moment $\vec{\mu}$ and angular momentum $\vec{L}$, we will rewrite Equation 31-39 in the form

$$\vec{\mu} = g\left(\frac{q}{2m}\vec{L}\right) \qquad (A\text{-}1)$$

where g is our fudge factor. It is the factor we have to introduce to make classical calculations give the correct results.

The value of g depends upon the kind of system we are talking about. If we are talking about the angular momentum of an electron in orbit about a nucleus, then g=1 and the classical equations work. If we are talking about the spin angular momentum of an electron, then g=2. For electron spin, the classical formulas are off by a factor of 2. For the proton, g has to have the value 2×2.79 in order to get the proton magnetic moment given in Equation 9. Since protons and neutrons are composite particles made from quarks it should not be surprising that g should have a peculiar value. This factor g is called either the **gyromagnetic ratio** or **Landé g factor**. Fudge factors sound better if you give them impressive names.

Combining our modified Equation A-1 for $\vec{\mu}$ with 31-37 for E, we get

$$E_{mag} = -\vec{\mu} \cdot \vec{B} = g\left(\frac{q}{2m}\right)\vec{L} \cdot \vec{B} \qquad (A\text{-}2)$$

as the semi classical formula for the magnetic energy of a particle in a magnetic field. We say semi classical because of the correction factor g.

If the particle has an angular momentum $\vec{L}$ and the magnetic field exerts a torque $\vec{\tau}$, the particle should precess like a gyroscope. Thus we can compare a bicycle wheel gyroscope subject to a gravitational torque to a current loop of magnetic moment $\vec{\mu}$ subject to a magnetic torque. To simplify the analysis, we are assuming that the magnetic moment $\vec{\mu}$ lies in the plane perpendicular to $\vec{B}$ so the magnetic torque $\vec{\tau} = \vec{\mu} \times \vec{B}$ has a magnitude μB.

Following the standard analysis of a gyroscope, we predict that the magnetic moment vector $\vec{\mu}$ should precess about the magnetic field vector $\vec{B}$ at a rate $\omega_{precession}$ given by

$$\omega_{precession} = \frac{\tau}{L} = \frac{\mu B}{L} \qquad (12\text{-}58)$$

a result we derived back in Chapter 12. Using our semi classical formula A1 for μ, we get

$$\omega_{precession} = g\left(\frac{qL}{2m}\right)\frac{B}{L}$$

The L's cancel, and we are left with

$$\omega_{precession} = g\left(\frac{q}{2m}\right)B \qquad (A\text{-}3)$$

The quantity $\omega_{precession}$ is the precessional frequency in radians per second. To convert this to cycles per second, we divide by 2π radians/cycle to get

$$f_{precession} = \frac{\omega_{precession}}{2\pi} = \frac{g}{2\pi}\left(\frac{q}{2m}\right)B \qquad (A\text{-}4)$$

as the precessional frequency of a charged orbiting or spinning particle in a magnetic field.

Applying Equation A-4 to the spin of a particle, we predict classically that if the particle is subject to an alternating electric and magnetic fields of frequency f, there will be a resonance and the particle can gain magnetic energy if the frequency f equals the precessional frequency $f_{precession}$.

To go to the quantum picture, multiply Equation A-4 through by Planck's constant h, to get

$$hf_{precession} = g\frac{h}{2\pi}\frac{q}{2m}B$$

$$= g\left(\frac{q\hbar}{2m}\right)B \qquad (A\text{-}5)$$

Applying this to an electron spin, setting q and m to the charge and mass of an electron, we get

$$hf_{precession} = g\mu_B B \qquad \begin{array}{c}\text{classical theory}\\ \text{with factor g}\end{array} \qquad (A\text{-}6)$$

where $\mu_B = e\hbar/2m$ is the Bohr magneton.

In the quantum picture, the electron gains magnetic energy if the photons in the radio wave have the right amount of energy to flip the spin of the electron. The Dirac equation gave the spin flip energy as

$$\Delta E = 2\mu_B B \qquad \begin{array}{c}\text{energy required}\\ \text{to flip the}\\ \text{electron spin}\end{array} \qquad (3\text{ repeated})$$

In Equation 7, we equated this energy to the photon energy to get

$$hf = 2\mu_B B \qquad \begin{array}{c}\text{set spin flip}\\ \text{energy equal to}\\ \text{photon energy}\end{array} \qquad (7\text{ repeated})$$

as the formula giving the frequency of the radio wave that can add magnetic energy to the electron. Comparing Equation (7) with the semi classical result (A-6), we see that the Landé g factor, the gyromagnetic ratio g, has to be set equal to 2 for the semi classical theory to agree with the Dirac equation

$$\boxed{g = 2 \qquad \begin{array}{c}\text{gyromagnetic ratio}\\ \text{for the electron spin}\end{array}} \qquad (A\text{-}7)$$

Chapter 40

Quantum Mechanics

That light had both a particle and a wave nature became apparent with Einstein's explanation of the photoelectric effect in 1905. One might expect that such a discovery would lead to a flood of publications speculating on how light could behave both as a particle and a wave. But no such response occurred. The particle wave nature was not looked at seriously for another 18 years, when de Broglie proposed that the particle wave nature of the electron was responsible for the quantized energy levels in hydrogen. Even then there was great reluctance to accept de Broglie's proposal as a satisfactory thesis topic.

Why the reluctance? Why did it take so long to deal with the particle-wave nature, first of photons then of electrons? What conceptual problems do we encounter when something behaves both as a particle and as a wave? How are these problems handled? That is the subject of this chapter.

TWO SLIT EXPERIMENT

Of all the experiments in physics, it is perhaps the 2 slit experiment that most clearly, most starkly, brings out the problems encountered with the particle-wave nature of matter. For this reason we will use the 2 slit experiment as the basis for much of the discussion of this chapter.

Let us begin with a review of the 2 slit experiment for water and light waves. Figure (1) shows the wave pattern that results when water waves emerge from 2 slits. The lines of nodes are the lines along which the waves from one slit just cancel the waves coming from the other. Figure (2) shows our analysis of the 2 slit pattern. The path length difference to the first minimum must be half a wavelength $\lambda/2$. This gives us the two similar triangles shown in Figure (2). If y_{min} is

much less than D, which it is for most 2 slit experiments, then the hypotenuse of the big triangle is approximately D and equating corresponding sides of the similar triangles gives us the familiar relationship

$$\frac{\lambda/2}{d} = \frac{y_{min}}{D}$$

$$\lambda = \frac{2y_{min}d}{D} \qquad (1)$$

Figure (3a) is the pattern we get on a screen if we shine a laser beam through 2 slits. To prove that the dark bands are where the light from one slit cancels the light from the other, we have in Figure (3b) moved a razor blade in front of one of the slits. We see that the dark bands disappear, and we are left with a one slit pattern. The dark bands disappear because there is no longer any cancellation of the waves from the 2 slits.

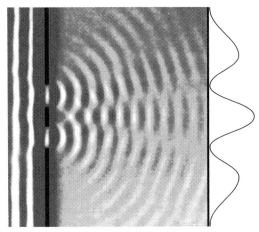

Figure 1
Water waves emerging from two slits.

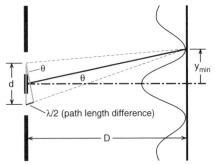

Figure 2
Analysis of the two slit pattern. We get a minimum when the path length difference is half a wavelength.

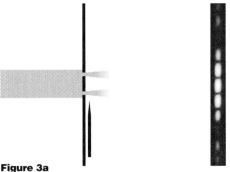

Figure 3a
Two slit interference pattern for light. The closely spaced dark bands are where the light from one slit cancels the light from the other.

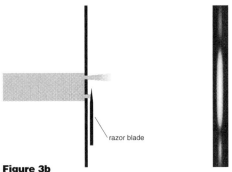

Figure 3b
Move a razor blade in front of one of the slits, and the closely spaced dark bands disappear. There is no more cancellation.

In 1961, Claus Jönsson did the 2 slit experiment using electrons instead of light, with the results shown in Figure (4). Assuming that the electron wavelength is given by the de Broglie formula $p = h/\lambda$, the dark bands are located where one would expect waves from the 2 slits to cancel. The 2 slit experiment gives the same result for light and electron waves.

The Two Slit Experiment from a Particle Point of View

In Figure (3a), the laser interference patterns were recorded on a photographic film. The pattern is recorded when individual photons of the laser light strike individual silver halide crystals in the film, producing a dark spot where the photon landed. Where the image shows up white in the positive print, many photons have landed close together exposing many crystal grains.

In a more modern version of the experiment one could use an array of photo detectors to count the number of photons landing in each small element of the array. The number of counts per second in each detector could then be sent to a computer and the image reconstructed on the computer screen. The result would look essentially the same as the photograph in Figure (3a).

The point is that the image of the two slit wave pattern for light is obtained by counting particles, not by measuring some kind of a wave height. When we look at the two slit experiment from the point of view of counting particles, the experiment takes on a new perspective.

Figure 4
Two slit experiment using electrons. (By C. Jönsson)

Imagine yourself shrunk down in size so that you could stand in front of a small section of the photographic screen in Figure (3a). Small enough that you want to avoid being hit by one of the photons on the laser beam. As you stand at the screen and look back at the slits, you see photons being sprayed out of both slits as if two machine guns were firing bullets at you, but you discover that there is a safe place to stand. There are these dark bands where the particles fired from one slit cancel the particles coming from the other.

Then one of the slits is closed, there is no more cancellation, the dark bands disappear as seen in Figure (3b). There is no safe place to stand when particles are being fired at you from only one slit. It is hard to imagine in our large scale world how it would be safe to have two machine guns firing bullets at you, but be lethal if only one is firing. It is hard to visualize how machine gun bullets could cancel each other. But the particle wave nature of light seems to require us to do so. No wonder the particle nature of light remained an enigma for nearly 20 years.

Two Slit Experiment—One Particle at a Time

You might object to our discussion of the problems involved in interpreting the two slit experiment. After all, Figure (1) shows water waves going through two slits and producing an interference pattern. The waves from one slit cancel the waves from the other at the lines of nodes. Yet water consists of particles—water molecules. If we can get a two slit pattern for water molecules, what is the big deal about getting a two slit pattern for photons ? Couldn't the photons somehow interact with each other the way water molecules do, and produce an interference pattern?

Photons do not interact with each other the way water molecules do. Two laser beams can cross each other with no detectable interaction, while two streams of water will splash off of each other. But one still might suspect that the cancellation in the two slit experiment for light is caused by some kind of interaction between the photons. This is even more likely in the case of electrons, which are strongly interacting charged particles.

a) 10 dots

b) 100 dots

c) 1000 dots

d) 10000 dots

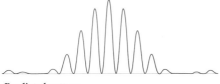

Predicted pattern

Experimental results by C. Jönsson

Figure 5
Computer simulation of the 2 slit electron diffraction experiment, as if the electrons had landed one at a time.

In an earlier text, we discussed the possibility of an experiment in which electrons would be sent through a two slit array, one electron at a time. The idea was to eliminate any possibility that the electrons could produce the two slit pattern by bouncing into each other or interacting in any way. Since the experiment had not yet been done, we drew a sketch of what the results should look like. That sketch now appears in a number of introductory physics texts.

When he saw the sketch, Lawrence Campbell of the Los Alamos Scientific Laboratories did a computer simulation of the experiment. We will first discuss Campbell's simulation, and then compare the simulation with the results of the actual experiment which was performed in 1991.

It is not too hard to guess some of the results of sending electrons through two slits, one at a time. After the first electron goes through you end up with one dot on the screen showing where the electron hit. The single dot is not a wave pattern. After two electrons, two dots; you cannot make much of a wave pattern out of two dots.

If, after many thousands of electrons have hit the screen, you end up with a two slit pattern like that shown in Figure (4), that means that none of the electrons land where there will eventually be a dark band. You know where the first dot, and the second dot cannot be located. Although two dots do not suggest a wave pattern, some aspects of the wave have already imposed themselves by preventing the dots from being located in a dark band.

To get a better idea of what is happening, let us look at Campbell's simulation in Figure (5). In (5a), and (5b) we see 10 dots and 100 dots respectively. In neither is there an apparent wave pattern, both look like a fairly random scatter of dots. But by the time there are 1000 dots seen in (5c), a fairly distinctive interference pattern is emerging. With 10,000 dots of (5d), we see a fairly close resemblance between Campbell's simulation and Jönsson's experimental results. Figure (5e) shows the wave pattern used for the computer simulation.

Although the early images in Figure (5) show nearly random patterns, there must be some order. Not only do the electrons not land where there will be a dark band, but they must also accumulate in greater numbers where the brightest bands will eventually be. If this were a roulette type of game in Las Vegas, you should put your money on the center of the brightest band as being the location most likely to be hit by the next electron.

Campbell's simulation was done as follows. Each point on the screen was assigned a probability. The probability was set to zero at the dark bands and to the greatest value in the brightest band. Where each electron landed was randomly chosen, but a randomness governed by the assigned probability.

How to assign a probability to a random event is illustrated by a roulette wheel. On the wheel, there are 100 slots, of which 49 are red, 49 black and 2 green. Thus where the ball lands, although random, has a 49% chance of being on red, 49% on black, 2% on green, and 0% on blue, there being no blue slots.

In the two slit simulation, the probability of the electron landing at some point was proportional to the intensity of the two slit wave pattern at that point. Where the wave was most intense, the electron is most likely to land. Initially the pattern looks random because the electrons can land with roughly equal probability in any of the bright bands. But after many thousands of electrons have landed, you see the details of the two slit wave pattern. The dim bands are dimmer than the bright ones because there was a lower probability that the electron could land there.

Figure (6) shows the two slit experiment performed in 1991 by Akira Tonomura and colleagues. The experiment involved a novel use of a superconductor for the two slits, and the incident beam contained so few electrons per second that no more than one electron was between the slits and the screen at any one time. The screen consisted of an array of electron detectors which recorded the time of arrival of each electron in each detector. From this data the researchers could reconstruct the electron patterns after 10 electrons (6a), 100 electrons (6b), 3000 electrons (6c), 20,000 electrons (6d) and finally after 70,000 electrons in Figure (6e). Just as in Campbell's simulation, the initially random looking patterns emerge into the full two slit pattern when enough electrons have hit the detectors.

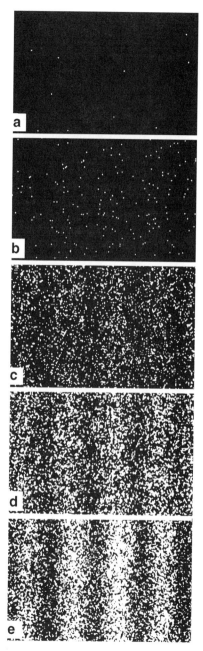

Figure 6
Experiment in which the 2 slit electron interference pattern is built up one electron at a time. (A. Tonomura, J. Endo, T. Matsuda, T. Kawasaki, American Journal of Physics, Feb. 1989. See also Physics Today, April 1990, Page 22.)

Born's Interpretation of the Particle Wave

In 1926, while calculating the scattering of electron waves, Max Born discovered an interpretation of the electron wave that we still use today. In Born's picture, the electron is actually a particle, but it is the electron wave that governs the behavior of the particle. The electron wave is a probability wave governing the probability of where you will find the electron.

To apply Born's interpretation to the two slit electron experiment, we do what Campbell did in the simulation of Figure (5). We first calculate what the wave pattern at the screen would be for a wave passing through the two slits. It is the two slit interference pattern we have seen for water waves, light waves and electron waves. *We then interpret the intensity of the pattern at some point on the screen as being proportional to the probability that the electron will land at that point.* We cannot predict where any given electron will actually land, any more than we can predict where the ball will end up on the roulette wheel. But we can predict what the pattern will look like after many electrons have landed. If we repeat the experiment, the electrons will not land in the same places, but eventually the same two slit pattern will result.

Exercise 1

Figure (36-16) reproduced here, shows the diffraction pattern produced when a beam of electrons is scattered by the atoms of a graphite crystal. Explain what you would expect to see if the electrons went through the graphite crystal one at a time and you could watch the pattern build up on the screen. Could you market this apparatus in Las Vegas, and if so, how would you use it?

Figure 36-16
Diffraction pattern produced by electrons passing through a graphite crystal.

Photon Waves

Both electrons and photons have a particle-wave nature related by the de Broglie formula $p = h/\lambda$, and both produce a two slit interference pattern. Thus one would expect that the same probability interpretation should apply to electron waves and light waves.

We have seen, however, that a light wave, according to Maxwell's equations, consists of a wave of electric and magnetic fields $\vec{E}$ and $\vec{B}$. These are vector fields that at each point in space have both a magnitude and a direction. Since probabilities do not point anywhere, we cannot directly equate $\vec{E}$ and $\vec{B}$ to some kind of probability.

To see how to interpret the wave nature of a photon, let us first consider something like a radio wave or a laser beam that contains many billions of photons. In our discussion of capacitors in Chapter 27, we saw that the energy density in a classical electric field was given by

$$\left.\begin{array}{c}\text{Energy}\\ \text{density}\end{array}\right\} = \frac{\varepsilon_0 E^2}{2} \quad \begin{array}{l}\text{energy density in}\\ \text{an electric field}\end{array} \quad (27\text{-}36)$$

where $E^2 = \vec{E} \cdot \vec{E}$. In an electromagnetic wave there are equal amounts of energy in the electric and the magnetic fields. Thus the energy density in a classical electromagnetic field is twice as large as that given y Equation 27-36, and we have

$$E\frac{\text{joules}}{\text{meter}^3} = \varepsilon_0 E^2 \quad \begin{array}{l}\text{energy density in an}\\ \text{electromagnetic wave}\end{array}$$

If we now picture the electromagnetic wave as consisting of photons whose energy is given by Einstein's photoelectric formula

$$E_{\text{photon}} = hf\frac{\text{joules}}{\text{photon}}$$

then the density of photons in the wave is given by

$$n = \frac{\varepsilon_0 E^2\,\text{joules}\,/\text{meter}^3}{hf\;\;\text{joules}\,/\text{photon}}$$

$$n = \frac{\varepsilon_0 E^2}{hf}\frac{\text{photons}}{\text{meter}^3} \quad \begin{array}{l}\text{density of photons}\\ \text{in an electromagnetic}\\ \text{wave of frequency}\,f\end{array} \quad (1)$$

where f is the frequency of the wave.

In Exercise 2, we have you estimate the density of photons one kilometer from the antenna of the student AM radio station at Dartmouth College. The answer turns out to be around .25 billion photons/cc, so many photons that it would be hard to detect them individually.

Exercise 2

To estimate the density of photons in a radio wave, we can, instead of calculating $\vec{E}$ for the wave, simply use the fact that we know the power radiated by the station. As an example, suppose that we are one kilometer away from a 1000 watt radio station whose frequency is $1.4 \times 10^6 Hz$. A 1000 watt station radiates 1000 joules of energy per second or 10^{-6} joules in a nanosecond. In one nanosecond the radiated wave moves out one foot or about 1/3 of a meter. If we ignore spatial distortions of the wave, like reflections from the ground, etc., then we can picture this 10^{-6} joules of energy as being located in a spherical shell 1/3 of a meter thick, expanding out from the antenna.

(a) What is the total volume of a spherical shell 1/3 of a meter thick and 1 kilometer in radius?

(b) What is the average density of energy, in joules/m³ of the radio wave 1 kilometer from the antenna

(c) What is the energy, in joules, of one photon of frequency $1.4 \times 10^6 Hz$?

(d) What is the average density of photons in the radio wave 1 kilometer from the station? Give the answer first in photons/m³ and then photons per cubic centimeter. (The answer should be about .25 billion photons/cm³.)

Now imagine that instead of being one kilometer from the radio station, you were a million kilometers away. Since the volume of a spherical shell 1/3 of a meter thick increases as r^2, (the volume being $(1/3) \times 4\pi r^2$) the density of photons would decrease as $1/r^2$. Thus if you were 10^6 times as far away, the density of photons would be 10^{-12} times smaller. At one million kilometers, the average density of photons in the radio wave would be

$$
\left.\begin{array}{l}\text{number of photons}\\ \text{per cubic centimeter}\\ \text{at 1 million kilometers}\end{array}\right\} = \frac{\text{number at 1km}}{10^{12}}
$$

$$
= \frac{.25 \times 10^9}{10^{12}}
$$

$$
= .00025 \frac{\text{photons}}{\text{cm}^3} \quad (2)
$$

In the classical picture of Maxwell's equations, the radio wave has a continuous electric and magnetic field even out at 1 million kilometers. You could calculate the value of $\vec{E}$ and $\vec{B}$ out at this distance, and the result would be sinusoidally oscillating fields whose structure is that shown back in Figure (32-23). But if you went out there and tried to observe something, all you would find is a few photons, on the order of .25 per liter (about one per gallon of space). If you look in 1 cubic centimeter of space, chances are you would not find a photon.

So how do you use Maxwell's equations to predict the results of an experiment to detect photons a million kilometers from the antenna? First you use Maxwell's equation to calculate $\vec{E}$ at the point of interest, then evaluate the quantity $(\varepsilon_0 \vec{E} \cdot \vec{E}/hf)$, and finally interpret the result as the probability of finding a photon in the region of interest. If, for example, we were looking in a volume of one cubic centimeter, the probability of finding a photon there would be about .00025 or .025%.

This is an explicit prescription for turning Maxwell's theory of electromagnetic radiation into a probability wave for photons. If the wave is intense, as it was close to the antenna, then $(\varepsilon_0 \vec{E} \cdot \vec{E}/hf)$ represents the density of photons. If the wave is very faint, then $(\varepsilon_0 \vec{E} \cdot \vec{E}/hf)$ becomes the probability of finding a photon in a certain volume of space.

Reflection and Fluorescence

An interesting example of the probability interpretation of light waves is provided by the phenomena of reflection and of fluorescence.

When a light beam is reflected from a metal surface, the angle of reflection, labeled θ_r in Figure (7a) is equal to the angle of incidence θ_i. The reason for this is seen in Figure (7b). The incident light wave is scattered by many atoms in the metal surface. The scattered waves add up to produce the reflected wave as shown in Figure (7b). Any individual photon in the incident wave must have an equal probability of being scattered by all of these atoms in order that the scattered probability waves add up to the reflected wave shown in (7b).

When you have a fluorescent material, you see a rather uniform eerie glow rather than a reflected wave. The light comes out in all directions as in Figure (8a).

The wavelength of the light from a fluorescent material is not the same wavelength as the incident light. What happens is that a photon in the incident beam strikes and excites an individual atom in the material. The excited atom then drops back down to the ground state radiating two or more photons to get rid of the excitation energy. (Ultraviolet light is often used in the incident beam, and we see the lower energy visible photons radiated from the fluorescing material.)

The reason that fluorescent light emerges in many directions rather than in a reflected beam is that an individual photon in the incident beam is absorbed by and excites one atom in the fluorescent material. There is no probability that it has struck any of the other atoms. The fluorescent light is then radiated as a circular wave from that atom, and the emerging photon has a more or less equal probability of coming out in all directions above the material.

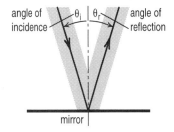

Figure 7a
When a light wave strikes a mirror, the angle of incidence equals the angle of reflection.

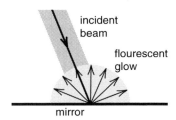

Figure 8a
When a beam of light strikes a fluorescent material, we see an eerie glow rather than a normal reflected light.

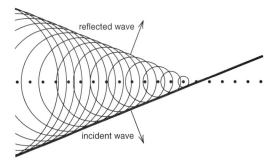

Figure 7b
The reflected wave results from the scattering of the incident wave by many atoms. If the incident wave contains a single photon, that photon must have an equal probability of being scattered by many atoms in order to emerge in the reflected wave.

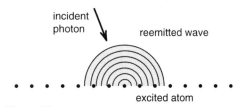

Figure 8b
Fluorescence occurs when an individual atom is excited and radiates its extra energy as two distinct photons. Since there is no chance that the radiation came from other atoms, the radiated wave emerges only from the excited atom.

A Closer Look at the Two Slit Experiment

While the probability interpretation of electron and photon waves provides a reasonable explanation of some phenomena, the interpretation is not without problems. To illustrate what these problems are, consider the following thought experiment.

Imagine that we have a large box with two slits at one end and a photographic film at the other, as shown in Figure (9). Far from the slits is an electron gun that produces a weak beam of electrons, so weak that on the average only one electron per hour passes through the slits and strikes the film. For simplicity we will assume that the electrons go through the slits on the hour, there being the 9:00 AM electron, the 10:00 AM electron, etc.

The electron gun is one of the simple electron guns we discussed back in Chapter 28. The beam is so spread out that there is no way it can be aimed at one slit or the other. Our beam covers both slits, meaning that each of the electrons has an equal chance of going through the top or bottom slit.

We will take the probability interpretation of the electron wave seriously. If the electron has an equal probability of passing through either slit, then an equally intense probability wave must emerge from both slits. When the probability waves get to the photographic film, there will be bands along which waves from one slit cancel waves from the other, and we should eventually build up a two slit interference pattern on the film.

Suppose that on our first run of the thought experiment, we do build up a two slit pattern after many hours and many electrons have hit the film.

We will now repeat the experiment with a new twist. We ask for a volunteer to go inside the box, look at the slits, and see which one each electron went through. John volunteers, and we give him a sheet of paper to write down the results. To make the job easier, we tell him to just look at the bottom slit on the hour to see if the electron went through that slit. If for example he sees an electron come out of the bottom slit at 9:00 AM, then the 9:00 AM electron went through the bottom slit. If he saw no electron at 10:00 AM, then the 10:00 AM electron must have gone through the upper slit.

If John does his job carefully, what kind of a pattern should build up on the film after many electrons have gone through? If the 9:00 AM electron was seen to pass through the bottom slit, then there is no probability that it went through the top slit. As a result, a probability wave can emerge only from the bottom slit, and there can be no cancellation of probability waves at the photographic film. Since the 10:00 AM electron did not go through the bottom slit, the probability wave must have emerged only from the top slit and there again can be no cancellation of waves at the photographic film.

If John correctly determines which slit each electron went through, there can be no cancellation of waves from the two slits, and we have to end up with a one slit pattern on the film. Just the knowledge of which slit each electron went through has to change the two slit pattern into a one slit pattern. With Born's probability interpretation of electron waves, just the **knowledge** of which slit the electrons go through **changes the result of the experiment**. Does this really happen, or have we entered the realm of metaphysics?

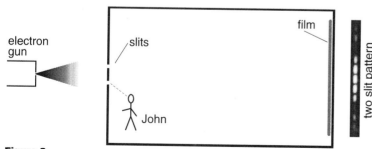

Figure 9
In this thought experiment, we consider the possibility that someone is looking at the two slits to see which slit each electron comes through.

Let us return to our thought experiment. John has been in the box for a long time now, so that a number of electrons have hit the film. We take the film out, develop it, and clearly see a two slit interference pattern emerging. There are the dark bands along which waves from one slit cancel the waves from the other slit.

Then we go over to the door on the side of the box, open it and let John out, asking to see his results. We look at his sheet of paper and nothing is written on it. "What were you doing all of that time," we ask. "What do you mean, what was I doing? How could I do anything? You were so careful sealing up the box from outside disturbances that it was dark inside. I couldn't see a thing and just had to wait until you opened the door. Not much of a fun experiment."

"Next time," John said, "give me a flashlight so I can see the electrons coming through the slits. Then I can fill out your sheet of paper."

"Better be careful," Jill interrupts, "about what kind of a flashlight you give John. A flashlight produces a beam of photons, and John can only see a passing electron if one of the flashlight's photons bounces off the electron."

"Remember that the energy of a photon is proportional to its frequency. If the photons from John's flashlight have too high a frequency, the photon hitting the passing electron will change the motion of the electron and mess up the two slit pattern. Give John a flashlight that produces low frequency, low energy photons, so he won't mess up the experiment."

"But," Bill responds, "a low frequency photon is a long wavelength photon. Remember that demonstration where waves were scattered from a tiny object? The scattered waves were circular, and contained no information about the shape of the object (Figure 36-1). You can't use waves to study details that are much smaller than the wavelength of the wave. That is why optical microscopes can't be used to study viruses that are smaller than a wavelength of visible light."

"If John's flashlight," Bill continues, "produced photons whose wavelength was longer than the distance between the two slits, then even if he hit the electron with one of the photons in the wave, John could not tell which slit the electron came through."

"Let us do some calculations," the professor says. "The most delicate way we can mess up the experiment is to hit an electron sideways, changing the electron's direction of motion so that if it were heading toward a maxima, it will instead land in a minima, filling up the dark bands and making the pattern look like a one slit pattern. Here is a diagram for the situation (Figure 10)."

"In the top sketch (10a), John is shining his flashlight at an electron that has just gone through the slit and is heading toward the central maximum. In the middle sketch (10b) the photon has knocked the electron sideways, so that it is now headed toward the first minimum in the diffraction pattern. Let us assume that all the photon's momentum $\vec{p}_{photon}$ has been transferred to the electron, so that the electron's new momentum is now

Figure 36-1 (reproduced)
If an object is smaller than a wavelength, the scattered waves are circular and do not contain information about the shape of the object.

Incident and scattered wave

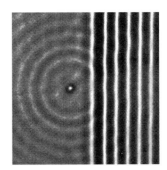

After incident wave has passed

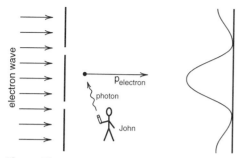

Figure 10a
In order to see the electron, John uses a flashlight, and strikes the electron with a photon.

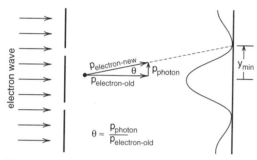

Figure 10b
Assume the photon's momentum has been absorbed by the electron. This could deflect the electron's path by an angle θ.

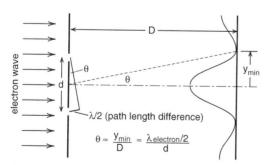

Figure 10c
Analysis of the two slit pattern. The angle to the first minimum is determined by using similar triangles. If the angle θ is small, then sinθ ≈ θ.

$$\vec{P}_{electron-new} = \vec{P}_{electron-old} + \vec{P}_{photon} \qquad (3)$$

The angle θ by which the electron is deflected is approximately given by

$$\theta \approx \frac{p_{photon}}{p_{electron}} = \frac{h/\lambda_{photon}}{h/\lambda_{electron}}$$

$$\theta \approx \frac{\lambda_{electron}}{\lambda_{photon}} \qquad (4)$$

where we used the de Broglie formula for the photon and electron momenta."

"In the bottom sketch we have the usual analysis of a two slit pattern. If the angle θ to the first minimum is small, which it usually is for a two slit experiment, then by similar triangles we have

$$\theta \approx \frac{y_{min}}{D} = \frac{\lambda_{electron}/2}{d} \qquad (5)$$

Equating the values of θ from equations (5) and (4), we get

$$\theta = \frac{\lambda_{electron}}{2d} = \frac{\lambda_{electron}}{\lambda_{photon}} \qquad (6)$$

"Look!" Bill says, "$\lambda_{electron}$ cancels and we are left with "

$$\boxed{\lambda_{photon} = 2d} \qquad (7)$$

"I told you," Jill interrupts, "that you had to be careful about what wavelength photons John could use. Here we see that if John's photons have a wavelength of 2d or less, his photons will carry enough of a punch, enough momentum to destroy the two slit pattern. Be sure John's photons have a wavelength longer than 2d so that they will be incapable of knocking an electron from a maxima to a minima."

"No way," responds Bill. "A wavelength of 2d is already too big. John cannot use photons with a wavelength any greater than the slit width d if he wants to see which slit the electron went through. And you want him to use photons with a wavelength greater than 2d!"

"That's the dilemma," the professor replies. "If John uses photons whose wavelength is short enough to see which slit the electron went through, he is likely to mess up the experiment and destroy the two slit pattern."

"It looks like the very act of getting information is messing up the experiment," Jill muses.

"It messes it up if we use photons," Bill responds." Let us work out a better experiment where we do a more delicate measurement to see which slit the electron went through. Do the experiment so delicately that we do not affect the motion of the electron, but accurately enough to see which slit the electron went through."

"How would you do that?" Jill asks.

"Maybe I would put a capacitor plate on one of the slits," Bill responds, "and record the capacitor voltage. If the electron went through that slit, the electric field of the electron should affect the voltage on the capacitor and leave a blip on my oscilloscope screen. If I don't see a blip, the electron went through the other slit."

"Would this measurement affect the motion of the electron?" Jill asks.

"I don't see why," Bill responds.

"Think about this," the professor interrupts. "We are now interpreting the electric and magnetic fields of a light wave as a probability wave for photons. In this view, all electric and magnetic phenomena are ultimately caused by photons. The electric and magnetic fields we worked with earlier in the course are now to be thought of as a way of describing the behavior of the underlying photons."

"That's crazy," Bill argues. "You mean, for example, that the good old $1/r^2$ Coulomb force law that holds the hydrogen atom together, is caused by photons? I don't see how."

"It's hard to visualize, but you can use a photon picture to explain every detail of the interaction between the electron and proton in the hydrogen atom. That calculation was actually done back in 1947. The modern view is that all electric and magnetic phenomena are caused by photons."

"If all electric and magnetic phenomena are caused by photons," Jill observes, "then Bill's capacitor plate and voltmeter, which uses electromagnetic phenomena, is based on photons. Since photons obey the de Broglie relationship, the photons in Bill's experiment should have the same effect as the photons from John's flashlight. If John's photons mess up the experiment, Bill's should too!"

"I have an idea," Bill says. "Aren't there such a thing as gravitational waves?"

"Yes," replies the professor. "They are very hard to make, and very hard to detect. We have not been able to make or detect them yet in the laboratory. But back in the 1970's Joe Taylor at the University of Massachusetts discovered a pair of binary neutron stars orbiting about each other. Since the stars eclipse each other, Taylor could accurately measure the orbital period."

"According to Einstein's theory of gravity, the orbiting neutron stars should radiate gravitational waves and lose energy. Joe Taylor has conclusively shown that the pair of stars are losing energy just as predicted by Einstein's theory. Taylor got the Nobel prize for this work in 1993."

"Is Einstein's theory a quantum theory?" Bill asks.

"What do you mean by that?" Jill asks.

"I mean," Bill responds, "in Einstein's theory, do gravitational waves have a particle wave nature like electromagnetic waves? Are there particles in a gravitational wave like there are photons in a light wave?"

"Not in Einstein's theory," the professor replies. "Einstein's theory is strictly a classical theory. No particles in the wave."

"Then if Einstein's theory is correct," Bill continues, "I should be able to make a gravitational wave with a very short wavelength and very little energy."

"Couldn't I then use this short wavelength, low energy, gravitational wave to see which slit the electron went through? I would make the wavelength much shorter than the slit spacing d so that there would be no doubt about which slit the electron went through. But I would use a very low energy, delicate wave so that I would not affect the motion of the electron."

"You could do that if Einstein's theory is right," the professor replies.

"But," Bill responds, "that allows me to tell which slit the electron went through without destroying the two slit pattern. What happens to the probability interpretation of the electron wave? If I know which slit the electron went through, the probability wave must have come from that slit, and we must get a one slit pattern. If John used gravitational waves instead of light waves in his flashlight, he could observe which slit the electron went through without destroying the two slit pattern."

"You have just stumbled upon one of the major outstanding problems in physics," the professor replies. "As far as we know there are four basic forces in nature. They are gravity, the electromagnetic force, the weak interaction, and the so-called gluon force that holds quarks together. I listed these in the order in which they were discovered."

"Now three of these forces, all but gravity, are known to have a particle-wave nature like light. All the particles obey the de Broglie relation $p = h/\lambda$."

"As a result, if we perform our two slit electron experiment, trying to see which slit the electron went through, and we use apparatus based on non gravitational forces, we run into the same problem we had with John's flashlight. The only chance we have for detecting which slit the electron went through without messing up the two slit pattern, is to use gravity."

"Could Einstein be wrong?" Jill asks. "Couldn't gravitational waves also have a particle nature? Couldn't the gravitational particles also obey the de Broglie relation?"

"Perhaps," the professor replies. "For years, physicists have speculated that gravity should have a particle-wave nature. They have even named the particle -- they call it a **graviton**. One problem is that gravitons should be very, very hard to detect. The only way we know that gravitational waves actually exist is from Joe Taylor's binary neutron stars. There are various experiments designed to directly observe gravitational waves, but no waves have yet been seen in these experiments."

"In the case of electromagnetism, we saw electromagnetic radiation -- i.e., light -- long before photons were detected in Hertz's photoelectric effect experiment. After gravitational waves are detected, then we will have to do the equivalent of a photoelectric effect experiment for gravity in order to see the individual gravitons. The main problem here is that the gravitational radiation we expect to see, like that from massive objects such as neutron stars, is very low frequency radiation. Thus we would be dealing with very low energy gravitons which would be hard to detect individually."

"And there is another problem," the professor continues, "no one has yet succeeded in constructing a consistent quantum theory of gravity. There are mathematical problems that have yet to be overcome. At the present time, the only consistent theory of gravity we have is Einstein's classical theory."

"It looks like two possibilities," Jill says. "If the probability interpretation of electron waves is right, then there has to be a quantum theory of gravity, gravitons have to exist. If Einstein's classical theory is right, then there is some flaw in the probability interpretation."

"That is the way it stands now," the professor replies.

THE UNCERTAINTY PRINCIPLE

We have just seen that, for the probability interpretation of particle-waves to be a viable theory, there can be no way we can detect which slit the electron went through without destroying the two slit pattern. Also we have seen that if every particle and every force have a particle wave nature obeying the de Broglie relationship $\lambda = h/p$, then there is no way we can tell which slit the electron went through without destroying the two slit pattern. Both the particle-wave nature of matter, and the probability interpretation of particle waves, lead to a basic limitation on our ability to make experimental measurements. This basic limitation was discovered by Werner Heisenberg shortly before Schrödinger developed his wave equation for electrons. Heisenberg called this limitation the ***uncertainty principle***.

When you cannot do something, when there is really no way to do something, physicists give the failure a name and call it a basic law of physics. We began the text with the observation that you cannot detect uniform motion. Michaelson and Morley thought they could, repeatedly tried to do so, and failed. This failure is known as the principle of relativity which Einstein used as the foundation of his theories of relativity. Throughout the text we have seen the impact of this simple idea. When combined with Maxwell's theory of light, it implied that light traveled at the same speed relative to all observers. That implied moving clocks ran slow, moving lengths contracted, and the mass of a moving object increased with velocity. This led to the relationship $E = mc^2$ between mass and energy, and to the connections between electric and magnetic fields. The simple idea that you cannot measure uniform motion has an enormous impact on our understanding of the way matter behaves.

Now, with the particle-wave nature of matter, we are encountering an equally universal restriction on what we can measure, and that restriction has an equally important impact on our understanding of the behavior of matter. Our discussion of the uncertainty principle comes at the end of the text rather than at the beginning only because it has taken a while to develop the concepts we need to explain this restriction. With the principle of relativity we could rely on the student's experience with uniform motion, clocks and meter sticks. For the uncertainty principle, we need some understanding of the behavior of particles and waves, and as we shall see, Fourier analysis plays an important role.

There are two forms of the uncertainty principle, one related to measurements of position and momentum, and the other related to measurements of time and energy. They are not separate laws, one can be derived from the other. The choice of which to use is a matter of convenience. Our discussion of the two slit experiment and the de Broglie relationship naturally leads to the position-momentum form of the law, while Fourier analysis naturally introduces the time-energy form.

POSITION-MOMENTUM FORM OF THE UNCERTAINTY PRINCIPLE

In our two slit thought experiment, in the attempt to see which slit the electron went through, we used a beam of photons whose momenta was related to their wavelength by $p = h/\lambda$. The wave nature of the photon is important because we cannot see details smaller than a wavelength λ when we scatter waves from an object. When we use waves of wavelength λ, the uncertainty in our measurement is at least as large as λ. Let us call the uncertainty in the position measurement Δx.

However when we use photons to locate the electron, we are slugging the electron with particles, photons of momentum $p_{photon} = h/\lambda$. Since we do not know where the photons are within a distance λ, we do not know exactly how the electron was hit and how much momentum it absorbed from the photon. The electron could have absorbed the full photon momentum p_{photon} or none of it. If we observe the electron, we make the electron's momentum uncertain by an amount at least as large as p_{photon}. Calling the uncertainty in the electron's momentum $\Delta p_{electron}$ we have

$$\Delta p_{electron} = p_{photon} = \frac{h}{\lambda} = \frac{h}{\Delta x} \qquad (8)$$

multiplying through by Δx gives

$$\Delta p \Delta x = h \qquad (9)$$

In Equation 9, Δp and Δx represent the smallest possible uncertainties we can have when measuring the position of the electron using photons. To allow for the fact that we could get much greater uncertainties using poor equipment or sloppy techniques, we will write the equation in the form

$$\boxed{\Delta p \Delta x \geq h} \quad \begin{array}{l} \textit{position–momentum} \\ \textit{form of the} \\ \textit{uncertainty principle} \end{array} \qquad (10)$$

indicating that the product of the uncertainties is at least as large as Planck's constant h.

If all forces have a particle nature, and all particles obey the de Broglie relationship, then the fact that we derived Equation 10 using photons makes no difference. We have to get the same result using any particle, in any possible kind of experiment. Thus Equation 10 represents a fundamental limitation on the measurement process itself!

Equation 10 is not like any formula we have previously dealt with in the text. It gives you an estimate, not an exact value. Often you will see the formula written $\Delta p \Delta x \geq \hbar$ with $\hbar = h/2\pi$, rather than h, appearing on the right side. Whether you use h or $\hbar$ depends upon how you wish to define the uncertainties Δp and Δx. But it is not necessary to be too precise. The important point is that the product $\Delta p \Delta x$ must be at least of the order of magnitude h. It cannot be h/100 or something smaller.

The gist of the uncertainty principle is that the more accurately you measure the position of the particle, the more you mess up the particle's momentum. Or, the more accurately you measure the momentum of a particle, the less you know about the particle's position.

Equation (10) is not quite right, because it turns out that an accurate measurement of the x position of a particle does not necessarily mess up the particle's y component of momentum, only its x component. A more accurate statement of the uncertainty principle is

$$\Delta p_x \Delta x \geq h \qquad (11a)$$

$$\Delta p_y \Delta y \geq h \qquad (11b)$$

where Δp_x is the uncertainty in the particle's x component of momentum due to a measurement of its x position, and Δp_y is the uncertainty in the y component of momentum resulting from a y position measurement. The quantities Δx and Δy are the uncertainty in the x and y measurements respectively.

Single Slit Experiment

In our two slit thought experiment, we measured the position of the electron by hitting it with a photon. Another way to measure the position of a particle is to send it through a slit. For example, suppose a beam of particles impinges on a slit of width (w) as illustrated in Figure (11). We know that any particle that makes it to the far side of the slit had, at one time, been within the slit. At that time we knew its y position to within an uncertainty Δy equal to the width (w) of the slit.

$$\Delta y = w \tag{12}$$

This is an example of a position measurement with a precisely known uncertainty Δy.

According to the uncertainty principle, the particle's y component of momentum is uncertain by an amount Δp_y given by Equation 11b as

$$\Delta p_y \geq \frac{h}{\Delta y} = \frac{h}{w} \tag{13}$$

Equation 13 tells us that the smaller Δy, i.e., the narrower the slit, the bigger the uncertainty Δp_y we create in the particle's y momentum. This is what happens if the particle's motion is governed by its wave nature.

In Figure (12a) we have a ripple tank photograph of a wave passing through a moderately narrow slit. The wave on the far side of the slit is seen to spread out a bit. We can calculate the amount of spread by noting that the beam is mostly contained within the central maximum of the single slit diffraction pattern.

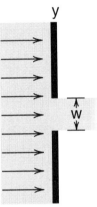

Figure 11
When a particle goes through the slit, its y position is known to within an uncertainty $\Delta y = w$.

Now suppose that this wave represented a beam of photons or electrons. On the right side of the slit, all the particles have a definite x component of momentum $p_x = h/\lambda$ and no y momentum. The uncertainty in the y momentum is zero.

Once the particle's have gone through the slit, the beam spreads out giving the particles a y momentum.

Since you do not know whether any given particle in the beam will go straight ahead, up or down, the spread of the beam introduces an uncertainty Δp_y in the particle's y momentum. This spread is illustrated in Figure (12b).

In Figure (13a), we see a wave passing through a narrower slit than the one in Figure (12a). With the narrower slit, we have made a more precise measurement of the particle's y position. We have reduced the uncertainty $\Delta y = w$. According to the uncertainty principle $\Delta p_y \geq h/\Delta y$, a decrease in Δy should increase the uncertainty Δp_y in the particle's y momentum. But an increase in Δp_y means that the beam should spread out more, which is what it does in Figure (13). In going from Figure (12) to Figure (13), we have cut the slit width about in half and about doubled the spread. I.e., cutting Δy in half doubles Δp_y as expected.

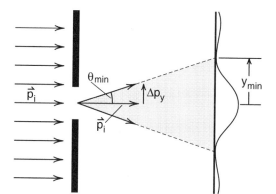

Figure 12a
Wave picture of the single slit experiment, wide slit.

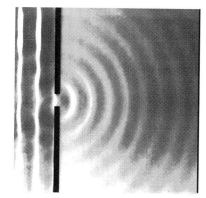

Figure 13a
Narrow the slit and the wave spreads out.

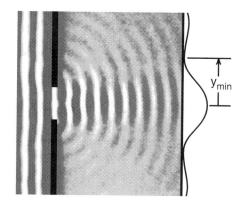

Figure 12b
Particle picture of the single slit experiment.

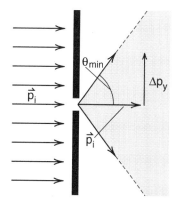

Figure 13b
With a narrower slit, Δp_y increases.

Example 1

We can use our analysis of the single slit pattern in Chapter 33 to show that Δp_y and Δy are related by the uncertainty principle. We saw that when a wave goes through a single slit of width w, the distance y_{min} to the first minimum is given by

$$y_{min} = \frac{\lambda D}{w} \tag{33-14}$$

where λ is the wavelength and D the distance to the screen as shown in Figure (14a). The angle to the first minimum is given by

$$\tan(\theta_{min}) = \frac{y_{min}}{D} = \frac{(\lambda D/w)}{D} = \frac{\lambda}{w} \tag{14}$$

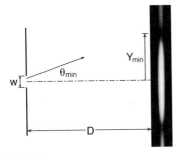

Figure 14a

After the beam emerges from the slit, the momenta of the particles spreads out through the same angle θ_{min} as indicated in Figure (14b). From that figure we have

$$\tan(\theta_{min}) = \frac{\Delta p_y}{p_x} \tag{15}$$

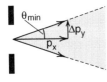

Figure 14b

where Δp_y represents the possible spread in the particle's y momenta. Equating values of $\tan(\theta_{min})$ from Equation 14 and Equation 15 gives

$$\frac{\lambda}{w} = \frac{\Delta p_y}{p_x} \tag{16}$$

The particles entered the slit as plane waves with only an x component of momentum given by de Broglie's formula

$$p_x = \frac{h}{\lambda} \tag{17}$$

Using this formula for p_x in Equation 16 gives

$$\frac{\lambda}{w} = \frac{\Delta p_y}{(h/\lambda)} = \frac{\lambda \Delta p_y}{h} \tag{18}$$

The λ's cancel, and we are left with

$$\left(\Delta p_y\right)w = h \tag{19}$$

But the slit width w is Δy, the uncertainty in the y measurement, thus

$$\Delta p_y \Delta y = h \tag{20}$$

There is an equal sign in Equation 20 because this particular measurement of the y position of the particle causes the least possible uncertainty in the particle's y component of momentum. (Note that the x component of the particle's momentum is more or less unaffected by the slit. The wave has the same wavelength λ before and after going through the slit. It is the y component of momentum that changed from zero on the left side to $\pm\Delta p_y$ on the right.)

Exercise 3

A microwave beam, consisting of $1.24 \times 10^{-4}\,eV$ photons impinges on a slit of width (w) as shown in Figure 14a.

(a) What is the momentum p_x of the photons in the laser beam before they get to the slit?

(b) When the photon passes through the slit, their y position is known to an uncertainty $\Delta y = w$, the slit width. Before the photons get to the slit, their y momentum has the definite value $p_y = 0$. Passing through the slit makes the photon's y momentum uncertain by an amount Δp_y. Using the uncertainty principle, calculate what the slit width (w) must be so that Δp_y is equal to the photon's original momentum p_x. How does w compare with the wavelength λ of the laser beam?

(c) If Δp_y becomes as large as the original momentum p_x, what can you say about the wave pattern on the right side of Figure (11)? Is this consistent with what you know about waves of wavelength λ passing through a slit of this width? Explain.

TIME-ENERGY FORM OF THE UNCERTAINTY PRINCIPLE

The second form of the uncertainty principle, which perhaps has an even greater impact on our understanding of the behavior of matter, involves the measurement of the energy of a particle, and the time available to make the measurement. *The shorter the time available, the less accurate the energy measurement is*. If ΔE is the uncertainty in the results of our energy measurement, and Δt the time we had to make the measurement, then ΔE and Δt are related by

$$\boxed{\Delta E \Delta t \geq h} \tag{21}$$

One can derive this form of the uncertainty principle from $\Delta p \Delta x \geq h$, but we can gain a better insight into the relationship by starting with an explicit example.

A device that has become increasingly important in research, particularly in the study of fast reactions in molecules and atoms, is the pulsed laser. The lasers we have used in various experiments are all continuous beam lasers. The beam is at least as long as the distance from the laser to the wall. If we had a laser that we could turn on and off in one nanosecond, the pulse would be 1 foot or 30 cm long and contain

$$\frac{30 cm}{6 \times 10^{-5} \dfrac{cm}{wavelength}} = 5 \times 10^5 wavelengths$$

Even a picosecond laser pulse which is 1000 times shorter, contains 500 wavelengths. Some of the recent pulsed lasers can produce a pulse 500 times shorter than that, only 2 femtoseconds (2×10^{-15} seconds) long. These lasers emit a pulse that is only one wavelength long.

For our example of the time-energy form of the uncertainty principle, we wish to consider the nature of the photons in a 2 femtosecond long laser pulse. If we want to measure the energy of the photons in such a pulse, we only have 2 femtoseconds to make the measurement because that is how long the pulse takes to go by us. In the notation of the uncertainty principle

$$\Delta t = 2 \times 10^{-15} sec \qquad \begin{array}{l} \textit{time available to} \\ \textit{measure the energy} \\ \textit{of the photons} \\ \textit{in our laser pulse} \end{array} \tag{22}$$
$$= 2 \text{ femtoseconds}$$

Let us suppose that the laser produces red photons whose wavelength is $6.2 \times 10^{-5} cm$, about the wavelength of the lasers we have been using. According to our usual formula for calculating the energy of the photons in such a laser beam we have

$$E_{photon} = \frac{12.4 \times 10^{-5} eV \; cm}{6.2 \times 10^{-5} cm} = 2 \text{ eV} \tag{23}$$

Now let us use the uncertainty principle in the form

$$\Delta E \geq \frac{h}{\Delta t} \tag{24}$$

to calculate the uncertainty in any measurement we would make the energy of the photons in the 2 femtosecond laser beam. We have

$$\Delta E \geq \frac{h}{\Delta t} \geq \frac{6.63 \times 10^{-27} erg \; sec}{2 \times 10^{-15} sec}$$
$$\geq 3.31 \times 10^{-12} ergs \tag{25}$$

Converting ΔE from ergs to electron volts, we get

$$\Delta E \geq \frac{3.31 \times 10^{-12} ergs}{1.6 \times 10^{-12} ergs/eV} \approx 2 \text{ eV} \tag{26}$$

The uncertainty in any energy measurement we make of these photons is as great as the energy itself! If we try to measure the energy of these photons, we expect the answers to range from $E - \Delta E = 0 \text{ eV}$ up to $E + \Delta E = 4 \text{ eV}$. Why does this happen? Why is the energy of the photons in this beam so uncertain? *Fourier analysis provides the answer*.

We can see why the energy of the photons in the 2 femtosecond pulse is so uncertain by comparing the Fourier transform of a long laser pulse with that of a pulse consisting of only one wavelength.

Figure (15) shows the Fourier transform of an infinitely long sine wave. You will recall that, in the design of the MacScope program, it is assumed that we are analyzing a repeated waveform. If you continuously repeat the waveform seen in the upper half of the diagram, you get an infinitely long cyclic wave which is a pure sine wave. (Sine waves are by definition infinitely long waves.) In effect we have in Figure (15) selected 16 cycles of the pure sine wave, and the Fourier analysis box shows that we have a pure 16th harmonic. This sine wave has a definite frequency f, and if this represented a laser beam, the photons in the beam would have a precise energy given by the formula $E = hf$. There is no uncertainty in the energy of this infinitely long sine wave. (It would take an infinite time Δt to make sure that the wave was infinitely long, with the result $\Delta E = h/\Delta t = h/\infty = 0$.)

In Figure (16) we are looking at a waveform consisting of a single pulse. This would accurately represent the output of a red laser that continuously emitted single wavelength pulses spaced 16 wavelengths apart. (Remember that our program assumes that the wave shape is repeated.) From the Fourier analysis box we see that there is a dramatic difference between the composition of a pure sine wave and of a single pulse. To construct a single pulse out of sine waves, we have to add up a slew of harmonics. The single pulse is more like a drum beat while the continuous wave is more like a flute. (In the appendix we show how the sine wave harmonics add up to produce a pulse.)

In Figure (16) we see that the dominant harmonic is still around the sixteenth, as it was for the continuous wave, but there is a spread of harmonics from near zero up to almost the 32nd. For a laser pulse to have this shape, it must consist of frequencies ranging from near zero up to twice the natural frequency. Each of these frequencies contains photons whose energy is given by Einstein's formula $E = hf$ where f is the frequency of the harmonic.

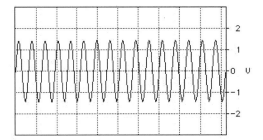

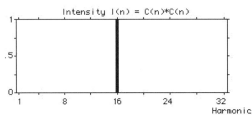

Intensity $I(n) = C(n)*C(n)$

Figure 15
A pure sine wave has a single frequency.

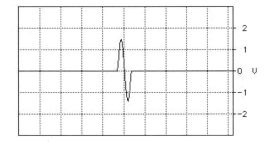

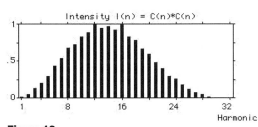

Intensity $I(n) = C(n)*C(n)$

Figure 16
One cycle of a wave is made up of a spread of harmonics

In Figure (17) we have reproduced the Fourier analysis box of Figure (16), but relabeled the horizontal axis in electron volts. We have assumed that the 16th harmonic represented 2 eV photons which would be the case if the wave were an infinitely long red laser pulse. Now the diagram represents the density of photons of different energies in the laser beam. While 2 eV is the most likely energy, there is a spread of energies ranging from nearly 0 eV up to nearly 4 eV. If we measure the energy of a photon in the beam, our answer is 2 eV with an uncertainty of 2 eV, just as predicted by the uncertainty principle $\Delta E \Delta t \geq h$.

In Figure (18) we analyze a pulse two wavelengths long. Now we see that the spread of frequencies required to reconstruct this waveform is only half as wide, ranging from the 8th to 24th harmonic, or from 1 eV to 3 eV. We have twice as long to study a 2 wavelength pulse, and the uncertainty in energy ΔE is only about 1 eV, or half as big.

Going to a 4 wavelength pulse in Figure (19) we see that by doubling the time available we again cut in half the uncertainty ΔE in energy. Now the energy varies from about 1.5 eV to 2.5 eV for a $\Delta E = .5$ eV. This is just what you expect from $\Delta E \Delta t \geq h$. You should now begin to see that the uncertainty principle is a simple rule evolving from the wave nature of particles. (By the way, it would be more accurate to write $\Delta E \Delta t = h$ for this discussion, because we are describing the very least uncertainty in energy.)

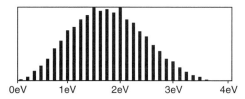

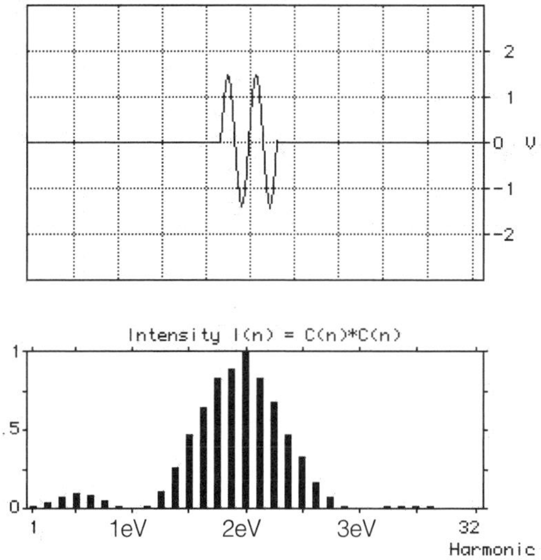

Figure 17
Photon energies in single wavelength pulse of a red laser beam.

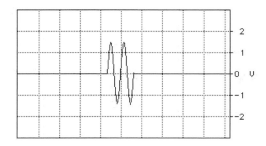

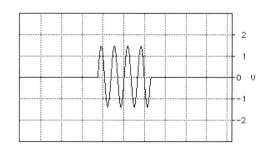

Figure 18
A two cycle wave has half the spread of harmonics.

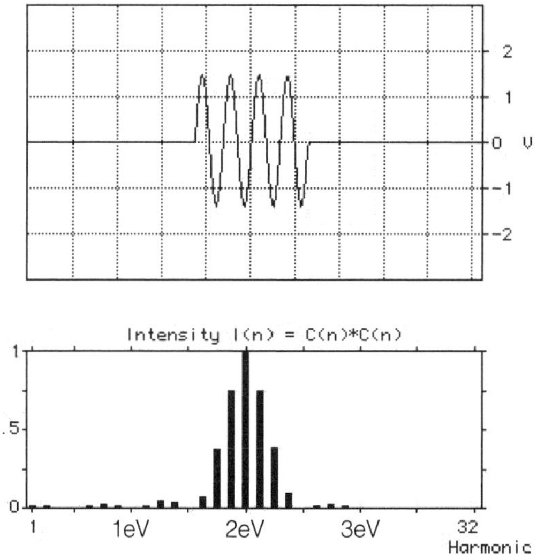

Figure 19
A four cycle wave has a fourth the spread of harmonics.

Probability Interpretation

We have interpreted Figure (17) as representing the spread in energies of the photons in a 2 femtosecond red laser pulse. What if the pulse consisted of only a single photon? Then how do we interpret this spread in energies? The answer is that we use a probability interpretation. The photon in the pulse has different probabilities of having different energies.

In our discussion of light waves, we saw that the energy density in a light wave was proportional to the square of the amplitude of the wave. This is reasonable because while the amplitude of a wave can be positive or negative, the square of the amplitude, which we call the *intensity* is always positive. Probabilities, like energy densities, also have to be positive, thus we should associate the probability of a photon as having a given frequency with the intensity or square of the amplitude of the wave of that frequency.

In Figure (20) we show the intensities (square of the amplitudes) of the harmonics that make up the single wavelength pulse. (This is plotted automatically by MacScope when we click on the button labeled Φ.) We see that squaring the amplitudes narrows the spread.

Figure (20) has the following interpretation when applied to pulses containing a single photon. If we measure the energy of the photon, we are most likely to get an answer close to 2 eV but there is a reasonable probability of getting an answer lower than 1 eV or even higher than 3 eV. The heights of the bars tell us the relative probability of measuring that energy for the photon.

Measuring Short Times

We have said that the new pulsed lasers produce pulses as short as 2 femtoseconds. How do we know that? Suppose we gave you the job of measuring the length of the laser pulse, and the best oscilloscope you had could measure times no shorter than a nanosecond. This is a million times too slow to see a femtosecond pulse. What do you do?

If you cannot measure the time directly, you can be sneaky and use the uncertainty principle. Send the laser pulse through a diffraction grating, and record the spread in wavelengths, i.e., the spread in energies of the photons in the pulse. If the line is very sharp, if they are all red photons of a single wavelength and energy, then you know that there is no measurable uncertainty ΔE in the photon energies, and the pulse must last a time Δt that is considerably longer than 2 femtoseconds. If, on the other hand, the line is spread out from the near infra red to violet, if the spread in energies is from 1 eV to 3 eV, and the spread is not caused by some other phenomena (like the Doppler effect), then from the uncertainty principle you know that the pulse is only about a femtosecond long. (You know, for example, it cannot be as long as 10 femtoseconds, or as short as a tenth of a femtosecond.)

Thus, with the uncertainty principle, you can use a diffraction grating rather than a clock or oscilloscope to measure very short times. Instead of being an annoying restriction on our ability to make experimental measurements, the uncertainty principle can be turned into an important scientific tool for measuring short times and, as we shall see, short distances.

Exercise 4

An electron is in an excited state of the hydrogen atom, either the second energy level at -3.40 eV or the third energy level at -1.51 eV. You wish to do an experiment to decide which of these two states the electron is in. What is the least amount of time you **must** take to make this measurement?

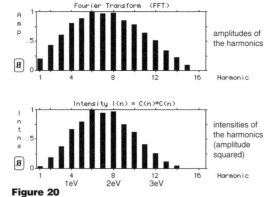

Figure 20
Intensities of the harmonics are proportional to the square of the amplitudes.

Short Lived Elementary Particles

We usually think of the rest energy of a particle as having a definite value. For example the rest energy of a proton is $938.2723 \times 10^6 \text{eV}$. The proton itself is a composite particle made of 3 quarks, and the number 938.2723 MeV represents the total energy of the quarks in the allowed wave pattern that represents a proton. This rest energy has a very definite value because the proton is a stable particle with plenty of time to settle into a precise wave pattern.

A rather different particle is the so called "$\Lambda(1520)$", which is another combination of 3 quarks, but very short lived. The name comes partly from the fact that the particle's rest mass energy is about 1520 million electron volts (MeV). As indicated in Figure (21), a $\Lambda(1520)$ can be created as a result of the collision between a K^- meson and a proton.

We are viewing the collision in a special coordinate system, where the total momentum of the incoming particles is zero. In this coordinate system, the resulting $\Lambda(1520)$ will be at rest. By conservation of energy, the total energy of the incoming particles should equal the rest mass energy of the $\Lambda(1520)$. Thus if we collide K^- particles with protons, we expect to create a $\Lambda(1520)$ particle only if the incoming particles have the right total energy.

Figure (22) shows the results of some collision experiments, where a K^- meson and a proton collided to produce a Λ and two π mesons. The probability of such a result peaked when the energy of the incoming particles was 1,520 MeV. This peak occurred because the incoming K^- meson and proton created a $\Lambda(1520)$ particle, which then decayed into the Λ and two π mesons, as shown in Figure (21). The $\Lambda(1520)$ was not observed directly, because its lifetime is too short.

Figure (22) shows that the energy of the incoming particles does not have to be exactly 1520 MeV in order to create a $\Lambda(1520)$. The peak is in the range from about 1510 to 1530 Mev, which implies that the rest mass energy of the $\Lambda(1520)$ is 1520 MeV plus or minus about 10 MeV. From one experiment to another, the rest mass energy can vary by about 20 MeV. (The experimentalists quoted a variation of 16 MeV.)

a) A K^-meson and a proton are about to collide. We are looking at the collision in a coordinate system where the total momentum is zero (the so called "center of mass" system).

b) *In the collision a $\Lambda(1520)$ particle is created. It is at rest in this center of mass system*

c) *The $\Lambda(1520)$ then quickly*
decays *into a lower energy Λ particle and two π mesons.*

Figure 21
A $\Lambda(1520)$ particle can be created if the total energy (in the center of mass system) of the incoming particles equals the rest mass energy of the $\Lambda(1520)$.

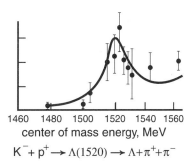

center of mass energy, MeV

$$K^- + p^+ \longrightarrow \Lambda(1520) \longrightarrow \Lambda + \pi^+ + \pi^-$$

Figure 22
The probability that a K^-meson and a proton collide to produce a Λ particle, and two π mesons peak at an energy of 1520 MeV. The peak results from the fact that a $\Lambda(1520)$ particle was created and quickly decayed into the Λ and two π mesons. The probability peaks at 1520 MeV, but can be seen to spread out over a range of about 16MeV. The small circles are experimental values, the vertical lines represent the possible error in the value. (Data from M.B. Watson et al., Phys. Rev. 131(1963).)

Why isn't the peak sharp? Why does the rest mass energy of the $\Lambda(1520)$ particle vary by as much as 16 to 20 MeV from one experiment to another? The answer lies in the fact that *the lifetime of the $\Lambda(1520)$ is so short, that the particle does not have enough time to establish a definite rest mass energy*. The 16 MeV variation is the uncertainty ΔE in the particles rest mass energy that results from the fact that the particle's lifetime is limited.

The uncertainty principle relates the uncertainty in energy ΔE to the time Δt available to establish that energy. To establish the rest mass energy, time Δt available is the particle's *lifetime*. Thus we can use the uncertainty principle to estimate the lifetime of the $\Lambda(1520)$ particle. With $\Delta E \times \Delta t \approx h$ we get

$$\Delta t \approx \frac{h}{\Delta E} = \frac{6.63 \times 10^{-27}\text{erg sec}}{16\text{ MeV} \times \left(1.6 \times 10^{-6}\dfrac{\text{erg}}{\text{MeV}}\right)}$$

$$\Delta t \approx 2.6 \times 10^{-22}\text{ seconds} \qquad (27)$$

The lifetime of the $\Lambda(1520)$ particle is of the order of 10^{-22} seconds! This is only about 10 times longer than it takes light to cross a proton! Only by using the uncertainty principle could we possibly measure such short times.

THE UNCERTAINTY PRINCIPLE AND ENERGY CONSERVATION

The fact that for short times the energy of a particle is uncertain, raises an interesting question about basic physical laws like the law of conservation of energy. If a particle's energy is uncertain, how do we know that energy is conserved in some process involving that particle? The answer is -- *we don't*.

One way to explain the situation is to say that nature will cheat if it can get away with it. Energy does not have to be conserved if we cannot do an experiment to demonstrate a lack of conservation of energy.

Consider the process shown in Figure (23). It shows a red, 2 eV photon traveling along in space. Suddenly the photon creates an positron-electron pair. The rest mass energy of both the positron and the electron are .51 MeV. Thus we have a 2 eV photon creating a pair of particles whose total energy is $1.02 \times 10^6\text{eV}$, a huge violation of the law of conservation of energy. A short time later the electron and positron come back together, annihilate, leaving behind a 2 eV photon. This is an equally huge violation of the conservation of energy.

But have we really violated the conservation of energy? During its lifetime, the positron-electron pair is a composite object whose total energy is uncertain. If the pair lived a long time, its total energy would be close to the expected energy of $1.02 \times 10^6\text{eV}$. But suppose the pair were in existence only for a very short time Δt, a time so short that the uncertainty in the energy could be as large as $1.02 \times 10^6\text{eV}$. Then there is some probability that the energy of the pair might be only 2 eV and the process shown in Figure (2) could happen.

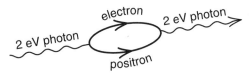

Figure 23
Consider a process where a 2 eV photon suddenly creates a positron-electron pair. A short time later the pair annihilates, leaving a 2 eV photon. In the long range, energy is conserved.

The length of time Δt that the pair could exist and have an energy uncertain by 1.02 MeV is

$$\Delta t = \frac{h}{\Delta E} = \frac{6.63 \times 10^{-27} \text{ erg sec}}{1.02 \text{ MeV} \times 1.6 \times 10^{-6} \frac{\text{erg}}{\text{MeV}}}$$

$$\Delta t = 4 \times 10^{-21} \text{sec} \qquad (32)$$

Another way to view the situation is as follows. Suppose the pair in Figure (20) lasted only 4×10^{-21} seconds or less. Even if the pair had an energy of $1.02 \times 10^6 \text{eV}$, the lifetime is so short that any measurement of the energy of the pair would be uncertain by at least $1.02 \times 10^6 \text{eV}$, and the experiment could not detect the violation of the law of conservation of energy. In this point of view, if we cannot perform an experiment to detect a violation of the conservation law, then the process should have some probability of occurring.

Does a process like that shown in Figure (23) actually occur? If so, is there any way that we can know that it does? The answer is yes, to both questions. It is possible to make extremely accurate studies of the energy levels of the electron in hydrogen, and to make equally accurate predictions of the energy using the theory of *quantum electrodynamics*. We can view the binding of the electron in hydrogen as resulting from the continual exchange of photons between the electron and proton. During this continual exchange, there is some probability that the photon creates a positron electron pair that quickly annihilates as shown in Figure (23). In order to predict the correct values of the hydrogen energy levels, the process shown in Figure (23) has to be included. Thus we have direct experimental evidence that for a short time the particle antiparticle pair existed.

QUANTUM FLUCTUATIONS AND EMPTY SPACE

We began the text with a discussion of the principle of relativity—that you could not detect your own motion relative to empty space. The concept of empty space seemed rather obvious—space with nothing in it. But the idea of empty space is not so obvious after all.

With the discovery of the cosmic background radiation, we find that all the space in this universe is filled with a sea of photons left over from the big bang. We can accurately measure our motion relative to this sea of photons. The earth is moving relative to this sea at a velocity of 600 kilometers per second toward the Vergo cluster of galaxies. While this measurement does not violate the principle of relativity, it is in some sense a measurement of our motion relative to the universe as a whole.

Empty space itself may not be empty. Consider a process like that shown in Figure (24) where a photon, an electron, and a positron are all created at some point in space. A short while later the three particles come back together with the positron and electron annihilating and the photon being absorbed.

One's first reaction might be that such a process is ridiculous. How could these three particles just appear and then disappear? To do this we would have to violate both the laws of conservation of energy and momentum.

But, of course, the uncertainty principle allows us to do that. We can, in fact, use the uncertainty principle to estimate how long such an object could last. The arguments would be similar to the ones we used in the analysis of the process shown in Figure (23).

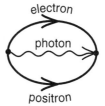

electron

photon

positron

Figure 24
Quantum fluctuation. The uncertainty principle allows such an object to suddenly appear, and then disappear.

In the theory of quantum electrodynamics, a completely isolated process like that shown in Figure (24) does not affect the energy levels of the hydrogen atom and should be undetectable in electrical measurements. But such a process might affect gravity. A gravitational wave or a graviton might interact with the energy of such an object. Some calculations have suggested that such interactions could show up in Einstein's classical theory of gravity as a correction to the famous *cosmological constant* we discussed in Chapter 21.

An object like that shown in Figure (23) is an example of what one calls a *quantum fluctuation*. Here we have something that appears and disappears in so-called empty space. If such objects can keep appearing and disappearing, then we have to revise our understanding of what we mean by empty.

The uncertainty principle allows us to tell the difference between a quantum fluctuation and a real particle. A quantum fluctuation like that in Figure (24) violates conservation of energy, and therefore cannot last very long. A real particle can last a long time because energy conservation is not violated.

However, there is not necessarily that much difference between a real object and a quantum fluctuation. To see why, let us take a closer look at the π meson. The π^+ is a particle with a rest mass energy of 140 MeV, that consists of a quark-antiquark pair. The quark in that pair is the so-called *up* quark that has a rest mass of roughly 400 MeV. The other is the *antidown* quark that has a rest mass of about 700 MeV. (Since we can't get at isolated quarks, the quark rest masses are estimates, but should not be too far off). Thus the two quarks making up the π meson have a total rest mass of about 1100 MeV. How could they combine to produce a particle whose rest mass is only 140 MeV?

The answer lies in the potential energy of the gluon force that holds the quarks together. As we have seen many times, the potential energy of an attractive force is negative. In this case the potential energy of the gluon force is almost as big in magnitude as the rest mass of the quarks, reducing the total energy from 1100 MeV to 140 MeV.

Suppose we had an object whose negative potential energy was as large as the positive rest mass energy. Imagine, for example, that the object consisted of a collection of point sized elementary particles so close together that their negative gravitational potential energy was the same magnitude as the positive rest mass and kinetic energy. Suppose such a collection of particles were created in a quantum fluctuation. How long could the fluctuation last?

Since such an object has no total energy, the violation ΔE of energy conservation is zero, and therefore the lifetime $\Delta t = h/\Delta E$ could be forever.

Suppose the laws of physics required that such a fluctuation rapidly expand, greatly increasing both the positive rest mass and kinetic energy, while maintaining the corresponding amount of negative gravitational potential energy. As long as ΔE remained zero, the expanding fluctuation could keep on going. Perhaps such a fluctuation occurred 14 billion years ago and we live in it now.

APPENDIX

HOW A PULSE IS FORMED
FROM SINE WAVES

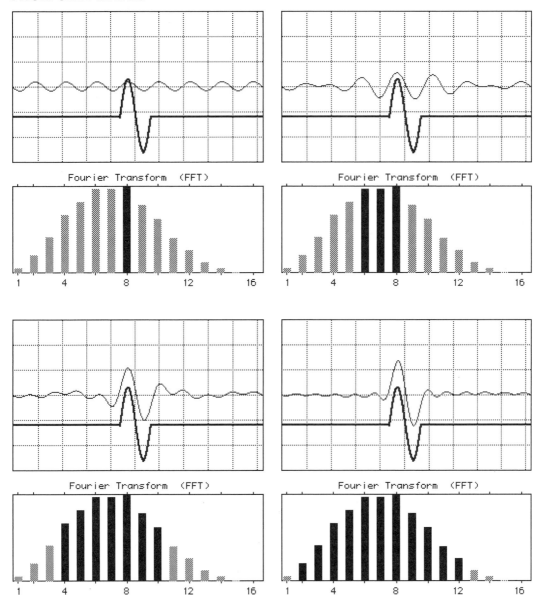

Figure A1
*By selecting more and more harmonics, you can
see how the sine waves add up to produce a pulse.*

Chapter on
Geometrical Optics

For over 100 years, from the time of Newton and Huygens in the late 1600s, until 1801 when Thomas Young demonstrated the wave nature of light with his two slit experiment, it was not clear whether light consisted of beams of particles as proposed by Newton, or was a wave phenomenon as put forward by Huygens. The reason for the confusion is that almost all common optical phenomena can be explained by tracing light rays. The wavelength of light is so short compared to the size of most objects we are familiar with, that light rays produce sharp shadows and interference and diffraction effects are negligible.

To see how wave phenomena can be explained by ray tracing, consider the reflection of a light wave by a metal surface. When a wave strikes a very small object, an object much smaller than a wavelength, a circular scattered wave emerges as shown in the ripple tank photograph of Figure (36-1) reproduced here. But when a light wave impinges on a metal surface consisting of many small atoms, represented by the line of dots in Figure (36-2), the circular scattered waves all add up to produce a reflected wave that emerges at an angle of reflection θ_r equal to the angle of incidence θ_i. Rather than sketching the individual crests and troughs of the incident wave, and adding up all the scattered waves, it is much easier to treat the light as a ray that reflected from the surface. This ray is governed by the law of reflection, namely $\theta_r = \theta_i$.

Figure 36-1
An incident wave passing over a small object produces a circular scattered wave.

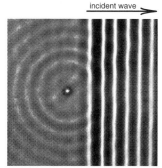

Light ray reflected from a mirror.

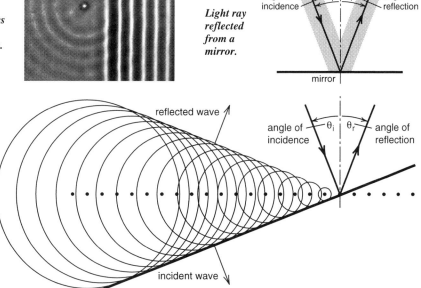

Figure 36-2
Reflection of light. In the photograph, we see an incoming plane wave scattered by a small object. If the object is smaller than a wavelength, the scattered waves are circular. When an incoming light wave strikes an array of atoms in the surface of a metal, the scattered waves add up to produce a reflected wave that comes out at an angle of reflection θ_r equal to the angle of incidence θ_i.

The subject of geometrical optics is the study of the behavior of light when the phenomena can be explained by ray tracing, where shadows are sharp and interference and diffraction effects can be neglected. The basic laws for ray tracing are extremely simple. At a reflecting surface $\theta_r = \theta_i$, as we have just seen. When a light ray passes between two media of different **indexes of refraction**, as in going from air into glass or air into water, the rule is $n_1 \sin \theta_1 = n_2 \sin \theta_2$, where n_1 and n_2 are constants called indices of refraction, and θ_1 and θ_2 are the angles that the rays made with the line perpendicular to the interface. This is known as **Snell's law**.

This entire chapter is based on the two rules $\theta_r = \theta_i$ and $n_1 \sin \theta_1 = n_2 \sin \theta_2$. These rules are all that are needed to understand the function of telescopes, microscopes, cameras, fiber optics, and the optical components of the human eye. You can understand the operation of these instruments without knowing anything about Newton's laws, kinetic and potential energy, electric or magnetic fields, or the particle and wave nature of matter. In other words, there is no prerequisite background needed for studying geometrical optics as long as you accept the two rules which are easily verified by experiment.

In most introductory texts, geometrical optics appears after Maxwell's equations and theory of light. There is a certain logic to this, first introducing a basic theory for light and then treating geometrical optics as a practical application of the theory. But this is clearly not an historical approach since geometrical optics was developed centuries before Maxwell's theory. Nor is it the only logical approach, because studying lens systems teaches you nothing more about Maxwell's equations than you can learn by deriving Snell's law. Geometrical optics is an interesting subject full of wonderful applications, a subject that can appear anywhere in an introductory physics course.

We have a preference not to introduce geometrical optics after Maxwell's equations. With Maxwell's theory, the student is introduced to the wave nature of one component of matter, namely light. If the focus is kept on the basic nature of matter, the next step is to look at the photoelectric effect and the particle nature of light. You then see that light has both a particle and a wave nature, which opens the door to the particle-wave nature of all matter and the subject of quantum mechanics. We have a strong preference not to interrupt this focus on the basic nature of matter with a long and possibly distracting chapter on geometrical optics.

REFLECTION FROM CURVED SURFACES

The Mormon Tabernacle, shown in Figure (1), is constructed in the shape of an ellipse. If one stands at one of the focuses and drops a pin, the pin drop can be heard 120 feet away at the other focus. The reason why can be seen from Figure (2), which is similar to Figure (8-28) where we showed you how to draw an ellipse with a pencil, a piece of string, and two thumbtacks.

The thumbtacks are at the focuses, and the ellipse is drawn by holding the string taut as shown. As you move the pencil point along, the two sections of string always make equal angles θ_i and θ_r to a line perpen-

Mormon Tabernacle under construction, 1866.

Mormon Tabernacle finished, 1871.

Mormon Tabernacle today.
Figure 1

dicular or normal to the part of the ellipse we are drawing. The best way to see that the angles θ_i and θ_r are always equal is to construct your own ellipse and measure these angles at various points along the curve.

If a sound wave were emitted from focus 1 in Figure (2), the part of the wave that traveled over to point A on the ellipse would be reflected at an angle θ_r equal to the angle of incidence θ_i, and travel over to focus 2. The part of the sound wave that struck point B on the ellipse, would be reflected at an angle θ_r equal to it's angle of incidence θ_i, and also travel over to focus 2. If you think of the sound wave as traveling out in rays, then all the rays radiated from focus 1 end up at focus 2, and that is why you hear the whisper there. We say that the rays are *focused* at focus 2, and that is why these points are called focuses of the ellipse. (Note also that the path lengths are the same, so that all the waves arriving at focus 2 are in phase.)

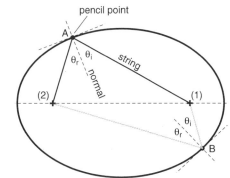

Figure 2
Drawing an ellipse using a string and two thumbtacks.

Figure 2a
A superposition of the top half of Figure 2 on Figure 1.

The Parabolic Reflection

You make a parabola out of an ellipse by moving one of the focuses very far away. The progression from a parabola to an ellipse is shown in Figure (3). For a true parabola, the second focus has to be infinitely far away.

Suppose a light wave were emitted from a star and traveled to a parabolic reflecting surface. We can think of the star as being out at the second, infinitely distant, focus of the parabola. Thus all the light rays coming in from the star would reflect from the parabolic surface and come to a point at the near focus. The rays from the star approach the reflector as a parallel beam of rays, thus a parabolic reflector has the property of focusing parallel rays to a point, as shown in Figure (4a).

If parallel rays enter a deep dish parabolic mirror from an angle off axis as shown in Figure (4b), the rays do not focus to a point, with the result that an off axis star would appear as a blurry blob. (This figure corresponds to looking at a star 2.5° off axis, about 5 moon diameters from the center of the field of view.)

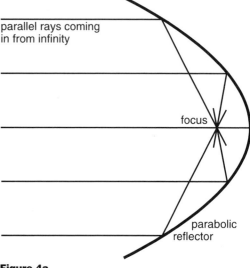

Figure 4a
Parallel rays, coming down the axis of the parabola, focus to a point.

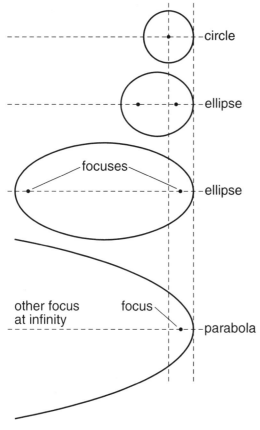

Figure 3
Evolution of an ellipse into a parabola. For a parabola, one of the focuses is out at infinity.

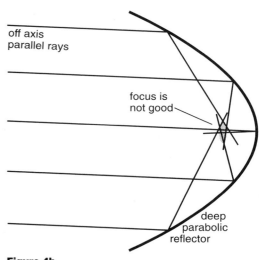

Figure 4b
For such a deep dish parabola, rays coming in at an angle of 2.5° do not focus well.

One way to get sharp images for parallel rays coming in at an angle is to use a shallower parabola as illustrated in Figure (4c). In that figure, the *focal length* (distance from the center of the mirror to the focus) is 2 times the mirror diameter, giving what is called an $f2$ mirror. In Figure (4d), you can see that rays coming in at an angle of 2.5° (blue lines) almost focus to a point. Typical amateur telescopes are still shallower, around $f8$, which gives a sharp focus for rays off angle by as much as 2° to 3°.

As we can see in Figure (4d), light coming from two different stars focus at two different points in what is called the *focal plane* of the mirror. If you placed a photographic film at the focal plane, light from each different star, entering as parallel beams from different angles, would focus at different points on the film, and you would end up with a photographic image of the stars. This is how distant objects like stars are photographed with what is called a *reflecting telescope*.

light from star on axis

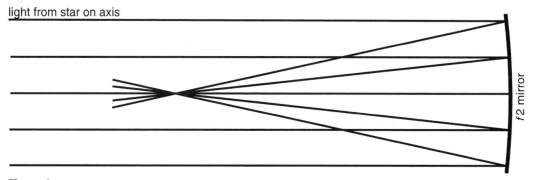

Figure 4c
A shallow dish is made by using only the shallow bottom of the parabola. Here the focal length is twice the diameter of the dish, giving us an f2 mirror. Typical amateur telescopes are still shallower, having a focal length around 8 times the mirror diameter (f8 mirrors). [The mirror in Figure 4b, that gave a bad focus, was f.125, having a focal length 1/8 the diameter of the mirror.]

light from star #2, 2.5° off axis

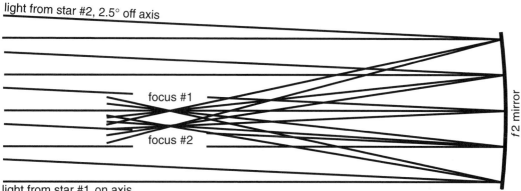

light from star #1, on axis

Figure 4d
We can think of this drawing as representing light coming in from a red star at the center of the field of view, and a blue star 2.5° (5 full moon diameters) away. Separate images are formed, which could be recorded on a photographic film. With this shallow dish, the off axis image is sharp (but not quite a point).

MIRROR IMAGES

The image you see in a mirror, although very familiar, is still quite remarkable in its reality. Why does it look so real? You do not need to know how your eye works to begin to see why.

Consider Figure (5a) where light from a point source reaches your eye. We have drawn two rays, one from the source to the top of the eye, and one to the bottom. In Figure (5b), we have placed a horizontal mirror as shown and moved the light source a distance h above the mirror equal to the distance it was below the mirror before the mirror was inserted. Using the rule that the angle of incidence equals the angle of reflection, we again drew two rays that went from the light source to

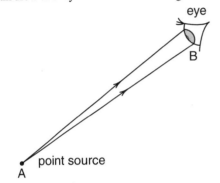

Figure 5a
Light from a point source reaching your eye.

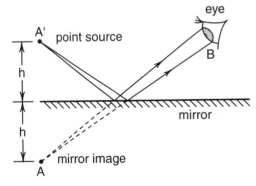

Figure 5b
There is no difference when the source is at point A, or at point A' and the light is reflected in a mirror.

the top and to the bottom of the eye. You can see that if you started at the eye and drew the rays back as straight lines, ignoring the mirror, the rays would intersect at the old source point A as shown by the dotted lines in Figure (5b).

To the eye (or a camera) at point B, there is no detectable difference between Figures (5a) and (5b). In both cases, the same rays of light, coming from the same directions enter the eye. Since the eye has no way of telling that the rays have been bent, we perceive that the light source is at the **image point** A rather than at the source point A'.

When we look at an extended object, its image in the mirror does not look identical to the object itself. In Figure (6), my granddaughter Julia is holding her right hand in front of a mirror and her left hand off to the side. The image of the right hand looks like the left hand. In particular, the fingers of the mirror image of the right hand curl in the opposite direction from those of the right hand itself. If she were using the right hand rule to find the direction of the angular momentum of a rotating object, the mirror image would look as if she were using a left hand rule.

It is fairly common knowledge that left and right are reversed in a mirror image. But if left and right are reversed, why aren't top and bottom reversed also? Think about that for a minute before you go on to the next paragraph.

Figure 6
The image of the right hand looks like a left hand.

To see what the image of an extended object should be, imagine that we place an arrow in front of a mirror as shown in Figure (7). We have constructed rays from the tip and the base of the arrow that reflect and enter the eye as shown. Extending these rays back to the image, we see that the image arrow has been reversed *front to back*. That is what a mirror does. The mirror image is reversed front to back, not left to right or top to bottom. It turns out that the right hand, when reversed front to back as in its image in Figure (6), has the symmetry properties of a left hand. If used to define angular momentum, you would get a left hand rule.

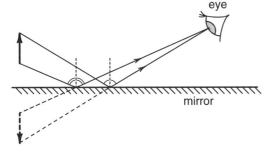

Figure 7
A mirror image changes front to back, not left to right.

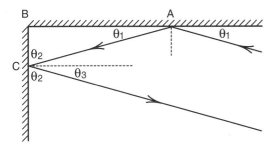

Figure 8a
With a corner reflector, the light is reflected back it the same direction from which it arrived.

The Corner Reflector

When two vertical mirrors are placed at right angles as shown in Figure (8a), a horizontal ray approaching the mirrors is reflected back in the direction from which it came. It is a little exercise in trigonometry to see that this is so. Since the angle of incidence equals the angle of reflection at each mirror surface, we see that the angles labeled θ_1 must be equal to each other and the same for the angles θ_2. From the right triangle ABC, we see that $\theta_1 + \theta_2 = 90°$. We also see that the angles $\theta_2 + \theta_3$ also add up to $90°$, thus $\theta_3 = \theta_1$, which implies the exiting ray is parallel to the entering one.

If you mount three mirrors perpendicular to each other to form the corner of a cube, then light entering this so called *corner reflector* from any angle goes back in the direction from which it came. The Apollo II astronauts placed the array of corner reflectors shown in Figure (8b) on the surface of the moon, so that a laser beam from the earth would be reflected back from a precisely known point on the surface of the moon. By measuring the time it took a laser pulse to be reflected back from the array, the distance to the moon could be measured to an accuracy of centimeters. With the distance to the moon known with such precision, other distances in the solar system could then be determined accurately.

Figure 8b
Array of corner reflectors left on the moon by the Apollo astronauts. A laser pulse from the earth, aimed at the reflectors, returns straight back to the laser. By measuring the time the pulse takes to go to the reflectors and back, the distance to that point on the moon and back can be accurately measured.

MOTION OF LIGHT THROUGH A MEDIUM

We are all familiar with the fact that light can travel through clear water or clear glass. With some of the new glasses developed for fiber optics communication, light signals can travel for miles without serious distortion. If you made a mile thick pane from this glass you could see objects through it.

From an atomic point of view, it is perhaps surprising that light can travel any distance at all through water or glass. A reasonable picture of what happens when a light wave passes over an atom is provided by the ripple tank photograph shown in Figure (36-1) reproduced here. The wave scatters from the atom, and since atoms are considerably smaller than a wavelength of visible light, the scattered waves are circular like those in the ripple tank photograph. The final wave is the sum of the incident and the scattered waves as shown in Figure (36-1a).

When light passes through a medium like glass or water, the wave is being scattered by a huge number of atoms. The final wave pattern is the sum of the incident wave and all of the many billions of scattered waves. You might suspect that this sum would be very complex, but that is not the case. At the surface some of the incident wave is reflected. Inside the medium, the *incident and scattered waves add up to a new wave* of the same frequency as the incident wave but which travels *at a reduced speed*. The speed of a light wave in water for example is 25% less than the speed of light in a vacuum.

incident wave →

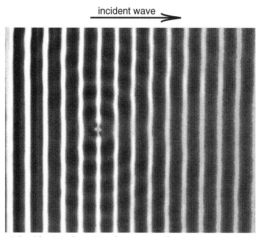

incident wave →

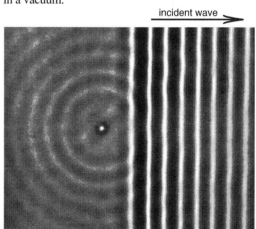

a) Incident and scattered wave together.

b) After incident wave has passed.

Figure 36-1
If the scattering object is smaller than a wavelength, we get circular scattered waves.

The optical properties of lenses are a consequence of this effective reduction in the speed of light in the lens. Figure (9) is a rather remarkable photograph of individual short pulses of laser light as they pass through and around a glass lens. You can see that the part of the wave front that passed through the lens is delayed by its motion through the glass. The thicker the glass, the greater the delay. You can also see that the delay changed the shape and direction of motion of the wave front, so that the light passing through the lens focuses to a point behind the lens. This is how a lens really works.

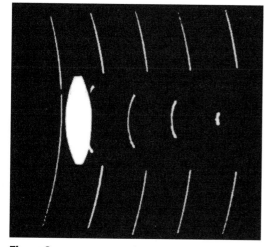

Figure 9
Motion of a wave front through a glass lens. The delay in the motion of the wave front as it passes through the glass changes the shape and direction of motion of the wave front, resulting in the focusing of light. (This photograph should not be confused with ripple tank photographs where wavelengths are comparable to the size of the objects. Here the wavelength of the light is about one hundred thousand times smaller than the diameter of the lens, with the result we get sharp shadows and do not see diffraction effects.)

In the 18/February/1999 issue of **Nature** *it was announced that a laser pulse travelled through a gas of supercooled sodium atoms at a speed of 17 meters per second! (You can ride a bicycle faster than that.) This means that the sodium atoms had an index of refraction of about 18 million, 7.3 million times greater than that of diamond!*

Index of Refraction

The amount by which the effective speed of light is reduced as the light passes through a medium depends both upon the medium and the wavelength of the light. There is very little slowing of the speed of light in air, about a 25% reduction in speed in water, and nearly a 59% reduction in speed in diamond. In general, blue light travels somewhat slower than red light in nearly all media.

It is traditional to describe the slowing of the speed of light in terms of what is called the ***index of refraction*** of the medium. The index of refraction n is defined by the equation

$$\left.\begin{array}{l}\text{speed of light}\\\text{in a medium}\end{array}\right\} \quad v_{light} = \frac{c}{n} \qquad (1)$$

The index n has to equal 1 in a vacuum because light always travels at the speed 3×10^8 meters in a vacuum. The index n can never be less than 1, because nothing can travel faster than the speed c. For yellow sodium light of wavelength $\lambda = 5.89 \times 10^{-5}$ cm (589 nanometers), the index of refraction of water at 20° C is n = 1.333, which implies a 25% reduction in speed. For diamond, n = 2.417 for this yellow light. Table 1 gives the indices of refraction for various transparent substances for the sodium light.

Vacuum	1.00000	exactly
Air (STP)	1.00029	
Ice	1.309	
Water (20° C)	1.333	
Ethyl alcohol	1.36	
Fuzed quartz	1.46	
Sugar solution (80%)	1.49	
Typical crown glass	1.52	
Sodium Chloride	1.54	
Polystyrene	1.55	
Heavy flint glass	1.65	
Sapphire	1.77	
Zircon	1.923	
Diamond	2.417	
Rutile	2.907	
Gallium phosphide	3.50	
Very cold sodium atoms	18000000	for laser pulse

Table 1
Some indices of refraction for yellow sodium light at a wavelength of 589 nanometers.

Exercise 1a

What is the speed of light in air, water, crown glass, and diamond. Express your answer in feet/nanosecond. (Take c to be exactly 1 ft/nanosecond.)

Exercise 1b

In one of the experiments announced in *Nature,* a laser pulse took 7.05 microseconds to travel .229 millimeters through the gas of supercooled sodium atoms. What was the index of refraction of the gas for this particular experiment? (The index quoted on the previous page was for the slowest observed pulse. The pulse we are now considering went a bit faster.)

CERENKOV RADIATION

In our discussion, in Chapter 1, of the motion of light through empty space, we saw that nothing, not even information, could travel faster than the speed of light. If it did, we could, for example, get answers to questions that had not yet been thought of.

When moving through a medium, the speed of a light wave is slowed by repeated scattering and it is no longer true that nothing can move faster than the speed of light in that medium. We saw for example that the speed of light in water is only 3/4 the speed c in vacuum. Many elementary particles, like the muons in the muon lifetime experiment, travel at speeds much closer to c. When a charged particle moves faster than the speed of light in a medium, we get an effect not unlike the sonic boom produced by a supersonic jet. We get a *shock wave of light* that is similar to a sound shock wave (sonic boom), or to the water shock wave shown in Figure (33-30) reproduced here. The light shock wave is called *Cerenkov radiation* after the Russian physicist Pavel Cerenkov who received the 1958 Nobel prize for discovering the effect.

In the muon lifetime picture, one observed how long muons lived when stopped in a block of plastic. The experiment was made possible by Cerenkov radiation. The muons that stopped in the plastic, entered moving faster than the speed of light in plastic, and as a result emitted a flash of light in the form of Cerenkov radiation. When the muon decayed, a charged positron and a neutral neutrino were emitted. In most cases the charged positron emerged faster than the speed of light in the plastic, and also emitted Cerenkov radiation. The two flashes of light were detected by the phototube which converted the light flashes to voltage pulses. The voltage pulses were then displayed on an oscilloscope screen where the time interval between the pulses could be measured. This interval represented the time that the muon lived, mostly at rest, in the plastic.

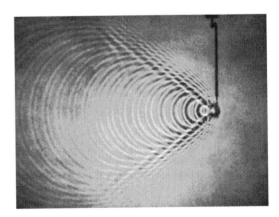

Figure 33-30
When the source of the waves moves faster than the speed of the waves, the wave fronts pile up to produce a shock wave as shown. This shock wave is the sonic boom you hear when a jet plane flies overhead faster than the speed of sound.

SNELL'S LAW

When a wave enters a medium of higher index of refraction and travels more slowly, the wavelength of the wave changes. The wavelength is the distance the wave travels in one period, and if the speed of the wave is reduced, the distance the wave travels in one period is reduced. (In most cases, the frequency or period of the wave is not changed. The exceptions are in fluorescence and nonlinear optics where the frequency or color of light can change.)

We can calculate how the wavelength changes with wave speed from the relationship

$$\lambda \frac{cm}{cycle} = \frac{v_{wave} \frac{cm}{sec}}{T \frac{sec}{cycle}}$$

Setting $v_{wave} = c/n$ for the speed of light in the medium, gives for the corresponding wavelength λ_n

$$\lambda_n = \frac{v_{wave}}{T} = \frac{c/n}{T} = \frac{1}{n}\frac{c}{T} = \frac{\lambda_0}{n} \qquad (2)$$

where $\lambda_0 = c/T$ is the wavelength in a vacuum. Thus, for example, the wavelength of light entering a diamond from air will be shortened by a factor of $1/2.42$.

What happens when a set of periodic plane waves goes from one medium to another is illustrated in the ripple tank photograph of Figure (10). In this photograph, the water has two depths, deeper on the upper part where the waves travel faster, and shallower in the lower part where the waves travel more slowly. You can see that the wavelengths are shorter in the lower part, but there are the same number of waves. (We do not gain or loose waves at the boundary.) The frequency, the number of waves that pass you per second, is the same on the top and bottom.

The only way that the wavelength can be shorter and still have the same number of waves is for the wave to bend at the boundary as shown. We have drawn arrows showing the direction of the wave in the deep water (the incident wave) and in the shallow water (what we will call the *transmitted* or *refracted* wave), and we see that the change in wavelength causes a sudden change in direction of motion of the wave. If you look carefully you will also see reflected waves which emerge at an angle of reflection equal to the angle of incidence.

Figure (11) shows a beam of yellow light entering a piece of glass. The index of refraction of the glass is 1.55, thus the wavelength of the light in the glass is only .65 times as long as that in air ($n \approx 1$ for air). You can see both the bending of the ray as it enters the glass and also the reflected ray. (You also see internal reflection and the ray emerging from the bottom surface.) You cannot see the individual wave crests, but otherwise Figures (10) and (11) show similar phenomena.

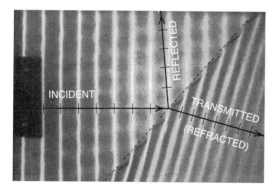

Figure 10
Refraction at surface of water. When the waves enter shallower water, they travel more slowly and have a shorter wavelength. The waves must travel in a different direction in order for the crests to match up.

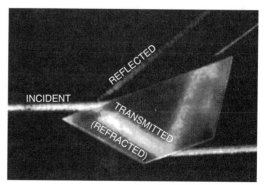

Figure 11
Refraction at surface of glass. When the light waves enter the glass, they travel more slowly and have a shorter wavelength. Like the water waves, the light waves must travel in a different direction in order for the crests to match up.

Derivation of Snell's Law

To calculate the angle by which a light ray is bent when it enters another medium, consider the diagram in Figure (12). The drawing represents a light wave, traveling in a medium of index n_1, incident on a boundary at an angle θ_1. We have sketched successive incident wave crests separated by the wavelength λ_1. Assuming that the index n_2 in the lower medium is greater than n_1, the wavelength λ_2 will be shorter than λ_1 and the beam will emerge at the smaller angle θ_2.

To calculate the angle θ_2 at which the transmitted or refracted wave emerges, consider the detailed section of Figure (12) redrawn in Figure (13a). Notice that we have labeled two apparently different angles by the same label θ_1. Why these angles are equal is seen in the construction of Figure (13b) where we see that the angles α and θ_1 are equal.

Exercise 2

Show that the two angles labeled θ_2 in Figure (13a) must also be equal.

Since the triangles ACB and ADB are right triangles in Figure (13a), we have

$$\lambda_1 = AB \sin(\theta_1) = \lambda_0/n_1 \tag{3}$$

$$\lambda_2 = AB \sin(\theta_2) = \lambda_0/n_2 \tag{4}$$

where AB is the hypotenuse of both triangles and λ_0 is the wavelength when $n_0 = 1$. When we divide Equation 4 by Equation 5, the distances AB and λ_0 cancel, and we are left with

$$\frac{\sin(\theta_1)}{\sin(\theta_2)} = \frac{n_2}{n_1}$$

or

$$\boxed{n_1 \sin(\theta_1) = n_2 \sin(\theta_2)} \quad \textit{Snell's law} \tag{5}$$

Equation 5, known as Snell's law, allows us to calculate the change in direction when a beam of light goes from one medium to another.

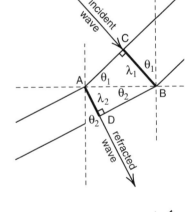

Figure 13a
The angles involved in the analysis.

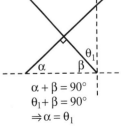

Figure 13b
Detail.

$$\alpha + \beta = 90°$$
$$\theta_1 + \beta = 90°$$
$$\Rightarrow \alpha = \theta_1$$

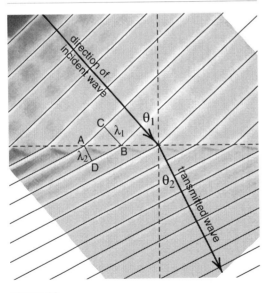

Figure 12
Analysis of refraction. The crests must match at the boundary between the different wavelength waves.

INTERNAL REFLECTION

Because of the way rays bend at the interface of two media, there is a rather interesting effect when light goes from a material of higher to a material of lower index of refraction, as in the case of light going from water into air. The effect is seen clearly in Figure (14). Here we have a multiple exposure showing a laser beam entering a tank of water, being reflected by a mirror, and coming out at different angles. The outgoing ray is bent farther away from the normal as it emerges from the water. We reach the point where the outgoing ray bends and runs parallel to the surface of the water. This is a critical angle, for if the mirror is turned farther, the ray can no longer get out and is completely reflected inside the surface.

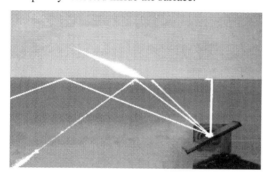

Figure 14
Internal reflection. We took three exposures of a laser beam reflecting off an underwater mirror set at different angles. In the first case the laser beam makes it back out of the water and strikes a white cardboard behind the water tank. In the other two cases, there is total internal reflection at the under side of the water surface. In the final exposure we used a flash to make the mirror visible.

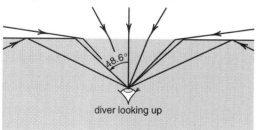

diver looking up

Figure 14a
When you are swimming under water and look up, you see the outside world through a round hole. Outside that hole, the surface is a silver mirror.

It is easy to calculate the critical angle θ_c at which this complete internal reflection begins. Set the angle of refraction, θ_2 in Figure (14), equal to 90° and we get from Snell's law

$$n_1 \sin\theta_c = n_2 \sin\theta_2 = n_2 \sin 90° = n_2$$

$$\sin\theta_c = \frac{n_2}{n_1} \; ; \quad \boxed{\theta_c = \sin^{-1}\frac{n_2}{n_1}} \qquad (6)$$

For light emerging from water, we have $n_2 \approx 1$ for air and $n_1 = 1.33$ for water giving

$$\sin^{-1}\theta_2 = \frac{1}{1.33} = .75$$

$$\theta_c = 48.6° \qquad (7)$$

Anyone who swims underwater, scuba divers especially, are quite familiar with the phenomenon of internal reflection. When you look up at the surface of the water, you can see the entire outside world through a circular region directly overhead, as shown in Figure (14a). Beyond this circle the surface looks like a silver mirror.

Exercise 3

A glass prism can be used as shown in Figure (15) to reflect light at right angles. The index of refraction n_g of the glass must be high enough so that there is total internal reflection at the back surface. What is the least value n_g one can have to make such a prism work? (Assume the prism is in the air where $n \approx 1$.)

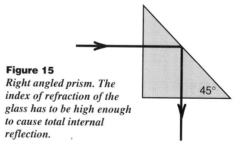

Figure 15
Right angled prism. The index of refraction of the glass has to be high enough to cause total internal reflection.

Fiber Optics

Internal reflection plays a critical role in modern communications and modern medicine through fiber optics. When light is sent down through a glass rod or fiber so that it strikes the surface at an angle greater than the critical angle, as shown in Figure (16a), the light will be completely reflected and continue to bounce down the rod with no loss out through the surface. By using modern very clear glass, a fiber can carry a light signal for miles without serious attenuation.

The reason it is more effective to use light in glass fibers than electrons in copper wire for transmitting signals, is that the glass fiber can carry information at a much higher rate than a copper wire, as indicated in Figure (16b). This is because laser pulses traveling through glass, can be turned on and off much more rapidly than electrical pulses in a wire. The practical limit for copper wire is on the order of a million pulses or bits of information per second (corresponding to a *baud rate* of one *megabit*). Typically the information rate is much slower over commercial telephone lines, not much in excess of 30 to 50 thousand bits of information per second (corresponding to 30 to 50 *kilobaud*). These rates are fast enough to carry telephone conversations or transmit text to a printer, but painfully slow for sending pictures and much too slow for digital television signals. High definition digital television will require that information be sent at a rate of about 3 million bits or pulses every 1/30 of a second for a baud rate of 90 million baud. (Compare that with the baud rate on your computer modem.) In contrast, fiber optics cables are capable of carrying pulses or bits at a rate of about a billion (10^9) per second, and are thus well suited for transmitting pictures or many phone conversations at once.

By bundling many fine fibers together, as indicated in Figure (17), one can transmit a complete image along the bundle. One end of the bundle is placed up against the object to be observed, and if the fibers are not mixed up, the image appears at the other end.

To transmit a high resolution image, one needs a bundle of about a million fibers. The tiny fibers needed for this are constructed by making a rather large bundle of small glass strands, heating the bundle to soften the glass, and then stretching the bundle until the individual strands are very fine. (If you have heated a glass rod over a Bunsen burner and pulled out the ends, you have seen how fine a glass fiber can be made this way.)

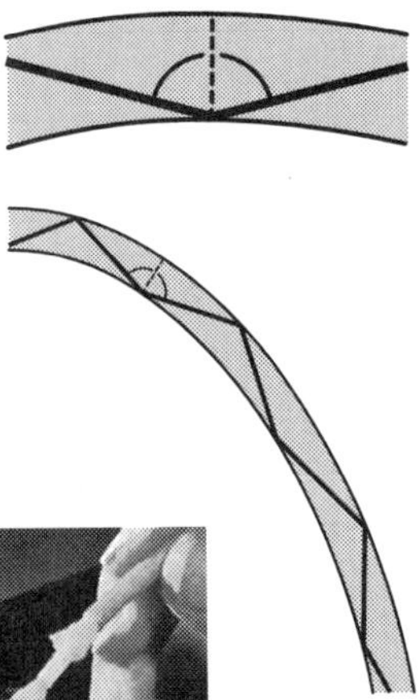

Figure 16a
Because of internal reflections, light can travel down a glass fiber, even when the fiber is bent.

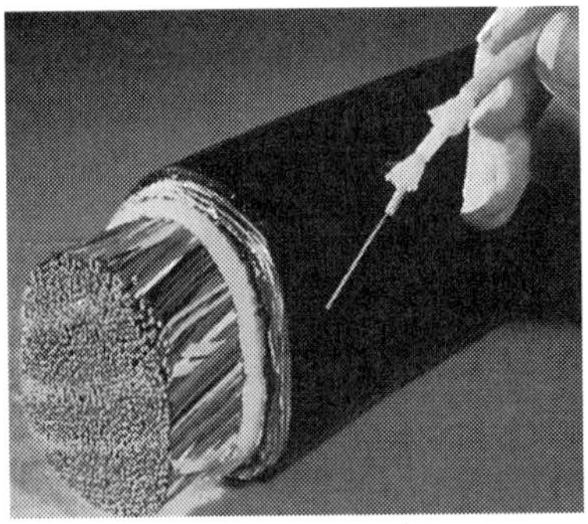

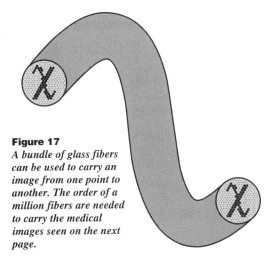

Figure 17
A bundle of glass fibers can be used to carry an image from one point to another. The order of a million fibers are needed to carry the medical images seen on the next page.

Figure 16b
A single glass fiber can carry the same amount of information as a fat cable of copper wires.

Medical Imaging

The use of fiber optics has revolutionized many aspects of medicine. It is an amazing experience to go down and look inside your own stomach and beyond, as the author did a few years ago. This is done with a flexible fiber optics instrument called a retroflexion, producing the results shown in Figure (18). An operation, such as the removal of a gallbladder, which used to require opening the abdomen and a long recovery period, can now be performed through a small hole near the navel, using fiber optics to view the procedure. You can see the viewing instrument and such an operation in progress in Figure (19).

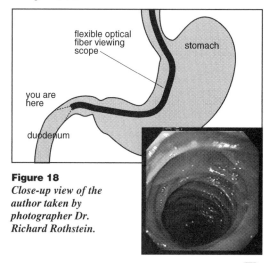

Figure 18
Close-up view of the author taken by photographer Dr. Richard Rothstein.

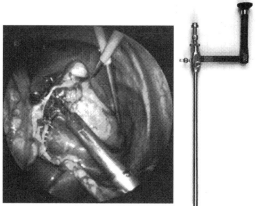

Figure 19
Gallbladder operation in progress, being viewed by the rigid laparoscope shown on the right. Such views are now recorded by high resolution television.

PRISMS

So far in our discussion of refraction, we have considered only beams of light of one color, one wavelength. Because the index of refraction generally changes with wavelength, rays of different wavelength will be bent at different angles when passing the interface of two media. Usually the index of refraction of visible light increases as the wavelength becomes shorter. Thus when white light, which is a mixture of all the visible colors, is sent through a prism as shown in Figure (20), the short wavelength blue light will be deflected by a greater angle than the red light, and the beam of light is separated into a rainbow of colors.

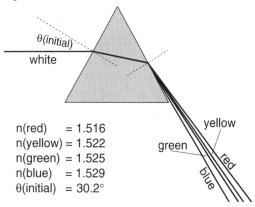

$n(red) = 1.516$
$n(yellow) = 1.522$
$n(green) = 1.525$
$n(blue) = 1.529$
$\theta(initial) = 30.2°$

Figure 20
When light is sent through a prism, it is separated into a rainbow of colors. In this scale drawing, we find that almost all the separation of colors occurs at the second surface where the light emerges from the glass.

Rainbows

Rainbows in the sky are formed by the reflection and refraction of sunlight by raindrops. It is not, however, particularly easy to see why a rainbow is formed. René Descartes figured this out by tracing rays that enter and leave a spherical raindrop.

In Figure (21a) we have used Snell's law to trace the path of a ray of yellow light that enters a spherical drop of water (of index $n = 1.33$), is reflected on the back side, and emerges again on the front side. (Only a fraction of the light is reflected at the back, thus the reflected beam is rather weak.) In this drawing, the angle θ_2 is determined by $\sin(\theta_1) = 1.33 \sin(\theta_2)$. At the back, the angles of incidence and reflection are equal, and at the front we have $1.33 \sin(\theta_2) = \sin(\theta_1)$ (taking the index of refraction of air = 1). Nothing is hard about this construction, it is fairly easy to do with a good drafting program like Adobe Illustrator and a hand calculator.

In Figure (21b) we see what happens when a number of parallel rays enter a spherical drop of water. (This is similar to the construction that was done by Descartes in 1633.) When you look at the outgoing rays, it is not immediately obvious that there is any special direction for the reflected rays. But if you look closely you will see that the ray we have labeled #11 is the one that comes back at the widest angle from the incident ray.

Ray #1, through the center, comes straight back out. Ray #2 comes out at a small angle. The angles increase up to Ray #11, and then start to decrease again for Rays #12 and #13. In our construction the maximum angle, that of Ray #11, was 41.6°, close to the theoretical value of 42° for yellow light.

What is more important than the fact that the maximum angle of deviation is 42° is the fact that the rays close to #11 emerge as more or less parallel to each other. The other rays, like those near #3 for example come out at diverging angles. That light is spread out. But the light emerging at 42° comes out as a parallel beam. When you have sunlight striking many raindrops, *more yellow light is reflected back at this angle of 42° than any other angle.*

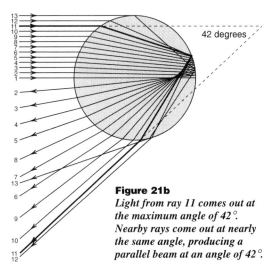

Figure 21b
Light from ray 11 comes out at the maximum angle of 42°. Nearby rays come out at nearly the same angle, producing a parallel beam at an angle of 42°.

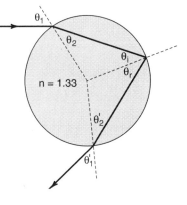

Figure 21a
Light ray reflecting from a raindrop.

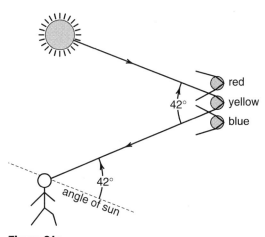

Figure 21c
You will see the yellow part of the rainbow at an angle of 42° as shown above. Red will be seen at a greater angle, blue at a lesser one.

Repeat the construction for red light where the index of refraction is slightly less than 1.33, and you find the maximum angle of deviation and the direction of the parallel beam is slightly greater than 42°. For blue light, with a higher index, the deviation is less.

If you look at falling raindrops with the sun at your back as shown in Figure (21c), you will see the yellow part of the rainbow along the arc that has an angle of 42° from the rays of sun passing you. The red light, having a greater angle of deviation will be above the yellow, and the blue will be below, as you can see in Figure (21d).

Sometimes you will see two or more rainbows if the rain is particularly heavy (we have seen up to 7). These are caused by multiple internal reflections. In the second rainbow there are two internal reflections and the parallel beam of yellow light comes out at an angle of 51°. Because of the extra reflection the red is on the inside of the arc and the blue on the outside.

Exercise 4

Next time you see a rainbow, try to measure the angle the yellow part of the arc makes with the rays of sun passing your head.

Figure 21d
Rainbow over Cook's Bay, Moorea. (The color version on the CD looks a lot better.)

The Green Flash

The so called green flash at sunset is a phenomenon that is supposed to be very rare, but which is easy to see if you can look at a distant sunset through binoculars. (Don't look until the very last couple of seconds so that you will not hurt your eyes.)

The earth's atmosphere acts as a prism, refracting the light as shown in Figure (22). The main effect is that when you look at a sunset, the sun has already set; only its image is above the horizon. But, as seen in Figure (20), the atmospheric prism also refracts the different colors in the white sunlight at different angles. Due to the fact that the blue light is refracted at a greater angle than the red light, the blue image of the sun is slightly higher above the horizon than the green image, and the green image is higher than the red image. We have over emphasized the displacement of the image in Figure (22). The blue image is only a few percent of the sun's diameter above the red image. Before the sun sets, the various colored images are more or less on top of each other and the sun looks more or less white.

If it is a very clear day, and you watch the sunset with binoculars, just as the sun disappears, for about 1/2 second, the sun turns a deep blue. The reason is that all the other images have set, and for this short time only this blue image is visible. We should call this the "blue flash".

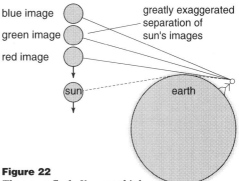

Figure 22
The green flash. You can think of the white sun as consisting of various colored disks that add up to white. The earth's atmosphere acts as a prism, diffracting the light from the setting sun, separating the colored disks. The blue disk is the last to set. Haze in the atmosphere can block the blue light, leaving the green disk as the last one seen.

If the atmosphere is not so clear, if there is a bit of haze or moisture as one often gets in the summer, the blue light is absorbed by the haze, and the last image we see setting is the green image. This is the origin of the green flash. With still more haze you get a red sunset, all the other colors having been absorbed by the haze.

Usually it requires binoculars to see the green or blue colors at the instant of sunset. But sometimes the atmospheric conditions are right so that this final light of the sun is reflected on clouds and can be seen without binoculars. If the clouds are there, there is probably enough moisture to absorb the blue image, and the resulting flash on the clouds is green.

Halos and Sun Dogs

Another phenomenon often seen is the reflection of light from hexagonal ice crystals in the atmosphere. The reflection is seen at an angle of 22° from the sun. If the ice crystals are randomly oriented then we get a complete halo as seen in Figure (23a). If the crystals are falling with their flat planes predominately horizontal, we only see the two pieces of the halo at each side of the sun, seen in Figure (24). These little pieces of rainbow are known as "sun dogs".

Figure 23
Halo caused by reflection by randomly oriented hexagonal ice crystals.

Figure 24
Sun dogs caused by ice crystals falling flat.

LENSES

The main impact geometrical optics has had on mankind is through the use of lenses in microscopes, telescopes, eyeglasses, and of course, the human eye. The basic idea behind the construction of a lens is Snell's law, but as our analysis of light reflected from a spherical raindrop indicated, we can get complex results from even simple geometries like a sphere.

Modern optical systems like the zoom lens shown in Figure (25) are designed by computer. Lens design is an ideal problem for the computer, for tracing light rays through a lens system requires many repeated applications of Snell's law. When we analyzed the spherical raindrop, we followed the paths of 12 rays for an index of refraction for only yellow light. A much better analysis would have resulted from tracing at least 100 rays for the yellow index of refraction, and then repeating the whole process for different indices of refraction, corresponding to different wavelengths or colors of light. This kind of analysis, while extremely tedious to do by hand, can be done in seconds on a modern desktop computer.

In this chapter we will restrict our discussion to the simplest of lens systems in order to see how basic instruments, like the microscope, telescope and eye, function. You will not learn here how to design a color corrected zoom lens like the Nikon lens shown below.

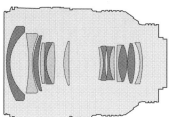

Figure 25
Nikon zoom lens.

Spherical Lens Surface

A very accurate spherical surface on a piece of glass is surprisingly easy to make. Take two pieces of glass, put a mixture of grinding powder and water between them, rub them together in a somewhat regular, somewhat irregular, pattern that one can learn in less than 5 minutes. The result is a spherical surface on the two pieces of glass, one being concave and the other being convex. The reason you get a spherical surface from this somewhat random rubbing is that only spherical surfaces fit together perfectly for all angles and rotations. Once the spheres have the desired radius of curvature, you use finer and finer grits to smooth out the scratches, and then jeweler's rouge to polish the surfaces. With any skill at all, one ends up with a polished surface that is perfectly spherical to within a fraction of a wavelength of light.

To see the optical properties of a spherical surface, we can start with the ray diagram we used for the spherical raindrop, and remove the reflections by extending the refracting medium back as shown in Figure (26a). The result is not encouraging. The parallel rays entering near the center of the surface come together—*focus*—quite a bit farther back than rays entering near the outer edge. This range of focal distances is not useful in optical instruments.

In Figure (26b) we have restricted the area where the rays are allowed to enter to a small region around the center of the surface. To a very good approximation all these parallel rays come together, focus, at one point. This is the characteristic we want in a simple lens, to bring parallel incoming rays together at one point as the parabolic reflector did.

Figure (26b) shows us that the way to make a good lens using spherical surfaces is to use only the central part of the surface. Rays entering near the axis as in Figure (26b) are deflected only by small angles, angles where we can approximate $\sin(\theta)$ by θ itself. When the angles of deflection are small enough to use small angle approximations, a spherical surface provides sharp focusing. As a result, in analyzing the small angle spherical lenses, we can replace the exact form of Snell's law

$$n_1 \sin(\theta_1) = n_2 \sin(\theta_2) \qquad \text{(5 repeated)}$$

by the approximate equation

$$n_1 \theta_1 = n_2 \theta_2 \quad \begin{array}{l} \textit{Snell's law} \\ \textit{for small} \\ \textit{angles} \end{array} \qquad (8)$$

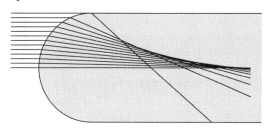

Figure 26a
Focusing properties of a spherical surface. (Not good!)

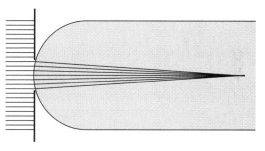

Figure 26b
We get a much better focus if we use only a small part of the spherical surface.

Focal Length of a Spherical Surface

Let us now use the simplified form of Snell's law to calculate the focal length f of a spherical surface, i.e., the distance behind the surface where entering parallel rays come to a point. Unless you plan to start making your own lenses, you do not really need this result, but the exercise provides an introduction to how focal lengths are related to the curvature of lenses.

Consider two parallel rays entering a spherical surface as shown in Figure (27). One enters along the axis of the surface, the other a distance h above it. The angle labeled θ_1 is the angle of incidence for the upper ray, while θ_2 is the refracted angle. These angles are related by Snell's law

$$n_1 \theta_1 = n_2 \theta_2$$

or

$$\theta_2 = \frac{n_1}{n_2} \theta_1 \qquad (9)$$

If you recall your high school trigonometry you will remember that the outside angle of a triangle, θ_1 in Figure (27a), is equal to the sum of the opposite angles, θ_2 and α in this case. Thus

$$\theta_1 = \theta_2 + \alpha$$

or using Equation 9 for θ_2

$$\theta_1 = \frac{n_1}{n_2} \theta_1 + \alpha \qquad (10)$$

Now consider the two triangles reproduced in Figures (27b) and (27c). Using the small angle approximation $\tan(\theta) \approx \sin(\theta) \approx \theta$, we have for Figure (27b)

$$\theta_1 \approx \frac{h}{r} \; ; \quad \alpha \approx \frac{h}{f} \qquad (11)$$

Substituting these values for θ_1 and α into Equation 10 gives

$$\frac{h}{r} = \frac{n_1}{n_2} \frac{h}{r} + \frac{h}{f} \qquad (12)$$

The height h cancels, and we are left with

$$\boxed{\frac{1}{f} = \frac{1}{r}\left(1 - \frac{n_1}{n_2}\right)} \qquad (13)$$

The fact that the height h cancels means that parallel rays entering at any height h (as long as the small angle approximation holds) will focus at the same point a distance f behind the surface. This is what we saw in Figure (26b).

Figure (26b) was drawn for $n_1 = 1$ (air) and $n_2 = 1.33$ (water) so that $n_1/n_2 = 1/1.33 = .75$. Thus for that drawing we should have had

$$\frac{1}{f} = \frac{1}{r}\left(1 - .75\right) = \frac{1}{r}\left(.25\right) = \frac{1}{r}\left(\frac{1}{4}\right)$$

or

$$f = 4r \qquad (14)$$

as the predicted focal length of that surface.

Figure 27
Calculating the focal length f of a spherical surface.

parallel rays

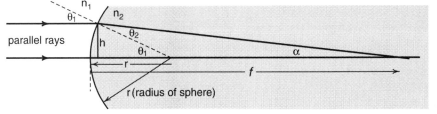

r (radius of sphere)

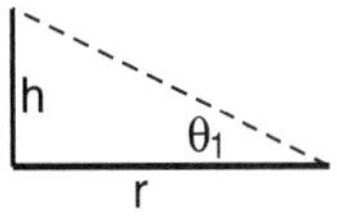

Figure 27b
$\theta_1 \approx h/r$

Figure 27a
$\theta_1 = \theta_2 + \alpha$

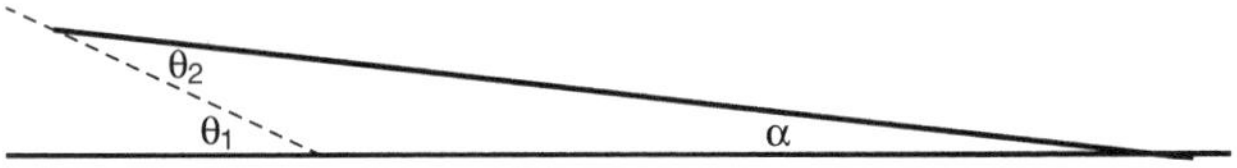

Figure 27c
$\alpha \approx h/f$

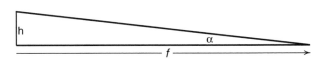

Exercise 5

Compare the prediction of Equation 14 with the results we got in Figure (26b). That is, what do you measure for the relationship between f and r in that figure?

Exercise 6

The index of refraction for red light in water is slightly less than the index of refraction for blue light. Will the focal length of the surface in Figure (26b) be longer or shorter than the focal length for red light?

Exercise 7

The simplest model for a fixed focus eye is a sphere of index of refraction n_2. The index n_2 is chosen so that parallel light entering the front surface of the sphere focuses on the back surface as shown in Figure (27d). What value of n_2 is required for this model to work when $n_1 = 1$? Looking at the table of indexes of refraction, Table 1, explain why such a model would be hard to achieve.

Aberrations

When parallel rays entering a lens do not come to focus at a point, we say that the lens has an *aberration*. We saw in Figure (26a) that if light enters too large a region of a spherical surface, the focal points are spread out in back. This is called *spherical aberration*. One cure for spherical aberration is to make sure that the diameter of any spherical lens you use is small in comparison to the radius of curvature of the lens surface.

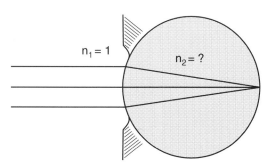

Figure 27d
A simple, but hard to achieve, model for an eye.

We get rainbows from raindrops and prisms because the index of refraction for most transparent substances changes with wavelength. As we saw in Exercise 6, this causes red light to focus at a different point than yellow or blue light, (resulting in colored bands around the edges of images). This problem is called *chromatic aberration*. The cure for chromatic aberration is to construct complex lenses out of materials of different indices of refraction. With careful design, you can bring the focal points of the various colors back together. Some of the complexity in the design of the zoom lens in Figure (25) is to correct for chromatic aberration.

Astigmatism is a common problem for the lens of the human eye. You get astigmatism when the lens is not perfectly spherical, but is a bit cylindrical. If, for example, the cylindrical axis is horizontal, then light from a horizontal line will focus farther back than light from a vertical line. Either the vertical lines in the image are in focus, or the horizontal lines, but not both at the same time. (In the eye, the cylindrical axis does not have to be horizontal or vertical, but can be at any angle.)

There can be many other aberrations depending upon what distortions are present in the lens surface. We once built a small telescope using a shaving mirror instead of a carefully ground parabolic mirror. The image of a single star stretched out in a line that covered an angle of about 30 degrees. This was an extreme example of an aberration called *coma*. That telescope provided a good example of why optical lenses and mirrors need to be ground very accurately.

What, surprisingly, does not usually cause a serious problem is a small scratch on a lens. You do not get an image of the scratch because the scratch is completely out of focus. Instead the main effect of a scratch is to scatter light and fog the image a bit.

Perhaps the most famous aberration in history is the spherical aberration in the primary mirror of the orbiting Hubble telescope. The aberration was caused by an undetected error in the complex apparatus used to test the surface of the mirror while the mirror was being ground and polished. The ironic part of the story is that the aberration could have easily been detected using the same simple apparatus all amateur telescope makers use to test their mirrors (the so called Foucault test), but such a simple minded test was not deemed necessary.

What saved the Hubble telescope is that the engineers found the problem with the testing apparatus, and could therefore precisely determine the error in the shape of the lens. A small mirror, only a few centimeters in diameter, was designed to correct for the aberration in the Hubble image. When this correcting mirror was inserted near the focus of the main mirror, the aberration was eliminated and we started getting the many fantastic pictures from that telescope.

Another case of historical importance is the fact that Issac Newton invented the reflecting telescope to avoid the chromatic aberration present in all lenses at that time. With a parabolic reflecting mirror, all parallel rays entering the mirror focus at a point. The location of the focal point does not depend on the wavelength of the light (as long as the mirror surface is reflecting at that wavelength). You also do not get spherical aberration either because a parabolic surface is the correct shape for focusing, no matter how big the diameter of the mirror is compared to the radius of curvature of the surface.

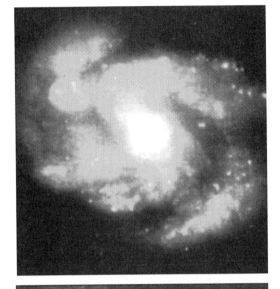

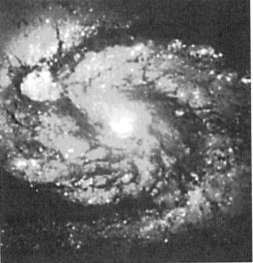

Figure 28
Correction of the Hubble telescope mirror. Top: before the correction. Bottom: same galaxy after correction. Left: astronauts installing correction mirror.

THIN LENSES

In Figure (29), we look at what happens when parallel rays pass through the two spherical surfaces of a lens. The top diagram (a) is a reproduction of Figure (26b) where a narrow bundle of parallel rays enters a new medium through a single spherical surface. By making the diameter of the bundle of rays much less than the radius of curvature of the surface, the parallel rays all focus to a single point. We were able to calculate where this point was located using small angle approximations.

In Figure (29b), we added a second spherical surface. The diagram is drawn to scale for indices of refraction n = 1 outside the gray region and n = 1.33 inside, and using Snell's law at each interface of each ray. (The drawing program Adobe Illustrator allows you to do this quite accurately.) The important point to note is that the parallel rays still focus to a point. The difference is that the focal point has moved inward.

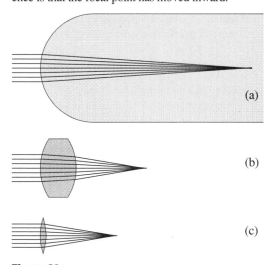

(a)

(b)

(c)

Figure 29
A two surface lens. Adding a second surface still leaves the light focused to a point, as long as the diameter of the light bundle is small compared to the radii of the lens surfaces.

In Figure (29c), we have moved the two spherical surfaces close together to form what is called a *thin lens*. We have essentially eliminated the distance the light travels between surfaces. If the index of refraction outside the lens is 1 and has a value n inside, and surfaces have radii of curvature r_1 and r_2, then the focal length f of the lens given by the equation

$$\frac{1}{f} = (n-1)\left(\frac{1}{r_1} + \frac{1}{r_2}\right)$$ *lens maker's equation* (15)

Equation 15, which is known as the **lens maker's equation**, can be derived in a somewhat lengthy exercise involving similar triangles.

Unless you are planning to grind your own lenses, the lens maker's equation is not something you will need to use. When you buy a lens, you specify what focal length you want, what diameter the lens should be, and whether or not it needs to be corrected for color aberration. You are generally not concerned with how the particular focal length was achieved—what combination of radii of curvatures and index of refraction were used.

Exercise 8

(a) See how well the lens maker's equation applies to our scale drawing of Figure (29c). Our drawing was done to a scale where the spherical surfaces each had a radius of $r_1 = r_2 = 37$ mm, and the distance f from the center of the lens to the focal point was 55 mm.

(b) What would be the focal length f of the lens if it had been made from diamond with an index of refraction n = 2.42?

The Lens Equation

What is important in the design of a simple lens system is where images are formed for objects that are different distances from the lens. Light from a very distant object enters a lens as parallel rays and focuses at a distance equal to the focal length f behind the lens. To locate the image when the object is not so far away, you can either use a simple graphical method which involves a tracing of two or three rays, or use what is called the *lens equation* which we will derive shortly from the graphical approach.

For our graphical work, we will use an arrow for the object, and trace out rays coming from the tip of the arrow. Where the rays come back together is where the image is formed. We will use the notation that the object is at a distance *(o)* from the lens, and that the image is at a distance *(i)* as shown in Figure (30).

In Figure (30) we have located the image by tracing three rays from the tip of the object. The top ray is parallel to the axis of the lens, and therefore must cross the axis at the focal point behind the lens. The middle

ray, which goes through the center of the lens, is undeflected if the lens is thin. The bottom ray goes through the focal point in front of the lens, and therefore must come out parallel to the axis behind the lens. (Lenses are symmetric in that parallel light from either side focuses at the same distance f from the lens.) The image is formed where the three rays from the tip merge. To locate the image, you only need to draw two of these three special rays.

Exercise 9

(a) Graphically locate the image of the object in Figure (31).

(b) A ray starts out from the tip of the object in the direction of the dotted line shown. Trace out this ray through the lens and show where it goes on the back side of the lens.

In Exercise 9, you found that, once you have located the image, you can trace out any other ray from the tip of the object that passes through the lens, because these rays must all pass through the tip of the image.

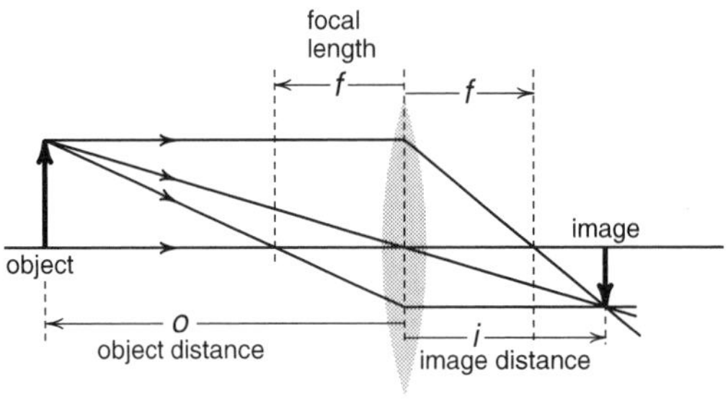

Figure 30
Locating the image using ray tracing. Three rays are easy to draw. One ray goes straight through the center of the lens. The top ray, parallel to the axis, intersects the axis where parallel rays would focus. A ray going through the left focus, comes out parallel to the axis. The image of the arrow tip is located where these rays intersect.

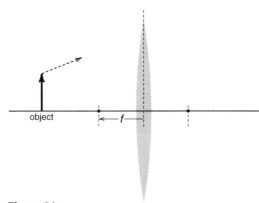

Figure 31
Locate the image of the arrow, and then trace the ray starting out in the direction of the dotted line.

There is a very, very simple relationship between the object distance o, the image distance i and the lens focal length f. It is

$$\frac{1}{o} + \frac{1}{i} = \frac{1}{f} \qquad \begin{array}{l} \textit{the lens} \\ \textit{equation} \end{array} \qquad (16)$$

Equation 16 is worth memorizing if you are going to do any work with lenses. It is the equation you will use all the time, it is easy to remember, and as you will see now, the derivation requires some trigonometry you are not likely to remember. We will take you through the derivation anyway, because of the importance of the result.

In Figure (32a), we have an object of height A that forms an inverted image of height B. We located the image by tracing the top ray parallel to the axis that passes through the focal point behind the lens, and by tracing the ray that goes through the center of the lens.

In Figure (32b) we have selected one of the triangles that appears in Figure (32a). The triangle starts at the tip of the object, goes parallel to the axis over to the image, and then down to the tip of the image. The length of the triangle is $(o + i)$ and the height of the base is (A+B). The lens cuts this triangle to form a smaller similar triangle whose length is o and base is (A). The ratio of the base to length of these similar triangles must be equal, giving

$$\frac{A}{o} = \frac{A+B}{(o+i)} \Rightarrow \frac{(A+B)}{A} = \frac{(o+i)}{o} \qquad (17)$$

In Figure (32c) we have selected another triangle which starts where the top ray hits the lens, goes parallel to the axis over to the image, and down to the tip of the image. This triangle has a length i and a base of height (A+B) as shown. This triangle is cut by a vertical line at the focal plane, giving a smaller similar triangle of length f and base (A) as shown. The ratio of the length to base of these similar triangles must be equal, giving

$$\frac{A}{f} = \frac{A+B}{i} \Rightarrow \frac{(A+B)}{A} = \frac{i}{f} \qquad (18)$$

Combining Equations 17 and 18 gives

$$\frac{i}{f} = \frac{o+i}{o} = 1 + \frac{i}{o} \qquad (19)$$

Finally, divide both sides by i and we get

$$\frac{1}{f} = \frac{1}{i} + \frac{1}{o} \qquad \begin{array}{l} \textit{lens} \\ \textit{equation} \end{array} \qquad (16)$$

which is the lens equation, as advertised.

Note that the lens equation is an exact consequence of the geometrical construction shown back in Figure (30). There is no restriction about small angles. However if you are using spherical lenses, you have to stick to small angles or the light will not focus to a point.

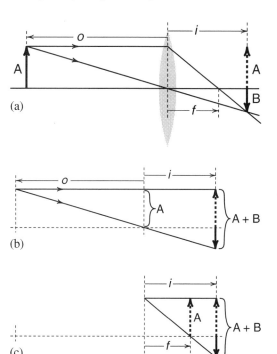

(a)

(b)

(c)

Figure 32
Derivation of the lens equation.

Negative Image Distance

The lens equation is more general than you might expect, for it works equally well for positive and negative distances and focal lengths. Let us start by seeing what we mean by a negative image distance. Writing Equation 15 in the form

$$\frac{1}{i} = \frac{1}{f} - \frac{1}{o} \tag{16a}$$

let us see what happens if $1/o$ is bigger than $1/f$ so that i turns out to be negative. If $1/o$ is bigger than $1/f$, that means that o is less than f and we have placed the object within the focal length as shown in Figure (33).

When we trace out two rays from the tip of the image, we find that the rays diverge after they pass through the lens. They diverge as if they were coming from a point behind the object, a point shown by the dotted lines. In this case we have what is called a ***virtual image***, which is located at a ***negative image distance*** (i). This negative image distance is correctly given by the lens equation (16a).

(We will not drag you through another geometrical proof of the lens equation for negative image distances. It should be fairly convincing that just when the image distance becomes negative in the lens equation, the geometry shows that we switch from a real image on the right side of the lens to a virtual image on the left.)

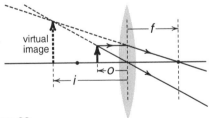

Figure 33
When the object is located within the focal length, we get a virtual image behind the object.

Negative Focal Length and Diverging Lenses

In Figure (33) we got a virtual image by moving the object inside the focal length. Another way to get a virtual image is to use a diverging lens as shown in Figure (34). Here we have drawn the three special rays, but the role of the focal point is reversed. The ray through the center of the lens goes through the center as before. The top ray parallel to the axis of the lens diverges outward as if it came from the focal point on the left side of the lens. The ray from the tip of the object headed for the right focal point, comes out parallel to the axis. Extending the diverging rays on the right, back to the left side, we find a virtual image on the left side.

You get diverging lenses by using concave surfaces as shown in Figure (34). In the lens maker's equation,

$$\frac{1}{f} = (n-1)\left(\frac{1}{r_1} + \frac{1}{r_2}\right) \qquad \begin{array}{l}\textit{lensmaker's}\\ \textit{equation}\end{array} \tag{15}$$

you replace $1/r$ by $-1/r$ for any concave surface. If $1/f$ turns out negative, then you have a diverging lens. Using this negative value of f in the lens equation (with $f = -|f|$) we get

$$\frac{1}{i} = -\left(\frac{1}{|f|} + \frac{1}{o}\right) \tag{16b}$$

This always gives a negative image distance i, which means that diverging lenses only give virtual images.

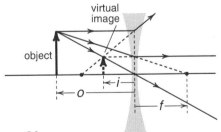

Figure 34
A diverging lens always gives a virtual image.

Exercise 10

You have a lens making machine that can grind surfaces, either convex or concave, with radius of curvatures of either 20 cm or 40 cm, or a flat surface. How many different kinds of lenses can you make? What is the focal length and the name of the lens type for each lens? Figure (35) shows the names given to the various lens types.

Negative Object Distance

With the lens equation, we can have negative image distances and negative focal lengths, and also negative object distances as well.

In all our drawings so far, we have drawn rays coming out of the tip of an object located at a positive object distance. A negative object distance means we have a virtual object where rays are converging toward the tip of the virtual object but don't get there. A comparison of the rays emerging from a real object and converging toward a virtual object is shown in Figure (36). The converging rays (which were usually created by some other lens) can be handled with the lens equation by assuming that the distance from the lens to the virtual object is negative.

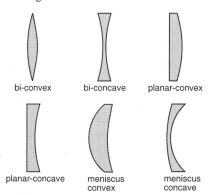

Figure 35
Various lens types. Note that eyeglasses are usually meniscus convex or meniscus concave.

As an example, suppose we have rays converging to a point, and we insert a diverging lens whose negative focal length $f = -|f|$ is equal to the negative object distance $o = -|o|$ as shown in Figure (37). The lens equation gives

$$\frac{1}{i} = \frac{1}{f} - \frac{1}{o} = \frac{1}{-|f|} - \frac{1}{-|o|} = \frac{1}{|o|} - \frac{1}{|f|} \quad (20)$$

If $|f| = |o|$, then $1/i = 0$ and the image is infinitely far away. This means that the light emerges as a parallel beam as we showed in Figure (37).

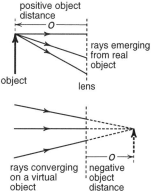

Figure 36
Positive and negative object distances.

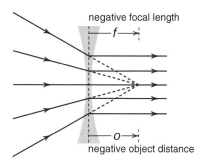

Figure 37
Negative focal length.

Multiple Lens Systems

Using the lens equation, and knowing how to handle both positive and negative distances and focal lengths, you can design almost any simple lens system you want. The idea is to work your way through the system, one lens at a time, where the image from one lens becomes the object for the next. We will illustrate this process with a few examples.

As our first example, consider Figure (38a) where we have two lenses of focal lengths $f_1 = 10$ cm and $f_2 = 12$ cm separated by a distance $D = 40$ cm. An object placed at a distance $o_1 = 17.5$ cm from the first lens creates an image a distance i_1 behind the first lens. Using the lens equation, we get

$$i_1 = \frac{1}{f_1} - \frac{1}{o_1} = \frac{1}{10} - \frac{1}{17.5} = \frac{1}{23.33} \qquad (21)$$

$$i_1 = 23.33 \text{ cm}$$

the same distance we got graphically in Figure (38a).

This image, which acts as the object for the second lens has an object distance

$$o_2 = D - i_1 = 40 \text{ cm} - 23.33 \text{ cm} = 16.67 \text{ cm}$$

This gives us a final upright image at a distance i_2 given by

$$\frac{1}{i_2} = \frac{1}{f_2} - \frac{1}{o_2} = \frac{1}{12} - \frac{1}{16.67} = \frac{1}{42.86} \qquad (22)$$

$$i_2 = 42.86 \text{ cm} \qquad (23)$$

which also accurately agrees with the geometrical construction.

In Figure (38b), we moved the second lens up to within 8 cm of the first lens, so that the first image now falls behind the second lens. We now have a negative object distance

$$o_2 = D - i_1 = 8 \text{ cm} - 23.33 \text{ cm} = -15.33 \text{ cm}$$

Using this negative object distance in the lens equation gives

$$\frac{1}{i_2} = \frac{1}{f_2} - \frac{1}{o_2} = \frac{1}{12} - \frac{1}{-15.33}$$

$$= \frac{1}{12} + \frac{1}{15.33} = \frac{1}{6.73}$$

$$i_2 = 6.73 \text{ cm} \qquad (24)$$

In the geometrical construction we find that the still inverted image is in fact located 6.73 cm behind the second image.

While it is much faster to use the lens equation than trace rays, it is instructive to apply both approaches for a few examples to see that they both give the same result. In drawing Figure (38b) an important ray was the one that went from the tip of the original object, down through the first focal point. This ray emerges from the first lens traveling parallel to the optical axis. The ray then enters the second lens, and since it was parallel to the axis, it goes up through the focal point of the second lens as shown. The second image is located by drawing the ray that passes straight through the second lens, heading for the tip of the first image. Where these two rays cross is where the tip of the final image is located.

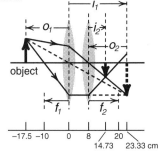

Figure 38b
We moved the second lens in so that the second object distance is negative. We now get an inverted image 6.73 cm from the second lens.

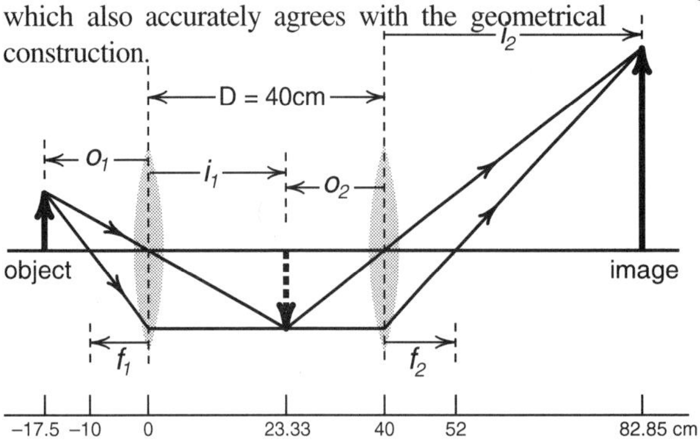

Figure 38a
Locating the image in a two lens system.

In Figure (38c) we sketched a number of rays passing through the first lens, heading for the first image. These rays are converging on the second lens, which we point out in Figure (36b) was the condition for a negative object distance.

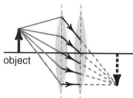

Figure 38c

Two Lenses Together

If you put two thin lenses together, as shown in Figure (39), you effectively create a new thin lens with a different focal length. To find out what the focal length of the combination is, you use the lens equation twice, setting the second object distance o_2 equal to minus the first image distance $-i_1$.

$$o_2 = -i_1 \qquad \begin{array}{l} \textit{for two lenses} \\ \textit{together} \end{array} \qquad (25)$$

From the lens equations we have

$$\frac{1}{i_1} = \frac{1}{f_1} - \frac{1}{o_1} \qquad (26)$$

$$\frac{1}{i_2} = \frac{1}{f_2} - \frac{1}{o_2} \qquad (27)$$

Setting $o_2 = -i_1$ in Equation 27 gives

$$\frac{1}{i_2} = \frac{1}{f_2} - \frac{1}{(-i_1)} = \frac{1}{f_2} + \frac{1}{i_1}$$

Using Equation 26 for $1/i_1$ gives

$$\frac{1}{i_2} = \frac{1}{f_2} + \frac{1}{f_1} - \frac{1}{o_1}$$

$$\frac{1}{o_1} + \frac{1}{i_2} = \frac{1}{f_1} + \frac{1}{f_2} \qquad (28)$$

Now o_1 is the object distance and i_2 is the image distance for the pair of lenses. Treating the pair of lenses as a single lens, we should have

$$\frac{1}{o_1} + \frac{1}{i_2} = \frac{1}{f} \qquad (29)$$

where f is the focal length of the combined lens.

Comparing Equations 28 and 29 we get

$$\boxed{\frac{1}{f} = \frac{1}{f_1} + \frac{1}{f_2}} \qquad \begin{array}{l} \textit{focal length of two} \\ \textit{thin lenses together} \end{array} \qquad (30)$$

as the simple formula for the combined focal length.

Exercise 11

(a) Find the image distances i_2 for the geometry of Figures (38), but with the two lenses reversed, i.e., with $f_1 = 12$ cm, $f_2 = 10$ cm. Do this for both length D = 40 cm and D = 8 cm.

(b) If the two lenses are put together (D = 0) what is the focal length of the combination?

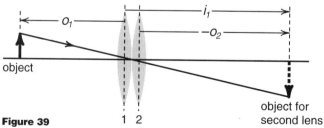

Figure 39
Two lenses together. Since the object for the second lens is on the wrong side of the lens, the object distance o_2 is negative in this diagram. If the lenses are close together, i_1 and $-o_2$ are essentially the same.

Magnification

It is natural to define the magnification created by a lens as the ratio of the height of the image to the height of the object. In Figure (40) we have reproduced Figure (38a) emphasizing the heights of the objects and images.

We see that the shaded triangles are similar, thus the ratio of the height B of the first image to the height A of the object is

$$\frac{B}{A} = \frac{i_1}{o_1} \qquad (31)$$

We could define the magnification in the first lens as the ratio of B/A, but instead we will be a bit tricky and include a - (minus) sign to represent the fact that the image is inverted. With this convention we get

$$\boxed{m_1 = \frac{-B}{A} = \frac{-i_1}{o_1}} \qquad \begin{array}{l} \textit{definition of} \\ \textit{magnification} \, m \end{array} \qquad (32)$$

Treating B as the object for the second lens gives

$$m_2 = \frac{-C}{B} = \frac{-i_2}{o_2} \qquad (33)$$

The total magnification m_{12} in going from the object A to the final image C is

$$m_{12} = \frac{C}{A} \qquad (34)$$

which has a + sign because the final image C is upright. But

$$\frac{C}{A} = \left(\frac{-C}{B}\right)\left(\frac{-B}{A}\right) \qquad (35)$$

Thus we find that the final magnification is the product of the magnifications of each lens.

$$\boxed{m_{12} = m_1 \, m_2} \qquad (36)$$

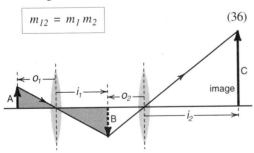

Figure 40
Magnification of two lenses.

Exercise 12

Figures (38) and (40) are scale drawings, so that the ratio of image to object sizes measured from these drawings should equal the calculated magnifications.

(a) Calculate the magnifications m_1, m_2 and m_{12} for Figure (38a) or (40) and compare your results with magnifications measured from the figure.

(b) Do the same for Figure (38b). In Figure (38b), the final image is inverted. Did your final magnification m_{12} come out negative?

Exercise 13

Figure (41a) shows a magnifying glass held 10 cm above the printed page. Since the object is inside the focal length we get a virtual image as seen in the geometrical construction of Figure (41b). Show that our formulas predict a positive magnification, and estimate the focal length of the lens. (Answer: about 17 cm.)

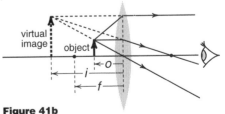

Figure 41a
Using a magnifying glass.

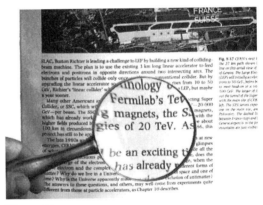

Figure 41b
When the magnifying glass is less than a focal length away from the object, we see an upright virtual image.

THE HUMAN EYE

A very good reason for studying geometrical optics is to understand how your own eye works, and how the situation is corrected when something goes wrong.

Back in Exercise 6 (p21), during our early discussion of spherical lens surfaces, we considered as a model of an eye a sphere of index of refraction n_2, where n_2 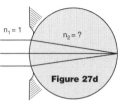 was chosen so that parallel rays which entered the front surface focused on the back surface as shown in Figure (27d). The value of n_2 turned out to be $n_2 = 2.0$. Since the only common substance with an index of refraction greater than zircon at $n = 1.923$ is diamond at $n = 2.417$, it would be difficult to construct such a model eye. Instead some extra focusing capability is required, both to bring the focus to the back surface of the eye, and to focus on objects located at various distances.

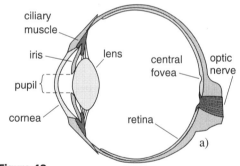

Figure 42
The human eye. The cornea and the lens together provide the extra focusing power required to focus light on the retina. (Photograph of the human eye by Lennart Nilsson.)

Figure (42a) is a sketch of the human eye and Figure 42b a remarkable photograph of the eye. As seen in (42a), light enters the *cornea* at the front of the eye. The amount of light allowed to enter is controlled by the opening of the *iris*. Together the cornea and *crystalline lens* focuses light on the retina which is a film of *nerve fibers* on the back surface of the eye. Information from the new fibers is carried to the brain through the optic nerve at the back. In the retina there are two kinds of nerve fibers, called *rods* and *cones*. Some of the roughly 120 million rods and 7 million cones are seen magnified about 5000 times in Figure (43). The slender ones, the rods, are more sensitive to dim light, while the shorter, fatter, cones, provide our color sensitivity.

In our discussion of the human ear, we saw how there was a mechanical system involving the basilar membrane that distinguished between the various frequencies of incoming sound waves. Information from nerves attached to the basilar membrane was then enhanced through processing in the local nerve fibers before being sent to the brain via the auditory nerve. In the eye, the nerve fibers behind the retina, some of which can be seen on the right side of Figure (43), also do a considerable amount of information processing before the signal travels to the brain via the optic nerve. The way that information from the rods and cones is processed by the nerve fibers is a field of research.

Returning to the front of the eye we have the surface of the cornea and the crystalline lens focusing light on the retina. Most of the focusing is done by the cornea. The shape, and therefore the focal length of the crystalline lens can be altered slightly by the *ciliary muscle* in order to bring into focus objects located at different distances.

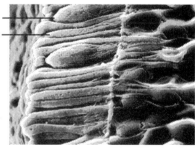

Figure 43
Rods and cones in the retina. The thin ones are the rods, the fat ones the cones.

In a normal eye, when the ciliary muscle is in its resting position, light from infinity is focused on the retina as shown in Figure (44a). To see a closer object, the ciliary muscles contract to shorten the focal length of the cornea-lens system in order to continue to focus light on the cornea (44b). If the object is too close as in Figure (44c), the light is no longer focused and the object looks blurry. The shortest distance at which the light remains in focus is called the *near point*. For children the near point is as short as 7 cm, but as one ages and the crystalline lens becomes less flexible, the near point recedes to something like 200 cm. This is why older people hold written material far away unless they have reading glasses.

Nearsightedness and Farsightedness

Not all of us have the so called *normal* eyes described by Figure (44). There is increasing evidence that those who do a lot of close work as children end up with a condition called *nearsightedness* or *myopia* where the eye is elongated and light from infinity focuses inside the eye as shown in Figure (45a). This can be corrected by placing a diverging lens in front of the eye to move the focus back to the retina as shown in Figure (45b).

The opposite problem, farsightedness, where light focuses behind the retina as shown in Figure (46a) is corrected by a converging lens as shown in Figure (46b).

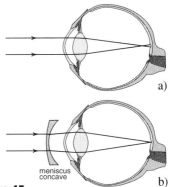

Figure 45
Nearsightedness can be corrected by a convex lens..

Figure 44a
Parallel light rays from a distant object are focuses on the retina when the ciliary muscles are in the resting position.

Figure 44b
The ciliary muscle contracts to shorten the focal length of the cornea-lens system in order to focus light from a more nearby object.

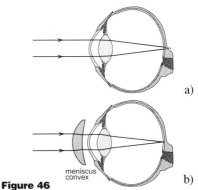

Figure 46
Farsightedness can be corrected by a convex lens

Figure 44c
When an object is to close, the light cannot be focused on the retina.

THE CAMERA

There are a number of similarities between the human eye and a simple camera. Both have an iris to control the amount of light entering, and both record an image at the focal plane of the lens. In a camera, the focus is adjusted, not by changing the shape of the lens as in the eye, but by moving the lens back and forth. The eye is somewhat like a TV camera in that both record images at a rate of about 30 per second, and the information is transmitted electronically to either the brain or a TV screen.

On many cameras you will find a series of numbers labeled by the letter f, called the f **number** or f **stop**. Just as for the parabolic reflectors in figure 4 (p5), the f number is the *ratio of the lens focal length to the lens diameter*. As you close down the iris of the camera to reduce the amount of light entering, you reduce the effective diameter of the lens and therefore increase the f number.

Exercise 14

The iris on the human eye can change the diameter of the opening to the lens from about 2 to 8 millimeters. The total distance from the cornea to the retina is typically about 2.3 cm. What is the range of f values for the human eye? How does this range compare with the range of f value on your camera? (If you have one of the automatic point and shoot cameras, the f number and the exposure time are controlled electronically and you do not get to see or control these yourself.)

Figure 47a
The Physics department's Minolta single lens reflex camera.

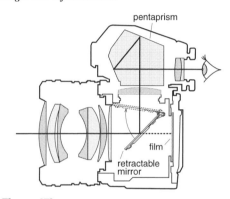

Figure 47b
The lens system for a Nikon single lens reflex camera. When you take the picture, the hinged mirror flips out of the way and the light reaches the film. Before that, the light is reflected through the prism to the eyepiece.

Depth of Field

There are three ways to control the exposure of the film in a camera. One is by the speed of the film, the second is the exposure time, and the third is the opening of the iris or f stop. In talking a picture you should first make sure the exposure is short enough so that motion of the camera and the subject do not cause blurring. If your film is fast enough, you can still choose between a shorter exposure time or a smaller f stop. This choice is determined by the **depth of field** that you want.

The concept of depth of field is illustrated in Figures (48a and b). In (48a), we have drawn the rays of light from an object to an image through an $f2$ lens, a lens with a focal length equal to twice its diameter. (The effective diameter can be controlled by a flexible diaphragm or iris like the one shown.) If you placed a film at the image distance, the point at the tip of the object arrow would focus to a point on the film. If you moved the film forward to position 1, or back to position 2, the image of the arrow tip would fill a circle about equal to the thickness of the three rays we drew in the diagram.

If the film were ideal, you could tell that the image at positions 1 or 2 was out of focus. But no film or recording medium is ideal. If you look closely enough there is always a graininess caused by the size of the basic medium like the silver halide crystals in black and white film, the width of the scan lines in an analog TV camera, or the size of the pixels in a digital camera. If the image of the arrow tip at position 1 is smaller than the grain or pixel size then you cannot tell that the picture is out of focus. You can place the recording medium anywhere between position 1 and 2 and the image will be as sharp as you can get.

In Figure (48b), we have drawn the rays from the same object passing through a smaller diameter $f8$ lens. Again we show by dotted lines positions 1 and 2 where the image of the arrow point would fill the same size circle as it did at positions 1 and 2 for the $f2$ lens above. Because the rays from the $f8$ lens fill a much narrower cone than those from the $f2$ lens, there is a much greater distance between positions 1 and 2 for the $f8$ lens.

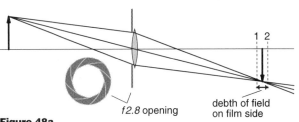

f2.8 opening   debth of field on film side

Figure 48a
A large diameter lens has a narrow depth of field.

Photograph taken at f 5.6.

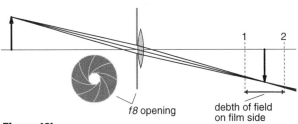

f8 opening   debth of field on film side

Figure 48b
Reducing the effective diameter of the lens increases the depth of field.

Photograph taken at f 22.

If Figures (48) represented a camera, you would not be concerned with moving the film back and forth. Instead you would be concerned with how far the image could be moved back and forth and still appear to be in focus. If the film were at the image position and you then moved the object in and out, you could not move it very far before it's image was noticeably out of focus with the $f2$ lens. You could move it much farther for the $f8$ lens.

This effect is illustrated by the photographs on the right side of Figures 48, showing a close-up tree and the distant tower on Baker Library at Dartmouth College. The upper picture taken at $f5.6$ has a narrow depth of field, and the tower is well out of focus. In the bottom picture, taken at $f22$, has a much broader depth of field and the tower is more nearly in focus. (In both cases we focused on the nearby tree bark.)

Camera manufacturers decide how much blurring of the image is noticeable or tolerable, and then figure out the range of distances the object can be moved and still be acceptably in focus. This range of distance is called the **depth of field**. It can be very short when the object is up close and you use a wide opening like $f2$. It can be quite long for a high f number like $f22$. The inexpensive fixed focus cameras use a small enough lens so that all objects are "in focus" from about 3 feet or 1 meter to infinity.

In the extreme limit when the lens is very small, the depth of field is so great that everything is in focus everywhere behind the lens. In this limit you do not even need a lens, a pinhole in a piece of cardboard will do. If enough light is available and the subject doesn't not move, you can get as good a picture with a pinhole camera as one with an expensive lens system. Our pinhole camera image in Figure (49) is a bit fuzzy because we used too big a pinhole.

(If you are nearsighted you can see how a pinhole camera works by making a tiny hole with your fingers and looking at a distant light at night without your glasses. Just looking at the light, it will look blurry. But look at the light through the hole made by your fingers and the light will be sharp. You can also see the eye chart better at the optometrists if you look through a small hole, but they don't let you do that.)

Figure 49a
We made a pinhole camera by replacing the camera lens with a plastic film case that had a small hole poked into the end.

Figure 49b
Photograph of Baker library tower, taken with the pinhole camera above. If we had used a smaller hole we would have gotten a sharper focus.

Figure 48c
Camera lens. This lens is set to f11, and adjusted to a focus of 3 meters or 10 ft. At this setting, the depth of field ranges from 2 to 5 meters.

Eye Glasses and a Home Lab Experiment

When you get a prescription for eyeglasses, the optometrist writes down number like -1.5, -1.8 to represent the *power* of the lenses you need. These cryptic number are the power of the lenses measured in ***diopters***. What a diopter is, is simply the reciprocal of the focal length $1/f$, where f is measured in meters. A lens with a power of 1 diopter is a converging lens with a focal length of 1 meter. Those of us who have lenses closer to –4 in power have lenses with a focal length of –25 cm, the minus sign indicating a diverging lens to correct for nearsightedness as shown back in Figure (45).

If you are nearsighted and want to measure the power of your own eyeglass lenses, you have the problem that it is harder to measure the focal length of a diverging lens than a converging lens. You can quickly measure the focal length of a converging lens like a simple magnifying glass by focusing sunlight on a piece of paper and measuring the distance from the lens to where the paper is starting to smoke. But you do not get a real image for a diverging lens, and cannot use this simple technique for measuring the focal length and power of diverging lenses used by the nearsighted.

As part of a project, some students used the following method to measure the focal length and then determine the power in diopters, of their and their friend's eyeglasses. They started by measuring the focal length f_0 of a simple magnifying glass by focusing the sun. Then they placed the magnifying glass and the eyeglass lens together, measured the focal length of the combination, and used the formula

$$\frac{1}{f} = \frac{1}{f_1} + \frac{1}{f_2} \qquad \text{(30 repeated)}$$

to calculate the focal length of the lens.

(Note that if you measure distances in meters, then $1/f_1$ is the power of lens 1 in diopters and $1/f_2$ that of lens 2. Equation 30 tells you that the power of the combination $1/f$ is the sum of the powers of the two lenses.

Exercise 15

Assume that you find a magnifying lens that focuses the sun at a distance of 10 cm from the lens. You then combine that with one of your (or a friends) eyeglass lenses, and discover that the combination focus at a distance of 15 cm. What is the power, in diopters, of

(a) The magnifying glass.

(b) The combination.

(c) The eyeglass lens.

Exercise 16 – Home Lab

Use the above technique to measure the power of your or your friend's glasses. If you have your prescription compare your results with what is written on the prescription. (The prescription will also contain information about axis and amount of astigmatism. That you cannot check as easily.

THE EYEPIECE

When the author was a young student, he wondered why you do not put your eye at the focal point of a telescope mirror. That is where the image of a distance object is, and that is where you put the film in order to record the image. You do not put your eye at the image because it would be like viewing an object by putting your eyeball next to it. The object would be hopelessly out of focus. Instead you look through an eyepiece.

The eyepiece is a magnifying glass that allows your eye to comfortably view an image or small object up close. For a normal eye, the least eyestrain occurs when looking at a distant object where the light from the object enters the eye as parallel rays. It is then that the ciliary muscles in the eye are in a resting position. If the image or small object is placed at the focal plane of a lens, as shown in Figure (50), light emerges from the lens as parallel rays. You can put your eye right up to that lens, and view the object or image as comfortably as you would view a distant scene.

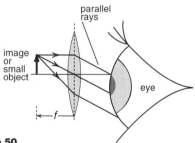

Figure 50
The eyepiece or magnifier. To look at small object, or to study the image produced by another lens or mirror, place the image or object at the focal plane of a lens, so that the light emerges as parallel rays that your eye can comfortably focus upon.

Exercise 17 - The Magnifying Glass

There are three distinct ways of viewing an object through a magnifying glass, which you should try for yourself. Get a magnifying glass and use the letters on this page as the object to be viewed.

(a) First measure the focal length of the lens by focusing the image of a distant object onto a piece of paper. A light bulb across the room or scene out the window will do.

(b) Draw some object on the paper, and place the paper at least several focal lengths from your eye. Then hold the lens about 1/2 a focal length above the object as shown in Figure (51a). You should now see an enlarged image of the object as indicated in Figure (51a). You are now looking at the virtual image of the object. Check that the magnification is roughly a factor of 2×.

(c) Keeping your eye in the same position, several focal lengths and at least 20 cm from the paper, pull the lens back toward your eye. The image goes out of focus when the lens is one focal length above the paper, and then comes back into focus upside down when the lens is farther out. You are now looking at the real image as indicated in Figure (51b). Keep your head far enough back that your eye can focus on this real image.

Hold the lens two focal lengths above the page and check that the inverted real image of the object looks about the same size as the object itself. (As you can see from Figure (51b), the inverted image should be the same size as the object, but 4 focal lengths closer.)

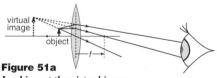

Figure 51a
Looking at the virtual image.

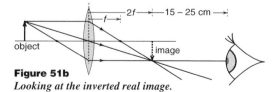

Figure 51b
Looking at the inverted real image.

(d) Now hold the lens one focal length above the page and put your eye right up to the lens. You are now using the lens as an eyepiece as shown in Figure (50). The letters will be large because your eye is close to them, and they will be comfortably in focus because the rays are entering your eye as parallel rays like the rays from a distant object. When you use the lens as an eyepiece you are not looking at an image as you did in parts (b) and (c) of this exercise, instead your eye is creating an image on your retina from the parallel rays.

(e) As a final exercise, hold the lens one focal length above a page of text, start with your eye next to the lens, and then move your head back. Since the light from the page is emerging from the lens as parallel rays, the size of the letters should not change as you move your head back. Instead what you should see is fewer and fewer letters in the magnifying glass as the magnifying glass itself looks smaller when farther away. This effect is seen in Figure (52).

The Magnifier

When jewelers work on small objects like the innards of a watch, they use what they call a *magnifier* which can be a lens mounted at one end of a tube as shown in Figure (53). The length of the tube is equal to the focal length of the lens, so that if you put the other end of the tube up against an object, the lens acts as an eyepiece and light from the object emerges from the lens as parallel rays. By placing your eye close to the lens, you get a close up, comfortably seen view of the object. You may have seen jewelers wear magnifiers like that shown in Figure (54).

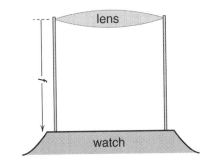

Figure 53
A magnifier.

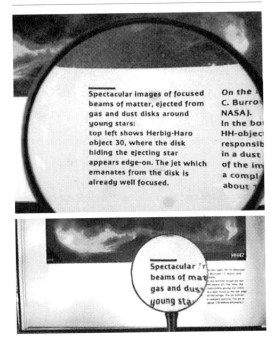

Figure 52
When the lens is one focal length from the page, the emerging rays are parallel. Thus the image letters do not change size as we move away. Instead the lens looks smaller, and we see fewer letters in the lens.

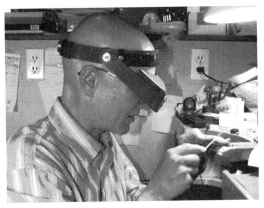

Figure 54
Jeweler Paul Gross with magnifier lenses mounted in visor.

Angular Magnification

Basically all the magnifier does is to allow you to move the object close to your eye while keeping the object comfortably in focus. It is traditional to define the *magnification* of the magnifier as the ratio of the size of the object as seen through the lens to the size of the object as you would see it without a magnifier. By size, we mean the angle the object subtends at your eye. This is often called the *angular magnification*.

The problem with this definition of magnification is that different people, would hold the object at different distances in order to look at it without a magnifier. For example, us nearsighted people would hold it a lot closer than a person with normal vision. To avoid this ambiguity, we can choose some standard distance like 25 cm, a standard near point, at which a person would normally hold an object when looking at it. Then the angular magnification of the magnifier is the ratio of the angle θ_m subtended by the object when using the magnifier, as shown in Figure (55a), to the angle θ_0 subtended by the object held at a distance of 25 cm, as shown in Figure (55b).

$$\frac{\text{angular}}{\text{magnification}} = \frac{\theta_m}{\theta_0} \quad \begin{matrix} \text{\textit{angles defined}} \\ \text{\textit{in Figure 55}} \end{matrix} \quad (37)$$

To calculate the angular magnification we use the small angle approximation $\sin\theta \approx \theta$ to get

$$\theta_m = \frac{y}{f} \qquad \text{\textit{from Figure 55a}}$$

$$\theta_0 = \frac{y}{25\ \text{cm}} \qquad \text{\textit{from Figure 55b}}$$

which gives

$$\frac{\text{angular}}{\text{magnification}} = \frac{y/f}{y/25\ \text{cm}} = \frac{25\ \text{cm}}{f} \quad (38)$$

Thus if our magnifier lens has a focal length of 5 cm, the angular magnification is 5×. Supposedly the object will look five times bigger using the magnifier than without it.

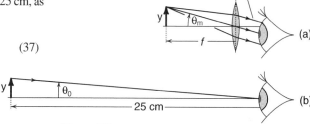

Figure 55
The angles used in defining angular magnification.

TELESCOPES

The basic design of a telescope is to have a large lens or parabolic mirror to create a bright real image, and then use an eyepiece to view the image. If we use a large lens, that lens is called an **objective lens**, and the telescope is called a **refracting telescope**. If we use a parabolic mirror, then we have a **reflecting telescope**.

The basic design of a refracting telescope is shown in Figure (56). Suppose, as shown in Figure (56a), we are looking at a constellation of stars that subtend an angle θ_0 as viewed by the unaided eye. The eye is directed just below the bottom star and light from the top star enters at an angle θ_0. In Figure (56b), the lens system from the telescope is placed in front of the eye, and we are following the path of the light from the top star in the constellation.

The parallel rays from the top star are focused at the focal length f_0 of the objective lens. We adjust the eyepiece so that the image produced by the objective lens is at the focal point of the eyepiece lens, so that light from the image will emerge from the eyepiece as parallel rays that the eye can easily focus.

As with the magnifier, we define the magnification of the telescope as the ratio of the size of (angle subtended by) the object as seen through the object to the size of (angle subtended by) the object seen by the unaided eye. In Figure (56) we see that the constellation subtends an angle θ_0 as viewed by the unaided eye, and an angle θ_i when seen through the telescope. Thus we define the magnification of the telescope as

$$m = \frac{\theta_i}{\theta_0} \qquad \begin{array}{l} \textit{magnification} \\ \textit{of telescope} \end{array} \qquad (39)$$

To calculate this ratio, we note from Figure (56c) that, using the small angle approximation $\sin\theta \approx \theta$, we have

$$\theta_0 = \frac{y_i}{f_0} \ ; \quad \theta_i = \frac{y_i}{f_e} \qquad (40)$$

where f_0 and f_e are the focal lengths of the objective and eyepiece lens respectively. In the ratio, the image height y_i cancels and we get

$$m = \frac{\theta_i}{\theta_0} = \frac{y_i/f_e}{y_i/f_0}$$

$$\boxed{m = \frac{f_0}{f_e}} \qquad (41)$$

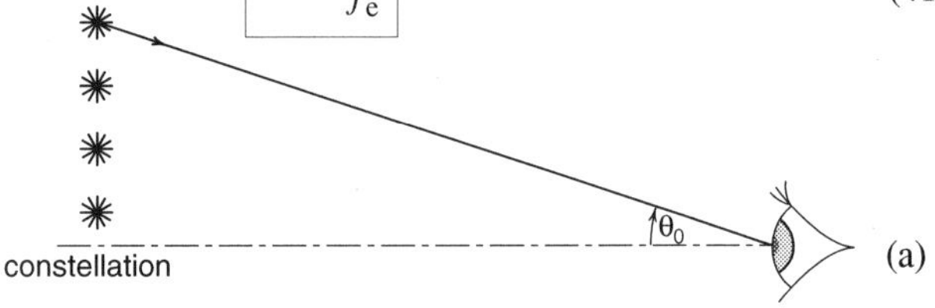

Figure 56a
The unaided eye looking at a constellation of stars that subtend an angle θ_o.

Figure 56b
Looking at the same constellation through a simple refracting telescope. The objective lens produces an inverted image which is viewed by the eyepiece acting as a magnifier. Note that the parallel light from the star focuses at the focal point of the objective lens. With the image at the focal point of the eyepiece lens, light from the image emerges as parallel rays that are easily focused by the eye.

Figure 56c
Relationship between the angles θ_0, θ_i, and the focal lengths.

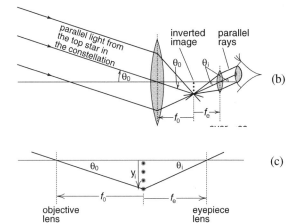

The same formula also applies to a reflecting telescope with f_0 the focal length of the parabolic mirror. Note that there is no arbitrary number like 25 cm in the formula for the magnification of a telescope because telescopes are designed to look at distant objects where the angle θ_0 the object subtends to the unaided eye is the same for everyone.

The first and the last of the important refracting telescopes are shown in Figures (57). The telescope was invented in Holland in 1608 by Hans Lippershy. Shortly after that, Galileo constructed a more powerful instrument and was the first to use it effectively in astronomy. With a telescope like the one shown in Figure (57a), he discovered the moons of Jupiter, a result that provided an explicit demonstration that heavenly bodies could orbit around something other than the earth. This countered the long held idea that the earth was at the center of everything and provided support for the Copernican sun centered picture of the solar system.

When it comes to building large refracting telescopes, the huge amount of glass in the objective lens becomes a problem. The 1 meter diameter refracting telescope at the Yerkes Observatory, shown in Figure (57b), is the largest refracting telescope ever constructed. That was built back in 1897. The largest reflecting telescope is the new 10 meter telescope at the Keck Observatory at the summit of the inactive volcano Mauna Kea in Hawaii. Since the area and light gathering power of a telescope is proportional to the area or the square of the diameter of the mirror or objective lens, the 10 meter Keck telescope is 100 times more powerful than the 1 meter Yerkes telescope.

Exercise 8

To build your own refracting telescope, you purchase a 3 inch diameter objective lens with a focal length of 50 cm. You want the telescope to have a magnification $m = 25\times$.

(a) What will be the f number of your telescope? (1 inch = 2.54 cm).

(b) What should the focal length of your eyepiece lens be?

(c) How far behind the objective lens should the eyepiece lens be located?

(d) Someone give you an eyepiece with a focal length of 10 mm. Using this eyepiece, what magnification do you get with your telescope?

(e) You notice that your new eyepiece is not in focus at the same place as your old eyepiece. Did you have to move the new eyepiece toward or away from the objective lens, and by how much?

(f) Still later, you decide to take pictures with your telescope. To do this you replace the eyepiece with a film holder. Where do you place the film, and why did you remove the eyepiece?

Figure 57b
The Yerkes telescope is the world's largest refracting telescope, was finished in 1897. Since then all larger telescopes have been reflectors.

Figure 57a
Galileo's telescope. With such an instrument Galileo discovered the moons of Jupiter.

Reflecting telescopes

In several ways, the reflecting telescope is similar to the refracting telescope. As we saw back in our discussion of parabolic mirrors, the mirror produces an image in the focal plane when the light comes from a distant object. This is shown in Figure (58a) which is similar to our old Figure (4). If you want to look at the image with an eyepiece, you have the problem that the image is in front of the mirror where, for a small telescope, your head would block the light coming into the scope. Issac Newton, who invented the reflecting telescope, solved that problem by placing a small, flat, 45° reflecting surface inside the telescope tube to deflect the image outside the tube as shown in Figure (58b). There the image can easily be viewed using an eyepiece. Newton's own telescope is shown in Figure (58d). Another technique, used in larger telescopes, is to reflect the beam back through a hole in the mirror as shown in Figure (58c).

The reason Newton invented the reflecting telescope was to avoid an effect called **chromatic aberration**. When white light passes through a simple lens, different wavelengths or colors focus at different distances behind the lens. For example if the yellow light is in focus the red and blue images will be out of focus. In contrast, all wavelengths focus at the same point using a parabolic mirror.

However, problems with keeping the reflecting surface shinny, and the development of lens combinations that eliminated chromatic aberration, made refracting telescopes more popular until the late 1800's. The invention of the durable silver and aluminum coatings on glass brought reflecting telescopes into prominence in the twentieth century.

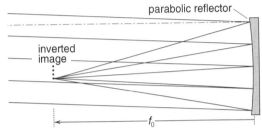

Figure 58a
A parabolic reflector focuses the parallel rays from a distant object, forming an image a distance f_0 in front of the mirror.

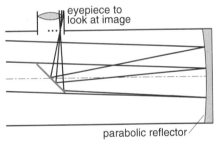

Figure 58b
Issac Newton's solution to viewing the image was to deflect the beam using a 45° reflecting surface so that the eyepiece could be outside the telescope tube.

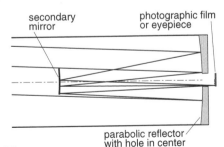

Figure 58c
For large telescopes, it is common to reflect the beam back through a hole in the center of the primary mirror. This arrangement is known as the Cassegrain design.

Figure 58d
Issac Newton's reflecting telescope.

Large Reflecting Telescopes.

The first person to build a really large reflecting telescope was William Hershel, who started with a two inch reflector in 1774 and by 1789 had constructed the four foot diameter telescope shown in Figure (59a). Among Hershel's accomplishments was the discovery of the planet Uranus, and the first observation a distant nebula. It would be another 130 years before Edwin Hubble, using the 100 inch telescope on Mt. Wilson would conclusively demonstrate that such nebula were in fact galaxies like our own milky way. This also led hubble to discover the expansion of the universe.

During most of the second half of the twentieth century, the largest telescope has been the 200 inch (5 meter) telescope on Mt. Palomar, shown in Figure (59b). This was the first telescope large enough that a person could work at the prime focus, without using a secondary mirror. Hubble himself is seen in the observing cage at the prime focus in Figure (59c).

Recently it has become possible to construct mirrors larger than 5 meters in diameter. One of the tricks is to cast the molten glass in a rotating container and keep the container rotating while the glass cools. A rotating liquid has a parabolic surface. The faster the rotation the deeper the parabola. Thus by choosing the right rotation speed, one can cast a mirror blank that has the correct parabola built in. The surface is still a bit rough, and has to be polished smooth, but the grinding out oh large amounts of glass is avoided. The 6.5 meter mirror, shown in Figures (59d and e), being installed on top of Mt. Hopkins in Arizona, was built this way. Seventeen tons of glass would have to have been ground out if the parabola had not been cast into the mirror blank.

Figures 59b,c
*The Mt. Palomar
200 inch telescope.
Below is Edwin
Hubble in the
observing cage.*

Figure 59a
*William Hershel's 4 ft diameter, 40 ft long
reflecting telescope which he completed in 1789.*

Figures 59d,e
*The 6.5 meter MMT
telescope atop Mt.
Hopkins. Above, the
mirror has not been
silvered yet. The blue
is a temporary
protective coating.
Below, the mirror is
being hoisted into
the telescope frame.*

Hubble Space Telescope

An important limit to telescopes on earth, in their ability to distinguish fine detail, is turbulence in the atmosphere. Blobs of air above the telescope move around causing the star image to move, blurring the picture. This motion, on a time scale of about 1/60 second, is what causes stars to appear to twinkle.

The effects of turbulence, and any distortion caused by the atmosphere, are eliminated by placing the telescope in orbit above the atmosphere. The largest telescope in orbit is the famous Hubble telescope with its 1.5 meter diameter mirror, seen in Figure (60). After initial problems with its optics were fixed, the Hubble telescope has produced fantastic images like that of the Eagle nebula seen in Figure (7-17) reproduced here.

With a modern telescope like the Keck (see next page), the effects of atmospheric turbulance can mostly be eliminated by having a computer can track the image of a bright star. The telescopes mirror is flexible enough that the shape of the mirror can then be be modified rapidly and by a tiny amount to keep the image steady.

Figure 7-17
The eagle nebula, birthplace of stars. This Hubble photograph, which apeared on the cover of Time magazine, is perhaps the most famous.

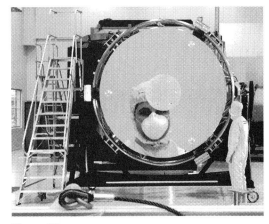

Figure 60a
The Hubble telescope mirror. How is that for a shaving mirror?

Figure 60b
Hubble telescope before launch.

Figure 60c
Hubble telescope being deployed.

World's Largest Optical Telescope

As of 1999, the largest optical telescope in the world is the Keck telescope located atop the Mauna Kea volcano in Hawaii, seen in Figure (61a). Actually there are two identical Keck telescopes as seen in the close-up, Figure (61b). The primary mirror in each telescope consists of 36 hexagonal mirrors fitted together as seen in Figure (61c) to form a mirror 10 meters in diameter. This is twice the diameter of the Mt. palomar mirror we discussed earlier.

The reason for building two Keck telescopes has to do with the wave nature of light. As we mentioned in the introduction to this chapter, geometrical optics works well when the objects we are studying are large compared to the wavelength of light. This is illustrated by the ripple tank photographs of Figures (33-3) and (33-8) reproduced here. In the left hand figure, we see we see a wave passing through a gap that is considerably wider than the wave's wavelength. On the other side of the gap there is a well defined beam with a distinct shadow. This is what we assume light waves do in geometrical optics.

In contrast, when the water waves encounter a gap whose width is comparable to a wavelength, as in the right hand figure, the waves spread out on the far side. This is a phenomenon called ***diffraction***. We can even see some diffraction at the edges of the beam emerging from the wide gap.

Diffraction also affects the ability of telescopes to form sharp images. The bigger the diameter of the telescope, compared to the light wavelength, the less important diffraction is and the sharper the image that can be formed. By combining the output from the two Keck telescopes, one creates a telescope whose effective diameter, for handling diffraction effects, is equal to the 90 meter separation of the telescopes rather than just the 10 meter diameter of one telescope. The great improvement in the image sharpness that results is seen in Figure (61d). On the left is the best possible image of a star, taken using one telescope alone. When the two telescopes are combined, they get the much sharper image on the right.

Figures 61 c
The 36 mirrors forming Keck's primary mirror. We have emphasized the outline of the upper 4 mirrors.

Figures 33-3,8
Unless the gap is wide in comparison to a wavelength, diffraction effects are important.

Figures 61 a,b
The Keck telescopes atop Mauna Kea volcano in Hawaii

Figures 62
Same star, photographed on the left using one scope, on the right with the two Keck telescopes combined.

Infrared Telescopes

Among the spectacular images in astronomy are the large dust clouds like the ones that form the Eagle nebula photographed by the Hubble telescope, and the famous Horsehead nebula shown in Figure (63a). But a problem is that astronomers would like to see through the dust, to see what is going on inside the clouds and what lies beyond.

While visible light is blocked by the dust, other wavelength's of electromagnetic radiation can penetrate these clouds. Figure (63b) is a photograph of the same patch of sky as the Horsehead nebula in (63a), but observed using infrared light whose wavelengths are about 3 times longer than the wavelengths of visible light. First notice that the brightest stars are at the same positions in both photographs. But then notice that the black cloud, thought to resemble a horses head, is missing in the infrared photograph. The stars in and behind the cloud shine through; their infrared light is not blocked by the dust.

Where does the infrared light come from? If you have studied Chapter 35 on the Bohr theory of hydrogen, you will recall that hydrogen atoms can radiate many different wavelengths of light. The only visible wavelengths are the three longest wavelengths in the Balmer series. The rest of the Balmer series and all of the Lyman series consist of short wavelength ultraviolet light. But all the other wavelengths radiated by hydrogen are infrared, like the Paschen series where the electron ends up in the third energy level. The infrared wavelengths are longer than those of visible light. Since hydrogen is the major constituent of almost all stars, it should not be surprising that stars radiate infrared as well as visible light.

A telescope designed for looking at infrared light is essentially the same as a visible light telescope, except for the camera. Figure (64) shows the infrared telescope on Mt. Hopkins used to take the infrared image of the Horsehead nebula. We enlarged the interior photograph to show the infrared camera which is cooled by a jacket of liquid nitrogen (essentially a large thermos bottle surrounding the camera).

a)
Visible light photograph

b)
Infrared light photograph

Figure 63
The horsehead nebula photographed in visible (a) and infrared light (b). The infrared light passes through the dust cloud.

Figure 64
Infrared telescope on Mt. Hopkins. Note that the infrared camera, seen in the blowup, is in a container cooled by liquid nitrogen. You do not want the walls of the camera to be "infrared hot" which would fog the image.

You might wonder why you have to cool an infrared camera and not a visible light camera. The answer is that warm bodies emit infrared radiation. The hotter the object, the shorter the wavelength of the radiation. If an object is hot enough, it begins to glow in visible light, and we say that the object is red hot, or white hot. Since you do not want the infrared detector in the camera seeing camera walls glowing "infrared hot", the camera has to be cooled.

Not all infrared radiation can make it down through the earth's atmosphere. Water vapor, for example is very good at absorbing certain infrared wavelengths. To observe the wavelengths that do not make it through, infrared telescopes have been placed in orbit. Figure 65 is an artist's drawing of the Infrared Astronomical Satellite (IRAS) which was used to make the infrared map of the entire sky seen in Figure (66). The map is oriented so that the Milky Way, our own galaxy, lies along the center horizontal plane. In visible light photographs, most of the stars in our own galaxy are obscured by the immense amount of dust in the plane

of the galaxy. But in an infrared photograph, the huge concentration of stars in the plane of the galaxy show up clearly.

At the center of our galaxy is a gigantic black hole, with a mass of millions of suns. For a visible light telescope, the galactic center is completely obscured by dust. But the center can be clearly seen in the infrared photograph of Figure (67), taken by the Mt. Hopkins telescope of Figure 64. This is not a single exposure, instead it is a composite of thousands of images in that region of the sky. Three different infrared wavelengths were recorded, and the color photograph was created by displaying the longest wavelength image as red, the middle wavelength as green, and the shortest wavelength as blue. In this photograph, you not only see the intense radiation from the region of the black hole at the center, but also the enormous density of stars at the center of our galaxy. (You do not see radiation from the black hole itself, but from nearby stars that may be in the process of being captured by the black hole.)

Figure 65
Artist's drawing of the infrared telescope IRAS in orbit.

Figure 66
Map of the entire sky made by IRAS. The center of the Milky Way is in the center of the map. This is essentially a view of our galaxy seen from the inside.

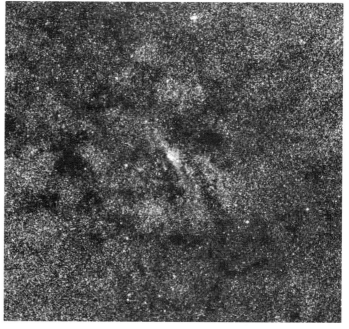

Figure 67
Center of our galaxy, where an enormous black hole resides. Not only is the galactic center rich in stars, but also in dust which prevents viewing this region in visible light.

Radio Telescopes

The earth's atmosphere allows not only visible and some infrared light from stars to pass through, but also radio waves in the wavelength range from a few millimeters to a good fraction of a meter. To study the radio waves emitted by stars and galaxies, a number of *radio telescopes* have been constructed.

For a telescope reflector to produce a sharp image, the surface of the reflector should be smooth and accurate to within about a fifth of a wavelength of the radiation being studied. For example, the surface of a mirror for a visible wavelength telescope should be accurate to within about 10^{-4} millimeters since the wavelength of visible light is centered around 5×10^{-4} millimeters. Radio telescopes that are to work with 5 millimeter wavelength radio waves, need surfaces accurate only to about a millimeter. Telescopes designed to study the important 21 cm wavelength radiation emitted by hydrogen, can have a rougher surface yet. As a result, radio telescopes can use sheet metal or even wire mesh rather than polished glass for the reflecting surface.

This is a good thing, because radio telescopes have to be much bigger than optical telescopes to order to achieve comparable images. The sharpness of an image, due to diffraction effects, is related to the ratio of the reflector diameter to the radiation wavelength. Since the radio wavelengths are at least 10^4 times larger than those for visible light, a radio telescope has to be 10^4 times larger than an optical telescope to achieve the same resolution.

The worlds largest radio telescope dish, shown in Figure (68), is the 305 meter dish at the Arecibo Observatory in Perto Rico. While this dish can see faint objects because of it's enormous size, and has been used to make significant discoveries, it has the resolving ability of an optical telescope about 3 centimeters in diameter, or a good set of binoculars .

As we saw with the Keck telescope, there is a great improvement in resolving power if the images of two or more telescopes are combined. The effective resolving power is related to the separation of the telescopes rather than to the diameter of the individual telescopes. Figure (69) shows the *Very Large Array (VLA)* consisting of twenty seven 25 meter diameter radio telescopes located in southern New Mexico. The dishes are mounted on tracks, and can be spread out to cover an area 36 kilometers in diameter. At this spacing, the resolving power is nearly comparable to a 5 meter optical telescope at Mt. Palomar.

Figures 69
The "Very Large Array" (VLA) of radio telescopes. The twenty seven telescopes can be spread out to a diameter of 36 kilometers.

Figure 68
Arecibo radio telescope. While the world's largest telescope dish remains fixed in the earth, the focal point can be moved to track a star.

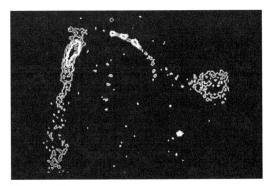

Figures 69b
Radio galaxy image from the VLA. Studying the radio waves emitted by a galaxy often gives a very different picture than visible light.

The Very Long Baseline Array (VLBA)

To obtain significantly greater resolving power, the *Very Long Baseline Array (VLBA)* was set up in the early 1990's. It consists of ten 25 meter diameter radio telescopes placed around the earth as shown in Figure (70). When the images of these telescopes are combined, the resolving power is comparable to an optical telescope 1000 meters in diameter (or an array of optical telescopes spread over an area one kilometer across).

The data from each telescope is recorded on a high speed digital tape with a time track created by a hydrogen maser atomic clock. The tapes are brought to a single location in Socorrow New Mexico where a high speed computer uses the accurate time tracks to combine the data from all the telescopes into a single image. To do this, the computer has to correct, for example, for the time difference of the arrival of the radio waves at the different telescope locations.

Because of it's high resolution, the VLBA can be used to study the structure of individual stars. In Figure 72 we see two time snapshots of the radio emission from the stellar atmosphere of a star 1000 light years away. With any of the current optical telescopes, the image of this star is only a point.

"Snapshots" of the Envelope of the Star TX Cam

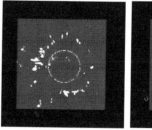
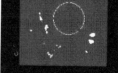

24 May 1997   29 October 1998

Figure 71
Very Long Baseline Array (VLBA) radio images of the variable star TX Cam which is located 1000 light years away. The approximate size of the star as it would be seen in visible light is indicated by the circle. The spots are silicon Monoxide (SiO) gas in the star's extended atmosphere. Motion of the these spots trace the periodic changes in the atmosphere of the star.
(Credit P.J. Diamond & A.J. Kembal, National Radio Astronomy, Associated Universities, Inc.)

Figure 70
The Very Long Baseline Array of radio antennas. They are located at a) Hancock New Hampshire b) Ft. Davis Texas c) Kitt Peak Arizona d) North Liberty Iowa e) St. Croix Virgin Islands f) Brewster Washington g) Mauna Kea Hawaii h) Pie Town New Mexico i) Los Alamos New Mexico j) Owen's Valley California.

MICROSCOPES

Optically, microscopes like the one seen in Figure (72), are telescopes designed to focus on nearby objects. Figure (73) shows the ray diagram for a simple microscope, where the objective lens forms an inverted image which is viewed by an eyepiece.

To calculate the magnification of a simple microscope, note that if an object of height y_0 were viewed unaided at a distance of 25 cm, it would subtend an angle θ_0 given by

$$\theta_0 = \frac{y_0}{25 \text{ cm}} \tag{42}$$

where throughout this discussion we will use the small angle approximation $\sin\theta \approx \tan\theta \approx \theta$.

A ray from the tip of the object (point A in Figure 73b), parallel to the axis, will cross the axis at point D, the focal point of the objective lens. Thus the height BC is equal to the height y_0 of the object, and the distance BD is the focal length f_0 of the objective, and the angle β is given by

$$\beta = \frac{y_0}{f_0} \quad \begin{array}{l} \textit{from triangle} \\ \textit{BCD} \end{array} \tag{43}$$

From triangle DEF, where the small angle at D is also β, we have

$$\beta = \frac{y_i}{L} \quad \begin{array}{l} \textit{from triangle} \\ \textit{DEF} \end{array} \tag{44}$$

where y_i is the height of the image and the distance L is called the **tube length** of the microscope.

Figure 72
Standard optical microscope, which my grandfather purchased as a medical student in the 1890's. Compare this with a microscope constructed 100 years later, seen in Figure (74) on the next page.

Equating the values of β in Equations 29 and 30 and solving for y_i gives

$$\beta = \frac{y_0}{f_0} = \frac{y_i}{L} \; ; \quad y_i = y_0 \frac{L}{f_0} \tag{45}$$

The eyepiece is placed so that the image of the objective is in the focal plane of the eyepiece lens, producing parallel rays that the eye can focus. Thus the distance EG equals the focal length f_0 of the eyepiece. From triangle EFG we find that the angle θ_i that image subtends as seen by the eye is

$$\theta_i = \frac{y_i}{f_e} \quad \begin{array}{l} \textit{angle subtended} \\ \textit{by image} \end{array} \tag{46}$$

Substituting Equation 45 for y_i in Equation 46 gives

$$\theta_i = \frac{L}{f_0} \frac{y_0}{f_e} \tag{47}$$

Finally, the magnification m of the microscope is equal to the ratio of the angle θ_i subtended by the image in the microscope, to the angle θ_0 the object subtends at a distance of 25 cm from the unaided eye.

$$m = \frac{\theta_i}{\theta_0} = \frac{L}{f_0} \frac{y_0}{f_e} \times \frac{1}{y_0/25 \text{ cm}} \tag{48}$$

where we used Equation 47 for θ_i and Equation 42 for θ_0. The distance y_0 cancels in Equation 48 and we get

$$\boxed{m = \frac{L}{f_0} \times \frac{25 \text{ cm}}{f_e}} \quad \begin{array}{l} \textit{magnification of a} \\ \textit{simple microscope} \end{array} \tag{49}$$

(We could have inserted a minus sign in the formula for magnification to indicate that the image is inverted.)

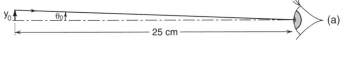

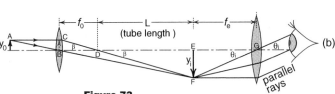

Figure 73
Optics of a simple microscope.

Scanning Tunneling Microscope

Modern research microscopes bear less resemblance to the simple microscope described above than the Hubble telescope does to Newton's first reflector telescope. In the research microscopes that can view and manipulate individual atoms, there are no lenses based on geometrical optics. Instead the surface to be studied is scanned, line by line, by a tiny probe whose operation is based on the particle-wave nature of electrons. An image of the surface is then reconstructed by computer and displayed on a computer screen. These microscopes work at a scale of distance much smaller than the wavelength of light, a distance scale where the approximations inherent in geometrical optics do not apply.

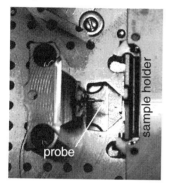

a) Probe and sample holder.

b) Vacuum chamber enclosing the probe and sample holder. Photograph taken in Geoff Nunes' lab at Dartmouth College.

Figure 74
Scanning Tunneling Microscope (STM). The tungsten probe seen in (a) has a very sharp point, about one atom across. With a couple of volts difference between the probe and the silicon crystal in the sample holder, an electric current begins to flow when the tip gets to within about fifteen angstroms (less than fifteen atomic diameters) of the surface. The current flows because the wave nature of the electrons allows them to "tunnel" through the few angstrom gap. The current increases rapidly as the probe is brought still closer. By moving the probe in a line sideways across the face of the silicon, while moving the probe in and out to keep the current constant, the tip of the probe travels at a constant height above the silicon atoms. By recording how much the probe was moved in and out, one gets a recording of the shape of the surface along that line. By scanning across many closely spaced lines, one gets a map of the entire surface. The fine motions of the tungsten probe are controlled by piezo crystals which expand or contract by tiny amounts when a voltage is applied to them. The final image you see was created by computer from the scanning data.

c) Surface (111 plane) of a silicon crystal imaged by this microscope. We see the individual silicon atoms in the surface

PHOTOGRAPH CREDITS

Figure 36-1, p1, p8; Scattered Wave
Education Development Center

Figure 33-30, p10; Shock wave
Education Development Center

Figures Optics-1 p3; Mormon Tabrenacle
The Church of Jesus Christ of Latter-day Saints,
Historical Department Archives

Figure Optics-8b p7; Corner reflectors
NASA

Figure Optics-9, p9; Wave through lens
Nils Abramson

Figure Optics-10, p11; Refraction, ripple tank
Education Development Center

Figure Optics-11, p9; Refraction, glass
©1990 Richard Megna/ Fundamental Photographs

Figure Optics-16b, *p14;* Glass Fiber
Foto Forum

Figure Optics-18, p15; Duodenum
Dr. Richard Rothstein

Figure Optics-19, p18; Halo
© Robert Greenler

Figure Optics-25, *p18;* Zoom lens
Nikon

Figure Optics-28, *p22;* Hubble Scope
NASA

Figure Optics-42, *p31;* Human eye
© Lennart Nilsson

Figure Optics-43, *p31;* Rods & cones
© Lennart Nilsson

Figure Optics-47b, *p33;* Single lens reflex
Adapted from Nikon drawing

Figure Optics-57a, *p41;* Galileo's telescope
© Institute and Museum of History of Science of
Florence Italy

Figure Optics-57b, *p41;* Yerkes telescope
University of Chicago

Figure Optics-58d, *p42; Newton's telescope*
Dorling Kindersley, Pockets Inventions

Figure Optics-59a, *p43; Hershel's telescope*
© Royal Astronomical Society Library

Figure Optics-59b,c, *p43; Palomar telescope*
Courtesy of The Archives, California Institute of
Technology

Figure Optics-59d,e, *p43; MMT telescope*
Lori Stiles, Universitu of Arizona News Service

Figure 7-17, p44; Eagle nebula
Space Telescope Science Institute/NASA

Figure Optics-60, *p44;* Hubble telescope
NASA

Figure Optics-61a,b,c, p45; Keck telescope
© Richard Wainscoat

Figure 33-3,8, *p45;* Wave through gap
Education Development Center

Figure Optics-62, p45; Sharpened star image
Keck

Figure Optics-63, p46; Horsehead nebula
Two Micron All Sky Survay (2MASS)

Figure Optics-64, p46; Infrared telescope
Two Micron All Sky Survay (2MASS)

Figure Optics-65, p47; IRAS drawing
Space Infrared Telescope Facility

Figure Optics-66, p47; Iras Milky Way
Space Infrared Telescope Facility

Figure Optics-67, p47; Center of galaxy
Two Micron All Sky Survay (2MASS)

Figure Optics-68, p48; Arecibo telescope
National Astronomy and Ionosphere Center

Figure Optics-69a, *p48; VLA radio telescopes*
Photo courtesy of NRAO/AUI.

Figure Optics-69b, *p48; Radio galaxy*
Photo courtesy of NRAO/AUI.

Figure Optics-70, *p49; Radio image of star*
Photo courtesy of NRAO/AUI.

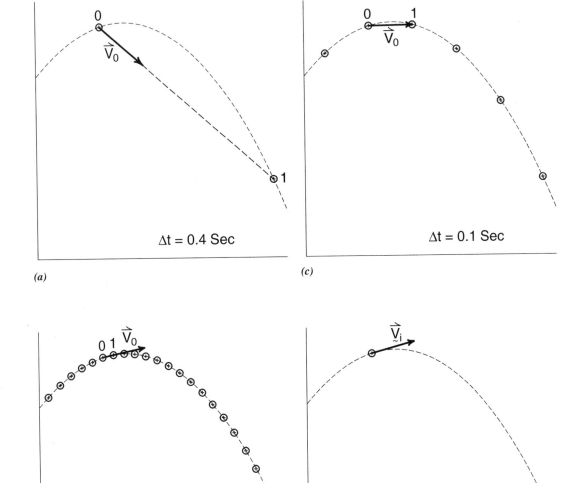

$\Delta t = 0.4$ Sec

(a)

$\Delta t = 0.1$ Sec

(c)

$\Delta t = 0.025$ Sec

(b)

instantaneous velocity

(d)

Figure 1
Transition to instantaneous velocity.

Calculus Chapter 1

Introduction to Calculus

This chapter, which replaces Chapter 4 in Physics 2000, is intended for students who have not had calculus, or as a calculus review for those whose calculus is not well remembered. If, after reading part way through this chapter, you feel your calculus background is not so bad after all, go back to Chapter 4 in Physics 2000, study the derivation of the constant acceleration formulas beginning on page 4-8, and work the projectile motion problems in the appendix to Chapter 4. Those who study all of this introduction to calculus should then proceed to the projectile motion problems in the appendix to Chapter 4 of the physics text.

In Chapter 3 of Physics 2000, we used strobe photographs to define velocity and acceleration vectors. The basic approach was to turn up the strobe flashing rate as we did in going from Figure (3-3) to (3-4) until all the kinks are clearly visible and the successive displacement vectors give a reasonable description of the motion. We did not turn the flashing rate too high, for the practical reason that the displacement vectors became too short for accurate work.

LIMITING PROCESS

In our discussion of instantaneous velocity we conceptually turned the strobe all the way up as illustrated in Figures (2-22a) through (2-22d), redrawn here in Figure (1). In these figures, we initially see a fairly large change in $\vec{v}_0$ as the strobe rate is increased and Δt reduced. But the change becomes smaller and it looks as if we are approaching some final value of $\vec{v}_0$ that does not depend on the size of Δt, provided Δt is small enough. It looks as if we have come close to the final value in Figure (1c).

The progression seen in Figure (1) is called a ***limiting process***. The idea is that there really is some true value of $\vec{v}_0$ which we have called the instantaneous velocity, and that we approach this true value for sufficiently small values of Δt. This is a calculus concept, and in the language of calculus, we are ***taking the limit as Δt goes to zero***.

The Uncertainty Principle

For over 200 years, from the invention of calculus by Newton and Leibnitz until 1924, the limiting process and the resulting concept of instantaneous velocity was one of the cornerstones of physics. Then in 1924 Werner Heisenberg discovered what he called the ***uncertainty principle*** which places a limit on the accuracy of experimental measurements.

LIMITING PROCESS

In Chapter 3 of Physics 2000, we used strobe photographs to define velocity and acceleration vectors. The basic approach was to turn up the strobe flashing rate, as we did in going from Figure (3-3) to (3-4) shown below. We turned the rate up until all the kinks are clearly visible and the successive displacement vectors give a reasonable description of the motion. We did not turn the flashing rate too high, for the practical reason that the displacement vectors became too short for accurate work.

In our discussion of instantaneous velocity we conceptually turned the strobe all the way up as illustrated in Figures (2-22a) through (2-22d), redrawn here in Figure (1). In these figures, we initially see a fairly large change in $\vec{v}_0$ as the strobe rate is increased and Δt reduced. But then the change becomes smaller, and it looks as if we are approaching some final value of $\vec{v}_0$ that does not depend on the size of Δt, provided Δt is small enough. It looks as if we have come close to the final value in Figure (1c).

The progression seen in Figure (1) is called a *limiting process*. The idea is that there really is some true value of $\vec{v}_0$ which we have called the *instantaneous velocity*, and that we approach this true value for sufficiently small values of Δt. This is a calculus concept, and in the language of calculus, we are *taking the limit as Δt goes to zero*.

THE UNCERTAINTY PRINCIPLE

For over 200 years, from the invention of calculus by Newton and Leibnitz until 1924, the limiting process and the resulting concept of instantaneous velocity was one of the cornerstones of physics. Then in 1924 Werner Heisenberg discovered what he called the *uncertainty principle* which places a limit on the accuracy of experimental measurements.

Heisenberg discovered something very new and unexpected. He found that the act of making an experimental measurement unavoidably affects the results of an experiment. This had not been known previously because the effect on large objects like golf balls is undetectable. But on an atomic scale where we study small systems like electrons moving inside an atom, the effect is not only observable, it can dominate our study of the system.

One particular consequence of the uncertainly principle is that the more accurately we measure the position of an object, the more we disturb the motion of the object. This has an immediate impact on the concept of instantaneous velocity. If we turn the strobe all the way up, reduce Δt to zero, we are in effect trying to measure the position of the object with infinite precision. The consequence would be an infinitely big disturbance of the motion of the object we are studying. If we actually could turn the strobe all the way up, we would destroy the object we were trying to study.

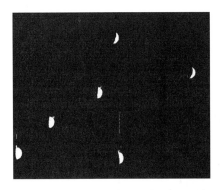 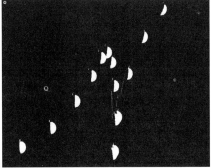

Figures 3-3 and 3-4 from Physics 2000
Strobe photographs of a moving object. In the first photograph, the time between flashes is so long that the motion is difficult to understand. In the second, the time between flashes was reduced and the motion is more easily understood.

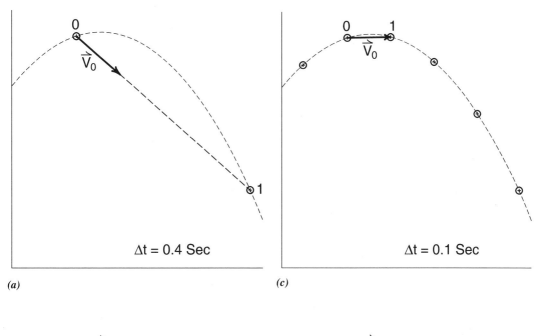

(a)   *(c)*

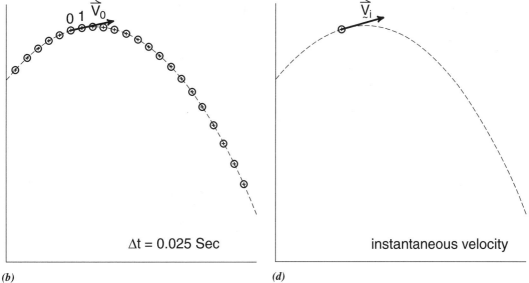

(b)   *(d)*

Figure 1
Transition to instantaneous velocity. As we reduce Δt,
there is less and less change in the vector $\vec{V}_0$. It looks
as if we are approaching an exact final value.

Heisenberg discovered something very new and unexpected. He found that the act of making an experimental measurement unavoidably affects the results of an experiment. This had not been known previously because the effect on large objects like golf balls is undetectable. But on an atomic scale where we study small systems like electrons moving inside an atom, the effect is not only observable, it can dominate our study of the system.

One particular consequence of the uncertainly principle is that the more accurately we measure the position of an object, the more we disturb the motion of the object. This has an immediate impact on the concept of instantaneous velocity. If we turn the strobe all the way up, reduce Δt to zero, we are in effect trying to measure the position of the object with infinite precision. The consequence would be an infinitely big disturbance of the motion of the object we are studying. If we actually could turn the strobe all the way up, we would destroy the object we were trying to study.

It turns out that the uncertainty principle can have a significant impact on a larger scale of distance than the atomic scale. Suppose, for example, that we constructed a chamber 1 cm on a side, and wished to study the projectile motion of an electron inside. Using Galileo's idea that objects of different mass fall at the same rate, we would expect that the motion of the electron projectile should be the same as more massive objects. If we took a strobe photograph of the electron's motion, we would expect get results like those shown in Figure (2). This figure represents projectile motion with an acceleration $g = 980 \, cm/sec^2$ and $\Delta t = .01 \, sec$, as the reader can easily check.

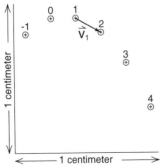

Figure 2
Hypothetical electron projectile motion experiment.

When we study the uncertainty principle in Chapter 40 of the physics text, we will see that a measurement which is accurate enough to show that Position (2) is below Position (1), could disturb the electron enough to reverse its direction of motion. The next position measurement could find the electron over where we drew Position (3), or back where we drew Position (0), or anywhere in the region in between. As a result we could not even determine what direction the electron is moving. This uncertainty would not be the result of a sloppy experiment, it is the best we can do with the most accurate and delicate measurements possible.

The uncertainty principle has had a significant impact on the way physicists think about motion. Because we now know that the measuring process affects the results of the measurement, we see that it is essential to provide experimental definitions to any physical quantity we wish to study. A conceptual definition, like turning the strobe all the way up to define instantaneous velocity, can lead to fundamental inconsistencies.

Even an experimental definition like our strobe definition of velocity can lead to inconsistent results when applied to something like the electron in Figure (2). But these inconsistencies are real. Their existence is telling us that the very concept of velocity is beginning to lose meaning for these small objects.

On the other hand, the idea of the limiting process and instantaneous velocity is very convenient when applied to larger objects where the effects of the uncertainty principle are not detectable. In this case we can apply all the mathematical tools of calculus developed over the past 250 years. The status of instantaneous velocity has changed from a basic concept to a useful mathematical tool. Those problems for which this mathematical tool works are called problems in *classical physics*; and those problems for which the uncertainty principle is important, are in the realm of what we call *quantum physics*.

CALCULUS DEFINITION OF VELOCITY

With the above perspective on the physical limitations on the limiting process, we can now return to the main topic of this chapter—the use of calculus in defining and working with velocity and acceleration.

In discussing the limiting process in calculus, one traditionally uses a special set of symbols which we can understand if we adopt the notation shown in Figure (3). In that figure we have drawn the coordinate vectors $\vec{R}_i$ and $\vec{R}_{i+1}$ for the i th and (i + 1) th positions of the object. We are now using the symbol $\overrightarrow{\Delta R}_i$ to represent the displacement of the ball during the i to i+1 interval. The vector equation for $\overrightarrow{\Delta R}_i$ is

$$\overrightarrow{\Delta R}_i = \vec{R}_{i+1} - \vec{R}_i \tag{1}$$

In words, Equation (1) tells us that $\overrightarrow{\Delta R}_i$ is the change, during the time Δt, of the position vector $\vec{R}$ describing the location of the ball.

The velocity vector $\vec{v}_i$ is now given by

$$\vec{v}_i = \frac{\overrightarrow{\Delta R}_i}{\Delta t} \tag{2}$$

This is just our old strobe definition $\vec{v}_i = \vec{s}_i/\Delta t$, but using a notation which emphasizes that the displacement $\vec{s}_i = \overrightarrow{\Delta R}_i$ is the *change in position* that occurs during the time Δt. The Greek letter Δ (delta) is used both to represent the idea that the quantity $\overrightarrow{\Delta R}_i$ or Δt is small, and to emphasize that both of these quantities change as we change the strobe rate.

The limiting process in Figure (1) can be written in the form

$$\underline{\vec{v}}_i \equiv \lim_{\Delta t \to 0} \frac{\overrightarrow{\Delta R}_i}{\Delta t} \tag{3}$$

where the word "limit" with $\Delta t \to 0$ underneath, is to be read as "limit as Δt goes to zero". For example we would read Equation (3) as "*the instantaneous velocity $\underline{\vec{v}}_i$ at position i is the limit, as Δt goes to zero, of the ratio $\overrightarrow{\Delta R}_i /\Delta t$.*"

For two reasons, Equation (3) is not quite yet in standard calculus notation. One is that in calculus, only the limiting value, in this case, the instantaneous velocity, is considered to be important. Our strobe definition $\vec{v}_i = \overrightarrow{\Delta R}_i /\Delta t$ is only a step in the limiting process. Therefore when we see the vector $\vec{v}_i$, we should assume that it is the limiting value, and no special symbol like the underline is used. For this reason we will drop the underline and write

$$\vec{v}_i = \lim_{\Delta t \to 0} \frac{\overrightarrow{\Delta R}_i}{\Delta t} \tag{3a}$$

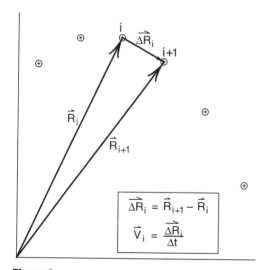

$$\overrightarrow{\Delta R}_i = \vec{R}_{i+1} - \vec{R}_i$$

$$\vec{V}_i = \frac{\overrightarrow{\Delta R}_i}{\Delta t}$$

Figure 3
Definitions of $\overrightarrow{\Delta R}_i$ and $\vec{v}_i$.

The second change deals with the fact that when Δt goes to zero we need an infinite number to time steps to get through our strobe photograph, and thus it is not possible to locate a position by counting time steps. Instead we measure the time t that has elapsed since the beginning of the photograph, and use that time to tell us where we are, as illustrated in Figure (4). Thus instead of using $\vec{v}_i$ to represent the velocity at position i, we write $\vec{v}(t)$ to represent the velocity at time t. Equation (3) now becomes

$$\vec{v}(t) = \lim_{\Delta t \to 0} \frac{\overrightarrow{\Delta R}(t)}{\Delta t} \tag{3b}$$

where we also replaced $\overrightarrow{\Delta R}_i$ by its value $\overrightarrow{\Delta R}(t)$ at time t.

Although Equation (3b) is in more or less standard calculus notation, the notation is clumsy. It is a pain to keep writing the word "limit" with a $\Delta t \to 0$ underneath. To streamline the notation, we replace the Greek letter Δ with the English letter d as follows

$$\lim_{\Delta t \to 0} \frac{\overrightarrow{\Delta R}(t)}{\Delta t} \equiv \frac{d\vec{R}(t)}{dt} \tag{4}$$

(The symbol $\equiv$ means *defined equal to*.) To a mathematician, the symbol $d\vec{R}(t)/dt$ is just shorthand

notation for the limiting process we have been describing. But to a physicist, there is a different, more practical meaning. Think of dt as a short Δt, short enough so that the limiting process has essentially occurred, but not too short to see what is going on. In Figure (1), a value of dt less than .025 seconds is probably good enough.

If dt is small but finite, then we know exactly what the $d\vec{R}(t)$ is. It is the small but finite displacement vector at the time t. It is our old strobe definition of velocity, with the added condition that dt is such a short time interval that the limiting process has occurred. From this point of view, dt is a real time interval and $d\vec{R}(t)$ a real vector, which we can work with in a normal way. The only thing special about these quantities is that when we see the letter d instead of Δ, we must remember that a limiting process is involved. In this notation, the calculus definition of velocity is

$$\vec{v}(t) = \frac{d\vec{R}(t)}{dt} \tag{5}$$

where $\vec{R}(t)$ and $\vec{v}(t)$ are the particle's coordinate vector and velocity vector respectively, as shown in Figure (5). Remember that this is just fancy shorthand notation for the limiting process we have been describing.

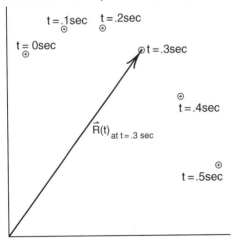

Figure 4
Rather than counting individual images, we can locate a position by measuring the elapsed time t. In this figure, we have drawn the displacement vector $\vec{R}(t)$ at time t = .3 sec.

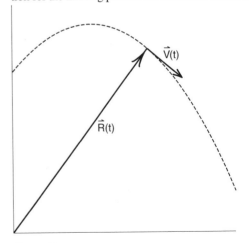

Figure 5
Instantaneous position and velocity at time t.

ACCELERATION

In the analysis of strobe photographs, we defined both a velocity vector $\vec{v}$ and an acceleration vector $\vec{a}$. The definition of $\vec{a}$, shown in Figure (2-12) reproduced here in Figure (6) was

$$\vec{a}_i \equiv \frac{\vec{v}_{i+1} - \vec{v}_i}{\Delta t} \qquad (6)$$

In our graphical work we replaced $\vec{v}_i$ by $\vec{s}_i/\Delta t$ so that we could work directly with the displacement vectors $\vec{s}_i$ and experimentally determine the behavior of the acceleration vector for several kinds of motion.

Let us now change this graphical definition of acceleration over to a calculus definition, using the ideas just applied to the velocity vector. First, assume that the ball reached position i at time t as shown in Figure (6). Then we can write

$$\vec{v}_i = \vec{v}(t)$$

$$\vec{v}_{i+1} = \vec{v}(t+\Delta t)$$

to change the time dependence from a count of strobe flashes to the continuous variable t. Next, define the vector $\overrightarrow{\Delta v}(t)$ by

$$\overrightarrow{\Delta v}(t) \equiv \vec{v}(t+\Delta t) - \vec{v}(t) \quad \left(= \vec{v}_{i+1} - \vec{v}_i \right) \qquad (7)$$

We see that $\overrightarrow{\Delta v}(t)$ is the change in the velocity vector as the time advances from t to $t+\Delta t$. The strobe definition of $\vec{a}_i$ can now be written

$$\vec{a}(t) \begin{pmatrix} strobe \\ definition \end{pmatrix} = \frac{\vec{v}(t + \Delta t) - \vec{v}(t)}{\Delta t} \equiv \frac{\overrightarrow{\Delta v}(t)}{\Delta t} \qquad (8)$$

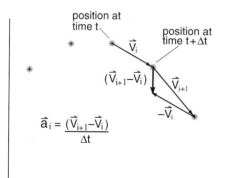

position at time t

position at time t+Δt

$\vec{V}_i$

$(\vec{V}_{i+1}-\vec{V}_i)$

$\vec{V}_{i+1}$

$-\vec{V}_i$

$$\vec{a}_i = \frac{(\vec{V}_{i+1}-\vec{V}_i)}{\Delta t}$$

Figure 6
Experimental definition of the acceleration vector.

Now go through the limiting process, turning the strobe up, reducing Δt until the value of $\vec{a}(t)$ settles down to its limiting value. We have

$$\vec{a}(t) \begin{pmatrix} calculus \\ definition \end{pmatrix} = \lim_{\Delta t \to 0} \frac{\vec{v}(t + \Delta t) - \vec{v}(t)}{\Delta t}$$

$$= \lim_{\Delta t \to 0} \frac{\overrightarrow{\Delta v}(t)}{\Delta t} \qquad (9)$$

Finally use the shorthand notation d/dt for the limiting process:

$$\boxed{\vec{a}(t) = \frac{d\vec{v}(t)}{dt}} \qquad (10)$$

Equation (10) does not make sense unless you remember that it is notation for all the ideas expressed above. Again, physicists think of dt as a short but finite time interval, and $d\vec{v}(t)$ as the small but finite change in the velocity vector during the time interval dt. It's our strobe definition of acceleration with the added requirement that Δt is short enough that the limiting process has already occurred.

Components

Even if you have studied calculus, you may not recall encountering formulas for the derivatives of vectors, like $d\vec{R}(t)/\Delta t$ and $d\vec{v}(t)/\Delta t$ which appear in Equations (5) and (10). To bring these equations into a more familiar form where you can apply standard calculus formulas, we will break the vector Equations (5) and (10) down into component equations.

In the chapter on vectors, we saw that any vector equation like

$$\vec{A} = \vec{B} + \vec{C} \qquad (11)$$

is equivalent to the three component equations

$$A_x = B_x + C_x$$
$$A_y = B_y + C_y \qquad (12)$$
$$A_z = B_z + C_z$$

The advantage of the component equations was that they are simply numerical equations and no graphical work or trigonometry is required.

The limiting process in calculus does not affect the decomposition of a vector into components, thus Equation (5) for $\vec{v}(t)$ and Equation (10) for $\vec{a}(t)$ become

$$\vec{v}(t) = d\vec{R}(t)/dt \tag{5}$$

$$v_x(t) = dR_x(t)/dt \tag{5a}$$
$$v_y(t) = dR_y(t)/dt \tag{5b}$$
$$v_z(t) = dR_z(t)/dt \tag{5c}$$

and

$$\vec{a}(t) = d\vec{v}(t)/dt \tag{10}$$

$$a_x(t) = dv_x(t)/dt \tag{10a}$$
$$a_y(t) = dv_y(t)/dt \tag{10b}$$
$$a_z(t) = dv_z(t)/dt \tag{10c}$$

Often we use the letter x for the x coordinate of the vector $\vec{R}$ and we use y for R_y and z for R_z. With this notation, Equation (5) assumes the shorter and perhaps more familiar form

$$v_x(t) = dx(t)/dt \tag{5a'}$$
$$v_y(t) = dy(t)/dt \tag{5b'}$$
$$v_z(t) = dz(t)/dt \tag{5c'}$$

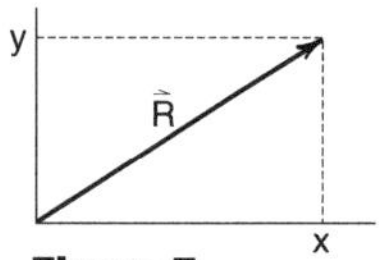

Figure 7

At this point the notation has become deceptively short. You now have to remember that x(t) stands for the x coordinate of the particle at a time t.

We have finally boiled the notation down to the point where it would be familiar from any calculus course. If we restrict our attention to one dimensional motion along the x axis. Then all we have to concern ourselves with are the x component equations

$$\boxed{\begin{aligned} v_x(t) &= \frac{dx(t)}{dt} \\[1em] a_x(t) &= \frac{dv_x(t)}{dt} \end{aligned}} \tag{10a}$$

INTEGRATION

When we worked with strobe photographs, the photograph told us the position $\vec{R}(t)$ of the ball as time passed. Knowing the position, we can then use Equation (5) to calculate the ball's velocity $\vec{v}(t)$ and then Equation (10) to determine the acceleration $\vec{a}(t)$. In general, however, we want to go the other way, and predict the motion from a knowledge of the acceleration. For example, imagine that you were in Galileo's position, hired by a prince to predict the motion of cannonballs. You know that a cannonball should not be much affected by air resistance, thus the acceleration throughout its trajectory should be the constant gravitational acceleration $\vec{g}$. You know that $\vec{a}(t) = \vec{g}$; how then do you use that knowledge in Equations (5) and (10) to predict the motion of the ball?

The answer is that you cannot with the equations in their present form. The equations tell you how to go from $\vec{R}(t)$ to $\vec{a}(t)$, while to predict motion you need to go the other way, from $\vec{a}(t)$ to $\vec{R}(t)$. The topic of this section is to see how to reverse the directions in which we use our calculus equations. Equations (5) and (10) involve the process called *differentiation*. We will see that when we go the other way the reverse of differentiation is a process called *integration*. We will see that integration is a simple concept, but a process that is sometimes hard to perform without the aid of a computer.

Prediction of Motion

In our earlier discussion, we have used strobe photographs to analyze motion. Let us see what we can learn from such a photograph for predicting motion. Figure (8) is our familiar projectile motion photograph showing the displacement $\vec{s}$ of a ball during the time the ball traveled from a position labeled (0) to the position labeled (4). If the ball is now at position (0) and each of the images is .1 seconds apart, then the vector $\vec{s}$ tells us where the ball will be at a time of .4 seconds from now. If we can predict $\vec{s}$, we can predict the motion of the ball. The general problem of predicting the motion of the ball is to be able to calculate $\vec{s}(t)$ for any time t.

From Figure (8) we see that $\vec{s}$ is the vector sum of the individual displacement vectors $\vec{s}_1$, $\vec{s}_2$, $\vec{s}_3$ and $\vec{s}_4$

$$\vec{s} = \vec{s}_1 + \vec{s}_2 + \vec{s}_3 + \vec{s}_4 \qquad (11)$$

We can then use the fact that $\vec{s}_1 = \vec{v}_1\Delta t$, $\vec{s}_2 = \vec{v}_2\Delta t$, etc. to get

$$\vec{s} = \vec{v}_1\Delta t + \vec{v}_2\Delta t + \vec{v}_3\Delta t + \vec{v}_4\Delta t \qquad (12)$$

Rather than writing out each term, we can use the *summation sign* Σ to write

$$\vec{s} = \sum_{i=1}^{4} \vec{v}_i\Delta t \qquad (12a)$$

Equation (12) is approximate in that the $\vec{v}_i$ are approximate (strobe) velocities, not the instantaneous veloci-

ties we want for a calculus discussion. In Figure (9) we improved the situation by cutting Δt to 1/4 of its previous value, giving us four times as many images and more accurate velocities $\vec{v}_i$.

We see that the displacement $\vec{s}$ is now the sum of 16 vectors

$$\vec{s} = \vec{s}_1 + \vec{s}_2 + \vec{s}_3 + ... + \vec{s}_{15} + \vec{s}_{16} \qquad (13)$$

Expressing this in terms of the velocity vectors $\vec{v}_1$ to $\vec{v}_{16}$ we have

$$\vec{s} = \vec{v}_1\Delta t + \vec{v}_2\Delta t + \vec{v}_3\Delta t + ... + \vec{v}_{15}\Delta t + \vec{v}_{16}\Delta t \quad (14)$$

or using our more compact notation

$$\vec{s} = \sum_{i=1}^{16} \vec{v}_i\Delta t \qquad (14a)$$

While Equation (14) for $\vec{s}$ looks quite different than Equation (12), the sum of sixteen vectors instead of four, the displacement vectors $\vec{s}$ in the two cases are exactly the same. Adding more intermediate images did not change where the ball was located at the time of t = .4 seconds. In going from Equation (12) to (14), what has changed in shortening the time step Δt, is that *the individual velocity vectors $\vec{v}_i$ become more nearly equal to the instantaneous velocity of the ball at each image.*

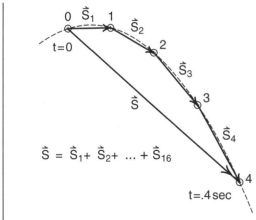

Figure 8
To predict the total displacement
$\vec{s}$, we add up the individual
displacements $\vec{s}_i$.

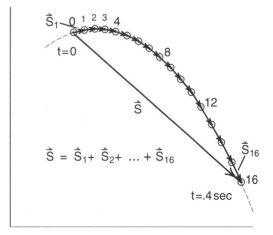

Figure 9
With a shorter time interval, we
add up more displacement vectors
to get the total displacement $\vec{s}$.

If we reduced Δt again by another factor of 1/4, so that we had 64 images in the interval $t = 0$ to $t = .4$ sec, the formula for $\vec{s}$ would become

$$\vec{s} = \sum_{i=1}^{64} \vec{v}_i \Delta t \tag{15a}$$

where now the $\vec{v}_i$ are still closer to representing the ball's instantaneous velocity. The more we reduce Δt, the more images we include, the closer each $\vec{v}_i$ comes to the instantaneous velocity $\vec{v}(t)$. While adding more images gives us more vectors we have to add up to get the total displacement $\vec{s}$, there is very little change in our formula for $\vec{s}$. If we had a million images, we would simply write

$$\vec{s} = \sum_{i=1}^{1000000} \vec{v}_i \Delta t \tag{16a}$$

In this case the $\vec{v}_i$ would be physically indistinguishable from the instantaneous velocity $\vec{v}(t)$. We have essentially reached a calculus limit, but we have problems with the notation. It is clearly inconvenient to label each $\vec{v}_i$ and then count the images. Instead we would like notation that involves the instantaneous velocity $\vec{v}(t)$ and expresses the beginning and end points in terms of the initial time t_1 and final time t_f, rather than the initial and final image numbers i.

In the calculus notation, we replace the summation sign Σ by something that looks almost like the summation sign, namely the ***integral sign*** $\int$. (The French word for integration is the same as their word for summation.) Next we replaced the individual $\vec{v}_i$ by the continuous variable $\vec{v}(t)$ and finally express the end points by the initial time t_i and the final time t_f. The result is

$$\vec{s} = \sum_{i=1}^{n} \vec{v}_i \Delta t \rightarrow \left(\begin{array}{l} as\ the\ number \\ n\ becomes \\ infinitely \\ large \end{array} \right) \int_{t_i}^{t_f} \vec{v}(t)dt \tag{17}$$

Calculus notation is more easily handled, or is at least more familiar, if we break vector equations up into component equations. Assume that the ball started at position i which has components $x_i = x(t_i)$ [read $x(t_i)$ as "x at time t_i"] and $y_i = y(t_i)$ as shown in Figure (10). The final position f is at $x_f = x(t_f)$ and $y_f = y(t_f)$.

Thus the displacement $\vec{s}$ has x and y components

$$s_x = x(t_f) - x(t_i)$$

$$s_y = y(t_f) - y(t_i)$$

Breaking Equation (17) into component equations gives

$$s_x = x(t_f) - x(t_i) = \int_{t_i}^{t_f} v_x(t)dt \tag{18a}$$

$$s_y = y(t_f) - y(t_i) = \int_{t_i}^{t_f} v_y(t)dt \tag{18b}$$

Here we will introduce one more piece of notation often used in calculus courses. On the left hand side of Equation (18a) we have $x(t_f) - x(t_i)$ which we can think of as the variable x(t) evaluated over the interval of time from t_i to t_f. We will often deal with variables evaluated over some interval and have a special notation for that. We will write

$$x(t_f) - x(t_i) \equiv x(t) \Big|_{t_i}^{t_f} \tag{19}$$

You are to read the symbol $x(t)\big|_{t_i}^{t_f}$ as "x of t evaluated from t_i to t_f". We write the initial time t_i at the bottom of the vertical bar, the final time t_f at the top.

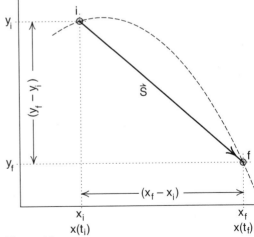

Figure 10
Breaking the vector $\vec{s}$ into components.

We use similar notation for any kind of variable, for example

$$f(x)\bigg|_{x_1}^{x_2} \equiv f(x_2) - f(x_1) \tag{19a}$$

Remember to subtract the variable when evaluated at the value at the bottom of the vertical bar.

With this notation, our Equation (18) can be written

$$s_x = x(t)\bigg|_{t_i}^{t_f} = \int_{t_i}^{t_f} v_x(t)dt \tag{18a'}$$

$$s_y = y(t)\bigg|_{t_i}^{t_f} = \int_{t_i}^{t_f} v_y(t)dt \tag{18b'}$$

Calculating Integrals

Equation (20) is nice and compact, but how do you use it? How do you calculate integrals? The key is to remember that an integral is just a fancy notation for a sum of terms, where we make the time step $\Delta(t)$ very small. Keeping this in mind, we will see that there is a very easy way to interpret an integral.

Figure 11a
Strobe photograph of ball moving at constant velocity in x direction.

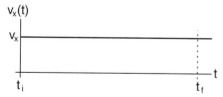

Figure 11b
Graph of $v_x(t)$ versus t for the ball of Figure 11a.

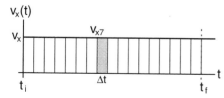

Figure 11c
Each $v_x \Delta t$ is the area of a rectangle.

To get this interpretation, let us start with the simple case of a ball moving in a straight line, for instance, the x direction, at a constant velocity v_x. A strobe picture of this motion would look like that shown in Figure (11a).

Figure (11b) is a graph of the ball's velocity $v_x(t)$ as a function of the time t. The vertical axis is the value of v_x, the horizontal axis is the time t. Since the ball is traveling at constant velocity, v_x has a constant value and is thus represented by a straight horizontal line. In order to calculate the distance that the ball has traveled during the time interval from t_i to t_f, we need to evaluate the integral

$$s_x = \int_{t_i}^{t_f} v_x(t)dt \quad\substack{\textit{distance ball}\\\textit{travels in}\\\textit{time interval}\\ t_i \textit{ to } t_f} \tag{18a}$$

To actually evaluate the integral, we will go back to our summation notation

$$s_x = \sum_{i_{initial}}^{i_{final}} v_{xi}\Delta t \tag{20}$$

and show individual time steps Δt in the graph of v_x versus t, as in Figure (11c).

We see that each term in Equation (20) is represented in Figure (11c) by a rectangle whose height is v_x and whose width is Δt. We have shaded in the rectangle representing the 7th term $v_{x7}\Delta t$. We see that $v_{x7}\Delta t$ is just the **area** of the shaded rectangle, and it is clear that the sum of all the areas of the individual rectangles is the total area under the curve, starting at time t_i and ending at time t_f. Here we are beginning to see that the process of integration is equivalent to finding the area under a curve.

With a simple curve like the constant velocity $v_x(t)$ in Figure (11c), we see by inspection that the total area from t_i to t_f is just the area of the complete rectangle of height v_x and width $(t_f - t_i)$. Thus

$$s_x = v_x \times (t_f - t_i) \tag{21}$$

This is the expected result for constant velocity, namely

$$\substack{\textit{distance}\\\textit{traveled}} = \textit{velocity} \times \textit{time} \quad\substack{\textit{for}\\\textit{constant}\\\textit{velocity}} \tag{21a}$$

To see that you are not restricted to the case of constant velocity, suppose you drove on a freeway due east (the x direction) starting at 9:00 AM and stopping for lunch at 12 noon. Every minute during your trip you wrote down the speedometer reading so that you had an accurate plot of $v_x(t)$ for the entire morning, a plot like that shown in Figure (12). From such a plot, could you determine the distance s_x that you had travelled?

Your best answer is to multiply each value v_i of your velocity by the time Δt to calculate the average distance traveled each minute. Summing these up from the initial time $t_i = 9{:}00\text{AM}$ to the final time $t_f = \text{noon}$, you have as your estimate

$$s_x \approx \sum_i v_{xi}\Delta t$$

(The symbol $\approx$ means **approximately equal**.)

To get a more accurate value for the distance traveled, you should measure your velocity at shorter time intervals Δt and add up the larger number of smaller rectangles. The precise answer should be obtained in the limit as Δt goes to zero

$$s_x = \lim_{\Delta t \to 0} \sum_i v_{xi}\Delta t = \int_{t_i}^{t_f} v_x(t)dt \qquad (22)$$

This limit is just the area under the curve that is supposed to represent the instantaneous velocity $v_x(t)$.

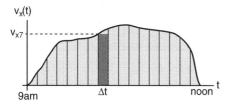

Figure 12
Plot of $v_x(t)$ for a trip starting at 9:00 AM and finishing at noon. The distance traveled is the area under the curve.

Thus we can interpret the integral of a curve as the area under the curve even when the curve is not constant or flat. Mathematicians concern themselves with curves that are so wild that it is difficult or impossible to determine the area under them. Such curves seldom appear in physics problems.

While the basic idea of integration is simple—just finding the area under a curve–in practice it can be quite difficult to calculate the area. Much of an introductory calculus course is devoted to finding the formulas for the areas under various curves. There are also books called **tables of integrals** where you look up the formula for a curve and the table tells you the formula for the area under that curve.

In Chapter 16 of the physics text, we will discuss a mathematical technique called **Fourier analysis**. This is a technique in which we can describe the shape of any continuous curve in terms of a sum of sin waves. (Why we want to do that will become clear then.) The process of Fourier analysis involves finding the area under some very complex curves, curves often involving experimental data for which we have no formula, only graphs. Such curves cannot be integrated by using a table of integrals, with the result that Fourier analysis was not widely used until the advent of the modern digital computer.

The computer made a difference, because we can find the area under almost any curve by breaking the curve into short pieces of length Δt, calculating the area $v_i\Delta t$ of each narrow rectangle, and adding up the area of the rectangles to get the total area. If the curve is so wild that we have to break it into a million segments to get an accurate answer, that might be too hard to do by hand, but it usually a very simple and rapid job for a computer. Computers can be much more efficient than people at integration.

The Process of Integrating

There is a language for the process of integration which we will now take you through. In each case we will check that the results are what we would expect from our summation definition, or the idea that an integral is the area under a curve.

The simplest integral we will encounter in the calculation of the area under a curve of unit height as shown in Figure (13). We have the area of a rectangle of height 1 and length $(t_f - t_i)$

$$\int_{t_i}^{t_f} 1 \, dt = \int_{t_i}^{t_f} dt = (t_f - t_i) \tag{22}$$

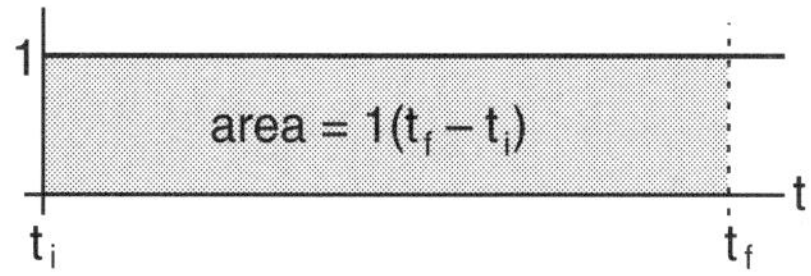

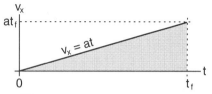

Figure 13
Area under a curve of unit height.

We will use some special language to describe this integration. We will say that the integral of dt is simply the time t, and that the integral of dt from t_i to t_f is equal to t evaluated from t_i to t_f. In symbols this is written as

$$\int_{t_i}^{t_f} dt = t \Big|_{t_i}^{t_f} = (t_f - t_i) \tag{23}$$

Recall that the vertical line after a variable means to evaluate that variable at the final position t_f (upper value), minus that variable evaluated at the initial position t_i (lower value). Notice that this prescription gives the correct answer.

The next simplest integral is the integral of a constant, like a constant velocity v_x over the interval t_i to t_f

$$\int_{t_i}^{t_f} v_x \, dt = v_x(t_f - t_i) \tag{24}$$

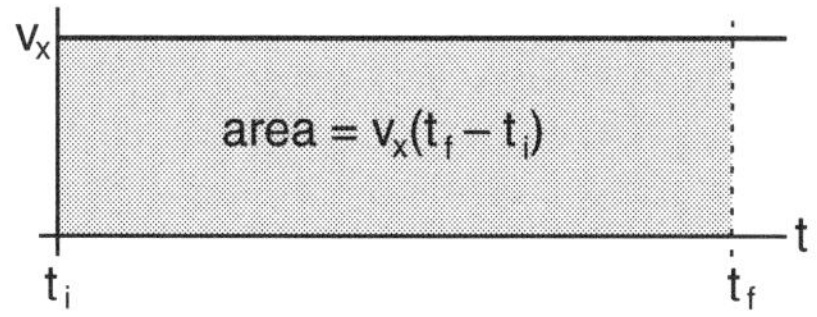

Figure 14
Area under the constant v_x curve.

Since $(t_f - t_i) = \int_{t_i}^{t_f} dt$, we can replace $(t_f - t_i)$ in Equation (24) by the integral to get

$$\int_{t_i}^{t_f} v_x \, dt = v_x \int_{t_i}^{t_f} dt \qquad v_x \, a \, constant \tag{25}$$

and we see that a constant like v_x can be taken outside the integral sign.

Let us try the simplest case we can think of where v_x is not constant. Suppose v_x starts at zero at time $t_i = 0$ and increases linearly according to the formula

$$v_x = at \tag{26}$$

Figure 15

When we get up to the time t_f the velocity will be (at_f) as shown in Figure (15). The area under the curve $v_x = at$ is a triangle whose base is of length t_f and height is at_f. The area of this triangle is one half the base times the height, thus we get for the distance s_x traveled by an object moving with this velocity

$$s_x = \int_0^{t_f} v_x \, dt = \tfrac{1}{2}(\text{base}) \times (\text{altitude})$$
$$= \tfrac{1}{2}(t_f)(at_f) = \tfrac{1}{2}at_f^2 \tag{27}$$

Now let us repeat the same calculation using the language one would find in a calculus book. We have

$$s_x = \int_0^{t_f} v_x \, dt = \int_0^{t_f} (at) \, dt \tag{28}$$

The constant (a) can come outside, and we know that the answer is $1/2 \, at_f^2$, thus we can write

$$s_x = a \int_0^{t_f} t \, dt = \tfrac{1}{2}at_f^2 \tag{29}$$

In Equation (29) we can cancel the a's to get the result

$$\int_0^{t_x} t \, dt = \tfrac{1}{2}t_f^2 \tag{30}$$

In a calculus text, you would find the statement that the integral $\int t\,dt$ is equal to $t^2/2$ and that the integral should be evaluated as follows

$$\int_0^{t_f} t\,dt = \frac{t^2}{2}\bigg|_0^{t_f} = \frac{t_f^2}{2} - \frac{0}{2} = \frac{t_f^2}{2} \qquad (31)$$

Indefinite Integrals

When we want to measure an actual area under a curve, we have to know where to start and stop. When we put these limits on the integral sign, like t_i and t_f, we have what is called a ***definite integral***. However there are times where we just want to know what the form of the integral is, with the idea that we will put in the limits later. In this case we have what is called an ***indefinite integral***, such as

$$\int t\,dt = \frac{t^2}{2} \quad \textit{indefinite integral} \qquad (32)$$

The difference between our definite integral in Equation (31) and the indefinite one in Equation (32) is that we have not chosen the limits yet in Equation (32). If possible, a table of integrals will give you a formula for the indefinite integral and let you put in whatever limits you want.

Integration Formulas

For some sets of curves, there are simple formulas for the area under them. One example is the set of curves of the form t^n. We have already considered the cases where $n = 0$ and $n = 1$.

$n = 0$

$$\int t^0 dt = \int dt = t$$

$n = 1$

$$\int t^1 dt = \int t\,dt = \frac{t^2}{2}$$

Some results we will prove later are

$n = 2$

$$\int t^2 dt = \frac{t^3}{3}$$

$n = 3$

$$\int t^3 dt = \frac{t^4}{4}$$

$$(33a,b,c,d)$$

Looking at the way these integrals are turning out, we suspect that the general rule is

$$\boxed{\int t^n dt = \frac{t^{n+1}}{n+1}} \qquad (34)$$

It turns out that Equation (34) is a general result for any value of n except $n = -1$. If $n = -1$, then you would have division by zero, which cannot be the answer. (We will shortly discuss the special case where $n = -1$.)

As long as we stay away from the $n = -1$ case, the formula works for negative numbers. For example

$$\int t^{-2} dt = \int \frac{dt}{t^2} = \frac{t^{(-2+1)}}{-2+1} = \frac{t^{-1}}{(-1)}$$

$$\boxed{\int \frac{dt}{t^2} = -\frac{1}{t}} \qquad (35)$$

In our discussion of gravitational and electrical potential energy, we will encounter integrals of the form seen in Equation (35).

Exercise 1

Using Equation (34) and the fact that constants can come outside the integral, evaluate the following integrals:

(a) $\displaystyle\int x\,dx$ *it does not matter whether we call the variable t or x*

(b) $\displaystyle\int_{x=1}^{x=2} x^5 dx$ *also sketch the area being evaluated*

(c) $\displaystyle\int_{t=1}^{t=2} \frac{dt}{t^2}$ *Show that you get a positive area.*

(d) $\displaystyle\int \frac{GmM}{r^2} dr$ *where G, m, and M are constants*

(e) $\displaystyle\int \frac{a}{y^{3/2}} dy$ *(a) is a constant*

NEW FUNCTIONS
Logarithms

We have seen that when we integrate a curve or function like t^2, we get a new function $t^3/3$. The functions t^2 and t^3 appear to be fairly similar; the integration did not create something radically different. However, the process of integration can lead to some curves with entirely different behavior. This happens, for example, in that special case $n = -1$ when we try to do the integral of t^{-1}.

It is certainly not hard to plot t^{-1}, the result is shown in Figure (16). Also there is nothing fundamentally difficult or peculiar about measuring the area under the t^{-1} curve from some t_i to t_f, as long as we stay away from the origin $t=0$ where t^{-1} blows up. The formula for this area turns out, however, to be the new function called the **natural logarithm**, abbreviated by the symbol *ln*. The area in Figure (16) is given by the formula

$$\int_{t_i}^{t_f} \frac{1}{t}\, dt = \ln(t_f) - \ln(t_i) \tag{36}$$

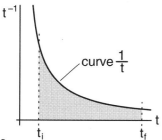

$\text{curve } \dfrac{1}{t}$

Figure 16
*Plot of t^{-1}. The area under this curve
is the natural logarithm ln.*

Two of the important but peculiar features of the natural logarithm are

$$\ln(ab) = \ln(a) + \ln(b) \tag{37}$$

$$\ln(\tfrac{1}{a}) = -\ln(a) \tag{38}$$

Thus we get, for example

$$\ln(t_f) - \ln(t_i) = \ln(t_f) + \ln\left(\frac{1}{t_i}\right)$$

$$= \ln\left(\frac{t_f}{t_i}\right) \tag{39}$$

Thus the area under the curve in Figure (16) is

$$\int_{t_i}^{t_f} \frac{dt}{t} = \ln\left(\frac{t_f}{t_i}\right) \tag{40}$$

While the natural logarithm has some rather peculiar properties it is easy to evaluate because it is available on all scientific calculators. For example, if $t_i = .5$ seconds and $t_f = 4$ seconds, then we have

$$\ln\left(\frac{t_f}{t_i}\right) = \ln\left(\frac{4}{.5}\right) = \ln(8) \tag{41}$$

Entering the number 8 on a scientific calculator and pressing the button labeled *ln*, gives

$$\ln(8) = 2.079 \tag{42}$$

which is the answer.

Exercise 2
Evaluate the integrals

$$\int_{.001}^{1000} \frac{dx}{x} \qquad \int_{.000001}^{1} \frac{dx}{x}$$

Why are the answers the same?

The Exponential Function

We have just seen that, while the logarithm function may have some peculiar properties, it is easy to evaluate using a scientific calculator. The question we now want to consider is whether there is some function that undoes the logarithm. When we enter the number 8 into the calculator and press ln, we get the number 2.079. Now we are asking if, when we enter the number 2.079, can we press some key and get back the number 8? The answer is, you press the key labeled e^x. The e^x key performs the **exponential function** which undoes the logarithm function. We say that the exponential function e^x is the **inverse** of the logarithm function ln.

Exponents to the Base 10

You are already familiar with exponents to the base 10, as in the following examples

$$10^0 = 1$$
$$10^1 = 10$$
$$10^2 = 100$$
$$10^6 = 1,000,000$$

$$10^{-1} = 1/10 = .1$$
$$10^{-2} = 1/100 = .01 \quad (43)$$
$$10^{-6} = .000001$$

The exponent, the number written above the 10, tells us how many factors of 10 are involved. A minus sign means how many factors of 10 we divide by. From this alone we deduce the following rules for the exponent to the base 10.

$$10^{-a} = \frac{1}{10^a} \quad (44)$$

$$10^a \times 10^b = 10^{a+b} \quad (45)$$

(Example $10^2 \times 10^3 = 100 \times 1000 = 100,000$.)

The inverse of the *exponent to the base 10* is the function called *logarithm to the base 10* which is denoted by the key labeled log on a scientific calculator. Formally this means that

$$\log\left(10^y\right) = y \quad (46)$$

Check this out on your scientific calculator. For example, enter the number 1,000,000 and press the log button and see if you get the number 6. Try several examples so that you are confident of the result.

The Exponential Function y^x

Another key on your scientific calculator is labeled y^x. This allows you to determine the value of any number y raised to the power (or exponent) x. For example, enter the number $y = 10$, and press the y^x key. Then enter the number $x = 6$ and press the $=$ key. You should see the answer

$$y^x = 10^6 = 1000000$$

It is quite clear that all exponents obey the same rules we saw for powers of 10, namely

$$y^a \times y^b = y^{a+b} \quad (47)$$

(Example $y^2 \times y^3 = \left(y \times y\right)\left(y \times y \times y\right) = y^5$.)

And as before

$$y^{-a} \equiv \frac{1}{y^a} \quad (48)$$

Exercise 3

Use your scientific calculator to evaluate the following quantities. (You should get the answers shown.)

(a) 10^6 (1000000)

(b) 2^3 (8)

(c) 23^0 (1)

(d) 10^{-1} (.1)

(To do this calculation, enter 10, then press y^x. Then enter 1, then press the +/– key to change it to –1, then press = to get the answer .1)

(e) $2^{-.5}$ $(1/\sqrt{2} = .707)$

(f) $\log(10)$ (1)

(g) $\ln(2.7183)$ (1) (very close to 1)

Try some other examples on your own to become completely familiar with the y^x key. (You should note that any positive number raised to the 0 power is 1. Also, some calculators, in particular the one I am using, cannot handle any negative values of y, not even $(-2)^2$ which is +4)

Euler's Number e = 2.7183...

We have seen that the function *log* on the scientific calculator undoes, is the inverse of, powers of 10. For example, we saw that

$$\log\left(10^x\right) = x \qquad\qquad \text{(46 repeated)}$$

Example: $\log\left(10^6\right) = 6$

Earlier we saw that the exponential function e^x was the inverse of the natural logarithm *ln*. This means that

$$\ln\left(e^x\right) = x \qquad\qquad (49)$$

The difference between the logarithm *log* and the natural logarithm *ln*, is that *log* undoes exponents of the number 10, while *ln* undoes exponents of the number e. This special number e, one of the fundamental mathematical constants like π, is known as **Euler's number**, and is always denoted by the letter e.

You can find the numerical value of Euler's number e on your calculator by evaluating

$$e^1 = e \qquad\qquad (50)$$

To do this, enter 1 into your calculator, press the e^x key, and you should see the result

$$e^1 = e = 2.718281828 \qquad\qquad (51)$$

We will run into this number throughout the course. You should remember that e is about 2.7, or you might even remember 2.718. (Only remembering e as 2.7 is as klutzy as remembering π as 3.1)

The terminology in math courses is that the function *log*, which undoes exponents of the number 10, is the *logarithm to the base 10*. The function *ln*, what we have called the *natural logarithm*, which undoes exponents of the number e, is the *logarithm to the base e*. You can have logarithms to any base you want, but in practice we only use base 10 (because we have 10 fingers) and the base e. The base e is special, in part because that is the logarithm that naturally arises when we integrate the function 1/x. We will see shortly that the functions *ln* and e^x have several more, very special features.

DIFFERENTIATION AND INTEGRATION

The scientific calculator is a good tool for seeing how the functions like *ln* and e^x are inverse of each other. Another example of inverse operations is integration and differentiation. We have seen that integration allows us to go the other way from differentiation [finding x(t) from v(t), rather than v(t) from x(t)]. However it is not so obvious that integration and differentiation are inverse operations when you think of integration as finding the area under a curve, and differentiation as finding limits of $\Delta x/\Delta t$ as Δt goes to zero. It is time now to make this relationship clear.

First, let us review our concept of a derivative. Going back to our strobe photograph of Figure (3), replacing $\vec{R}_i$ by $\vec{R}(t)$ and $\vec{R}_{i+1}$ by $\vec{R}(t+\Delta t)$, as shown in Figure (3a), our strobe velocity was then given by

$$\vec{v}(t) = \frac{\vec{R}(t+\Delta t) - \vec{R}(t)}{\Delta t} \qquad (52)$$

The calculus definition of the velocity is obtained by reducing the strobe time interval Δt until we obtain the instantaneous velocity $\vec{v}$.

$$\vec{v}_{calculus} = \lim_{\Delta t \to 0} \frac{\vec{R}(t + \Delta t) - \vec{R}(t)}{\Delta t} \qquad (53)$$

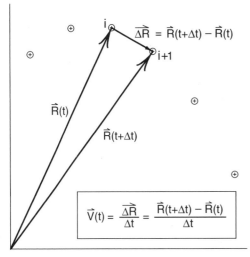

$$\vec{\Delta R} = \vec{R}(t+\Delta t) - \vec{R}(t)$$

$$\vec{V}(t) = \frac{\vec{\Delta R}}{\Delta t} = \frac{\vec{R}(t+\Delta t) - \vec{R}(t)}{\Delta t}$$

Figure 3a
Defining the strobe velocity.

While Equation (53) looks like it is applied to the explicit case of the strobe photograph of projectile motion, it is easily extended to cover any process of differentiation. Whatever function we have [we had $\vec{R}(t)$, suppose it is now f(t)], evaluate it at two closely spaced times, subtract the older value from the newer one, and divide by the time difference Δt. Taking the limit as Δt becomes very small gives us the derivative

$$\frac{df(t)}{dt} \equiv \lim_{\Delta t \to 0} \frac{f(t + \Delta t) - f(t)}{\Delta t} \qquad (54)$$

The variable with which we are differentiating does not have to be time t. It can be any variable that we can divide into small segments, such as x;

$$\frac{d}{dx} f(x) \equiv \lim_{\Delta x \to 0} \frac{f(x + \Delta x) - f(x)}{\Delta x} \qquad (55)$$

Let us see how the operation defined in Equation (55) is the inverse of finding the area under a curve.

Suppose we have a curve, like our old $v_x(t)$ graphed as a function of time, as shown in Figure (17). To find out how far we traveled in a time interval from t_i to some later time T, we would do the integral

$$x(T) = \int_{t_i}^{T} v_x(t) \, dt \qquad (56)$$

The integral in Equation (56) tells us how far we have gone at any time T during the trip. The quantity x(T) is a function of this time T.

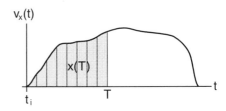

Figure 17
The distance traveled by the time T is the area under the velocity curve up to the time T.

Now let us differentiate the function x(T) with respect to the variable T. By our definition of differentiation we have

$$\frac{d}{dT}x(T) = \lim_{\Delta T \to 0} \frac{x(T + \Delta T) - x(T)}{\Delta T} \qquad (57)$$

Figure (17) shows us the function x(T). It is the area under the curve v(t) starting at t_i and going up to time t = T. Figure (18) shows us the function x(T + ΔT). It is the area under the same curve, starting at t_i but going up to t = T + Δt. When we subtract these two areas, all we have left is the area of the slender rectangle shown in Figure (19).

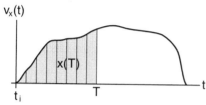

Figure 17 repeated
The distance x(T) traveled by the time T

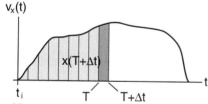

Figure 18
The distance x(T+Δt) traveled by the time T+Δt.

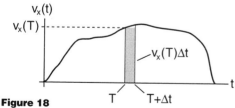

Figure 18
The distance x(T+Δt) − x(T)
traveled during the time Δt.

The rectangle has a height approximately v(T) and a width ΔT for an area

$$x(T + \Delta T) - x(T) = v_x(T)\Delta t \qquad (58)$$

Dividing through by ΔT gives

$$v_x(T) = \frac{x(T + \Delta T) - x(T)}{\Delta T} \qquad (59)$$

The only approximation in Equation (59) is at the top of the rectangle. If the curve is not flat, $v_x(T + \Delta T)$ will be different from $v_x(T)$ and the area of the sliver will have a value somewhere between $v_x(T)\Delta t$ and $v_x(T + \Delta t)\Delta t$. But if we take the limit as ΔT goes to zero, the value of $v_x(T + \Delta T)$ must approach $v_x(T)$, and we end up with the exact result

$$v_x(T) = \lim_{\Delta t \to 0} \frac{x(T + \Delta t) - x(T)}{\Delta t} \qquad (60)$$

This is just the derivative dx(t)/dt evaluated at t = T.

$$\boxed{v_x(T) = \left.\frac{dx(t)}{dt}\right|_{t = T}} \qquad (61a)$$

where we started from

$$\boxed{x(T) = \int_{t_i}^{T} v_x(t)\, dt} \qquad (61b)$$

Equations (61a) and (61b) demonstrate explicitly how differentiation and integration are inverse operations. The derivative allowed us to go from x(t) to $v_x(t)$ while the integral took us from $v_x(t)$ to x(t). This inverse is not as simple as pushing a button on a calculator to go from ln to e^x. Here we have to deal with limits on the integration and a shift of variables from t to T. But these two processes do allow us to go back and forth.

A Fast Way to go Back and Forth

We introduced our discussion of integration by pointing out that equations

$$v_x(t) = \frac{dx(t)}{dt}; \qquad a_x(t) = \frac{dv_x(t)}{dt} \qquad (62a,b)$$

went the wrong way in that we were more likely to know the acceleration $a_x(t)$ and from that want to calculate the velocity $v_x(t)$ and distance traveled $x(t)$. After many steps, we found that integration was what we needed.

We do not want to repeat all those steps. Instead we would like a quick and simple way to go the other way around. Here is how you do it. Think of the dt in (62a) as a small but finite time interval. That means we can treat it like any other number and multiply both sides of Equation (62a) through by it.

$$v_x(t) = \frac{dx(t)}{dt}$$

$$dx(t) = v_x(t)dt \qquad (63)$$

Now integrate both sides of Equation (63) from some initial time t_i to a final time T. (If you do the same thing to both sides of an equation, both sides should still be equal to each other.)

$$\int_{t_i}^{T} dx(t) = \int_{t_i}^{T} v_x(t)\,dt \qquad (64)$$

If dt is to be thought of as a small but finite time step, then dx(t) is the small but finite distance we moved in the time dt. The integral on the left side of Equation (64) is just the sum of all these short distances moved, which is just the total distance moved during the time from t_i to T.

$$\int_{t_i}^{T} dx(t) = x(t)\Big|_{t_i}^{T} = x(T) - x(t_i) \qquad (65)$$

Thus we end up with the result

$$x(t)\Big|_{t_i}^{T} = \int_{t_i}^{T} v_x(t)dt \qquad (66)$$

Equation (66) is a little more general than (62b) for it allows for the fact that $x(t_i)$ might not be zero. If,

however, we say that we started our trip at $x(t_i) = 0$, then we get the result

$$x(T) = \int_{t_i}^{T} v_x(t)dt \qquad (67)$$

representing the distance traveled since the start of the trip.

Constant Acceleration Formulas

The constant acceleration formulas, so well known from high school physics courses, are an excellent application of the procedures we have just described.

We will begin with motion in one dimension. Suppose a car is traveling due east, in the x direction, and for a while has a constant acceleration a_x. The car passes us at a time $t_i = 0$, traveling at a speed v_{x0}. At some later time T, if the acceleration a_x remains constant, how far away from us will the car be?

We start with the equation

$$a_x(t) = \frac{dv_x(t)}{dt} \qquad (68)$$

Multiplying through by dt to get

$$dv_x(t) = a_x(t)dt$$

then integrating from time $t_i = 0$ to time $t_f = T$, we get

$$\int_{0}^{T} dv_x(t) = \int_{0}^{T} a_x(t)dt \qquad (69)$$

Since the integral $\int dv_x(t) = v_x(t)$, we have

$$\int_{0}^{T} dv_x(t) = v_x(t)\Big|_{0}^{T} = v_x(T) - v_x(0) \qquad (70)$$

where $v_x(0)$ is the velocity v_{x0} of the car when it passed us at time $t = 0$.

While we can always do the left hand integral in Equation (69), we cannot do the right hand integral until we know $a_x(t)$. For the constant acceleration problem, however, we know that $a_x(t) = a_x$ is constant, and we have

$$\int_{0}^{T} a_x(t)dt = \int_{0}^{T} a_x dt \qquad (71)$$

Since constants can come outside the integral sign, we get

$$\int_0^T a_x dt = a_x \int_0^T dt = a_x t \Big|_0^T = a_x T \qquad (72)$$

where we used $\int dt = t$. Substituting Equations (70) and (72) in (69) gives

$$v_x T - v_{x0} = a_x T \qquad (73)$$

Since Equation (73) applies for any time T, we can replace T by t to get the well known result

$$\boxed{v_x(t) = v_{x0} + a_x t} \qquad (a_x \text{ constant}) \qquad (74)$$

Equation (74) tells us the speed of the car at any time t after it passed us, as long as the acceleration remains constant.

To find out how far away the car is, we start with the equation

$$v_x(t) = \frac{dx(t)}{dt} \qquad (62a)$$

Multiplying through by dt to get

$$dx(t) = v_x(t) \, dt$$

then integrating from time t = 0 to time t = T gives (as we saw earlier)

$$\int_0^T dx(t) = \int_0^T v_x(t) \, dt \qquad (75)$$

The left hand side is

$$\int_0^T dx(t) = x(t) \Big|_0^T = x(T) - x(0) \qquad (76)$$

If we measure along the x axis, starting from where we are (where the car was at t = 0) then x(0) = 0.

In order to do the right hand integral in Equation (75), we have to know what the function $v_x(t)$ is. But for constant acceleration, we have from Equation (74) $v_x(t) = v_{x0} + a_x t$, thus

$$\int_0^T v_x(t) \, dt = \int_0^T (v_{x0} + a_x t) \, dt \qquad (77)$$

One of the results of integration that you should prove for yourself (just sketch the areas) is the rule

$$\int_i^f \big[a(x) + b(x)\big] dx = \int_i^f a(x) dx + \int_i^f b(x) dx \qquad (78)$$

thus we get

$$\int_0^T (v_{x0} + a_x t) dt = \int_0^T v_{x0} dt + \int_0^T a_x t \, dt \qquad (79)$$

Since constants can come outside the integrals, this is equal to

$$\int_0^T (v_{x0} + a_x t) dt = v_{x0} \int_0^T dt + a_x \int_0^T t \, dt \qquad (80)$$

Earlier we saw that

$$\int_0^T dt = t \Big|_0^T = T - 0 = T \qquad (23)$$

$$\int_0^T t \, dt = \frac{t^2}{2} \Big|_0^T = \frac{T^2}{2} - 0 = \frac{T^2}{2} \qquad (30)$$

Thus we get

$$\int_0^T (v_{x0} + a_x t) dt = v_{x0} T + \frac{1}{2} a_x T^2 \qquad (81)$$

Using Equations (76) and (81) in (75) gives

$$x(T) - x_0 = v_{x0} T + \frac{1}{2} a_x T^2$$

Taking $x_0 = 0$ and replacing T by t gives the other constant acceleration formula

$$\boxed{x(t) = v_{x0} t + \frac{1}{2} a_x t^2} \qquad (a_x \text{ constant}) \qquad (82)$$

You can now see that the factor of $t^2/2$ in the constant acceleration formulas comes from the integral $\int t \, dt$.

Exercise 4

Find the formula for the velocity v(t) and position x(t) for a car moving with constant acceleration a_x, that was located at position x_i at some initial time t_i.

Start your calculation from the equations

$$v_x(t) = \frac{dx(t)}{dt}$$

$$a_x(t) = \frac{dv_x(t)}{dt}$$

and go through all the steps that we did to get Equations (74) and (82). See if you can do this without looking at the text.

If you have to look back to see what some steps are, then finish the derivation looking at the text. Then a day or so later, clean off your desk, get out a blank sheet of paper, write down this problem, put the book away and do the derivation. Keep doing this until you can do the derivation of the constant acceleration formulas without looking at the text.

Constant Acceleration Formulas in Three Dimensions

To handle the case of motion with constant acceleration in three dimensions, you start with the separate equations

$$v_x(t) = \frac{dx(t)}{dt} \qquad a_x(t) = \frac{dv_x(t)}{dt}$$

$$v_y(t) = \frac{dy(t)}{dt} \qquad a_y(t) = \frac{dv_y(t)}{dt}$$

$$v_z(t) = \frac{dz(t)}{dt} \qquad a_z(t) = \frac{dv_z(t)}{dt} \qquad (83)$$

Then repeat, for each pair of equations, the steps that led to the constant acceleration formulas for motion in the x direction. The results will be

$$x(t) = v_{x0}t + \frac{1}{2}a_x t^2 \qquad v_x(t) = v_{x0} + a_x t$$

$$y(t) = v_{y0}t + \frac{1}{2}a_y t^2 \qquad v_y(t) = v_{y0} + a_y t \quad (84)$$

$$z(t) = v_{z0}t + \frac{1}{2}a_z t^2 \qquad v_z(t) = v_{z0} + a_z t$$

The final step is to combine these six equations into the two vector equations

$$\boxed{\vec{x}(t) = \vec{v}_0 t + \frac{1}{2}\vec{a}t^2 \; ; \quad \vec{v}(t) = \vec{v}_0 + \vec{a}t} \quad (85)$$

These are the equations we analyzed graphically in Chapter 3 of the physics text, in Figure (3-34) and Exercise (3-9). (There we wrote $\vec{s}$ instead of $\vec{x}(t)$, and $\vec{v}_i$, rather than v_0.)

In many introductory physics courses, considerable emphasis is placed on solving constant acceleration problems. You can spend weeks practicing on solving these problems, and become very good at it. However, when you have done this, you have not learned very much physics because most forms of motion are not with constant acceleration, and thus the formulas do not apply. The formulas were important historically, for they were the first to allow the accurate prediction of motion (of cannonballs). But if too much emphasis is placed on these problems, students tend to use them where they do not apply. For this reason we have placed the exercises using the constant acceleration equations in an appendix at the end of chapter 4 of the physics text. There are plenty of problems there for all the practice you will need with these equations. Doing these exercises requires only algebra, there is no practice with calculus. To get some experience with calculus, be sure that you can confidently do Exercise 4.

MORE ON DIFFERENTIATION

In our discussion of integration, we saw that the basic idea was that the integral of some curve or function f(t) was equal to the area under that curve. That is an easy enough concept. The problems arose when we actually tried to find the formulas for the areas under various curves. The only areas we actually calculated were the rectangular area under f(t) = constant and the triangular area under f(t) = at. It was perhaps a surprise that the area under the simple curve 1/t should turn out to be a logarithm.

For differentiation, the basic idea of the process is given by the formula

$$\frac{df(t)}{dt} = \lim_{\Delta t \to 0} \frac{f(t + \Delta t) - f(t)}{\Delta t} \qquad \text{(54 repeated)}$$

Equation (54) is short hand notation for a whole series of steps which we introduced through the use of strobe photographs. The basic idea of differentiation is more complex than integration, but, as we will now see, it is often a lot easier to find the derivative of a curve than its integral.

Series Expansions

An easy way to find the formula for the derivative of a curve is to use a series expansion. We will illustrate the process by using the binomial expansion to calculate the derivative of the function x^n where n is any constant.

We used the binomial expansion, or at least the first two terms, in Chapter 1 of the physics text. That was during our discussion of the approximation formulas that are useful in relativistic calculations. As we mentioned in Exercise (1-5), the binomial expansion is

$$(x + \alpha)^n = x^n + n\alpha x^{n-1} + \frac{n(n-1)}{2!} \alpha^2 x^{n-2} \cdots$$

$$\text{(86)}$$

When α is a number much smaller than 1 ($\alpha << 1$), we can neglect α^2 compared to α (if $\alpha = .01$, $\alpha^2 = .0001$), with the result that we can accurately approximate $(x + \alpha)^n$ by

$$\boxed{(x + \alpha)^n \approx x^n + n\alpha x^{n-1}} \qquad \alpha << 1 \qquad \text{(87)}$$

Equation (87) gives us all the approximation formulas found in Equations (1-20) through (1-25) on page 1-28 of the physics text.

As an example of Equation (87), just to see that it works, let us take x = 5, n = 7 and $\alpha = .01$ to calculate $(5.01)^7$. From the calculator we get

$$(5.01)^7 = 79225.3344 \qquad \text{(88)}$$

(To do this enter 5.01, press the y^x button, then enter 7 and press the = button.) Let us now see how this result compares with

$$(x + \alpha)^n \approx x^n + n\alpha x^{n-1}$$
$$(5 + .01)^7 \approx 5^7 + 7(.01)5^6 \qquad \text{(89)}$$

We have

$$5^7 = 78125 \qquad \text{(90)}$$

$$7 \times .01 \times 5^6 = 7 \times .01 \times 15625 = 1093.75 \quad \text{(91)}$$

Adding the numbers in (90) and (91) together gives

$$5^7 + 7(.01)5^6 = 79218.75 \qquad \text{(92)}$$

Thus we end up with 79218 instead of 79225, which is not too bad a result. The smaller α is compared to one, the better the approximation.

Derivative of the Function x^n

We are now ready to use our approximation formula (87) to calculate the derivative of the function x^n. From the definition of the derivative we have

$$\frac{d(x^n)}{dx} = \lim_{\Delta x \to 0} \frac{(x + \Delta x)^n - x^n}{\Delta x} \qquad (93)$$

Since Δx is to become infinitesimally small, we can use our approximation formula for $(x + \alpha)^n$. We get

$$(x + \alpha)^n \approx x^n + n(\alpha)x^{n-1} \qquad (\alpha \ll 1)$$

$$(x + \Delta x)^n \approx x^n + n(\Delta x)x^{n-1} \qquad (\Delta x \ll 1) \;(94)$$

Using this in Equation (93) gives

$$\frac{d(x^n)}{dx} = \lim_{\Delta x \to 0} \left[\frac{\left(x^n + n(\Delta x)x^{n-1}\right) - x^n}{\Delta x} \right] \qquad (95)$$

We used an equal sign rather than an approximately equal sign in Equation (95) because our approximation formula (94) becomes exact when Δx becomes infinitesimally small.

In Equation (95), the terms x^n cancel and we are left with

$$\frac{d(x^n)}{dx} = \lim_{\Delta x \to 0} \left[\frac{n(\Delta x)x^{n-1}}{\Delta x} \right] \qquad (96)$$

At this point, the factors Δx cancel and we have

$$\frac{d(x^n)}{dx} = \lim_{\Delta x \to 0} \left[nx^{n-1} \right] \qquad (97)$$

Since no Δx's remain in our formula, we end up with the exact result

$$\boxed{\frac{d(x^n)}{dx} = nx^{n-1}} \qquad (98)$$

Equation (98) is the general formula for the derivative of the function x^n.

In our discussion of integration, we saw that a constant could come outside the integral. The same thing happens with a derivative. Consider, for example,

$$\frac{d}{dx}\left[a\,f(x)\right] = \lim_{\Delta x \to 0} \left[\frac{a\,f(x + \Delta x) - a\,f(x)}{\Delta x} \right]$$

Since the constant a has nothing to do with the limiting process, this can be written

$$\frac{d}{dx}\left[a\,f(x)\right] = a \lim_{\Delta x \to 0} \left[\frac{f(x + \Delta x) - f(x)}{\Delta x} \right]$$

$$= a\,\frac{df(x)}{dx} \qquad (99)$$

Exercise 5

Calculate the derivative with respect to x (i.e., d/dx) of the following functions. (When negative powers of x are involved, assume x is not equal to zero.)

(a) x

(b) x^2

(c) x^3

(d) $5x^2 - 3x$

(Before you do part (d), use the definition of the derivative to prove that $\frac{d}{dx}\left[f(x) + g(x)\right] = \frac{df(x)}{dx} + \frac{dg(x)}{dx}$)

(e) x^{-1}

(f) x^{-2}

(g) $\sqrt{x}$

(h) $1/\sqrt{x}$

(i) $3x^{.73}$

(j) $7x^{-.2}$

(k)   1

(In part (k) first show that this should be zero from the definition of the derivative. Then write $1 = x^0$ and show that Equation (98) also works, as long as x is not zero.)

(l)   5

The Chain Rule

There is a simple trick called the **chain rule** that makes it easy to differentiate a wide variety of functions. The rule is

$$\boxed{\frac{df[y(x)]}{dx} = \frac{df(y)}{dy}\frac{dy}{dx}} \qquad \textit{chain rule} \quad (100)$$

To see how this rule works, consider the function

$$f(x) = \left(x^2\right)^n \qquad (101)$$

We know that this is just $f(x) = x^{2n}$, and the derivative is

$$\frac{df(x)}{dx} = \frac{d}{dx}\left(x^{2n}\right) = 2nx^{2n-1} \qquad (102)$$

But suppose that we did not know this trick, and therefore did not know how to differentiate $(x^2)^n$. We do, however, know how to differentiate powers like x^2 and y^n. The chain rule allows us to use this knowledge in order to figure out how to differentiate the more complex function $(x^2)^n$.

We begin by defining $y(x)$ as

$$y(x) = x^2 \qquad (103)$$

Then our function $f(x) = (x^2)^n$ can be written in terms of y as follows

$$f(x) = (x^2)^n = \left(y(x)\right)^n = (y)^n = f(y)$$

$$f(y) = (y)^n \qquad (104)$$

Differentiating (103) and (104) gives

$$\frac{dy(x)}{dx} = \frac{d}{dx}\left(x^2\right) = 2x \qquad (105)$$

$$\frac{df(y)}{dy} = \frac{d}{dy}\left(y^n\right) = ny^{n-1} \qquad (106)$$

Using (104) and (105) in the chain rule (100) gives

$$\frac{df(y)}{dx} = \frac{df}{dy}\times\frac{dy}{dx} = \left(ny^{n-1}\right)\times\left(2x\right)$$

$$= 2ny^{n-1}x$$

$$= 2n\left(x^2\right)^{n-1}x$$

$$= 2n\left(x^{2(n-1)}\right)x$$

$$= 2n\left(x^{(2n-2)}\right)x \qquad (107)$$

$$= 2n\left(x^{(2n-2)+1}\right)$$

$$= 2nx^{2n-1}$$

which is the answer we expect.

In our example, using the chain rule was more difficult than differentiating directly because we already knew how to differentiate x^{2n}. But we will shortly encounter examples of new functions that we do not know how to differentiate directly, but which can be written in the form $f[y(x)]$; and where we know df/dy and dy/dx. We can then use the chain rule to evaluate the derivative df/dx. We will give you practice with the chain rule when we encounter these functions.

Remembering The Chain Rule

The chain rule can be remembered by thinking of the dy's as cancelling as shown.

$$\boxed{\frac{df(y)}{dy}\frac{dy}{dx} = \frac{df(y)}{dx}} \qquad \begin{array}{l}\textit{remembering}\\ \textit{the chain rule}\end{array} \quad (108)$$

Partial Proof of the Chain Rule (optional)

The proof of the chain rule is closely related to cancellation we showed in Equation (108). A partial proof of the rule proceeds as follows.

Suppose we have some function f(y) where y is a function of the variable x. As a result f[y(x)] is itself a function of x and can be differentiated with respect to x.

$$\frac{d}{dx} f[y(x)] = \lim_{\Delta x \to 0} \frac{f[y(x + \Delta x)] - f[y(x)]}{\Delta x} \quad (123)$$

Now define the quantity Δy by

$$\Delta y \equiv y(x + \Delta x) - y(x) \quad (124)$$

so that

$$y(x + \Delta x) = y(x) + \Delta y$$

$$f[y(x + \Delta x)] = f(y + \Delta y)$$

and Equation (123) becomes

$$\frac{d}{dx} f[y(x)] = \lim_{\Delta x \to 0} \frac{f(y + \Delta y) - f(y)}{\Delta x} \quad (125)$$

Now multiply (125) through by

$$1 = \frac{\Delta y}{\Delta y} = \frac{y(x + \Delta x) - y(x)}{\Delta y} \quad (126)$$

to get

$$\frac{d}{dx} f[y(x)]$$

$$= \lim_{\Delta x \to 0} \left| \frac{f(y + \Delta y) - f(y)}{\Delta x} \times \frac{y(x + \Delta x) - y(x)}{\Delta y} \right|$$

$$= \lim_{\Delta x \to 0} \left| \frac{f(y + \Delta y) - f(y)}{\Delta y} \times \frac{y(x + \Delta x) - y(x)}{\Delta x} \right|$$

$$(127)$$

where we interchanged Δx and Δy in the denominator.

(We call this a partial proof for the following reason. For some functions $y(x)$, the quantity $\Delta y = y(x + \Delta x) - y(x)$ may be identically zero for a small range of Δx. In that case we would be dividing by zero (the $1/\Delta y$) even before we took the limit as Δx goes to zero. A more complete proof handles the special cases separately. The resulting chain rule still works however, even for these special cases.)

Since $\Delta y = y(x + \Delta x) - y(x)$ goes to zero as Δx goes to zero, we can write Equation (127) as

$$\frac{d}{dx} f[y(x)]$$

$$= \left| \lim_{\Delta y \to 0} \frac{f(y + \Delta y) - f(y)}{\Delta y} \right|$$

$$\times \left| \lim_{\Delta x \to 0} \frac{y(x + \Delta x) - y(x)}{\Delta x} \right|$$

$$= \left| \frac{df(y)}{dy} \right| \left| \frac{dy}{dx} \right| \quad \text{(100 repeated)}$$

This rule works as long as the derivatives df/dy and dy/dx are meaningful, i.e., we stay away from kinks or discontinuities in f and y.

INTEGRATION FORMULAS

Knowing the formula for the derivative of the function x^n, and knowing that integration undoes differentiation, we can now use Equation (98)

$$\frac{dx^n}{dx} = nx^{n-1} \qquad \text{(98 repeated)}$$

to find the integral of the function x^n. We will see that this trick works for all cases except the special case where $n = -1$, i.e., the special case where the integral is a natural logarithm.

To integrate x^n, let us go back to our calculation of the distance s_x or $x(t)$ traveled by an object moving in the x direction at a velocity v_x. This was given by Equations (19) or (56) as

$$x(t)\Big|_{t_i}^{T} = \int_{t_i}^{T} v_x(t)\, dt \qquad (128)$$

where the instantaneous velocity $v_x(t)$ is defined as

$$v_x(t) = \frac{dx(t)}{dt} \qquad (129)$$

Suppose $x(t)$ had the special form

$$x(t) = t^{n+1} \qquad \text{(a special case)} \qquad (130)$$

then we know from our derivative formulas that

$$v(t) = \frac{dx(t)}{dt} = \frac{dt^{(n+1)}}{dt} = (n+1)t^n \qquad (131)$$

Substituting $x(t) = t^{n+1}$ and $v(t) = (n+1)t^n$ into Equation (128) gives

$$x(t)\Big|_{t_i}^{T} = \int_{t_i}^{T} v_x(t)\, dt \qquad (128)$$

$$\begin{aligned} t^{n+1}\Big|_{t_i}^{T} &= \int_{t_i}^{T} (n+1)t^n dt \\ &= (n+1)\int_{t_i}^{T} t^n dt \end{aligned} \qquad (132)$$

Dividing through by $(n+1)$ gives

$$\boxed{\int_{t_i}^{T} t^n dt = \frac{1}{n+1} t^{n+1}\Big|_{t_i}^{T}} \qquad (133)$$

If we choose $t_i = 0$, we get the simpler result

$$\int_{0}^{T} t^n dt = \frac{T^{n+1}}{n+1} \qquad (134)$$

and the indefinite integral can be written

$$\boxed{\int t^n dt = \frac{t^{n+1}}{n+1}} \qquad \text{(135) (also 34)}$$

This is the general rule we stated without proof back in Equation (34). Note that this formula says nothing about the case $n = -1$, i.e., when we integrate $t^{-1} = 1/t$, because $n+1 = -1+1 = 0$ and we end up with division by zero. But for all other values of n, we now have derived a general formula for finding the area under any curve of the form x^n (or t^n). This is a rather powerful result considering the problems one encounters actually finding areas under curves. (If you did not do Exercise 1, the integration exercises on page 14, or had difficulty with them, go back and do them now.)

Derivative of the Exponential Function

The previous work shows us that if we have a series expansion for a function, it is easy to obtain a formula for the derivative of the function. We will now apply this technique to calculate the derivative and integral of the exponential function e^x.

There is a series expansion for the function e^x that works for any value of α in the range -1 to $+1$.

$$e^\alpha \approx 1 + \alpha + \frac{\alpha^2}{2!} + \frac{\alpha^3}{3!} + \cdots \qquad (136)$$

where $2! = 2 \times 1$, $3! = 3 \times 2 \times 1 = 6$, etc. (The quantities $2!$, $3!$ are called *factorials*. For example $3!$ is called *three factorial*.)

To see how well the series (136) works, consider the case $\alpha = .01$. From the series we have, up to the α^3 term

$$\alpha = .01$$

$$\alpha^2 = .0001 ; \qquad \alpha^2/2 = .00005$$

$$\alpha^3 = .000001 ; \qquad \alpha^3/6 = .000000167$$

Giving us the approximate value

$$1 + \alpha + \frac{\alpha^2}{2!} + \frac{\alpha^3}{3!} = 1.010050167 \qquad (137)$$

When we enter .01 into a scientific calculator and press the e^x button, we get exactly the same result. Thus the calculator is no more accurate than including the α^3 term in the series, for values of α equal to .01 or less.

Let us now see how to use the series 136 for calculating the derivative of e^x. We have, from the definition of a derivative,

$$\frac{d}{dx}f(x) \equiv \lim_{\Delta x \to 0} \frac{f(x + \Delta x) - f(x)}{\Delta x} \qquad \text{(56 repeat)}$$

If $f(x) = e^x$, we get

$$\frac{d(e^x)}{dx} = \lim_{\Delta x \to 0}\left[\frac{e^{x + \Delta x} - e^x}{\Delta x}\right] \qquad (138)$$

To do this calculation, we have to evaluate the quantity $e^{x + \Delta x}$. First, we use the fact that for exponentials

$$e^{a+b} = e^a e^b$$

(Remember that $10^{2+3} = 10^2 \times 10^3 = 10^5$.) Thus

$$e^{x + \Delta x} = e^x e^{\Delta x} \qquad (139)$$

Now use the approximation formula (136), setting $\alpha = \Delta x$ and throwing out the α^2 and α^3 and higher terms because we are going to let Δx go to zero

$$e^{\Delta x} \approx 1 + \Delta x \qquad (140)$$

Substituting (140) in (139) gives

$$e^{x+\Delta x} \approx e^x(1 + \Delta x)$$

$$= e^x + e^x \Delta x \qquad (141)$$

Next use (141) in (138) to get

$$\frac{d(e^x)}{dx} = \lim_{\Delta x \to 0}\left[\frac{(e^x + e^x \Delta x) - e^x}{\Delta x}\right] \qquad (142)$$

The e^x terms cancel and we are left with

$$\frac{d(e^x)}{dx} = \lim_{\Delta x \to 0}\left[\frac{e^x \Delta x}{\Delta x}\right] = \lim_{\Delta x \to 0} e^x \qquad (143)$$

Since the Δx's cancelled, we are left with the exact result

$$\boxed{\frac{d(e^x)}{dx} = e^x} \qquad (144)$$

We see that the exponential function e^x has the special property that it is its own derivative.

We will often want to know the derivative, not just of the function e^x but of the slightly more general result e^{ax} where a is a constant. That is, we want to find

$$\frac{d}{dx}e^{ax} \qquad (a = constant) \qquad (145)$$

Solving this problem provides us with our first meaningful application of the chain rule

$$\frac{df(y)}{dx} = \frac{df(y)}{dy}\frac{dy}{dx} \qquad (100 \text{ repeated})$$

If we set

$$y = ax \qquad (146)$$

then we have

$$\frac{de^{ax}}{dx} = \frac{de^y}{dy}\frac{dy}{dx} \qquad (147)$$

Now

$$\frac{de^y}{dy} = e^y \qquad (148)$$

$$\frac{dy}{dx} = \frac{d}{dx}(ax) = a\frac{dx}{dx} = a \times 1 = a \qquad (149)$$

Using (148) and (149) in (147) gives

$$\frac{de^{ax}}{dx} = \left(e^y\right)(a) = \left(e^{ax}\right)(a) = ae^{ax}$$

Thus we have

$$\boxed{\frac{d}{dx}e^{ax} = ae^{ax}} \qquad (150)$$

This result will be used so often it is worth memorizing.

Exercise 6

For further practice with the chain rule, show that

$$\frac{de^{ax^2}}{dx} = 2axe^{ax^2}$$

Do this by choosing $y = ax^2$, and then do it again by choosing $y = x^2$.

Integral of the Exponential Function

To calculate the integral of e^{ax}, we will use the same trick as we used for the integral of x^n, but we will be a bit more formal this time. Let us start with Equation (128) relating position x(t) and velocity v(t) = dx(t)/dt go get

$$x(t)\Big|_{t_i}^{t_f} = \int_{t_i}^{t_f} v_x(t)\,dt = \int_{t_i}^{t_f} \frac{dx(t)}{dt}\,dt \qquad (128)$$

Since Equation (128) holds for any function x(t) [we did not put any restrictions on x(t)], we can write Equation (128) in a more abstract way relating any function f(x) to its derivative df(x)/dx;

$$\boxed{f(x)\Big|_{x_i}^{x_f} = \int_{x_i}^{x_f} \frac{df(x)}{dx}\,dx} \qquad (151)$$

To calculate the integral of e^{ax}, we set $f(x) = e^{ax}$ and $df(x)/dx = ae^{ax}$ to get

$$e^{ax}\Big|_{x_i}^{x_f} = \int_{x_i}^{x_f} ae^{ax}\,dx \qquad (152)$$

Dividing (157) through by (a) gives us the definite integral

$$\boxed{\int_{x_i}^{x_f} e^{ax}\,dx = \frac{1}{a}e^{ax}\Big|_{x_i}^{x_f}} \qquad (a = constant) \quad (153)$$

The corresponding indefinite integral is

$$\boxed{\int e^{ax}dx = \frac{e^{ax}}{a}} \qquad (a = constant) \qquad (154)$$

Exercise 7

The natural logarithm is defined by the equation

$$\ln(x) = \int\left(\frac{1}{x}\right)dx \qquad \text{(see Equations 33-40)}$$

Use Equation (151) to show that

$$\boxed{\frac{d}{dx}(\ln x) = \frac{1}{x}} \qquad (155)$$

(Hint—integrate both sides of Equation (155) with respect to x.)

DERIVATIVE AS THE SLOPE OF A CURVE

Up to now, we have emphasized the idea that the derivative of a function f(x) is given by the limiting process

$$\frac{df(x)}{dx} = \lim_{\Delta x \to 0} \left[\frac{f(x + \Delta x) - f(x)}{\Delta x} \right] \quad \text{(55 repeated)}$$

We saw that this form was convenient when we had an explicit way of calculating $f(x + \Delta x)$, as we did by using a series expansion. However, a lot of words are required to explain the steps involved in doing the limiting process indicated in Equation (55). In contrast, the idea of an integral as being the area under a curve is much easier to state and visualize. Now we will provide an easy way to state and interpret the derivative of a curve.

Consider the function f(x) graphed in Figure (20). At a distance x down the x axis, the curve had a height f(x) as shown. Slightly farther down the x axis, at $x + \Delta x$, the curve has risen to a height $f(x + \Delta x)$.

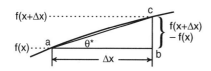

Figure 20

Two points on a curve, a distance Δx apart.

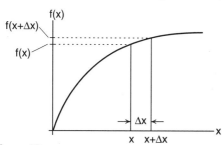

Figure 20a

At this point, the curve is tilted by approximately an angle θ.*

Figure (20a) is a blowup of the curve in the region between x and $x + \Delta x$. If the distance Δx is sufficiently small, the curve between x and $x + \Delta x$ should be approximately a straight line and that part of the curve should be approximately the hypotenuse of the right triangle abc seen in Figure (20a). Since the side opposite to the angle θ^* is $f(x + \Delta x) - f(x)$, and the adjacent side is Δx, we have the result that the tangent of the angle θ^* is

$$\tan \left(\theta^* \right) = \frac{f(x + \Delta x) - f(x)}{\Delta x} \quad (156)$$

When we make Δx smaller and smaller, take the limit as $\Delta x \to 0$, we see that the angle θ^* becomes more nearly equal to the angle θ shown in Figure (21), the angle of the curve when it passes through the point x. Thus

$$\tan \theta = \lim_{\Delta x \to 0} \frac{f(x + \Delta x) - f(x)}{\Delta x} \quad (157)$$

The tangent of the angle at which the curve passes through the point x is called the ***slope of the curve at the point x.*** Thus from Equation (157) we see that the slope of the curve is equal to the derivative of the curve at that point. We now have the interpretation that the derivative of a curve at some point is equal to the slope of the curve at that point, while the integral of a curve is equal to the area under the curve up to that point.

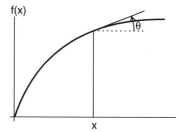

Figure 21

The tangent of the angle θ at which the curve passes through the point x is called the slope of the curve at that point.

Negative Slope

In Figure (22) we compare the slopes of a rising and a falling curve. In (22a), where the curve is rising, the quantity $f(x + \Delta x)$ is greater than $f(x)$ and the derivative or slope

$$\frac{df(x)}{dx} = \lim_{\Delta x \to 0} \left[\frac{f(x + \Delta x) - f(x)}{\Delta x} \right]$$

is a positive number.

In contrast, for the downward curve of Figure (22b), $f(x + \Delta x)$ is less than $f(x)$ and the slope is negative. For a curve headed downward, we have

$$\frac{df(x)}{dx} = -\tan(\theta) \quad \begin{array}{l} \textit{downward heading} \\ \textit{curve} \end{array} \quad (158)$$

(For this case you can think of θ as a negative angle, so that $\tan(\theta)$ would automatically come out negative. However it is easier simply to remember that the slope of an upward directed curve is positive and that of a downward directed cure is negative.)

Exercise 8

Estimate the numerical value of the slope of the curve shown in Figure (23) at points (a), (b), (c), (d) and (e). In each case do a sketch of $\left[f(x + \Delta x) - f(x) \right]$ for a small Δx, and let the slope be the ratio of $\left[f(x + \Delta x) - f(x) \right]$ to Δx. Your answers should be roughly 1, 0, –1, $+\infty$, $-\infty$.

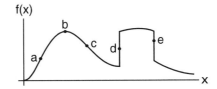

Figure 23
Estimate the slope at the various points indicated.

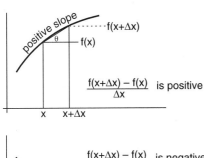

$$\frac{f(x+\Delta x) - f(x)}{\Delta x} \quad \text{is positive}$$

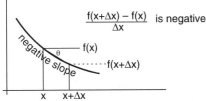

$$\frac{f(x+\Delta x) - f(x)}{\Delta x} \quad \text{is negative}$$

Figure 22
Going uphill is a positive slope, downhill is a negative slope.

THE EXPONENTIAL DECAY

A curve that we will encounter several times during the course is the function e^{-ax} shown in Figure (24), which we call an exponential decay. Since exponents always have to be dimensionless numbers, we are writing the constant (a) in the form $1/x_0$ so that the exponent x/x_0 is more obviously dimensionless.

The function e^{-x/x_0} has several very special properties. At $x = 0$, it has the numerical value 1 ($e^0 = 1$). When we get up to $x = x_0$, the curve has dropped to a value

$$e^{-x/x_0} = e^{-1} = \frac{1}{e} \quad (at \; x = x_0)$$
$$\approx \frac{1}{2.7} \tag{159}$$

When we go out to $x = 2x_0$, the curve has dropped to

$$e^{-2x_0/x_0} = e^{-2} = \frac{1}{e^2} \tag{160}$$

Out at $x = 3x_0$, the curve has dropped by another factor of e to $(1/e)(1/e)(1/e)$. This decrease continues indefinitely. It is the characteristic feature of an exponential decay.

Muon Lifetime

In the muon lifetime experiment, we saw that the number of muons surviving decreased with time. At the end of two microseconds, more than half of the original 648 muons were still present. By 6 microsec-

onds, only 27 remained. The decay of these muons is an example of an exponential decay of the form

$$\begin{pmatrix} \text{number of} \\ \text{surviving} \\ \text{muons} \end{pmatrix} = \begin{pmatrix} \text{number of} \\ \text{muons at} \\ \text{time } t = 0 \end{pmatrix} \times e^{-t/t_0} \tag{161}$$

where t_0 is the time it takes for the number of muons remaining to drop by a factor of $1/e = 1/2.7$. That time is called the muon *lifetime*.

We can use Equation (161) to estimate the muon lifetime t_0. In the movie, the number of mesons at the top of the graph, reproduced in Figure (25), is 648. That is at time $t = 0$. Down at time $t = 6$ microseconds, the number surviving is 27. Putting these numbers into Equation (161) gives

$$27 \; _{\text{muons}}^{\text{surviving}} = 648 \; _{\text{muons}}^{\text{initial}} \times e^{-6/t_0}$$

$$e^{-6/t_0} = \frac{27}{648} = .042 \tag{162}$$

Take the natural logarithm ln of both sides of Equation (162), [remembering that $ln(e^x) = x$] gives

$$ln\left(e^{-6/t_0}\right) = \frac{-6}{t_0} = ln(.042) = -3.17$$

where we entered .042 on a scientific calculator and pressed the ln key. Solving for t_0 we get

$$t_0 = \frac{6}{3.17} = 1.9 \; \text{microseconds} \tag{163}$$

This is close to the accepted value of $t_0 = 2.20$ microseconds which has been determined from the study of many thousands of muon decays.

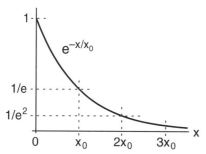

Figure 24
As we go out an additional distance x_0, the exponential curve drops by another factor of 1/e.

Figure 25
The lifetime of each detected muon is represented by the length of a vertical line. We can see that many muons live as long as 2 microseconds (2μs), but few live as long as 6 microseconds.

Half Life

The exponential decay curve e^{-t/t_0} decays to $1/e = 1/2.7$ of its value at time t_0. While $1/e$ is a very convenient number from a mathematical point of view, it is easier to think of the time $t_{1/2}$ it takes for half of the muons to decay. This time $t_{1/2}$ is called the **half life** of the particle.

From Figure (26) we can see that the half life $t_{1/2}$ is slightly shorter than the lifetime t_0. To calculate the half life from t_0, we have

$$e^{-t/t_0}\bigg|_{t=t_{1/2}} = e^{-t_{1/2}/t_0} = \frac{1}{2} \qquad (164)$$

Again taking the natural logarithm of both sides of Equation (164) gives

$$ln\left(e^{-t_{1/2}/t_0}\right) = \frac{-t_{1/2}}{t_0} = ln\left(\frac{1}{2}\right) = -.693$$

$$\boxed{t_{1/2} = .693\, t_0} \qquad (165)$$

From Equation (165) you can see that a half life $t_{1/2}$ is about .7 of the lifetime t_0. If the muon lifetime is 2.2μsec (we will abbreviate microseconds as μsec), and you start with a large number of muons, you would expect about half to decay in a time of

$$\left(t_{1/2}\right)_{muon} = .693 \times 2.2\mu sec = 1.5\,\mu sec$$

The basic feature of the exponential decay curve e^{-t/t_0} is that for every time t_0 that passes, the curve decreases by another factor of $1/e$. The same applies to the half life $t_{1/2}$. After one half life, e^{-t/t_0} has decreased to half its value. After a second half life, the curve is down to $1/4 = 1/2 \times 1/2$. After 3 half lives it is down to $1/8 = 1/2 \times 1/2 \times 1/2$ as shown in Figure (27).

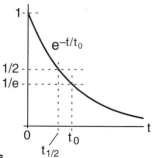

Figure 26
Comparison of the lifetime t_0 and the half-life $t_{1/2}$.

To help illustrate the nature of exponential decays, suppose that you started with a million muons. How long would you expect to wait before there was, on the average, only one left?

To solve this problem, you would want the number e^{-t/t_0} to be down by a factor of 1 million

$$e^{-t/t_0} = 1 \times 10^{-6}$$

Taking the natural logarithm of both sides gives

$$ln\left(e^{-t/t_0}\right) = \frac{-t}{t_0} = ln\left(1\times10^{-6}\right) = -13.8 \quad (166)$$

(To calculate $ln\left(1\times10^{-6}\right)$, enter 1, then press the *exp* key and enter 6, then press the +/– key to change it to –6. Finally press = to get the answer –13.8.)

Solving Equation (166) for t gives

$$t = 13.8\, t_0 = 13.8 \times 2.2\,\mu sec$$

$$\boxed{t = 30\ microseconds} \qquad (167)$$

That is the nature of an exponential decay. While you have nearly half a million left after around 2 microseconds, they are essentially all gone by 30 microseconds.

Exercise 9

How many factors of 1/2 do you have to multiply together to get approximately 1/1,000,000? Multiply this number by the muon half-life to see if you get about 30 microseconds.

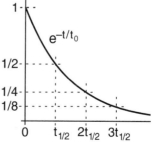

Figure 27
After each half-life, the curve decreases by another factor of 1/2.

Measuring the Time Constant from a Graph

The idea that the derivative of a curve is the slope of the curve, leads to an easy way to estimate a lifetime t_0 from an exponential decay curve e^{-t/t_0}.

The formula for the derivative of an exponential curve is

$$\frac{de^{at}}{dt} = ae^{at} \qquad \text{(150 repeated)}$$

Setting $a = -1/t_0$ gives

$$\frac{d}{dt}\left(e^{-t/t_0}\right) = -\frac{1}{t_0}e^{-t/t_0} \qquad (168)$$

Since the derivative of a curve is the slope of the curve, we set the derivative equal to the tangent of the angle the curve makes with the horizontal axis.

$$\frac{d}{dt}\left(e^{-t/t_0}\right) = -\frac{1}{t_0}e^{-t/t_0} = \tan\theta \qquad (168a)$$

The minus sign tells us that the curve is headed down.

In Figure (28), we have drawn a line tangent to the curve at the point $t = T$. This line intersects the (t) axis (the axis where e^{-t/t_0} goes to zero) at a distance (x) down the t axis.

The height (y) of the point where we drew the tangent curve is just the value of the function e^{-T/t_0}. The tangent of the angle θ is the opposite side (y) divided by the adjacent side (x)

$$\tan\theta = \frac{y}{x} = \frac{e^{-T/t_0}}{x} \qquad (169)$$

Equating the two magnitudes of $\tan\theta$ in Equations (169) in (168a) gives us

$$\frac{1}{t_0}e^{-T/t_0} = \frac{1}{x}e^{-T/t_0}$$

which requires that

$$x = t_0 \qquad (170)$$

Equation (170) tells us that the distance (x), the distance down the axis where the tangent lines intersect the axis, is simply the time constant t_0.

The result gives us a very quick way of determining the time constant t_0 of an exponential decay curve. As illustrated in Figure (29), choose any point on the curve, draw a tangent to the curve at that point and measure the distance down the axis where the tangent line intersects the axis. That distance will be the time constant t_0. We will use this technique in several laboratory exercises later in the course.

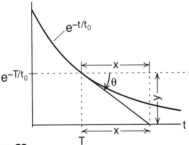

Figure 28
A line, drawn tangent to the exponential decay curve at some point T, intersects the axis a distance x down the axis. We show that this distance x is equal to the time constant t_0. This is true no matter what point T we start with.

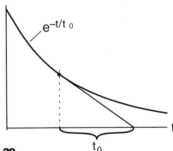

Figure 29
A quick way to estimate the time constant t_0 for an exponential decay curve is to draw the tangent line as shown.

THE SINE AND COSINE FUNCTIONS

The final topic in our introduction to calculus will be the functions $\sin\theta$ and $\cos\theta$ and their derivatives and integrals. We will need these functions when we come to rotational motion and wave motion.

The definition of $\sin\theta$ and $\cos\theta$, which should be familiar from trigonometry, are

$$\sin\theta = \frac{a}{c} \qquad \left(\frac{opposite}{hypotenuse}\right) \qquad (171a)$$

$$\cos\theta = \frac{b}{c} \qquad \left(\frac{adjacent}{hypotenuse}\right) \qquad (171b)$$

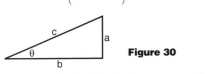

Figure 30

where θ is an angle of a right triangle as sown in Figure (30), (a) is the length of the side opposite to θ, (b) the side adjacent to θ and (c) the hypotenuse.

The formulas are simplified if we consider a right triangle whose hypotenuse is of length $c = 1$ as in Figure (31). Then we have

$$\sin\theta = a \qquad (172a)$$

$$\cos\theta = b \qquad (172b)$$

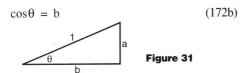

Figure 31

We can then fit our right triangle inside a circle of radius 1 as shown in Figure (32).

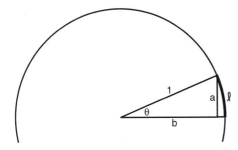

Figure 32
Fitting our right triangle inside a unit radius circle.

Radian Measure

We are brought up to measure angles in degrees, but physicists and mathematicians usually measure angles in *radians*. The angle θ measured in radians is defined as the *arc length* ℓ subtended by the angle θ on a circle of unit radius, as shown in Figure (32).

$$\theta_{radians} = \ell \qquad \begin{matrix} arc\ length\ subtended \\ by\ \theta\ on\ a\ unit\ circle \end{matrix} \qquad (173)$$

(If we had a circle of radius c, then we would define $\theta_{radians} = \ell/c$, a dimensionless ratio. In the special case $c = 1$, this reduces to $\theta_{radians} = \ell$.)

Since the circumference of a unit circle is 2π, we see that θ for a complete circle is 2π radians, which is the same as 360 degrees. This tells us how to convert from degrees to radians. We have the conversion factor

$$\frac{360\ degrees}{2\pi\ radians} = 57.3\ \frac{degrees}{radian} \qquad (174)$$

As an example of using this conversion factor, suppose we want to convert 30 degrees to radians. We would have

$$\frac{30\ degrees}{57.3\ degrees/radian} = .52\ radians \qquad (175)$$

To decide whether to divide by or multiply by a conversion factor, use the dimensions of the conversion factor. For example, if we had multiplied 30 degrees by our conversion factor, we would have gotten

$$30\ degrees \times 57.3\ \frac{degrees}{radian} = 1719\ \frac{degrees^2}{radian}$$

This answer may be correct, but it is useless.

The numbers to remember in using radians are the following:

$$\begin{matrix} 90° &=& \pi/2\ radians \\ 180° &=& \pi\ radians \\ 270° &=& 3\pi/2\ radians \\ 360° &=& 2\pi\ radians \end{matrix} \qquad (176)$$

The other values you can work out as you need them.

The Sine Function

In Figure (33) we have started with a circle of radius 1 and, in a somewhat random way, labeled 10 points around the circle. The arc length up to each of these points is equal to the angle, in radian measure, subtended by that point. The special values are:

$$\theta_0 = 0 \text{ radians}$$
$$\theta_4 = \pi/2 \text{ radians } (90°)$$
$$\theta_6 = \pi \text{ radians } (180°)$$
$$\theta_8 = 3\pi/2 \text{ radians } (270°)$$
$$\theta_{10} = 2\pi \text{ radians } (360°)$$

In each case the $\sin\theta$ is equal to the height (a) at that point. For example

$$\sin\theta_1 = a_1$$
$$\sin\theta_2 = a_2$$
$$\cdots\cdots$$
$$\sin\theta_{10} = a_{10}$$

We see that the height (a) starts out at $a_0 = 0$ for θ_0, increases up to $a_4 = 1$ at the top of the circle, drops back down to $a_6 = 0$ at $\theta_6 = \pi$, goes negative, down to $a_8 = -1$ at $\theta_8 = 3\pi/2$, and returns to $a_{10} = 0$ at $\theta_{10} = 2\pi$.

Our next step is to construct a graph in which θ is shown along the horizontal axis, and we plot the value of $\sin\theta = (a)$ on the vertical axis. The result is shown in Figure (34). The eleven points, representing the heights a_0 to a_{10} at θ_0 to θ_{10} are shown as large dots in Figure (34). We have also sketched in a smooth curve through these points, it is the curve we would get if we had plotted the value of (a) for every value of θ from $\theta = 0$ to $\theta = 2\pi$. The smooth curve is a graph of the function *sin θ*.

Exercise 10

Using the fact that the cosine function is defined as

$$\cos\theta = b \qquad \text{(b is defined in Figures 31, 32)}$$

plot the values of $b_0, b_1, \cdots, b_{10}$ on a graph similar to Figure (34), and show that the cosine function $\cos\theta$ looks like the curve shown in Figure (35).

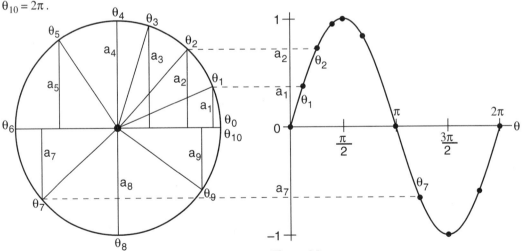

Figure 33
The heights a_i at various points around a unit circle.

Figure 34
Graph of the function $\sin\left(\theta\right)$.

There is nothing that says we have to stop measuring the angle θ after we have gone around once. On the second trip around, θ increases from 2π up to 4π, and the curve sinθ repeats itself. If we go around several times, we get a result like that shown in Figure (36).

Several cycles of the curve cosθ are shown in Figure (37). You can see that the only difference between a sine and a cosine curve is where you set θ = 0. If you move the origin of the cosine axis back (to the left) 90° (π/2), you get a sine wave.

Amplitude of a Sine Wave

A graph of the function y(θ) = c sinθ looks just like the curve in Figure (36), except the curve goes up to a height c and down to –c as shown in Figure (38). We would get the curve of Figure (38) by plotting points around a circle as in Figure (33), but using a circle of radius c. We call this factor c the *amplitude* of the sine wave. The function sinθ has an amplitude 1, while the sine wave in Figure (38) has an **amplitude c** (its values range from +c to –c).

c sinθ

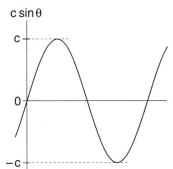

Figure 38
A sine wave of amplitude c.

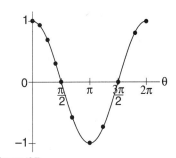

Figure 35
The cosine function.

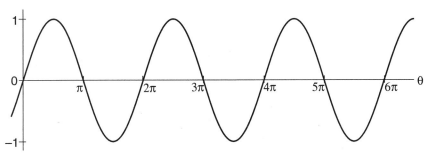

Figure 36
Several cycles of the curve sin(θ).

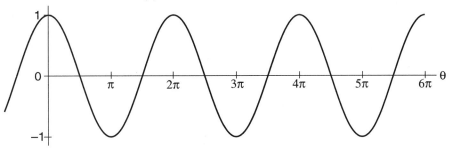

Figure 37
Several cycles of the curve cos(θ).

Derivative of the Sine Function

Since the sine and cosine functions are smooth curves, we should be able to calculate the derivatives and integrals of them. We will do this by first calculating the derivative, and then turning the process around to find the integral, just as we did for the functions x^n and e^x.

The derivative of the function $\sin\theta$ is defined as usual by

$$\frac{d(\sin\theta)}{d\theta} = \lim_{\Delta\theta\to 0}\left[\frac{\sin(\theta+\Delta\theta) - \sin\theta}{\Delta\theta}\right] \quad (177)$$

where $\Delta\theta$ is a small change in the angle θ.

The easiest way to evaluate this limit is to go back to the unit circle of Figure (25) and construct both $\sin\theta$ and $\sin(\theta+\Delta\theta)$ as shown in Figure (39). We see that $\sin\theta$ is the height of the triangle with an angle θ, while $\sin(\theta+\Delta\theta)$ is the height of the triangle whose center angle is $(\theta+\Delta\theta)$. What we have to do is calculate the difference in heights of these two triangles.

In Figure (40) we start by focusing our attention on the slender triangle abc with an angle $\Delta\theta$ at (a) and long sides of length 1 (since we have a unit circle). Since the angle $\Delta\theta$ is small, the short side of this triangle is essentially equal to the arc length along the circle from point (b) to point (c). And since we are using radian measure, this arc length is equal to the angle $\Delta\theta$.

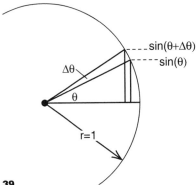

Figure 39
Triangles for the $\sin\theta$ and the $\sin(\theta+\Delta\theta)$.

Now draw a line vertically down from point (c) and horizontally over from point (b) to form the triangle bcd shown in Figure (40). The important point is that the angle at point (c) in this tiny triangle is the same as the angle θ at point (a). To prove this, consider the sketch in Figure (41). A line bf is drawn tangent to the circle at point (b), so that the angle abf is a right angle. That means the other two angles in the triangle add up to 90°, the total angle in any triangle being 180°

$$\theta + \varphi = 90° \quad (178)$$

Since the angle at (e) in triangle bef is also a right angle, the other two angles in the triangle bef, must also add up to 90°.

$$\alpha + \varphi = 90° \quad (179)$$

For both Equations (178) and (179) to be true, we must have $\alpha = \theta$.

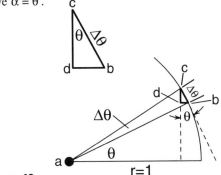

Figure 40
The difference between $\sin\theta$ and $\sin(\theta + \Delta\theta)$ is equal to the height of the side cdb of the triangle cdb.

$$\theta + \phi = 90°$$
$$\alpha + \phi = 90°$$
$$\therefore \alpha = \theta$$

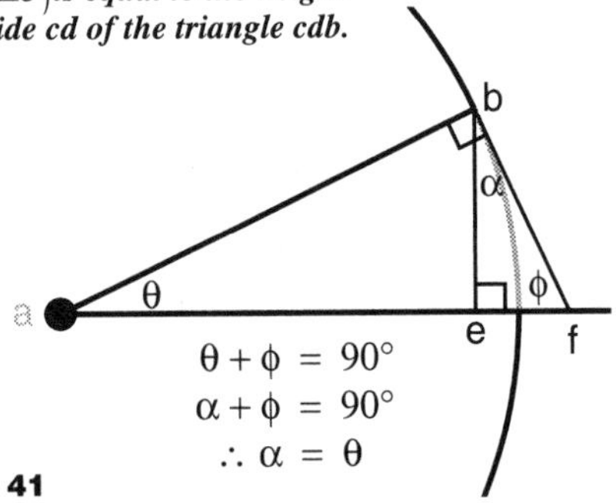

Figure 41
Demonstration that the angle α equals the angle θ.

The final step is to note that when $\Delta\theta$ in Figure (40) is very small, the side cb of the very small triangle is essentially tangent to the circle, and thus parallel to the side bf in Figure (41). As a result the angle between cb and the vertical is also the same angle θ.

Because the tiny triangle, shown again in Figure (42) has a hypotenuse $\Delta\theta$ and a top angle θ, the vertical side, which is equal to the difference between $\sin\theta$ and $\sin(\theta + \Delta\theta)$ has a height $(\cos\theta)\Delta\theta$. Thus we have

$$\sin(\theta + \Delta\theta) - \sin\theta = (\cos\theta)\Delta\theta \qquad (180)$$

Equation (180) becomes exact when $\Delta\theta$ becomes an infinitesimal angle.

We can now evaluate the derivative

$$\frac{d(\sin\theta)}{d\theta} = \lim_{\Delta\theta \to 0}\left[\frac{\sin(\theta + \Delta\theta) - \sin\theta}{\Delta\theta}\right]$$

$$= \lim_{\Delta\theta \to 0}\left[\frac{(\cos\theta)\Delta\theta}{\Delta\theta}\right]$$

$$= \lim_{\Delta\theta \to 0} \cos\theta$$

Thus we get the exact result

$$\frac{d}{d\theta}(\sin\theta) = \cos\theta \qquad (181)$$

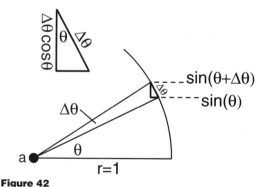

Figure 42
The difference between $\sin\theta$ and $\sin(\theta + \Delta\theta)$ is equal to $\Delta\theta\cos\theta$.

Exercise 11
Using a similar derivation, show that

$$\frac{d}{d\theta}(\cos\theta) = -\sin\theta \qquad (182)$$

Exercise 12
Using the chain rule for differentiation, show that

$$\left.\begin{array}{l}\dfrac{d}{d\theta}(\sin a\theta) = a\cos a\theta \\[2mm] \dfrac{d}{d\theta}(\cos a\theta) = -a\sin a\theta\end{array}\right\} (a = constant) \quad (183)$$

(Hint—if you need to, look at Equation (145) through (150).

Exercise 13
Using the fact that integration reverses differentiation, as we did in integrating the function e^x (Equations (151) through (154), show that

$$\int_{\theta_i}^{\theta_f}(\cos a\theta)d\theta = \frac{1}{a}\sin a\theta\Big|_{\theta_i}^{\theta_f} \qquad (184a)$$
$$(a = constant)$$
$$\int_{\theta_i}^{\theta_f}(\sin a\theta)d\theta = -\frac{1}{a}\cos a\theta\Big|_{\theta_i}^{\theta_f} \qquad (184b)$$

Use sketches of the integrals from $\theta_i = 0$ to $\theta_f = \pi/2$ to show that Equations (184a) and (184b) have the correct numerical sign. (Explicitly explain the minus sign in (184b).

P2000 Index